Metal Forming

Technology and Process Modelling

Metal Forming

Technology and Process Modelling

Edited by

Uday Shanker Dixit

Professor, Department of Mechanical Engineering
IIT Guwahati

R. Ganesh Narayanan

Asst. Professor, Department of Mechanical Engineering
IIT Guwahati

McGraw Hill Education (India) Private Limited

NEW DELHI

McGraw Hill Education Offices

New Delhi New York St Louis San Francisco Auckland Bogotá Caracas
Kuala Lumpur Lisbon London Madrid Mexico City Milan Montreal
San Juan Santiago Singapore Sydney Tokyo Toronto

McGraw Hill Education (India) Private Limited

Published by McGraw Hill Education (India) Private Limited,
P-24, Green Park Extension, New Delhi 110 006

Metal Forming: Technology and Process Modelling

ISBN (13): 978-1-25-905889-9
ISBN (10): 1-25-905889-1

Vice President and Managing Director: *Ajay Shukla*
Publishing Manager—Professional: *Mitadru Basu*
Editorial Researcher—Science, Technology & Computing: *Irene Rajalingam*
DGM—Sales and Business Development—Professional: *S Girish*
Deputy Marketing Manager—Science, Technology & Computing: *Rekha Dhyani*
General Manager—Production: *Rajender P Ghansela*
Manager—Production: *Reji Kumar*

Typeset at Shubham Composers, WZ-437 Madipur Village, New Delhi 110063 and printed at Sanat Printers, 312 EPIP, HSIDC, Kundli, Sonepat, Haryana.

Cover Printer: Sanat Printers

Cover Designer: Kapil Gupta

RLYYCRCOXRAZZ

Dedicated
To
Teaching Fraternity

Preface

Metal forming is one of the oldest and widely used manufacturing processes. Its advantages include no or very little wastage of material during processing, enhancement of the mechanical and metallurgical properties of the material and its suitability for mass production. Metal forming has progressed from being a craftsman based technology to a technology requiring highly scientific inputs. Last two-three decades have seen the extensive application of computational techniques in this area. However, in academic institutions, there is still a dearth of suitable courses on the scientific aspects of metal forming. It is felt that a suitable course should be offered to final or pre-final year undergraduate student and first year graduate students. At the same time, practicing engineers should also be provided an opportunity to keep them abreast with the latest developments in this area. With this viewpoint in mind and considering the shortage of advanced level books on metal forming, we have decided to bring out an edited book on the subject.

The present book discusses both fundamentals and advanced topics in both bulk and sheet metal forming. In this form, the book can be used as a textbook as well as a reference book. Chapters have been written by 18 authors from three countries associated with academia and industry. The chapters have been arranged in a sequence of "from fundamental to advanced topics".

Chapter 1 provides an introduction to metal forming technology and is descriptive in nature. Chapter 2 discusses the theory of plasticity, which serves as a foundation of metal forming research as well as processes involving plastic deformation. Chapter 3 describes modeling techniques for metal forming. It includes finite element method and soft computing tools. Chapter 4 highlights the role of friction and ways to manage it. First four chapters are of fundamental nature. Chapter 5 discusses both the fundamental and advanced aspects of metal extrusion. Chapter 6 discusses the advanced topics in die designing particularly for extrusion. Chapter 7 emphasizes the experimental and theoretical research in metal forming with the help of illustrative examples. Chapter 8 introduces sheet metal forming in detail. Chapter 9 discusses flange wrinkling in deep drawing, a commonly used sheet metal forming. Chapter

10 discusses forming of tailor welded blanks and chapter 11 focuses on the forming aspects pertaining to tailor welded blanks fabricated by friction stir welding process. Chapter 12 describes probabilistic and mesoscale approaches to model interfacial phenomena during hot flat rolling. Chapter 13 is the concluding chapter of the book, which discusses some of the nascent metal forming technologies and suggests direction for future research.

We are very grateful to authors for finding time for writing the chapters in spite of their busy schedule. We are thankful to Prof. M.S. Shunmugam of IIT Madras, Dr. P.M. Pandey of IIT Delhi, Prof. J. Paulo Davim of University of Aveiro, Portugal, Prof. Masayoshi Akiyama of Kyoto Institute of Technology, Japan, for their comments and suggestions about the chapters. We also thank Peter Jones, Iain McGregor, Yuguo An, Clive Sutton, Yi Gao, Ray Long, Nico Langerak, Henk Vegter, Eisso Atzema, Matthijs Toose, Jon King of Tata Steel for reviewing all the chapters and suggesting improvements.

We have taken utmost care to avoid any errors and not to omit any relevant reference and apologise in advance for any inadvertent errors and omissions. We request our readers to provide valuable suggestions for improving this book further. Any feedback can be sent directly to us at our e-mail addresses uday@iitg.ac.in and ganu@iitg.ac.in. We expect that students and practicing engineers will find this book useful.

Uday S. Dixit
(uday@iitg.ac.in)
R. Ganesh Narayanan
(ganu@iitg.ac.in)

Contents

CHAPTER 1 Metal Forming Process

R. Ganesh Narayanan, Uday S. Dixit
Department of Mechanical Engineering
Indian Institute of Technology Guwahati 781 039, India

1.1 INTRODUCTION

Metal forming is a manufacturing process, in which a metallic workpiece is plastically deformed by the application of forces. When the forces are applied on the workpiece, first the workpiece deforms elastically and then the plastic deformation takes place. If the load is removed during elastic deformation, the workpiece returns to its original state without causing any permanent set (deformation). In metal forming operations, usually the elastic deformation is much smaller compared with plastic deformation and is often neglected for finding out the total load required for deforming the material. However, it must be considered if the residual stress or spring back needs to be calculated.

The stress–strain behavior of a typical metal is shown schematically in Fig. 1.1(a). A sharp yield point is shown at which the state of metal changes from elastic to plastic. Most of the real metals will not display a sharp yield point. Figure 1.1(b) shows the stress–strain behavior of a hypothetical rigid-plastic material neglecting the elastic deformation. In practice, no material is rigid-plastic, but to simplify the analysis of metal forming process, it may be assumed that in a uniaxial tensile or compressive test, the material remains rigid till the stress reaches the yield strength, after which the plastic deformation starts. The stress–strain behavior of the material greatly influences the metal forming, but it is not the only important factor. The controllable process parameters, lubrication, contact surface condition, and the design of the metal forming setup also affect the metal forming process to a greater extent.

A brief description of stress–strain behavior and plastic deformation properties of a typical metal are given here. A typical engineering stress–strain behavior of metals is shown in Fig. 1.2. In annealed low carbon steel and some other materials, the stress–strain behavior is characterized by a sharp upper yield point. The subsequent drop in yield stress is followed by a yield plateau, in which barring small oscillations the yield stress remains constant. This phenomenon is known as the yield point phenomenon. Here one can clearly distinguish between elastic and plastic deformation. In many metallic materials, the transition between elastic and plastic parts is smooth as shown in Fig. 1.2. The deformation starts with linear elastic deformation till point B. Hooke's law of stress being linearly proportional to strain is obeyed in the region OB. In the region BA, the material is non-linearly elastic. Usually, the region BA is negligibly small and it is very difficult to locate the point A accurately. It is customary to take the yield point as a point C at which the permanent deformation is equivalent to an offset strain of 0.002. In Fig. 1.2, OD is equal to the offset strain. Note that the line DC is parallel to the line OB. After the onset of plastic deformation, stress increases non-linearly with further deformation due to strain hardening. At point E, corresponding to ultimate strength, necking starts. Finally, the necking is localized and material fails at point F. The elongation attained by the material till necking is called uniform elongation and the elongation till failure is called total elongation of the material.

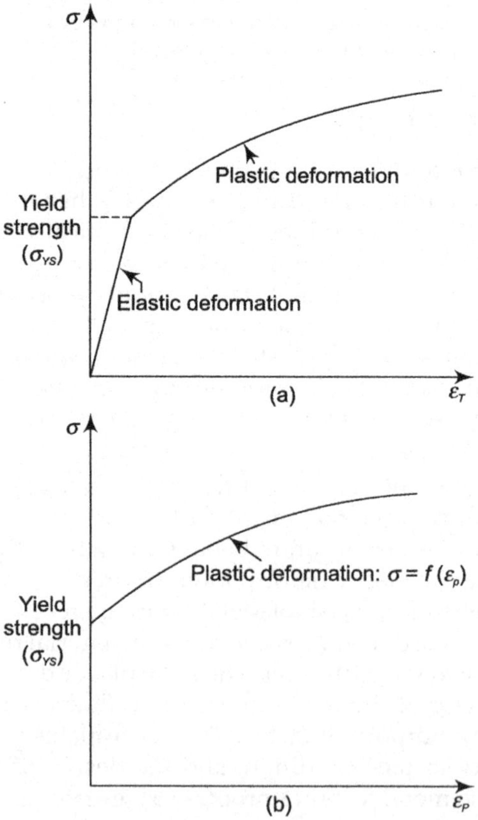

Fig. 1.1 Schematic diagram of stress–strain behavior (a) elastic-plastic material (b) rigid-plastic material

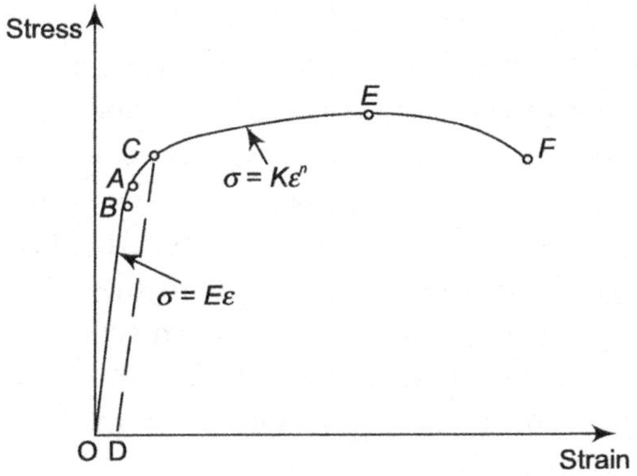

Fig. 1.2 Stress–strain behavior of a typical metal

It is possible to obtain some of the important forming properties from the stress–strain behavior of a material. The strain hardening behavior during plastic deformation can be described by strain hardening relations like Hollomon relation $\sigma = K\varepsilon^n$, where n is the strain hardening exponent, and K is the strength coefficient. It is to be noted that σ is the true stress and ε is the true plastic strain (total strain, if elastic strain is negligible) in Hollomon relation. The engineering stress–strain behavior can be transformed into true stress–strain behavior by volume constancy principle during plastic deformation.

There are many metal forming processes for manufacturing engineering products. For example, rolling is used to convert a thick slab into thin sheets having thickness of even in the sub-millimeter range. These sheets are then used to make automotive, aerospace, electronic components, and various consumer products by stamping, deep drawing, stretch forming, tube forming, bending, shearing or spinning, to name just a few sheet metal forming processes. Extrusion is used to make rods, seamless tubes, and products with various cross-sectional profiles that are used for engineering and construction industries. It should be noted that many a times the metal forming and other manufacturing processes are combinedly used to manufacture engineering products. For instance, (i) bolts and screws are made by cold heading the rod and thread rolling, (ii) cooking gas cylinders are made by deep drawing of upper and lower halves and welding, (iii) metal wash basins are made by deep drawing and hole making process. The main motivation of the metal forming studies is to understand the change in structural properties, design the process efficiently, and model it to have accurate performance prediction.

In general, the metal forming processes are categorized into (i) bulk metal forming, and (ii) sheet metal forming processes. The processes like forging, extrusion, rolling, wire drawing, heading, ironing, etc., can be included in bulk forming processes; and deep drawing, stretch forming, bending, spinning, hydro forming, etc., can be included in sheet forming processes. Bulk forming processes exhibit high volume to surface area ratio. The workpiece sheet used in sheet forming processes has a low volume to surface area ratio. Both the bulk forming and sheet forming processes can be performed at room temperature, high temperature, and medium temperature, depending on the workability of the materials. If a metal forming process is carried out at a temperature more than the recrystallization temperature, it is termed as hot working. The metal forming process performed without heating the work-piece is termed as cold working process. In between hot working and cold working is the warm working, in which the workpiece is heated but there is no recrystallization. In the following sections, some important bulk forming and sheet forming processes are described.

1.2 BULK METAL FORMING PROCESS

In bulk forming, generally, the input raw material is in the form of rod, slab or billet. The raw material undergoes plastic deformation and a shape and size change is witnessed in the output product. Other than the material properties, the working temperature, strain rate, and friction play a vital role in deciding the workability of the material. The important bulk forming processes like rolling, forging, extrusion, and wire drawing are described here.

1.2.1 Rolling

In rolling process, the semi-finished parts like slab, billet or bloom are rolled into plates, sheets, strips, rods, tubes by rollers rotating in opposite directions. The semi-finished products are also made mainly by hot rolling of cast ingots. Hot rolling is performed above recrystallization temperature, where large reduction in thickness, with less forming pressures is possible. In the case of sheet, strip or foil rolling, the deformation in width direction is negligible when compared to thickness and length direction. This type of deformation is known as plane strain deformation. This is generally seen in sheets whose deformation zone width is more than 10–20 times larger than the length of the zone. In the case of rolling of thick sheets and slabs, there exists considerable deformation along the width (called spread), length (called elongation), and thickness (called compression).

The rolls are classified based on the number of rolls, number of stand, type of products produced, size of the rolling mill, etc. Some important classifications are described here. Figures 1.3–1.5 show the schematic description of two-high, three-high, and four-high rolling mills. The two-high mill [Figs 1.3(a) and (b)] is the simplest one, and the slab is rolled by two rolls rotating in opposite direction— one in clockwise (CW) and another in counter-clockwise (CCW) directions. Once the rolling is completed, the rolled sheet is taken to the same entry location, and rolling is done again for further thickness reduction. The two-high mill can be a reversing mill as shown in Fig. 1.3(b), in which the rollers employed can rotate both in CW and CCW directions after each stage of thickness reduction. Hence, after first stage of rolling, the sheet can be inserted from the exit location itself for next rolling stage. The reversing mill is found to be more productive and efficient when compared with ordinary two-high rolling mill.

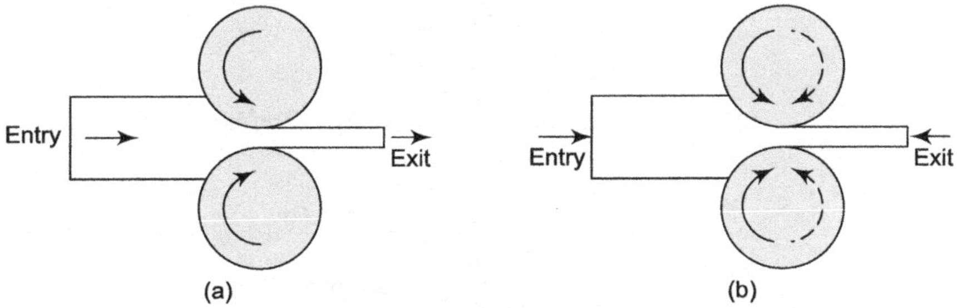

Fig. 1.3 Schematic diagram of two high rolling mill (a) without reversing facility, (b) with reversing facility

The three-high rolling (Fig. 1.4) mill contains three rollers— the top and bottom rollers rotate in same direction, while the middle one rotates in opposite direction. First, rolling is done between bottom and middle rolls in a direction and the same sheet is rolled in the opposite direction between top and middle rolls in the second operation for more thickness reduction. There is no reversing of roll direction in this case. In four-high roll mill (Fig. 1.5), the small work roll is involved in sheet rolling whereas large back-up rolls support the working roll throughout its length, so that large reductions in sheet thickness is possible without excessive deflection of working roll. Huge rolling operation involving large thickness reductions are possible in special roll mills like cluster mills and planetary mills. In a typical 20-high Sendzimir mill (Fig. 1.6), the small working rolls are supported by two intermediate back-up rolls which in turn are supported by three back-up rolls that in turn are supported by four back-up rolls throughout their length. All these rolls are arranged within a rigid housing. A large thickness reduction is possible with this rolling mill, without roller deflection. A planetary mill (Fig. 1.7) contains back-up rolls surrounded by small work rolls that are used for large thickness reductions.

For rolling operations involving different stages, rolls are arranged in "stands" as shown in Fig. 1.8. In each stand, uniform and preset thickness reduction occurs and finally the sheets are coiled up in a "windup reel". Since different thickness reduction is taken at each stand, the strip will move with different velocities between successive stands. The speed of each set of rolls is controlled.

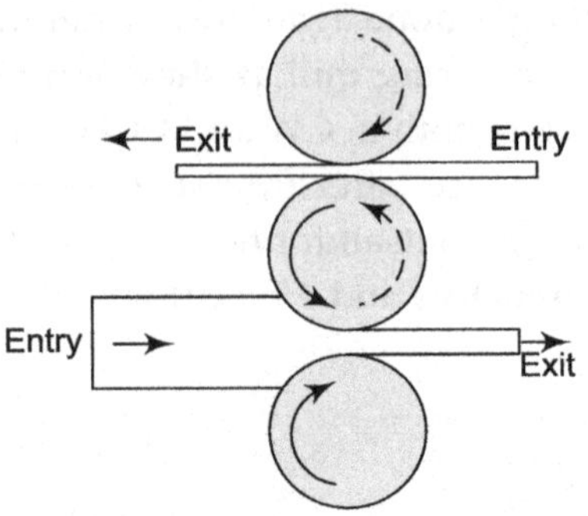

Fig. 1.4 Schematic diagram of three high rolling mill

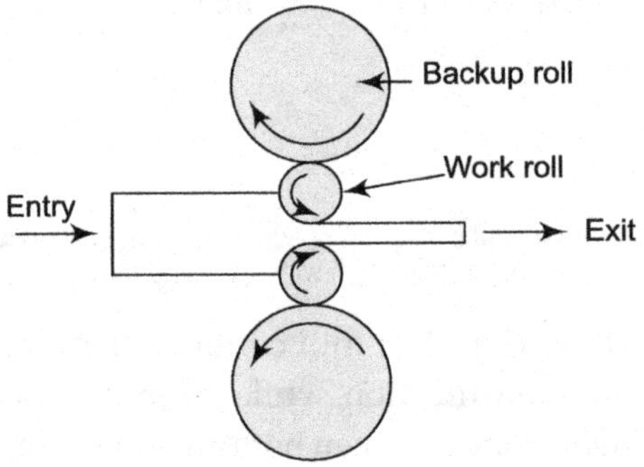

Fig. 1.5 Schematic diagram of four high rolling mill

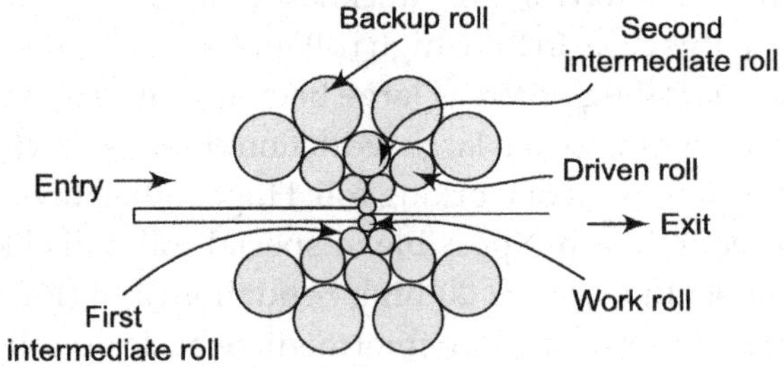

Fig. 1.6 Schematic diagram of Sendzimir rolling mill

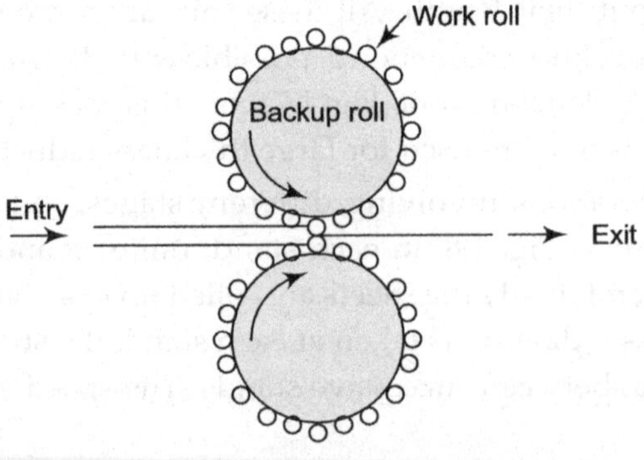

Fig. 1.7 Schematic diagram of planetary roll mill

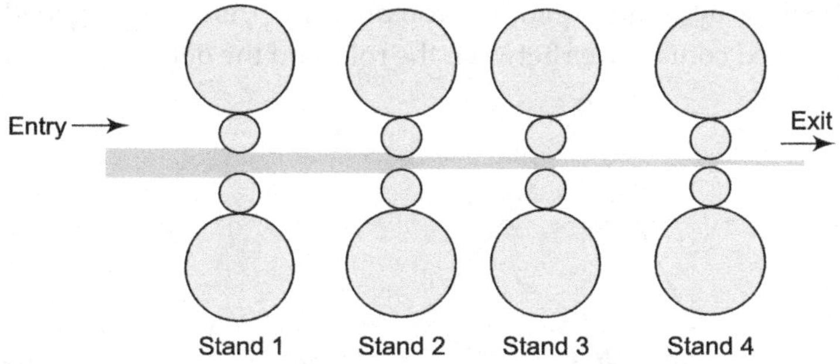

Fig. 1.8 Rolling mill with four roll stands

The rolls are important components in the rolling machine as they control and plastically deform the sheet for significant amount of thickness reduction. Selection of roll material is crucial for the performance of the process. Cast iron (as-cast and stress relieved), heat treated high alloy iron, cast steel, hardened forged steel (typical composition: 0.85% C, 0.30% Mn, 0.30% Si, 1.75% Cr, 0.10% V) are generally used to make rolls [1]. Recently high speed steels and semi-high speed steels are also used widely for hot strip rolling . These rolls are more resistant to wear than conventional high chromium rolls, which make them suitable for long rolling schedules and productivity. More details of roll materials and manufacturing can be found in the handbook edited by Semiatin [1].

The Schematic diagram of forces acting during strip rolling is shown in Fig. 1.9. A sheet material with the initial thickness of h_1 enters the rolling mill from entry side with a velocity of v_1. The sheet is rolled between two rolls. It has reduced thickness of h_2 at the exit side and leaves with the velocity of v_2. The sheet has intermediate thickness of h and velocity of v at a typical location in deformation. In strip rolling, the strip width w does not vary considerably. The volume constancy principle provides

$$wh_1v_1 = whv = wh_2v_2 \tag{1.1}$$

As per this equation, the exit velocity v_2 is greater than entry velocity v_1. In fact, the velocity of the rolled sheet continuously increases from entry to exit. There is a point or zone on the contact surface between the roll and the sheet where the roll surface velocity v_r is equal to the sheet velocity. This point is called neutral point, shown as point N in the Fig. 1.9. Many times, instead of a clear cut neutral point, there may be a neutral zone. At a typical point F, the radial force, F_r, and tangential friction force, F_f, are shown. The interface friction force acts towards the neutral point from the entry and the exit sides due to the velocity difference between the sheet and the roll surface before and after the neutral point. Before neutral point, the sheet moves slower than the roll surface and after the neutral point, the sheet moves faster than the roll surface. The vertical component of radial force is called the rolling load, F_{rv}. This is also termed as roll separating force, as the pressing force acting on the sheet by the roll and the force given by the sheet to the rolls for separating them apart are equal. The roll pressure p_r is the rolling load divided by the projected contact area between the rolls and the deforming material, i.e.,

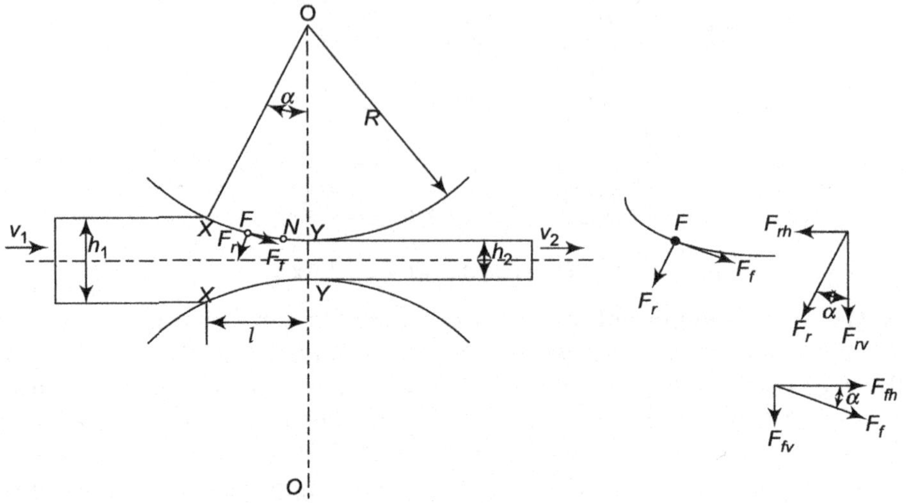

Fig. 1.9 Schematic diagram of forces acting during strip rolling

$$p_r = \frac{F_{rv}}{wl}, \tag{1.2}$$

where wl is the product of strip width and the projected length of the contact area. As described elaborately in [2], the projected length, l, is given by

$$l \approx [R(h_1 - h_2)]^{1/2} = \sqrt{R\Delta h}, \tag{1.3}$$

where 'Δh' is called draft. The experimental work done by many researchers, for example, Al-Salehi *et al.* [3] indicates that the roll pressure rises to the maximum at the neutral point and then decreases as shown schematically in Fig. 1.10. The angle between the entry sections to centerline of the rollers is called "angle of contact" or "angle of bite (α)". From Fig. 1.9, for the sheet metal to enter into the rolls, the horizontal component of friction force ($F_{fh} = F_f \cos \alpha$) that acts towards the roll gap, should be equal to or greater than the horizontal component of F_r, i.e., $F_r \sin \alpha$. In the limiting case,

$$F_f \cos \alpha = F_r \sin \alpha$$

or,

$$\tan \alpha = \frac{\sin \alpha}{\cos \alpha} = \frac{F_f}{F_r}.$$

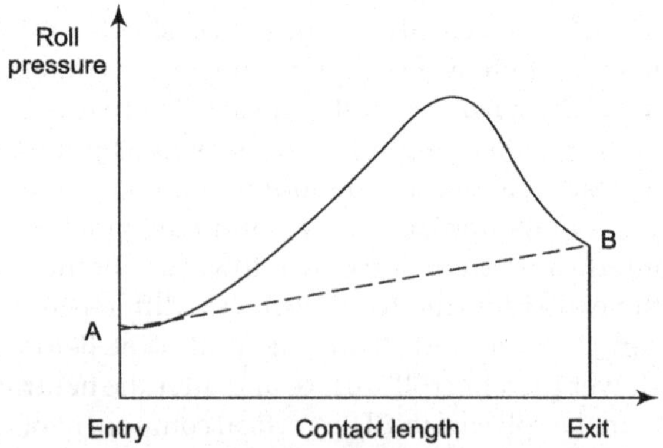

Fig. 1.10 Variation of roll pressure with contact length during rolling

Thus,

$$F_f = F_r \tan \alpha = F_r \mu;$$

providing

$$\mu = \tan \alpha \tag{1.4}$$

From Eq. (1.4), if $\mu = 0$, rolling cannot occur. If $\tan \alpha$ is greater than μ, then the sheet cannot be drawn into the rolls.

From Fig. 1.9, one can write

$$\tan \alpha = \frac{l}{\left[R - \left(\frac{\Delta h}{2}\right)\right]} \approx \frac{\sqrt{R\Delta h}}{\left[R - \frac{\Delta h}{2}\right]} \approx \sqrt{\frac{\Delta h}{R}}$$

From Eq. (1.4),

$$\mu = \tan \alpha \approx \sqrt{\frac{\Delta h}{R}};$$

$$(\Delta h)_{\max} = \mu^2 R \tag{1.5}$$

The maximum possible draft during cold rolling operation can be obtained from Eq. (1.5). Figures 1.11(a) and (b) depict the effect of friction coefficient and roll radius on maximum draft based on Eq. (1.5). With increase in friction coefficient from 0.08 to 0.1, there is a substantial increase in maximum draft for same roll radius [Fig. 1.11(a)]. Similarly by increasing the roll radius from 300 to 400 mm, for the same friction coefficient, considerable increase in maximum draft is observed [Fig. 1.11(b)].

An approximate model for evaluating the roll separating force can be developed by approximating the deformation zone (Fig. 1.9) with the homogeneous plane-strain upsetting process. The model to find the load per unit width of the strip [1, 2, 4] is given by

$$L = \frac{2\bar{\sigma}}{\sqrt{3}}\left(1 + \frac{ml_a}{4H}\right)l_a \tag{1.6}$$

where $\bar{\sigma}$ is the flow stress of the material, m is the friction factor (ratio of frictional traction to shear yield stress), H is the average strip thickness $\{H = 0.5\,(h_1 + h_2)\}$, and l_a is the approximate contact length ($l_a = R\alpha$).

The roll pressure during strip rolling with and without back and front tension is shown in Fig. 1.12 [7]. The maximum stress occurs at the neutral point. With the front and back tension, the neutral point shifts from its initial position. With the application of back tension, the neutral point shifts towards the roll exit. A very high back tension will eventually shift the neutral point to the roll exit. In this situation, the rolls slide over the sheet and they move faster than the sheet. The neutral point will shift towards the roll entry with the application of front tension (Fig. 1.12).

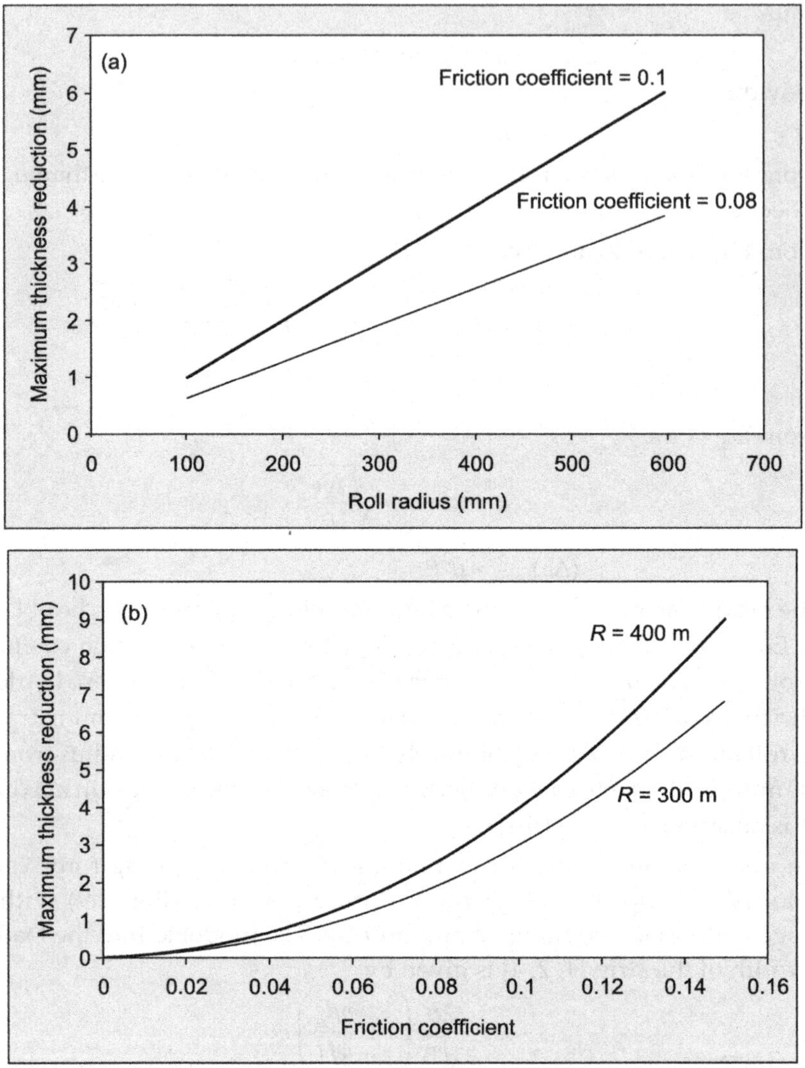

Fig. 1.11 Effect of friction coefficient and roll radius on maximum height reduction

There are various process and roll parameters that affect the strip rolling performance. Some of the important parameters are friction coefficient, front and back tension, strip width, roll bending force, roll roughness, lubrication viscosity, percentage reduction, and roll radius. The important rolling performances monitored are rolling pressure, rolling torque, forward slip, strip roughness, etc. The variation of friction and effect of friction on the rolling parameters were studied by many researchers including Tieu and Liu [5], Jeswiet [6], Dixit and Dixit [7], Gudur *et al.* [8], Jiang *et al.* [9], Gao *et al.* [10],

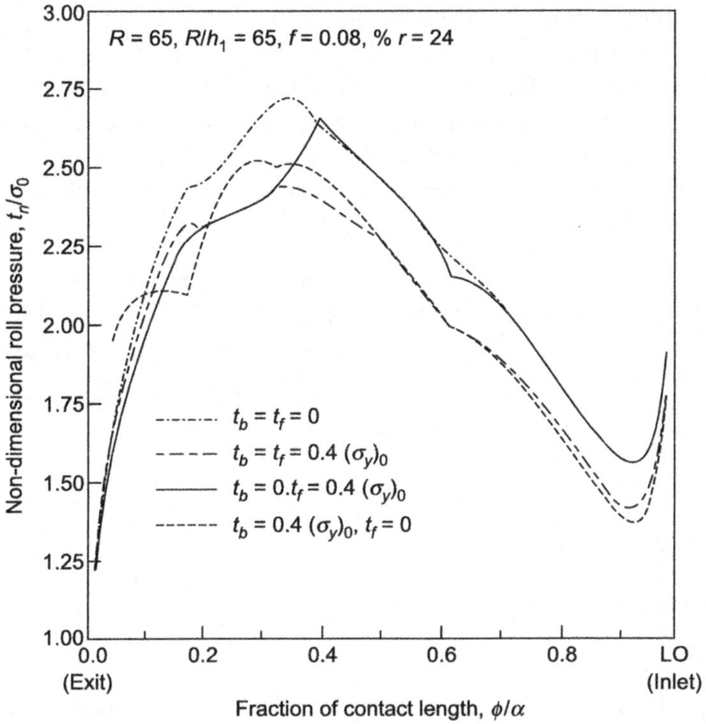

Fig. 1.12 Stress distribution (or roll pressure) during strip rolling with and without back and front tension (t_b: back tension; t_f: front tension; $(\sigma_y)_0$: yield stress in tension; R: undeformed roll radius; h_1: inlet thickness of strip; f: friction factor; r: reduction ratio; φ: angular position of a point on the interface; α: angle of contact). With permission from [7] copyright [1996] Elsevier

and Pawelski [11]. For example, Tieu and Liu [5] showed that the friction coefficient is not a constant and it varies with time and roll bite length, from entry to exit. Similar behavior is witnessed for different rolling reductions (37.01% and 40.38%) and rolling speed (5 and 30 rpm). Likewise, it was shown by Jeswiet [6] that friction force decreases along the rolling direction, tested for a reduction of 24.2%. Dixit and Dixit [7] used Wanheim and Bays friction model in their work to analyze the flat rolling process using finite elements. They have also applied fuzzy set theory. In their analysis, the initial yield stress, hardening coefficients, and friction coefficients are considered as the fuzzy process parameters. As a result of fuzzy input parameters, the output parameters also acquire fuzzy characteristics, thereby giving a more realistic simulation of experimental results which always have a scatter. Gudur *et al.* [8] proposed a method for the estimation of friction coefficient based on the asymmetric rolling operation. Using slab method, the curvature of the rolled sheet was estimated and the friction coefficient was evaluated by solving an inverse problem. It was found that as the coefficient of friction increases the curvature also goes on increasing (Fig. 1.13). Similarly curvature was found to increase with increase in speed ratio (Fig.1.14) [8]. An inverse problem of finding the friction coefficient by estimating the curvature was proposed and validated by authors. Jiang *et al.* [9] modeled cold strip rolling taking into account the friction along the strip rolling and transverse directions. A 3D rigid plastic finite element method was used. Simulation results show that the friction in both directions has a significant effect on the rolling mechanics, strip shape and profile.

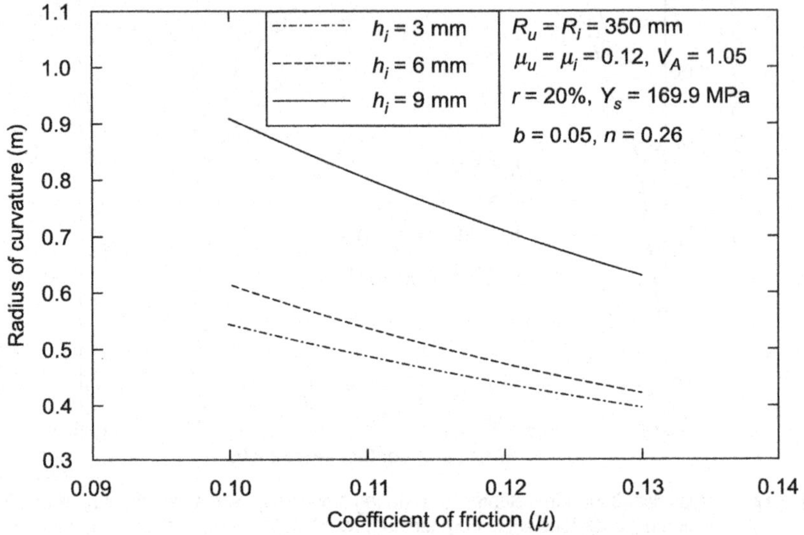

Fig. 1.13 Influence of friction coefficient on the radius of curvature. With permission from [8] copyright [2008] Elsevier

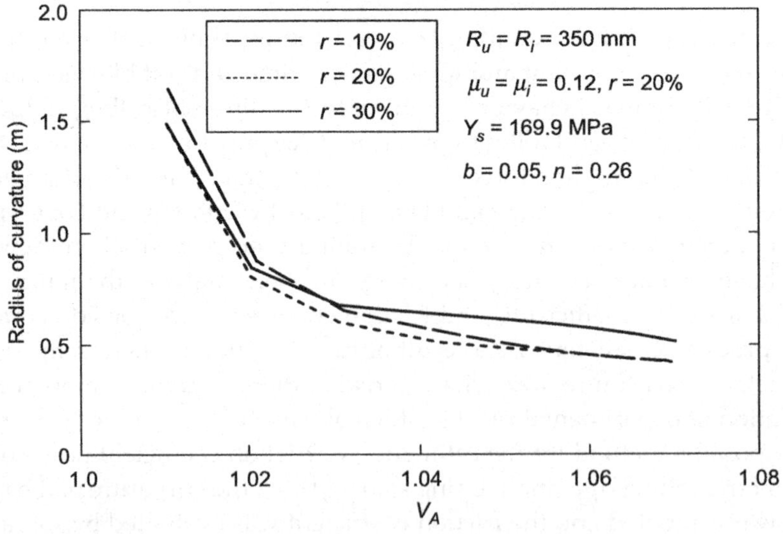

Fig. 1.14 Influence of speed ratio on the radius of curvature. With permission from [8] copyright [2008] Elsevier

An interesting work was done by Gao *et al.* [10] to understand the effect of varying friction coefficient on the rolling pressure and force. As described by authors, there are three different asymmetrical rolling processes, one with varying roll radii and constant angular velocity, second one with varying angular velocity and same roll radii and third one with changing friction coefficient by keeping the other two parameters constant. The effect of different friction coefficients on the rolling performance was studied by monitoring the roll pressure, force and torque, using slab method of analysis. The effect of back and front tension during asymmetrical rolling is also observed. The variable friction coefficient is quantified by friction coefficient ratio (μ_1/μ_2). It is observed that with increase in friction coefficient ratio, rolling pressure increases considerably. With the application of back and front tension, the rolling pressure is found to reduce. It was also shown that with more reduction per pass, the rolling pressure is found to increase. An interesting result is that the summation of friction coefficients has got more effect on the rolling pressure than the friction coefficient ratio. The friction coefficient ratio affects the shear deformation region considerably. The total rolling torque is found to decrease with increase in friction coefficient ratio.

The effect of roll roughness on the roll separating force and strip roughness can be understood from work done by Dick and Lenard [12]. With increase in roll speed, under dry condition, the roll force is either constant or decreases and in the mean time, with the increase of roll surface roughness, roll force is found to increase substantially. The authors also studied rolling in presence of three different lubricants. The strip roughness is found to increase with roll roughness. The lubricant viscosity had insignificant effect on the roll separating force.

As discussed earlier, Gudur *et al.* [8] proposed an inverse method for estimating friction coefficient by measuring the rolled sheet curvature during asymmetric rolling operation. The strip rolling is expected to show profound effect on the mechanical properties after rolling. Shen *et al.* [13] had proved that by tailoring the microstructure using rolling, one can obtain metals with high strength and high ductility. They demonstrated the effect of cold rolling on the tensile properties of interstitial free steel. In their work, it is observed that with increase in reduction from 13 to 75%, the ultimate strength increased significantly, with a substantial increase in ductility [13]. Das *et al.* [14] investigated the change in tensile properties of SS316L steel after cold rolling by ball indentation technique.

The *elastic flattening* of rolls will eventually result in more deformation of the rolls than the sheets. Thus there will be a minimum sheet thickness, below which the sheets cannot be rolled. The thinner gage sheets can be rolled by smaller radius rolls efficiently.

There are some rolling defects that affect the quality of sheets produced and hence their performance is affected. The *roll gap* must be uniform.If the roll deflection is excessive, edges of the sheet will be decreased more in thickness than the center region, and since volume and width remains constant, the sheet edges will elongate more than the center region, resulting in bow sheets. *Flatness* of the sheets should be maintained. Sometimes *waviness* develops in the sheet. If the rolls deflect in a convex fashion, the edges of the sheet will be elongated more in the longitudinal direction as compared to the center region. Since continuity has to be maintained between the sheet regions, the sheet edges will be compressed, while sheet center will be stretched in tension. This results in *wavy edge* or edge buckling. Rolls having a ground camber or crown are usually used for preventing this. *Edge cracking* occurs due to the tensile stretching of rolled sheet at the edges and compression at the center region. This occurs mainly because of the lateral expansion of the sheet. Under severe conditions, edge cracking can lead to *center split* of the sheet. *Alligatoring* (splitting of sheet across the thickness) type of fracture is also seen during rolling. Other defects like scratches during cold rolling and internal fissures are also witnessed during sheet rolling.

1.2.2 Forging

Forging is a separate class of bulk forming process wherein useful products can be obtained by shape change using hammering or pressing force. Generally the forging process is carried out in hot forming conditions (hot forging) because of less load requirements, though sometimes metals undergo forging at room temperatures (cold forging). When hammers are used, large impact blow is delivered on the material, while slow incremental compression is used to deform metal in the case of forging using a press. The hammers are used to deform the material by dissipating the kinetic energy of the ram. The mechanical presses are stroke-restricted machines since the length of the stroke and load availability at each level of the stroke determines the capability. The hydraulic presses are load-restricted machines and the maximum tonnage of the machines chiefly decides the ability and products produced by the machine.

The forging processes are divided into two categories namely "open die forging" and "closed die forging". In open die forging, the metal can flow freely without any restriction from die impression. Generally, the open die forging is used when (a) the quantity required are small to justify the cost of making dies, (b) the properties obtained after open die forging are satisfactory, and (c) the forging is too large to be produced by closed die forging. A simple example of open die forging is the "upsetting" operation. In this operation, a cylindrical billet is compressed between two flat dies. The billet bulges as shown in Fig. 1.15, and this depends on the interface friction between the flat dies and the billet. More friction will not permit the upper and lower surfaces to expand, with unrestricted expansion at the center. This forms "bulged" component. By increasing the lubrication, the upper and lower surfaces can be made to move freely equivalent to that of billet center, which will result in homogeneous upsetting with "un-bulged" sample. The examples discussed in [1], like making cylindrical work piece from square stock, forging of a gear blank and hub in open dies, four diameter spindle in open dies, etc., indicate the use of open die forging in making complex shapes with auxiliary equipment. In the case of closed die forging, the final product takes the shape of the die impression. The impression can be in one half of the die or made in two halves of the die. An example of closed die forging is shown in Fig. 1.16. The extra material is stored as "flash". The flash also controls the outward flow of metal so that the corners can be filled properly. The schematic diagram of forging load–stroke behavior during closed-die forging is shown in Fig. 1.17. The forging pressure increases with decreasing flash thickness and increasing

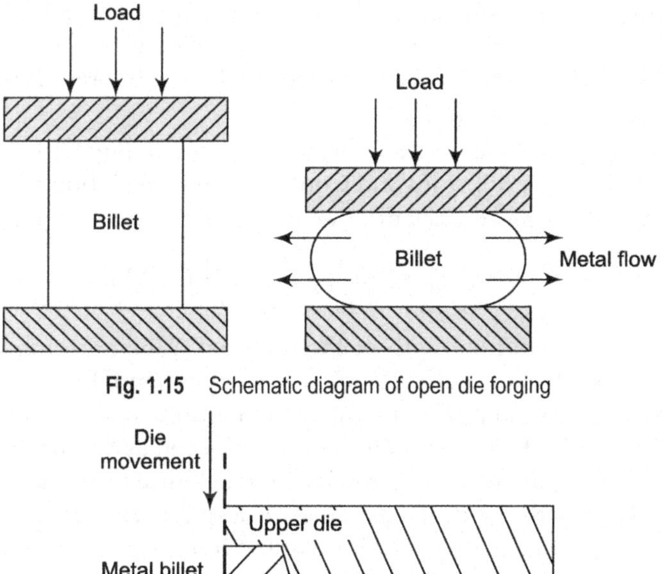

Fig. 1.15 Schematic diagram of open die forging

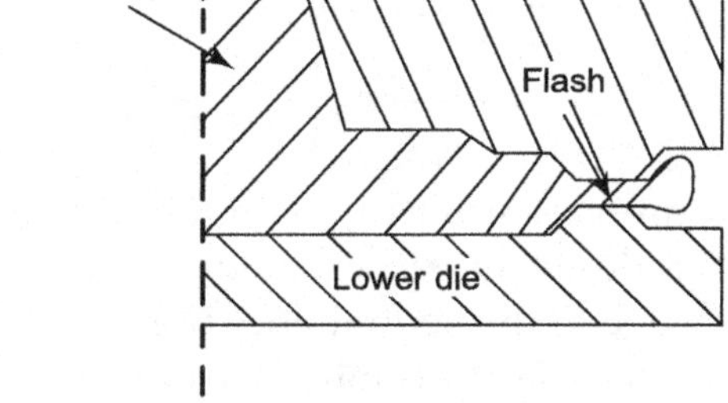

Fig. 1.16 Schematic diagram of closed die forging

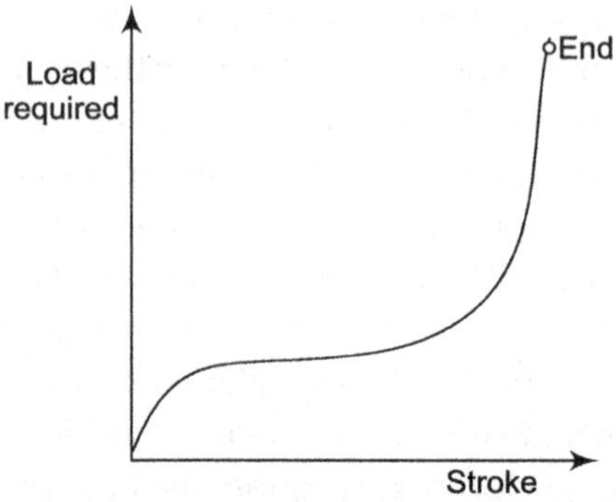

Fig. 1.17 Forging load–stroke behavior during closed-die forging

flash land width. The ideal design is to have minimum flash required to make the product. In closed die forging, heavy reductions are possible with close tolerances as compared to open die forging. Minimal post-forging machining is required in the case of closed die forging. Better mechanical properties are achievable. It is suitable for both low volume and high volume production. Aluminum alloys are easily forgeable at temperatures in the range 400–500°C and superalloys and tungsten alloys are difficult to forge even at high temperatures (1200–1300°C). Stainless steels show moderate forgeability at high temperatures (1100–1250°C).

A simple analysis of plane strain forging is shown in Fig. 1.18. By assuming the forging to be in equilibrium and Coulomb's law of sliding friction ($\tau = \mu p$ where 'μ' is the friction coefficient), the forging pressure is given by

$$p = \frac{2}{\sqrt{3}} \sigma_0 \exp\left(\frac{2\mu}{h}(l - x)\right) \tag{1.7}$$

Here 'σ_0' is the yield strength of the material. When forging pressure is plotted with the stroke , the pressure varies exponentially as per Eq (1.7). At the center of the plate the pressure is maximum which is illustrated as "friction hill" in Fig. 1.18. The center line can be seen as "neutral line" in this simple upsetting. The metal is stationary in the neutral line whereas it moves outward away from the neutral line. The same forging pressure can be obtained by assuming sticking friction condition at the interface, i.e., interfacial shear stress is equal to yield shear stress giving

$$p = \frac{2}{\sqrt{3}} \sigma_0 \left(1 + \frac{l - x}{h}\right) \tag{1.8}$$

The forging pressure varies linearly with the distance 'x' which is also shown in Fig. 1.18. Practically the friction conditions will change from sliding to sticking from the edge to the neutral line. The friction will be of sticking nature close to the neutral line. The combined friction hill is shown in Fig. 1.18 where sticking zone is visible near the neutral line showing a linear variation in pressure.

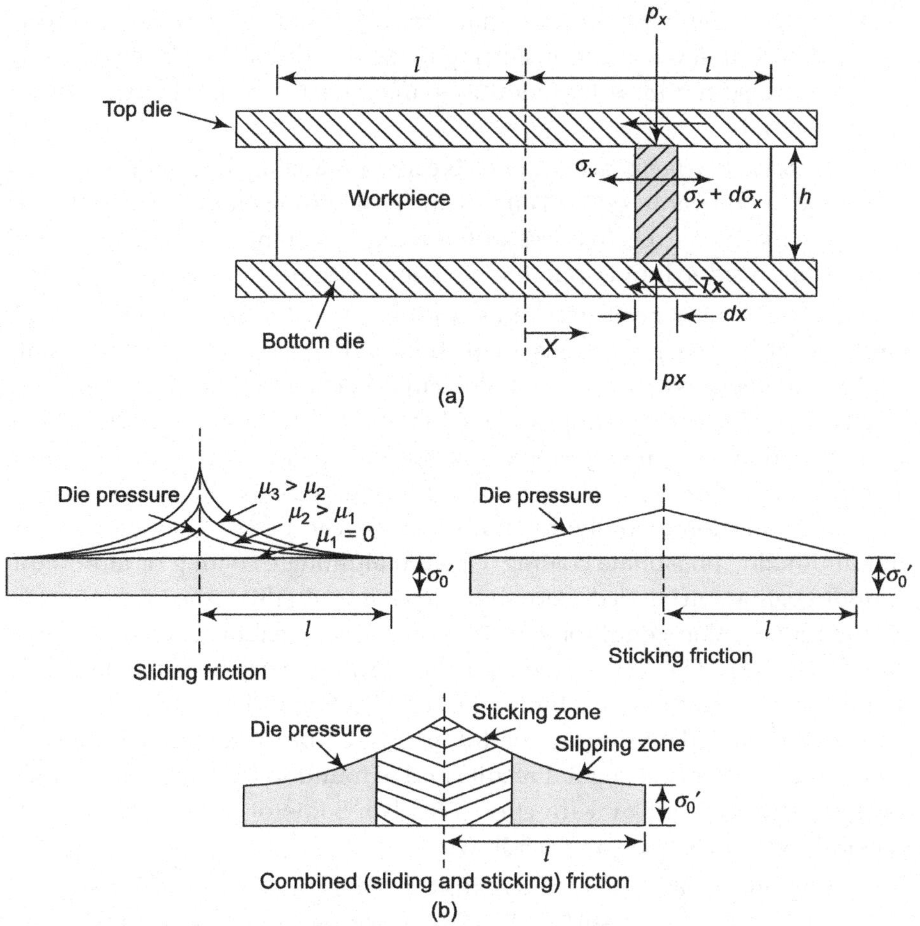

Fig. 1.18 (a) Simple analysis of plane strain forging, (b) Pressure variation at different friction conditions

There are varieties of lubricants that are used in cold and hot forging of metals. The selection of lubricants depends on the forging materials, forging operation, working temperatures, interface pressures and sliding speed. However, there are several functions that are common to lubricants for any processes and conditions [1]. Some of important functions are as follows:

- the *lubricating function* that optimizes the metal flow and controls the die filling. This is very crucial in case of closed-die forging that involves severe forming conditions requiring "hard surface coating" for successful operation
- the *tribological function* that reduces the wear and tear of the die spots that has high temperatures, sliding speed and forging pressures
- the *release function* that facilitates the easy removal of forged part from dies without adhesion during cold and hot forging

- the *heat protection function* that removes the heat generated during deformation without transferring to the dies. This if not done properly will raise the local temperatures affecting the forgeability of many materials
- the *safety function* that is responsible for minimizing the pollution caused due to lubricants, odor ill effects, fire risk, wrong handling of materials and machines, reactiveness with man and machines, tool corrosion and wear, residue build up in the die after continuous usage, etc.

Some of the important lubricants used for cold forging are, mineral oil, synthetic oil, and soap solutions. In the hot forging operations, the lubricants used are molten glass, graphite contained in water and molybdenum disulfide. Often conversion coatings are used as a lubricant carrier. Conversion coatings are the coatings obtained by the conversion of surface of the workpiece by chemical or electrochemical means. For forging of steels, phosphate coating with soap solution or molybdenum disulfide can be used. For forging of aluminum zinc phosphate coating, calcium aluminate coating or aluminum fluoride coating can be used. Ganesh Narayanan *et al.* [15] studied the influence of different friction conditions on the hardness distribution during a cold forging process. They also predicted the hardness variation during a cold forging-extrusion operation by an empirical relation developed by simple upsetting for varied lubricating conditions. The empirical equation prediction is validated with experimental results and is found to be consistent. Similar work was carried out by Kim *et al.* [16] also. Some of the defects that are generally seen in forging are as follows:

(i) *Surface layer deformation* – this defect is mainly because of the deformation being restricted to surface layer only without much happening in the interior of the forging. Instead of using light weight, rapid hammering operation, a heavy hammering operation can be used to minimize it.

(ii) *Surface cracking/defects* caused mainly by the over working of the surface at too low temperatures or as a result of hot shortness. A high sulfur content in the furnace atmosphere can cause hot shortness in steel and nickel.

(iii) *Flash cracking* during closed-die forging is another defect that is of concern as it penetrates into the forged component even after the flash is trimmed. This can be avoided by having a thicker flash or relocating the flash to a non-critical region of the forging.

(iv) *Cold shut or fold*— common in closed-die forging, is a discontinuity that is produced due to the folding or overlapping of one region of the metal into another. This occurs when the metal flows past the region that is already filled partially because of sharp die radius, excessive friction, etc. Generally sharp radius should be avoided in dies.

(v) *Underfilling*— occurs when the material is not flown fully into the corners, edges, etc. This can be avoided by proper cleaning of forging setup so that the residue accumulation and scale built-up can be minimized.

(vi) *Internal cracks*— generally seen in open-die forging like upsetting of cylindrical billets because of circumferential tensile stresses. This can be minimized by proper, continuous lubrication and using concave dies. This is not seen in closed-die forging as the lateral compressive stresses are generated by the reaction of deforming billet with the die wall in this case.

1.2.3 Extrusion

Extrusion is the process of plastically deforming a metal block through dies to reduce its area of cross-section. The metal billet is compressed by the ram against the die opening through which the metal gets extruded taking the shape of the opening. The cylindrical rods and tubes are generally made through extrusion process. During extrusion, the metal undergoes severe compressive deformation and the initial ingot microstructure is broken to obtain a part which is free of internal defects. Many a times, extrusion is used to process difficult to deform materials like stainless steel, high strength steels, nickel based alloys and materials deformable at high temperatures. Like any other forming processes, the extrusion can be cold, warm or hot.

The extrusion process can be classified into forward extrusion, backward extrusion, and lateral extrusion, depending on the metal flow with respect to ram displacement. Many of the extrusion processes are combination of these. In forward extrusion, the metal flows in the same direction as that of punch displacement. The billet is pushed through the die opening for the area reduction. In this case, there is a relative movement between billet and the die wall. Friction exists between the billet and the die wall and the billet–punch interface. In backward extrusion, the metal flows in the opposite direction as that of ram displacement, through a hollow ram. During backward extrusion, there is no relative motion between the billet and the die wall and hence there exists less friction during forming. This eventually results in less load requirement in backward extrusion when compared to forward extrusion (Fig. 1.19). Sometimes both backward and forward extrusions are combined to produce formed parts. The double cup extrusion test (DCET) is an example of this type. In *lateral extrusion*, the metal flow is perpendicular to the punch displacement. The billet is placed in the die container and pushed sideward through die opening. In *Hooker extrusion*, a variation of the forward extrusion process, a tube billet is pressed through a forward extrusion die that acts a pusher and mandrel, such that the outer diameter is reduced significantly and the tube is elongated. Similarly in the *ironing operation*, the wall thickness of the tubes is reduced by forcing it under tension/compression force into the forward extrusion die. The Schematic diagram of above described processes are shown in Fig. 1.20. *Impact extrusion,* is predominantly used to make short hollow products, like collapsible toothpaste tubes. As the name implies, the process is done on a high speed mechanical press, deforming the material by impact pressure. The process is restricted to softer materials like lead, aluminum, copper. Some of the advantages of cold extrusion over hot extrusion are dimensional control, superior surface finish, high production rates, clean and hygienic work environment, and net shaped formed components. The disadvantages are the manufacture of less complex parts, high load requirement, lubrication usage and cost, and limited deformation.

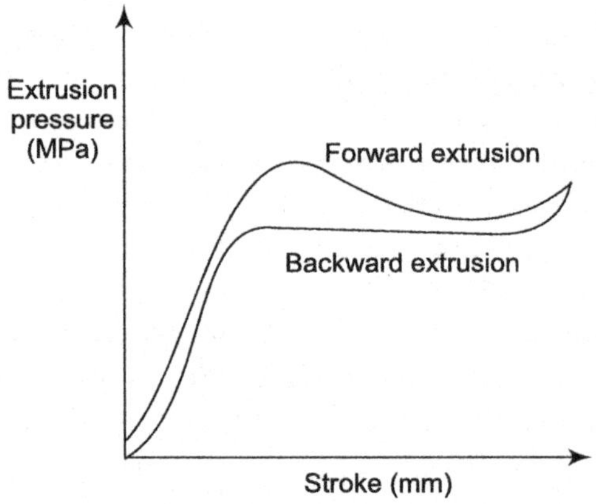

Fig. 1.19 Variation of extrusion pressure with stroke for backward and forward extrusion

The extrusion process is generally carried out in mechanical or hydraulic presses. The machines are expected to have high rigidity, accurate alignment, and long working strokes. The mechanical presses are suitable for high production rates, production volume, and large batch sizes. The extrusion process can be performed in single or multiple press stations depending on the machine capacity and reductions. The capacity of such machines varies from 1000 to 2000 tons. There are horizontal mechanical presses with coil feeds and billet shearing facility. The drive mechanism of the mechanical press also varies like crank, knuckle, link, and eccentric [1]. It is important to analyze the force–displacement curve of the press and of the extrusion process to ensure that sufficient deformation energy is available during the cycle. In multi-press station transfer presses, careful study of time–displacement curves of the press, transfer, part, and ejector is essential to ensure proper transfer. The hydraulic presses are generally suitable for low volume production, long working strokes and large extrusions. The tool design depends on the part dimension, alignment, friction, lubrication, tool setup and assembly, load distribution, and tool elastic deflection. The ram , die and shrink rings are important parts of an extrusion setup. The tools are made of AISI tool steels M2 and M4 heat treated to hardness of 60–70 HRC. Tungsten carbide tools are also used for extruding hard materials requiring large extrusion ratios. There is an optimum design of punch for favorable working of extrusion. A tapered punch with 170° included angle is found to be optimal to avoid lubricant escape during forming and prevent failures.

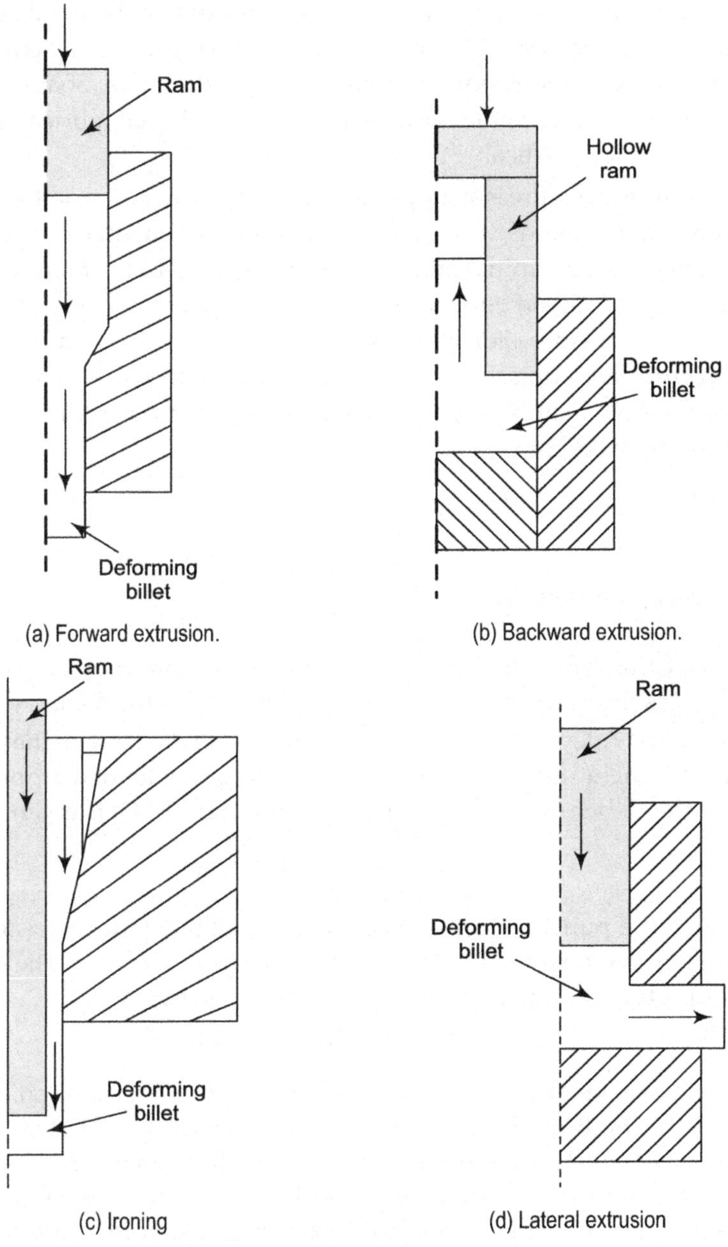

Fig. 1.20 Schematic description of different extrusion processes

The die extrusion angle (half angle) that is dependent on extrusion ratio varies between 5 and 30°. Lower angles are preferred for higher reductions. If higher angles are used, care should be taken that chevrons (arrow shaped internal defects) are not formed during multiple extrusion. An angle of 30–60° is used for hollow parts. A soap lubricant formulated to react with zinc phosphate coating, resulting in a layer of water insoluble metal soap (zinc stearate) on the surfaces of the billets are used that gives excellent lubricity even for high volume production and heavy reductions. Soaps with and without additive, polymer lubricants, oil and water based lubricants are also used in industrial practice.

The extrusion pressure is an important parameter which is monitored during extrusion and it depends on the amount of deformation, material properties, die and punch design, initial billet geometry and tool–billet interface friction. There are analytical and empirical equations that were developed to predict the extrusion pressure. For instance, Onuh *et al.* [17] developed empirical equations for predicting the normalized extrusion pressure as a function of die reduction (r), loading rate (V), and die angle for Al and Pb alloys. The equations are of the form:

For die reduction: $$\frac{P_{max}}{\sigma_0} = A + B \ln\left(\frac{1}{1-r}\right), \text{ and} \tag{1.9}$$

For loading rate: $$\frac{P_{max}}{\sigma_0} = C + D \ln (V) \tag{1.10}$$

The values for A, B, C and D are different for different materials and conditions. Here 'P_{max}' is the maximum extrusion pressure, 'σ_0' is the mean yield stress.

Ajiboye and Adeyemi [18] analyzed the impact of die land on the extrusion pressure by including the ironing effect through extended upper bound method. The dimensionless extrusion pressure [Eq. (1.11)] is given by,

$$P_{ave} = \frac{P^*}{A_0 V_0} \tag{1.11}$$

where V_0 is the punch velocity, A_0 is the original billet area, σ_0 is the mean yield strength of the material and P^* is the extrusion pressure. The extrusion power (or total powers of deformation, P^*) is given by,

$$P^* = W_i + \sum W_S + \sum W_f \tag{1.12}$$

In the above equation, W_i is the internal power of deformation, W_S is the shear loss at the boundaries of velocity discontinuity, and W_f is the total friction power which is the summation of power dissipated in ironing, friction dissipated in punch–metal interface and billet–die interface. It is observed from their work that the normalized extrusion pressure increases almost linearly with die land dimension. The theoretical calculations agree well with the experimental results. Similar work was done by the authors for varied die opening geometries like square, rectangular, I-shaped and T-shaped sections [19]. Analytical equations were developed for three different power losses that constitute the extrusion pressure by considering ironing effect for all die opening sections. The normalized extrusion pressure is found to vary linearly with the die land length.

Kudo [20] has developed the following expressions based on upper bound analyses for normalized extrusion pressure for smooth and rough square dies:

For smooth tools: $$\frac{P}{\sigma_0} = 0.88 - 1.30 \ln(1 - R) \tag{1.13}$$

For rough tools: $$\frac{P}{\sigma_0} = 1.06 - 1.55 \ln(1 - R) \tag{1.14}$$

These equations predict error less than ±5% for $R = 0.3 - 0.97$, where R is the extrusion ratio (ratio of initial cross-section area of the billet to the final area of cross-section after extrusion).

Most of the time, the extrusion is combined with forging operations to make parts. Choi *et al.* [21] studied the influence of different parameters like gap height, die corner radius, and friction factor on the forming load and volume ratio. It is observed that as compared with friction factor, gap height and die corner radius showed profound effect on the forming load and volume ratio. Tube extrusion from a rod or another tube is of interest to research community and is analyzed through analytical methods, finite element methods and experiments. It is known that during the manufacture of tubular products employing extrusion, the extrusion pressure, the state of the stresses and the flow pattern are influenced by parameters such as, reduction ratio, the shape of the die and the mandrel, the frictional conditions on the tool–metal interfaces and the strength and flow properties of the material.

A generalized slab analysis was presented by Chitkara and Aleem [22] for extrusion–piercing process through a shaped die using a general shaped mandrel starting from a hollow or solid circular billet. As per their analysis, the total extrusion force (F_e) is given by,

$$F_e = \int_{R_0}^{R_1} 2\sigma_z \pi r dr \tag{1.15}$$

where σ_z, is the axial stress component. The mean extrusion pressure on the punch/ram is estimated as,

$$P_e = \frac{F_e}{\pi(R_o^2 - R_i^2)} \tag{1.16}$$

Here R_i and R_o are inner radius and outer radius of the hollow billet, respectively. The analyses were performed for different mandrel types.

Mandrel diameters were varied as 12.8, 16.5, and 19.0 mm, for the conical mandrels and had the included cone angles of 70°, 90° and 120°. An ogival-shaped mandrel and another curved-shaped mandrel of 16.5 mm shank diameter were also fabricated. On each of the mandrel shanks, relief of 0.1–0.12 mm and land length of 3 mm was also provided. Besides these, three parallel shank mandrels of diameters 12.8, 16.5 and 19 mm were also made. Tellurium lead specimens (size: 12.5 mm diameter and 15 mm height) are used for experiments. The empirical equations (developed for extrusion pressure are as follows:

For extrusion of tubes with straight mandrel

$$\frac{P_{\text{avg}}}{\sigma_0} = 0.18 + 1.62 \ln R; \tag{1.17}$$

where R is the extrusion ratio,
and for extrusion–piercing of solid billet through shaped dies/mandrels

$$\frac{P_{\text{avg}}}{\sigma_0} = 0.6 + 1.63 \ln R; \tag{1.18}$$

An upper bound analysis on tube extrusion process can be seen in [23] also. In this work, a kinematically admissible velocity field is developed and the effect of process variables on the relative extrusion pressure is investigated. Based on this model, the equation for optimum die angle, dead zone cone angle, and relative extrusion pressure are also derived. The extrusion behavior analysis and prediction for bimetallic tubes [24–26] and composites [27–29] were presented by many researchers. The visio-plasticity method of analysis has been used to study the grid distortion, deformation zone analysis, velocity profiles, etc. A simple analysis of bimetallic rod after extrusion process can be obtained from work done by Berski *et al.* [24]. The work done by Avitzur *et al.* [25, 26] describes the prediction of fracture in the drawing of bimetallic rods and wires. The visio-plastic analysis of layered composites for different component ratio (volume of sleeve and core) using various dies is conducted by Irena and Andrzej [27]. The composite material consisting of the soft lead and hard lead are used for testing. The grid patterns were used to obtain the optimized die angle for efficient extrusion of composite material. The influence of die angle on the extrusion load is also studied. The simulation analysis of composite extrusion is studied by Schikorra and Kleiner [28], while analytical model using stream functions that can satisfy the velocity boundary conditions is proposed by Hwang and Hwang [29] for evaluating the deformation of composite rods consisting of copper sleeve layer and aluminum core layer. Experiments are conducted to validate the analytical method.

In order to increase the production rate of small components, the multi-hole extrusion process was developed. Multi-hole extrusion process is the extrusion process carried out through a die having more than one hole. The requirement of ram force is lesser in this process compared to single-hole extrusion process. The research on multi-hole extrusion process is several decades old. Dodeja and Johnson [30] carried out experiments to extrude pure lead, tellurium lead, pure tin and super pure aluminum at room temperature through square dies containing up to four holes arranged in varied fashion. They provided empirical expressions for the calculation of the ram force. Keife [31] has carried out upper bound analysis of extrusion through two die openings. Their theoretical work was validated by extruding model material like plasticine. Ulysee and Johnson [32] presented analytical and semi-analytical upper bound solutions for plane-strain extrusion through a die containing eccentric hole and unsymmetrical multi-holes. They have validated the results from upper bound analyses with the literature results and with the finite element method results. Recently Sinha *et al.* [33, 34] proposed simplified models to evaluate the single-hole and multi-hole extrusion process behavior. The ram and die pressures obtained from this are used to design a multi-hole extrusion process. Later the model proposed was validated by many experiments using lead and candle wax through various die configurations.

There are a few defects that occur during extrusion because of the way the metal flows and this depends on the friction and lubrication at the billet–tool interface. For a well lubricated container, the metal flows smoothly along the die wall and equivalent amount of deformation is seen at the billet center and the billet–die interface. The case with high friction (or low lubrication) between the billet and the die wall is characterized by a *dead zone* of stagnant material with little strain. The elements near to die wall undergo shear deformation resulting in redundant work which is not related to final extruded product dimensions. The high friction at the die wall concentrates the metal flow towards the center and an internal shear plane develops. Due to sticking friction condition, the metal shears internally along the shear plane and a thin layer of metal will remain in the container. The dead metal zone extends along the extruded outer surface. After about 75% of the extrusion is completed, the surface layer moves towards the billet center and deforms near the axis of the rod. Since the outer surface contains oxide layer, this metal flow forms internal oxide stringers and it looks like an annular ring of oxide. This defect is generally seen as *internal pipe*. This defect formation increases with increase in sticking friction condition. The usage of follower block (has slightly less diameter than the internal diameter of die) pushes the billet forward, leaving the thin oxide layer in the container itself. *Center burst or chevron cracking* is an arrow shaped defect that is formed during forward extrusion internally because of low extrusion ratio or high die angle. *Surface cracking* (Fig. 1.21) is formed on the extrude billet surface in the transverse direction by the longitudinal tensile stresses generated when the billet passes through the die. In cold extrusion, the defect is understood to occur because of the intermittent sticking of the billet to the die land resulting in sudden shoot up of load, and then breakaway.

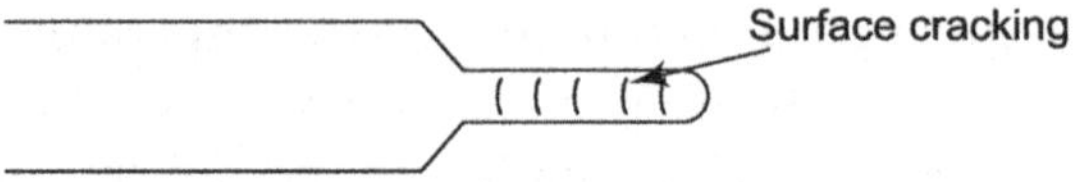

Fig. 1.21 Surface cracking during extrusion

1.2.4 Wire, Rod and Tube Drawing

Drawing operations involve pulling metal through a die hole by means of tensile force applied at the exit side of the die. This is just opposite to extrusion where metal billet is pushed through the die hole from the entry side. The area of cross-section/shape of the billet is changed during drawing operations. The metal is plastically deformed by tensile and compressive forces generated by the reaction of deforming billet with the die wall. The drawing operations are carried out in multi-stages causing heavy reductions of the input material. Generally the drawn material takes the shape and size of the die hole. Mostly the operation is done at room temperature, while some materials like tungsten (to make incandescent lamp filaments) are drawn at warm temperatures. In rod drawing, a rod is pulled through shaped die hole. The initial shape of the rod is round and is converted into another round rod or to other shapes like square, triangle, etc. In the case of tube drawing, the tube billet is drawn over a mandrel or without using a mandrel. In tube drawing without mandrel (known as tube sinking), the outer diameter of the tube is reduced, with increase in length and tube thickness. The increase in tube wall thickness and length depends on the flow stress of the material, friction conditions and die geometry. In tube drawing with fixed mandrel, generally used for drawing large to medium diameter straight tubes, the mandrel when pushed into the deforming zone, is pulled in the forward direction due to the friction generated by the translation of tube during forming. Sometimes while making long, small diameter tubes, the mandrel will also elongate and even undergo failure. In order to avoid this, floating mandrels are used for tube drawing. The Schematic diagram of rod and tube drawing (with, without mandrel) is shown in Figs 1.22 (a)–(d).

The rod and wire are not strictly distinguished. Generally the term "wire" refers to a small diameter product less than 5 mm. The rod and wire drawing operations are generally designed such that uncoiling, drawing and recoiling occur continuously at different stages. Rod coils, when ready for drawing, are usually butt welded for continuous drawing operation. The rod or wire drawing involves intermittent annealing in order to reduce the effect of strain hardening and flow stress increase during drawing.

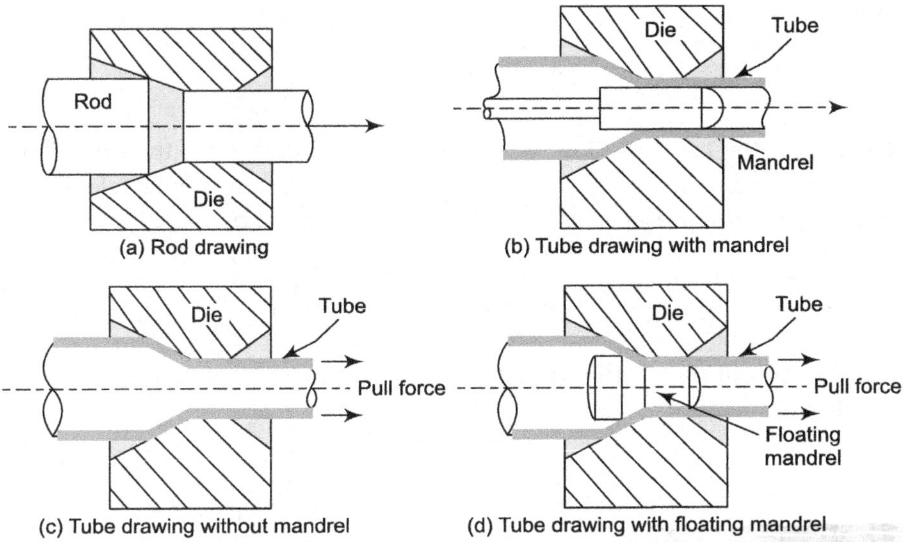

Fig. 1.22 Schematic diagram of rod and tube drawing (with, without mandrel)

The selection of tool materials for cold drawing of raw billet into rod, bar, and tube is dependent on the size, composition, shape, stock tolerance, and quantity of metal drawn. The dies used for drawing undergo severe abrasion. Dies used for drawing are made of hardened steel, tungsten carbide or diamond. The tube mandrels are usually fitted with carbide nibs. Lubrication is required for wire, bar, and tube drawing operations for reducing friction and wear. It should also dissipate heat during drawing, especially in high speed forming. It is important in tube drawing operations with fixed mandrels also (lubrication is required in floating mandrel any way). In dry drawing, the dry soap powder is used as lubricant that is placed in a die box through which the material will be passed before it is drawn in the die . This is used for drawing a steel wire of diameter 0.5–1 mm. The conversion coating consisting of sulfate or oxalate is also used. For drawing titanium, polymers are used as lubricant. For wet drawing, the lubricants are of oil-base or aqueous nature. This wet drawing practice is typical of all non-ferrous metals and steel wires less than 1 mm in diameter. The commonly used lubricant is sodium stearate. The average friction coefficient ranges from 0.01 to 0.1 in a practical wire drawing process.

A highly approximate analysis of cold drawing a wire (rigid, perfectly plastic, incompressible material) through *frictionless* conical die with the reduction in cross-sectional area from A_0 to A_1 delivered the wire drawing force 'F' and stress 'σ_w' as,

$$F = A_1 \sigma_0 \ln\left(\frac{A_0}{A_1}\right),$$

and

$$\sigma_W = \sigma_0 \ln\left(\frac{A_0}{A_1}\right) = \sigma_0 \ln\left(\frac{l_1}{l_0}\right), \tag{1.19}$$

where σ_0 is the constant flow stress. Here σ_w should be less than σ_0 so that wire does not break. The drawing stress in a die with semi-cone angle (θ) and friction (μ) in the presence of back tension (σ_b) is given by

$$\sigma_w = \frac{\sigma_0 (1+\beta)}{\beta}\left[1 - \left(\frac{D_1}{D_0}\right)^{2\beta}\right] + \sigma_b \left(\frac{D_1}{D_0}\right)^{2\beta} \tag{1.20}$$

In the absence of σ_b, the drawing stress, and force to give a reduction of 'r' where

$r = (A_0 - A_1)/A_0$ is given by

$$\sigma_w = \sigma_0 \left(1 + \frac{1}{\beta}\right)[1 - (1-r)^{\beta}], \tag{1.21}$$

and

$$F = A_1 \sigma_w \tag{1.22}$$

where

$$\beta = \frac{\mu}{\tan\theta}.$$

Similar equation like Eq. (1.21) was used by Rubio *et al.* [35] to investigate and compare the results with finite element results and experiments. Finite element modeling and analyses of wire drawing was performed by Dixit and Dixit [36] to investigate the influence of parameters like reduction ratio, semi-die angles, friction, back tension, and strain hardening. With increase in semi-die angle, the normalized drawing stress decreases to an optimum value and then increases. It is found that the optimum-die angle increase with increasing reduction for a given friction coefficient. The drawing stress increases with increase in percentage reduction. It is also seen that the optimum die angle increases with increasing coefficient of friction from zero to 0.1. Further, the variation of drawing stress with coefficient of friction is more significant at smaller die angles in comparison to that at larger die angles (Fig. 1.23). The drawing stress is observed to increase with increasing back tension. It is observed that the pattern of die-pressure distribution does not change greatly with strain-hardening. However, its value is greater with strain-hardening than without strain-hardening, along the whole taper length of the die. The influence of above said parameters on die pressure and separating force were also studied for three different materials viz., steel, aluminum, and copper.

Similar analyses were performed by Vega *et al.* [37] by comparing the experimental results with finite element simulations. They have quoted Siebel's equation for normalized drawing stress of wire drawing as,

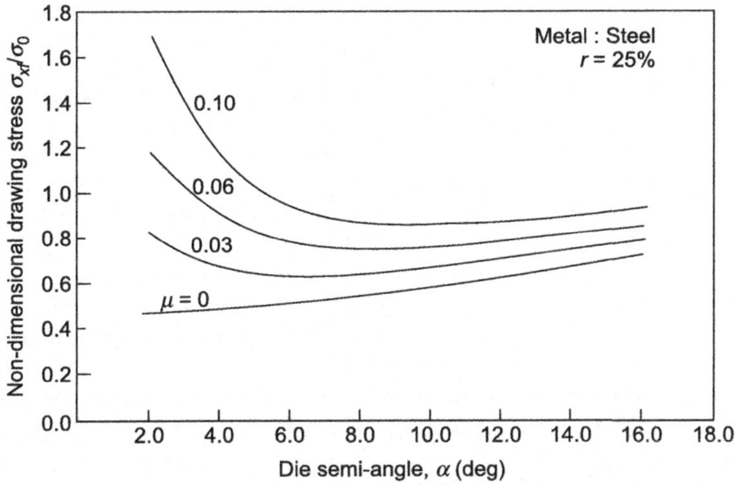

Fig. 1.23 Variation of drawing stress with die angle at different coefficients of friction during wire drawing [36]. With permission from [36] copyright [1995] Elsevier

$$\frac{\sigma_f}{\sigma_0} = \ln\left(\frac{R_i}{R_f}\right) + \mu \cot\theta \ln\left(\frac{R_i}{R_f}\right) + \frac{2\theta}{3} \tag{1.23}$$

where σ_f is the drawing stress, σ_0 is the yield stress, R_i is the initial wire radius, R_f is the wire radius after deformation, θ is the die semi-angle, and μ is the friction coefficient. An interesting behavior is the effect of semi die angle on the equivalent plastic strain distribution across the radius of the wire from center to surface, both at the entry and exit side of the die. It is observed at the entry side that with decrease in semi-die angle, the plastic strain is uniformly distributed. The strain distribution is uniform along the wire radius for semi-die angle of 7° and 8°, whereas it is not uniform at higher angles 14° and 18°. Similar behavior is observed at the exit side of the die also.

Avitzur [38] has proposed the normalized drawing stress as,

$$\frac{\sigma_f}{\sigma_0} = \frac{\left[\begin{array}{l}\dfrac{\sigma_b}{\sigma_0} + 2f(\theta)\ln\left(\dfrac{R_i}{R_f}\right) + \dfrac{2}{\sqrt{3}}\left(\dfrac{\theta}{\sin^2\theta} - \cot\theta\right) \\ + 2\mu\left(\cot(\theta)\left(1 - \dfrac{\sigma_b}{\sigma_0} - \ln\left(\dfrac{R_i}{R_f}\right)\right)\ln\left(\dfrac{R_i}{R_f}\right) + \dfrac{P}{R_f}\right)\end{array}\right]}{1 + 2\mu\dfrac{P}{R_f}} \tag{1.24}$$

where σ_b is the back tension and P is the bearing length.

In the case of tube drawing, the plug or mandrel decides the plastic deformation of tube wall. Under plane strain conditions, for tube drawing with a plug, the draw stress [2] is given by,

$$\left.\begin{aligned}\sigma_{xa} &= \sigma_0' \frac{1+B}{B}\left[1-\left(\frac{h_a}{h_b}\right)^B\right]\\ \text{where}\qquad & \\ B &= \frac{\mu_1+\mu_2}{\tan\alpha-\tan\beta}\ \text{(for fixed mandrel)}\end{aligned}\right\}\qquad(1.25)$$

and

$$B = \frac{\mu_1-\mu_2}{\tan\alpha-\tan\beta}\ \text{(for floating mandrel)}$$

Here μ_1 is the friction coefficient between tube and the die wall, μ_2 is the friction coefficient between tube and plug, α is the semi-cone angle of die, β is the semi-cone angle of plug (= 0° for cylindrical plug).

There are a few defects that are generated during wire drawing process. A comprehensive analysis of the surface defects can be found in Shinohara and Yoshida [39]. The important defects are scab, transverse crack, burnt, scratch, overlap, scale, rolled-in material, over filled, slip mark, rust. The experimental and finite element simulations were compared to study the evolution of various flaws during stainless steel wire drawing. Similarly the prediction of center burst, an arrow shaped internal defects, is studied by Ko and Kim [40] for extrusion and wire drawing. In this they have used Cockcroft–Latham criterion to predict the center burst defects and compared it with experiments and criterion given by Avitzur. Avitzur's criterion is based on an upper bound analysis assuming that when the central burst occurs, energy required for the flow with defect will be lesser than that for a defect-less flow By finite element analysis, Dixit and Dixit [36] observed that during the occurrence of central burst, the plastic zone narrows down at the center and strain-rates near to the die surface are much higher compared to the center of the wire.

1.3 SHEET FORMING PROCESS

The sheet metal forming processes involve plastic deformation of metallic sheets with the help of tools like punch, die, blank holder, draw bead to make useful sheet products. The material properties (like *n*, *m*, *r*); process parameters like friction between the tools and sheet, blank holding force, working temperature, strain rate; die and punch geometry influence the formability of sheet parts in a compounding fashion. Formability is nothing but the ability of material to be formed under the action of forces till occurrence of neck/failure. The important sheet forming operations like deep drawing, stretching (or stretch forming), bending, spinning are described here.

1.3.1 Deep Drawing

Deep drawing of sheet materials is used to make sheet products by pushing the sheet at the center portion into the die hole while holding it with the help of blank holder. The sheet is drawn in, to make a cup like component without wrinkling and failure. The blank holder is used to avoid or minimize wrinkling during this operation. The deep drawing operation is schematically shown in Fig. 1.24 for making simple cylindrical cup. The deformation is complex with tension in wall and bottom, compression and friction at the flange region (un-deformed region of sheet lying on the die surface), straightening in the die wall and bending at the punch and die radius. The process is utilized to make beverage cans, washbasins, cooking utensils, cylinders, automotive body parts, pressure vessels, etc. The clearance between the punch and die wall is decided as per the sheet thickness such that ironing is prevented. Some times ironing is also included as part of forming process which is performed subsequently. Optimum radii in punch bottom to side wall, and approach to die opening has to be provided to have safe drawing operation without tearing. The radii should be smooth and designed such that strain localization does not occur in those regions.

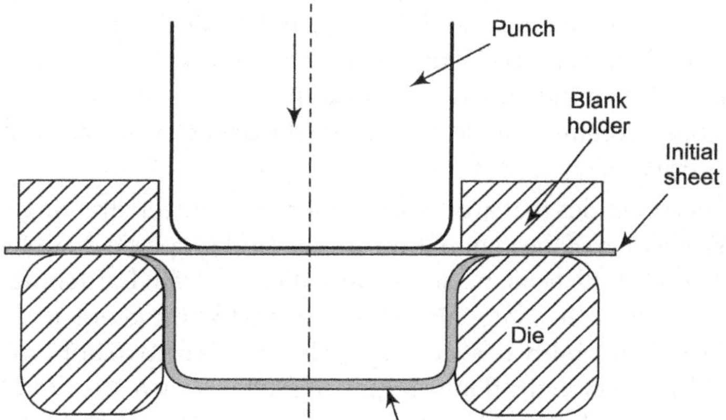

Fig. 1.24 Deep drawing of cylindrical cup

In most of the deep drawing operations, the cup is partially drawn, the flange region of the sheet is trimmed if not required, and the straight cup region is used. A desirable, but suitable, blank holding force (BHF) is applied on the sheet such that neither tearing (near the holding region) not wrinkling occurs during the deep drawing operation. Wrinkling occurs because of the compressive hoop stresses developed in the flange region. Two broad approaches are used to analyze the flange wrinkling: (i) *energy method*, and (ii) *bifurcation method*. These methods predict the critical values of certain process parameters at the initiation of wrinkling. In the energy method, a certain wave-form of the flange is assumed and the critical values are determined from the condition that this wave-form must be in equilibrium. In the bifurcation method, the critical values are obtained from the condition at which the solution of the (non-linear) governing equations becomes non-unique. These approaches will be described in detail later in the book.

In some cases, the cup shape is fully drawn into the die opening, and a straight wall cup shape is ejected through the die opening. Both mechanical and hydraulic presses are used for deep drawing operations. Hydraulic presses are generally preferred because of the controllability of punch movement. This operation can be performed in single or in multiple stages, with or without annealing in between stages.

At the start of the drawing process, the bottom of the punch touches the sheet surface. The sheet wraps around the punch bottom corner and on the die opening corner. The punch pushes the sheet into the die cavity that forms a straight side wall with some sheet part in the flange region. If the metal has a high resistance to thinning and thickening, the bottom radius and the upper sidewall areas remain close to their original thickness, and the radial and circumferential strains obtained in the drawn-in flange are increased to accommodate the deep-drawing process. Once the bottom is formed, the clearance between the punch and die is such that the metal in the cup side is free to move without excessive rubbing on the die walls. It is observed that slight roughening of the punch radius and minimizing the lubrication of this area contribute to better drawability, while the die opening should be smooth and well lubricated.

The sheet properties that are important for proper deep drawing operations are, the composition of the material; the mechanical properties like elongation—obtained from tensile tests, plastic strain ratio (r), strain hardening exponent (n), yield strength of the material; and properties like die geometry, and elastic modulus. The drawability of sheet material depends on two important factors: (i) the ability of sheet in the flange region to move freely in the radial direction in the plane of the sheet under shear deformation, (ii) the ability of sidewall material to resist deformation in the thickness direction. In order to have good deep drawability, maximum material flow is desirable in the plane of the sheet and maximum resistance to material flow in a direction perpendicular to the plane of the sheet. Since it is difficult to evaluate the flow strength in the thickness direction, plastic strain ratio (r) is used to compare strengths in the plane and thickness directions by determining true strains in these directions in a tension test. The plastic strain ratio is defined as the ratio of true width strain (ε_w) to true thickness strain (ε_t) for a material strained in longitudinal (perpendicular to width and thickness) direction, i.e.,

$$r = \frac{\varepsilon_w}{\varepsilon_t} \tag{1.26}$$

The sheet materials are anisotropic in nature, i.e., the properties of the sheet are different in different directions. Therefore, it is necessary to use the average of the strain ratios measured along 0°, 90°, and 45° to the rolling direction of the sheet to obtain an average strain ratio, $\bar{r}$, which is expressed as

$$\bar{r} = \frac{r_0 + r_{90} + 2r_{45}}{4} \tag{1.27}$$

where r_0, r_{45}, r_{90} are plastic strain ratios in 0, 45, and 90° to the rolling direction of the sheet. Here $\bar{r}$ is the measure of normal anisotropy of the sheet. For isotropic material, $\bar{r} = 1$. For higher resistance to thinning, $\bar{r} > 1$. The higher the normal anisotropy value, higher the resistance to thinning and hence deeper draws can be achieved. For many types of steel, for $\bar{r} > 1$, the flow stress in thickness direction will be more than the in-plane flow stress. The reverse is the case for $\bar{r} < 1$. However, for some commercial aluminum alloy and steel sheets an anomalous behavior is observed in which the flow stress in thickness direction is greater than in-plane flow stress, even when $\bar{r} < 1$.

There is another measure known as planar anisotropy (Δr) expressed as

$$\Delta r = \frac{r_0 + r_{90} - 2r_{45}}{2} \tag{1.28}$$

which is a measure of how different the 45° directions are from the symmetry axes. Drawability can also be expressed in terms of limiting draw ratio (LDR) or percentage of reduction based on cup tests. The limiting draw ratio (LDR) is the ratio of the diameter *D* of the largest blank that can be successfully drawn to the diameter of the punch *d*, i.e.,

$$\text{LDR} = \frac{D}{d} \tag{1.29}$$

The percentage reduction is then defined as

$$\text{percentage reduction} = \frac{100\,(D - d)}{D}. \tag{1.30}$$

The drawing operation is carried out either in hydraulic or mechanical presses. A double action hydraulic press is favored so that a uniform blank holding force can be applied over the entire stroke of the deep drawing operation. Double action hydraulic presses with die cushion arrangement are also preferred because of their constant drawing speed, stroke adjustment, and uniform application of clamping pressure.

The process parameters that affect the deep drawing behavior include punch and die radii, punch to die clearance, press speed, lubrication, and type of restraint of metal flow used. Material properties such as sheet thickness and anisotropy also affect deep drawing. The deep drawing force (maximum value and load–stroke behavior) and failure analyses are the important drawing behaviors that are predicted for a successful deep drawing operation. A typical punch force (or load)–stroke diagram during deep drawing operation is shown in Fig. 1.25. The following equation was developed through equilibrium analyses of deep drawn cup to evaluate the total punch force [2]:

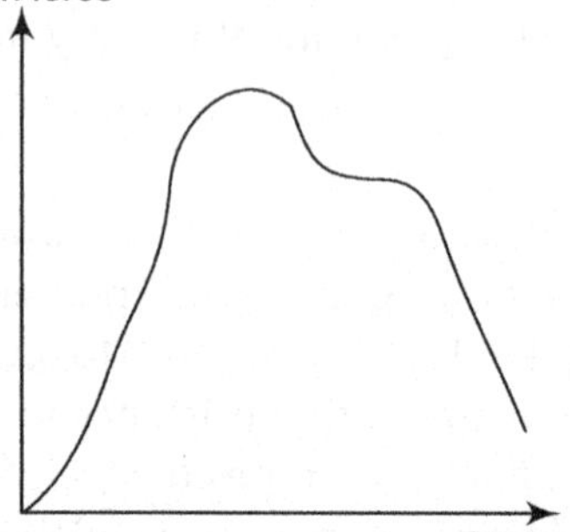

Fig. 1.25 Punch force–stroke diagram during deep drawing

$$P_T = \left[\pi D_p h\,(1.1\,\sigma_0)\ln\frac{D_0}{D_p} + \mu\left(2P_b\,\frac{D_p}{D_0}\right)\right]e^{\mu\pi/2} + B \quad (1.31)$$

where P_T is the total punch load, σ_0 is the average flow stress, D_p and D_0 are punch and blank diameters, P_b is the blank holding force, h is the wall thickness, μ is the friction coefficient, and B is the force required to bend and re-straighten blank. The load in deep drawing is applied to the cup bottom and then it is transmitted to the die wall. Generally failure occurs in the annular region just above the punch radius. This region undergoes tensile stretching without any bending and radial drawing. Failure occurs through necking, later converted into tearing, at a stress approximately equal to tensile strength, increased by a plane strain factor [2]. This is given by

$$P_{max} = \frac{2}{\sqrt{3}}\,F_u \pi D_p h, \quad (1.32)$$

where F_u is the tensile strength.

Also the theoretical upper limit of LDR is given by

$$\text{LDR} \approx \left(\frac{D_0}{D_P}\right)_{max} \approx e^{\eta}, \quad (1.33)$$

where η is an efficiency term accounted for friction losses. For $\eta = 1$, LDR = 2.7, while LDR ≈ 2 for $\eta = 0.7$. This agrees with the observation of how difficult it is to draw a cup with height more than its diameter even for a ductile material. More comprehensive analyses of deep drawing for evaluating stress, strains can be seen in the work done by Budiansky and Wang [41] and Woo [42].

The selection of *punch to die clearance* is important in deciding the smooth metal flow without burnishing of side wall and punching out of cup bottom. This can be guaranteed with a clearance per side of 7–15% of sheet thickness [1]. The drawing force will be minimal when the clearance per side is 15–20%

of sheet thickness and the cupped portions of the part are not in contact with the walls of the punch and die. The drawing force increases as the clearance decreases. It should be noted that a secondary peak occurs on the load–stroke curve where the sheet metal thickness is slightly greater than the clearance provided and thereafter ironing starts. The clearance between the punch and die for a rectangular cup (at the sidewalls and ends) is same as that for a circular cup. Clearance at the corners may be as much as 50% of sheet thickness to avoid ironing in these regions of the cup and to increase drawability. Redrawing operations require greater punch to die clearance as compared to first draw in order to compensate for thickening during cupping.

The *restraint given to the metal flow* during deep drawing is provided by the sheet thickness and the die radius. Additionally for successful operation, the blank holding force is also applied that prevents wrinkling and controls the metal flow. The ratio (l/t) between the supported length (the length from the edge of the blank to the die cavity) and the sheet thickness is a critical parameter to decide the usage of blank holding during drawing operation. The usage of blank holder prevents wrinkling, which if present, would deteriorate the deep drawability of the sheet. Much greater drawability is possible when a blank holder is used. The blank holders can be used in single action or double action presses. In double action presses, the blank holding is done first and then the punch is moved down to deform the sheet. The holding process is continued till the process is completed. By using die cushion arrangement and inverted die, similar action can be obtained in a single action press. The flat holders are used in many cases that restrict the movement just because of the friction between the sheet and the holder. Some times the blank holder can be used with drawbead facility that gives more restriction to the sheet movement required for making a product. The usage of drawbead will provide more stretching component rather than pure drawing. The drawbeads are used to lock the sheets in which case only stretching is present without any drawing mode of forming. There are flexible blank holders that are used to control the metal flow locally in specific locations. For example, Kinsey *et al.* [43] developed a novel method to control the weld line movement locally and this delivered a Tailor Welded Blank (TWB) product with increased formability.

The *press speed* for deep drawing operation is dependent on the sheet material formability, lubrication, toolings, and blank holder pressure. Speed is of greater significance in drawing stainless steels and heat resistant alloys than in drawing softer, more formable materials. It is observed that excessive press speeds have caused cracking and more wall-thinning in drawing a stronger, less ductile metals. By decreasing the speed, the stretching part is reduced and gives the metal enough time to undergo plastic deformation.

As discussed in other metal forming operations, *lubrication* significantly affects the performance of deep drawing operation. In deep drawing operation, lubrication serves two purposes. Firstly it reduces friction and prevent/ minimize galling (pressure welding) of tools and sheet material, wrinkling, and tearing; secondly, it should not enhance corrosion. Water based and oil based (like fatty oil, mineral oil, mill oil, etc.) lubricants are commonly used for drawing applications.

Material properties like *anisotropy and sheet thickness* also affect the deep drawing behavior. Planar anisotropy (defines the variation of properties in sheet plane), if very high, causes earing defect, which is nothing but the result of differential drawability of regions in the sheet plane. The thicker metal in the valleys between the ears forces the die open against the blankholder pressure, allowing the metal in the relatively thin areas near the ears to wrinkle. Sheet thickness affects the drawability through pressure requirement and wrinkling. Thinner blanks are more susceptible to wrinkling than thicker ones. Hence the blank holding pressure requirement for thicker sheets is minimum or even less than that required for thinner sheets. The finite element simulation by Padmanabhan *et al.* [44] suggests that by using variable blank holding force (BHF) and friction during drawing of LPG bottles, the metal flow and minimum thickness achieved can be enhanced. It is indicated by them that a low constant BHF in the starting stage prevents necking failure between punch and die radius and it plays a prominent role in the thickness distribution of the drawn part. When the punch force remains constant, an increasing blank holder force restrains the wrinkling tendency and enables a smooth flow of material into the die cavity [44].

Other than wrinkling, tearing, earing, the defects like orange peeling and stretcher strains also destroy the structural integrity of the formed parts. *Orange peeling* is the generation of higher surface roughness in the region of sheet that has undergone appreciable deformation. This is seen generally in large grain size material. This is due to the fact that the individual grains tend to deform independently of each other, and hence the grains stand out in relief on the deforming surface. This is better minimized by using finer grain sized material so that the grains deform as a whole. The *stretcher strains* are depressions on the sheet surface, mostly seen in low carbon steel sheets. This is directly related to presence of yield point in the stress–strain curve and the non-uniform deformation arising from the yield point elongation. This is eliminated by small amount of cold reduction (2%) in thickness. The temper rolling or skin rolling treatment cold works the material to eliminate the stretcher strains.

1.3.2 Stretch Forming

Stretch forming of sheets (or stretching) is a sheet forming process in which the sheet is formed over the die by keeping it under tension. The sheet is held at certain locations with the help of jaws or drawbeads, and then is wrapped around the form block or die to give a permanent shape change to the part. It is primarily used for making automotive door panels, automotive roof, aircraft skin panels, window frames, etc. Some of the advantages of stretch forming over other conventional forming operations are reduced springback, less likelihood of wrinkling, less force requirement, increase in strength, and hardness of material, lower residual stresses, no requirement of female die part, and usage of less expensive die materials for room temperature operations. Some of the limitations are as follows: the operation is not suitable for high volume production; it is best suited for shallow or flat components; not suited to progressive or transfer operations etc. The stretch forming is categorized into four different categories namely stretch draw forming, stretch wrapping, compression forming, and radial forming. The difference between these categories lies in the manner by which the sheets are stretched.

In *stretch draw forming*, the sheet is held in tension by grippers at the ends. A die or form block is moved against the sheet and is stretched taking the shape of the form block surface. Alternatively the form block is kept stationary; the grippers are moved such that the sheet is stretched over the block. The stretch draw forming can be used to make small contour shapes with the help of mating dies. In *stretch wrapping*, sufficient tension is applied to one end of a workpiece, while the form block revolves into the workpiece with the help of a turning table. The other end of the workpiece is clamped in a table gripper to the end of the form block.

The sheet stretching over the die is influenced by material parameters like strain hardening exponent (n), strain rate sensitivity (m), sheet anisotropy (r); friction conditions, and tool geometries. The stretching is used in many standard formability tests like limit dome height test (LDH) in which the sheet is gripped using drawbead and stretched using a hemispherical punch till necking occurs. The height (or strain called limit strain) at which failure/ necking occurs is the measure of formability of sheets. In a simple manner, the elongation at failure during tensile testing is the measure of ductility of sheet metals. The necking phenomenon in sheet stretching operation occurs in two steps — *diffused necking* and *localized necking*. In diffused necking, the neck region is extended over larger sheet region. For diffused necking, Considere has proposed a criterion during deformation of thin sheet, as early as 1885, in which the condition of instability resulting in localized deformation is defined by $dP = 0$. By assuming volume constancy relationship, the tensile instability condition is given by,

$$\frac{d\sigma}{d\varepsilon} = \sigma \tag{1.34}$$

In terms of engineering strain, the necking criterion can be expressed as,

$$\frac{d\sigma}{de} = \frac{\sigma}{1+e} \tag{1.35}$$

A simple form of the criterion is derived as below. From the definition of 'n',

$$n = \frac{d(\ln \sigma)}{d(\ln \varepsilon)} = \frac{\varepsilon}{\sigma}\frac{d\sigma}{d\varepsilon} \Rightarrow \frac{d\sigma}{d\varepsilon} = n\frac{\sigma}{\varepsilon} \tag{1.36}$$

By further substitution, for instability or necking to start,

$$\varepsilon_u = n \tag{1.37}$$

where 'ε_u' is the true uniform strain.

For localized necking during tensile deformation, $\frac{d\sigma}{d\varepsilon} = \frac{\sigma}{2}$. (1.38)

The localized necking is the case when the neck is restricted over a small, narrow band with a width equal to sheet thickness, inclined at $\theta = \pm 55°$ (approximately) to the major straining direction in the case of isotropic material. In the case of anisotropic material,

$$\theta = \arctan\left(\sqrt{\frac{(R+1)}{R}}\right) \tag{1.39}$$

where $R = \frac{d\varepsilon_2}{d\varepsilon_3}$, ε_2 is the strain along width direction, and ε_3 is the strain along thickness direction [45].

Similarly for a balanced biaxial tension condition of sheet metal, the instability occurs when, $\bar{\varepsilon} = 2n$, by assuming power law ($\bar{\sigma} = k\bar{\varepsilon}^n$) as strain hardening law. In the case of thin walled sphere under internal pressure, instability occurs when, $\bar{\varepsilon} = \frac{2}{3}n$ [2].

The impact of strain hardening exponent (n) and strain rate sensitivity (m) is significant during sheet stretching operation. The strain hardening exponent basically decides the strain gradient in the sheet material during forming. The strain gradient arises whenever the deformation is not uniform. The deformation becomes more uniform (reduced strain gradient) with increase in strain hardening exponent (larger n). This is expressed as Eq. (1.40) [2] by assuming Hollomon strain hardening law, i.e.,

$$\frac{d\varepsilon}{dr} = -\frac{(\varepsilon/r)}{n} \tag{1.40}$$

Once the material region is strain hardened, the load is passed on to the neighboring regions. This will make the material to strain more and in doing so, the strain gradient is reduced. The uniform strain gradient will result in increased formability of sheet material, i.e., sheet materials with larger n value exhibit better formability when compared with sheets with lower n value. The influence of strain rate sensitivity (m) on the forming extent is similar to the effect of n value. The m value governs the post-necking behavior. The conversion of diffused necking to localized necking is characterized by this value. The higher the *m* value, the more the diffused necking, and greater the deformation beyond the maximum load before fracture occurs. Thus, higher the *m* value, larger will be the forming limit.

During stretching operation, the formability of sheet metal is best described by *forming limit diagram* (*FLD*). In this diagram, *forming limit curve* (*FLC*), plotted between major strain (in Y-axis) and minor strain (in X-axis), is the index that says the amount of safe strains that can be incorporated into the sheet metal. A typical FLC is shown in Fig. 1.26. The FLC is the locus of all the limit strains in different strain paths of the sheet material. The plane-strain condition possesses the least forming limit, when compared with deep drawing and stretching strain paths. The influence of *n*, *m*, and *r* values on the predicted forming limit by Marciniak and Kuczynski (M-K) model is described in [46, 47]. It is seen that the positive '*m*' value (= 0.012) improves the forming limit as compared to negative '*m*' value (= – 0.012) by delaying the conversion of diffused necking to localized necking and then to failure [47]. The influence of '*r*' value on the forming limit curve is dependent on the yield criterion used for modeling. By using Hosford non-quadratic yield criterion, it is shown that '*r*' value has got negligible effect on the forming limit strains of sheet metal. On the other hand, by assuming planar isotropy, it is predicted that on decreasing the '*r*' value the level of the FLC is increased on the right hand side of FLD [46, 47]. This is explained by Sowerby and Duncan [47] that on increasing the '*r*' value from 0.5 to 3.0, the forming limit strain is found to decrease considerably. Like the effect of '*m*' value, on increasing '*n*' value from 0.1 to 0.4, the forming limit is found to increase considerably [47].

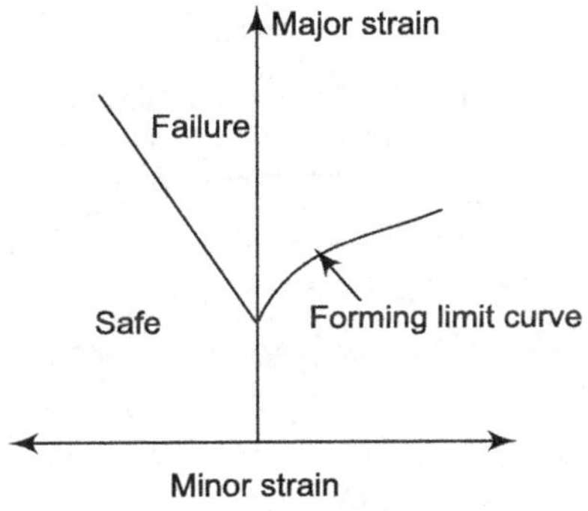

Fig. 1.26 Forming limit diagram (FLD) of sheets

1.3.3 Sheet Bending

Bending of sheets is a common sheet forming operation in which the deformation is concentrated in a small curved region, resulting in a non-uniform deformation across the thickness and curvature of the sheet. The different bending operations like V-bending, L-bending and U-bending are shown in Fig. 1.27. These operations are distinguished based on the dies used, with U-bending resulting in a channel type component. The sheet outer surface undergoes tension and inner surface undergoes compression during bending as shown in Fig. 1.28. When the metal is deformed beyond the elastic limit, permanent set occurs. However, some elastic deformation usually remains, so that the metal tends to *spring back* as the loading is released. This springback is compensated by overbending. If failure occurs during bending, it will occur on the outer surface, and wrinkling is witnessed on the inner surface. The friction forces act in the opposite direction to the sliding direction of sheet during bending and hence proper lubrication is required for the efficient running of the process. The identification of neutral axis in a bending operation is crucial in bending analyses. The neutral axis is the line of zero stress and strain in a bend with tension on one side and compression on the other side of the line. Before bending, the outer surface, inner surface lengths and neutral axis lengths are same as that of sheet length. After bending, the outer surface length is increased, while inner surface length is decreased. But the neutral axis length remains the same and hence it is considered as a reference for blank development calculations.

When the sheet is bent for the first time, the neutral axis is at the center of the sheet thickness. As bending progresses, the neutral axis shifts toward the inside, or compression side, of the bend.

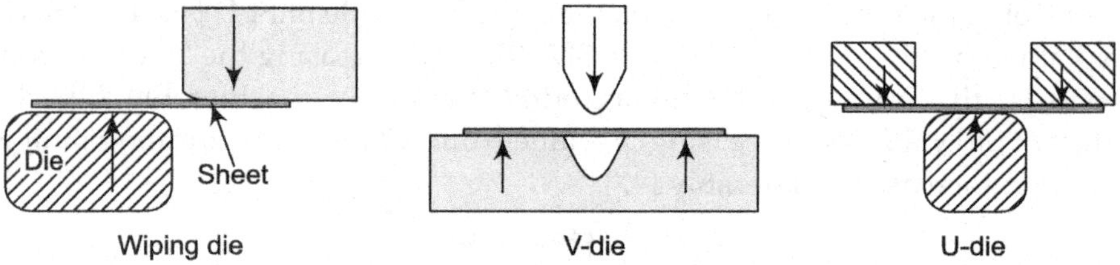

Fig. 1.27 Bending operations.

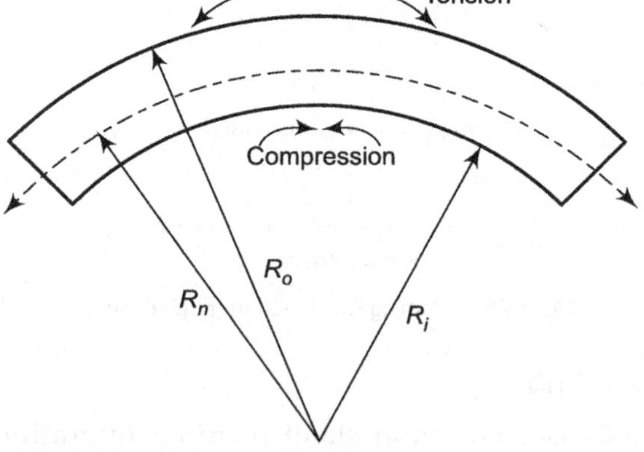

Fig. 1.28 Sheet bending showing tension and compression on the surfaces

In bending a thin sheet, it may be assumed that a plane normal section in the sheet will remain plane and normal and converge on the center of curvature as shown in Fig. 1.29. The line A_0B_0 at the middle surface may change its length to *AB*, if the sheet is under stretching during bending. The original length l_o becomes, $l_s = \rho\theta$. A line C_0D_0 at a distance y from the middle surface will deform to a length,

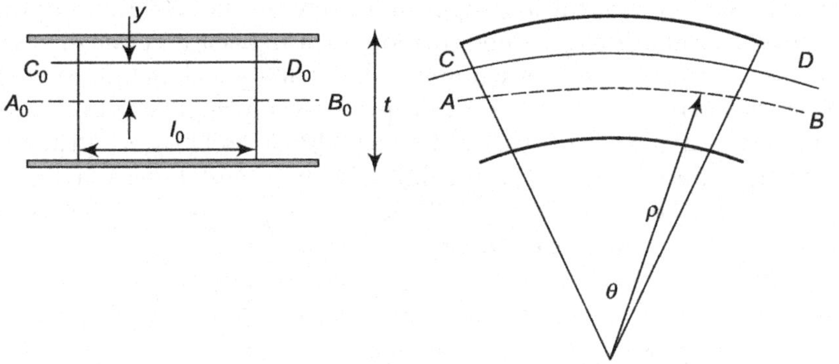

Fig. 1.29 Deformation of longitudinal fibres during bending and stretching

$$l = \theta(\rho + y) = \rho\theta\left(1 + \frac{y}{\rho}\right) = l_s\left(1 + \frac{y}{\rho}\right), \quad (1.41)$$

where ρ is the radius of curvature.

The axial strain of the fiber *CD* is

$$\varepsilon_1 = \ln\frac{l}{l_0} = \ln\frac{l_s}{l_0} + \ln\left(1 + \frac{y}{\rho}\right) = \varepsilon_a + \varepsilon_b \quad (1.42)$$

where 'ε_a' and 'ε_b' are the strains at the middle surface and bending strain, respectively. In the case of bending with radius of curvature larger compared to the thickness, the bending strain is approximated as

$$\varepsilon_b = \ln\left(1 + \frac{y}{\rho}\right) \approx \frac{y}{\rho} \quad (1.43)$$

The strain distribution is linear [48] across the thickness. For the strain distribution given by Eq. (1.43), the stress distribution can be evaluated by following a stress–strain law. This stress–strain law is usually approximated to have several material models like elastic bending, rigid-perfectly plastic bending, elastic-perfectly plastic bending, strain hardening sheet, etc. [48]. The strain also depends on the *bend ratio,* which is defined as the ratio of the radius of curvature to sheet thickness, ρ/t. Bending without tension is done by pure moment without any tension being applied. The neutral axis is at the mid-thickness in this case.

Spring back is considered as an important phenomenon during bending operations. The tensile stress, A, above the neutral line cause plastic deformation and it is less than ultimate tensile strength [Fig. 1.30(a)]. Upon load removal after first bending, the elastic band near the neutral axis [Fig. 1.30(b)] tries to return to the original flat condition but cannot, due to the restriction given by the plastic deformed regions. Some return occurs which is named as *springback*. Increase in elastic limit/yield strength, increases the springback and hence stronger sheets have greater degrees of springback. Springback is lower, when elastic modulus is reduced and plastic strain in increased. The springback prediction will be helpful in proper and accurate die design, minimizing springback, standardizing methods to reduce springback, accurate formability prediction, etc. There are many numerical and analytical methods by which springback can be predicted. More information on springback prediction can be obtained from the work done by Mullan [49], Gau and Kinzel [50] including Bauschinger effect, Li *et al.* [51] using finite element simulations, Wagoner *et al.* [52] simulating the effect of through thickness integration points, Bahloul *et al.* [53] using response surface methodology and analyses done by many others. The die design prediction by including springback errors can be understood from the work of Karafillis and Boyce [54] and Gan and Wagoner [55].

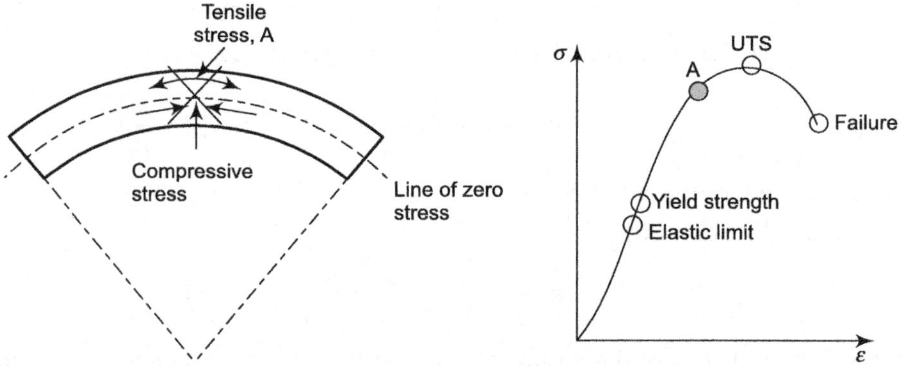

Fig. 1.30 (a) Changing stress patterns in a bend

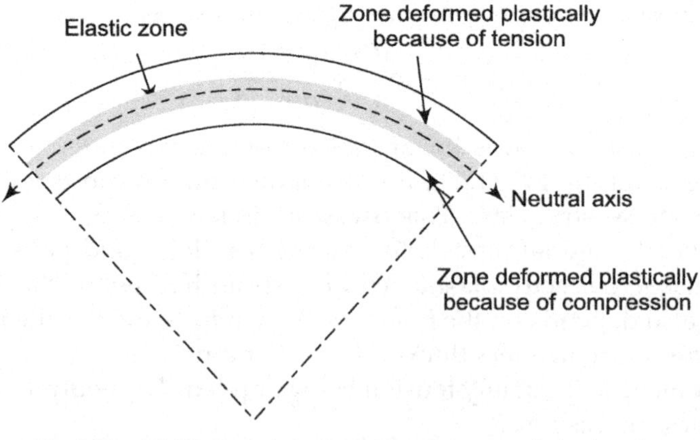

Fig. 1.30 (b) Elastic and plastic deformation zones during bending

1.3.4 Spinning

In spinning, the blank of certain thickness is held by a mandrel of certain shape, while the tool is made to press the blank against the mandrel. At the end of spinning process, blank takes the shape of the mandrel that is axi-symmetric. There are three types of metal spinning operations— manual spinning, power spinning, and tube spinning. In manual spinning, the tool used for spinning operation is manually or hydraulically operated as shown in Fig. 1.31. Generally the manual spinning is carried out at room temperatures. However, at elevated temperature metal spinning is performed for components with thick sections or for alloys with low ductility. It is meant for low volume production (say less than 1000 pieces). Larger volume parts are usually produced at lower cost by power spinning. Metal spinning is used to make components made of aluminum alloys, stainless steels, super alloys, brass, copper, and platinum. Power spinning can be carried out at high temperatures also. Since high shear stresses are involved in this, the heat generated can be removed only with proper lubrication. The surface roughness produced in this is of the order of 0.5–1.5 μm. Buckling, wrinkling, and formation of circumferential cracks are some of the defects seen during spinning.

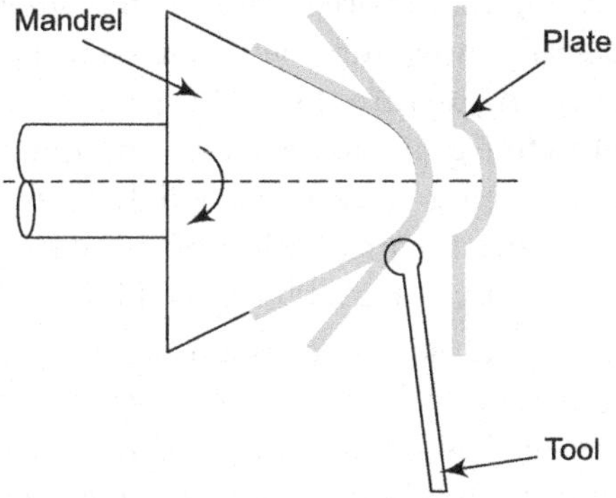

Fig. 1.31 Manual spinning operation

There are process variables like feed ratio, roller path, roller design, spinning ratio affecting the performance and ability to make successful spun components [56]. *Feed ratio* is defined as the ratio of the roller feed rate to the spindle speed. By keeping the feed ratio as constant, the roller feed and the spindle speed can be varied without any considerable influence on the quality of the product. High feed ratios generate higher forces, resulting in cracking. On the other hand, too low a feed ratio reduces the workability and severe thinning occurs. There are different *roller paths* such as linear, concave, convex, involute and quadratic, and they show significant influence on the deformation of the blank. With correct roller path, the formation of wrinkles and cracks can be avoided. The *roller design* is found to show profound influence mainly on the thickness uniformity of the blank. Smaller the nose roller radius, poor thickness uniformity is witnessed. *Spinning ratio* is defined as the ratio of blank diameter to mandrel diameter. The larger the spinning ratio, more difficult is the spinning process. If the spinning ratio is too large, circumferential splitting along the transition from the flange to the wall occurs [56]. A review on the mechanics of spinning process can be seen in Music *et al.* [57].

1.4 SOME OTHER FORMING PROCESS

In this section, some of the recently developed and researched metal forming processes like Equal Channel Angular Extrusion (ECAE), microforming, ring rolling, hydroforming, incremental forming, rubber pad forming, and age (or creep) forming are described briefly.

In *equal channel angular extrusion*, the deformable material undergoes severe plastic deformation creating large plastic strains by the application of simple shear. The tool used for ECAE contains two intersecting channels of identical cross-sections. A lubricated billet is pressed into the first channel and extruded through the second channel. The deformed billet can be removed from the end of second channel. Since the billet dimensions are same, the process can be repeated few times, changing the orientation of the billet between each pass. ECAE processed materials show improved mechanical properties, extremely fine grain sizes, and varying textures [58]. This process is very suitable for the production of bulk ultra fine grained materials. Though the application of this process is restricted, it is being researched widely by many researchers. The suitability of this technique for processing of tubular specimen geometries has also been investigated by Nagasekhar *et al.* [58].

The microforming is the process of manufacturing a part or feature by plastic deformation, whose at least one orthogonal view can be enclosed in a square of 1 mm size. These parts are used mostly in making electronic parts and in micro-electro mechanical systems (MEMS). Typical examples are microscrews, fasteners, etc. [59]. Microforming can be done with both bulk and sheet materials. In the bulk metal forming, a typical micro forming process is wire bending for the production of filaments and springs as used in medical applications and electronic industry. In this process, the influence of size effect on the flow behavior is significant. With decreasing specimen size, in general, a decrease in flow stress is observed. According to the surface layer model, the grains located at the surface of a tensile or upsetting specimen are less restricted than grains inside the material. Dislocations moving through the grains during deformation pile up at grain boundaries but not at the free surface. This results in less hardening and lower resistance against deformation of the surface grains. With decreasing specimen size and a size invariant microstructure, the share of surface grains increases, which leads to lower flow stress. Details about micro deep drawing of thin sheets can be obtained from the work of Saotome *et al.* [60].

In *ring rolling*, seamless ring shaped components are produced. The input material for the ring rolling process is of do-nut shape (shaped like a thick ring) made by hydraulic or mechanical presses or hammers. The donut shaped blank is placed over the undriven mandrel and the mandrel is forced with pressure against the rotating main roll. Once contact occurs between the rotating roll and the blank and the mandrel, since the main roll is rotating, the blank and the mandrel will also rotate in the opposite direction. The mandrel is then moved towards the main roll, shaping the ring, and reducing the wall thickness. Simultaneously the diameter is increased by circumferential extrusion that occurs during this process. The ring height is controlled by applying pressure through top and bottom rolls on the ring surface. This technology can be used in the manufacture of bearings, ring gears, final drive gears, transmission components, clutch components, and wheel blanks. Parts used in aerospace sector include rotating and non-rotating rings for fans, engine casings, and engine disks.

In *hydroforming,* the parts are made by plastically deforming the sheet or tube by the application of fluid pressure through movable punch. The sheet is deformed and it takes the shape of the die cavity. In tube hydroforming, the axial force is applied to the tube as well as internal pressure, which creates compressive stress in one direction so that the tube is deformed without thinning and necking is delayed. Some of the advantages of this process are requirement of simpler die construction and being economical for small number of parts. Disadvantages include presence of very high pressures and greater cycle time than for stamping in mechanical presses.

In *incremental sheet forming,* the sheet is clamped at the edge and a small tool is moved on the sheet surface. If an intermediate support is provided during forming, it is called two point incremental forming, otherwise it is called single point incremental forming. Both symmetric and asymmetric shapes are manufactured by this technique. A review by Emmens *et al.* [61] details the variants of incremental sheet forming process.

Rubber pad forming, also called as flexible die forming, involves forming of sheet metal with one of the tools made of flexible diaphragm or rubber pad, with another tool like that used in conventional forming operations. The flexible tool could be a punch or die cavity. The rubber pad acts like hydraulic fluid in exerting nearly equal pressure on all workpiece surfaces as it is pressed around the form block [1]. When compared with conventional forming or deep drawing, thinning of the sheet metal is reduced considerably, and excellent surface finish is obtained by using this method.

Age forming (or creep forming) is a shaping process mainly for heat treatable aluminum alloys that has gained popularity in recent times. In this process, parts are given aging treatment while simultaneously being deformed by the application of mechanical loads. It is used mainly for making large parts that have to be imparted a curvature. The part shape is obtained due to the creep that occurs at the aging temperatures. Since the part shape is achieved by creep, age forming is sometimes referred to as creep forming. The most common application of age forming is the shaping of upper wing skins in the aerospace industry. During age forming, the forming loads are usually less than the yield strength of the material, and the part shape is obtained due to the low temperature creep that occurs during the aging process. Hence, there is less non-uniformity in the microstructure as compared to cold formed parts. The age formed components have lower residual stresses and consequently better stress corrosion resistance.

1.5 EVALUATION OF FRICTION COEFFICIENT IN METAL FORMING

Friction conditions play a vital role in deciding the forging behavior of billets. There are various methods by which the friction factor m can be evaluated. Some of the methods like ring compression test (RCT), double cup extrusion test (DCE), forward extrusion, bucket cup extrusion test, spike test, upsetting sliding test, injection-upsetting method, and T-shape compression test are explained briefly.

In *ring compression test*, a ring shaped billet of size outer diameter : inner diameter : height–6:3:2, is compressed between two platens (upper and lower) using a hydraulic press. The variation in internal diameter with height reductions is plotted with specimen height reduction for different friction conditions. Depending on the friction between the ring sample and the platens, the inner diameter will decrease or increase with plastic reduction in height. Under low friction conditions, the inner diameter is found to increase (or bulge outwards), while under high friction conditions, the inner diameter is observed to decrease (or bulge inwards) (Fig. 1.32). The experimental "friction calibration curves" is plotted between percentage reduction in internal diameter and percentage reduction in height. Lots of analysis was performed by Male and Cockcroft [62], Yamada *et al.* [63], Carter and Lee [64], Robinson *et al.* [65] and many others. The RCT was conducted on porous metals to evaluate the interfacial friction factor. Tabata and Masaki [66] investigated the friction curves based on RCT on copper powder metals. The overall review of the ring compression test is presented by Rao and Sivaram [67]. The influence of sample geometry, barreling, changing friction conditions, strain rate, and temperature can be understood from the available literature. Recently, some researchers have observed that material properties also influence RCT. In RCT, the die geometry and metal flow are simple, the forging pressure is relatively low, and the new surface generated during deformation is small. Therefore, in a ring test the frictional characteristics of the forging lubricant cannot be adequately evaluated. Also, the friction factor evaluated from RCT can be used mainly for open die forging operations, where degree of deformation is less. In order to overcome the problem involved in RCT, a double cup extrusion (DCE) test was developed.

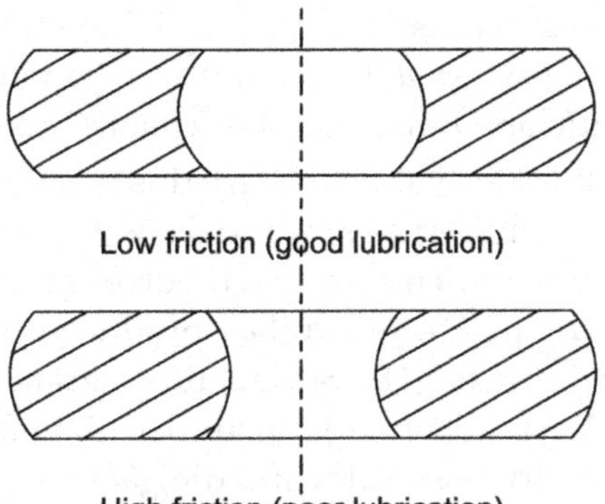

Fig. 1.32 Sample bulging during RCT

In *double cup extrusion test,* a cylindrical billet is plastically deformed between two punches, in which the bottom punch is kept stationary. The top punch which is attached to the ram of the hydraulic press is moved downwards. A combined forward–backward extrusion is performed on the billet and finally "H-shaped" sample is produced. The ratio of backward (h_1) to forward (h_2) extrusion cup heights is controlled by the friction conditions at the container–billet interface. The friction calibration curves can be obtained by plotting the height ratio (h_1/h_2) with stroke or height reduction for varied friction conditions (or friction factors, lubricants). The DCET is found to have following advantages over RCT.

(i) The test simulates severe deformation and friction conditions as seen in actual cold forging operations
(ii) The test is more suitable for closed-die forging operations, and
(iii) The test setup can be fabricated easily and reliable ranking of lubricants can be done

Like RCT, lots of work has been done on DCET for friction evaluation. The work done by Ghobrial *et al.* [68] shows that the different reductions in area have significant effect on the height ratio (h_1/h_2). In turn, the influence on friction factor '*m*' is also significant. The FE simulations performed also agreed well with the experiments for different area reductions. Barcellona *et al.* [69] from their work suggested that the '*m*' value obtained from RCT is more suitable for simple upsetting process simulation, whereas the '*m*' value from DCET is applicable for closed-die forging and extrusion processes. A critical evaluation of DCET was performed by Schrader *et al.* [70] in which the effect of different tool geometry and process parameters on the cup height ratio and contact pressure were analyzed. The impact of strain hardening exponent, strength coefficient, prior work hardening of the billet, billet height, extrusion ratio (d^2_{punch}/d^2_{billet}), punch die land, punch friction on the cup height ratio, contact pressure are studied. It is found that almost all the parameters show profound effect on the friction characteristics during DCET. Similar investigation was carried out by Buschhausen *et al.* [71] and they recommended the optimum usage of reduction ratio and billet height during DCE test.

Another important test to evaluate the friction is *spike forging test,* in which the cylindrical billet is compressed by a top die with a central bore, while the bottom die is kept stationary. This process involves both backward extrusion and radial forging. The height of "spike" produced describes the friction between the billet and the die. The Schematic diagram of spike forging test is shown in Fig. 1.33. Xu and Rao [72] studied the effect of ram velocity, specimen temperature, friction, billet and die geometry on the spike height of the deformed billet. The ram velocity and specimen temperature showed insignificant effect on the spike height of the formed billet. The interface friction and die geometry (like fillet radius, vertical draft) showed significant effect on the spike height. The optimum billet height for successful and efficient evaluation of friction conditions at the tool–billet interface has been suggested. The metal flow is more in the axial direction than in the radial direction, when the friction is large. Similar spike forging tests were conducted by Im *et al.* [73] to investigate the effect of heat loss on the metal flow through simulations in a Ti alloy. The comparison between simulation and experimental results is done for spike forging tests by Rao *et al.* [74] to investigate the impact of friction on the strain and strain rate distribution.

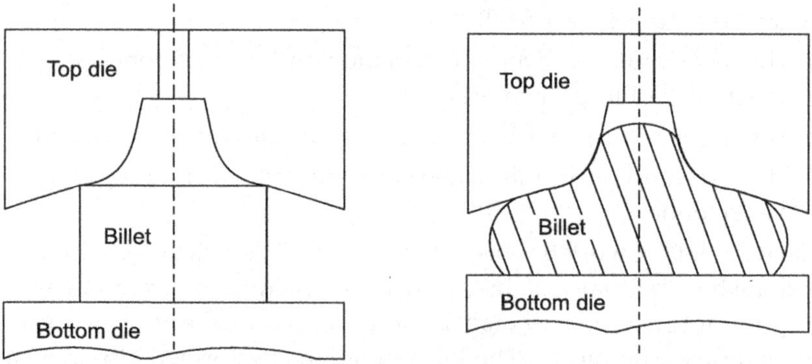

Fig. 1.33 Spike forging test

In *bucket cup extrusion test,* a cylindrical billet is compressed by a punch in a closed die. The billet undergoes backward extrusion and the final product is in "bucket form". The forging load during deformation and the bucket thickness quantifies friction between the tool and the billet. Shen *et al.* [75] compared the RCT and bucket test for their validity to evaluate the friction factors for four different lubricants. They found that the bucket test is advantageous over RCT in the following ways: the bucket test is complex representing large surface generation that can be seen in industrial forging operations, and for the same sample thickness, the forging pressure is more in the bucket test that in RCT that is generally seen in practical forging operations. They identified a new factor *"load multiplied by bucket thickness"* which is sensitive to changing friction and hence used to quantify friction in bucket cup test. They also compared the experimental results with simulations done in DEFORM, a finite element code.

Recently a *T-shape compression test* is developed by Zhang *et al.* [76] in which a cylindrical billet is compressed in a die with 'V' shaped groove. The curved surface of the billet touches the groove before forming. The billet undergoes extrusion in the V-groove and compression radially resulting in a T-shaped formed product. The height of deformation inside the V-groove determines the friction characteristics of the lubricant (schematic shown in Fig. 1.34). The authors have identified the slope of the load curve as a sensitive parameter to quantify friction factor and coefficient. The effect of die geometries like V-groove angle and corner radius on the height of the extruded part is analyzed. Through FE simulations, the friction factor and friction coefficient of different lubricants (oil, solid, mixed) were evaluated. It can be seen from their work that the friction factor 'm' is nearly 2.3 times the friction coefficient 'μ' for the same slope of load curve. Therefore, the friction coefficient can be easily obtained by knowing the friction factor of the lubricant. The amount of bulging in open die forging can also be used as a measure of coefficient of friction at the die-workpiece interface [77].

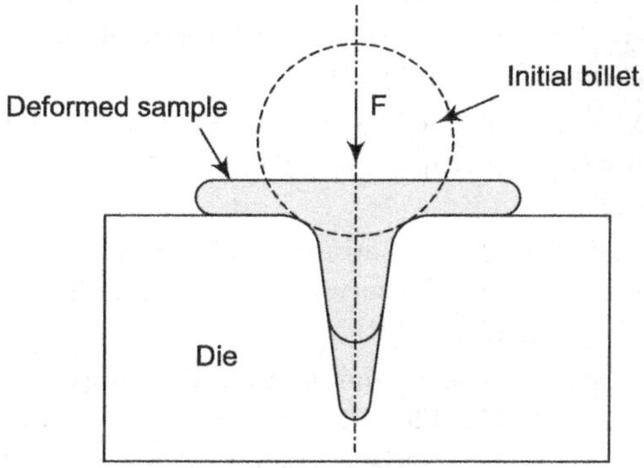

Fig. 1.34 *T*-shape compression test

1.6 CONCLUSION

Metal forming processes are used to make automotive, aerospace, household components, etc. through plastic deformation of billets and sheets. In this chapter, an overview of different bulk and sheet metal forming processes has been presented. Subsequent chapters will discuss various aspects of metal forming in detail. The information presented in this chapter has been compiled from a number of research papers and industry catalogues. For more information, about technological aspects of metal forming, the references cited in this chapter can be referred. It is also advisable to consult the latest data from industrial sources.

REFERENCES

1. Semiatin, S. L. (ed.) (2005). ASM Handbook, Metal working: Bulk forming (#06957G), Vol. 14A, *ASM International.*
2. Dieter, G. E., *Mechanical Metallurgy,* SI Metric Edition, McGraw-Hill, ISBN 0071004068.
3. Al-Salehi, F. A. R., Firbank and T. C., Lancaster, P. R. (1973). An experimental determinations of the roll pressure distributions in cold rolling, *International Journal of Mechanical Sciences,* **15**; 693.
4. Altan, T., Oh, S. I. and Gegel, H. L. (1983), Metal forming: Fundamentals and applications, in Gegel, H. L. (ed.), *American Society for Metals,* Metals park, OH, USA, ISBN 0871701677.
5. Tieu, A. K. and Liu, Y. J. (2004). Friction variation in the coldrolling process, *Tribology International,* **37**; 177.
6. Jeswiet, J. (1998). A comparison of friction coefficients in cold rolling, *Journal of Materials Processing Technology,* **80–81**; 239.
7. Dixit, U. S. and Dixit, P. M. (1996). A finite element analysis of flat rolling and application of fuzzy set theory, *International Journal of Machine Tools and Manufacture,* **36** (8); 947.
8. Gudur, P. P., Salunkhe, M. A. and Dixit, U. S. (2008). A theoretical study on the application of asymmetric rolling for the estimation of friction, *International Journal of Mechanical Sciences,* **50**; 315.
9. Jiang, Z. Y., Xiong, S. W., Tieu, A. K. and Wang, Q. J. (2008). Modelling of the effect of friction on cold strip rolling, *Journal of Materials Processing Technology,* **201**; 85.
10. Gao, H., Ramalingam, S. C., Barber, G. C. and Chen, G. (2002). Analysis of asymmetrical cold rolling with varying coefficients of friction, *Journal of Materials Processing Technology,* **124**; 178.
11. Pawelski, H., Friction inhomogeneities in cold rolling (2002). *Journal of Materials Processing Technology,* **125–126**; 392.
12. Dick, K. and Lenard, J. G. (2005). The effect of roll roughness and lubricant viscosity on the loads on the mill during cold rolling of steel strips, *Journal of Materials Processing Technology,* **168**; 16.
13. Shen, Y. F., Xue, W. Y., Wang, Y. D., Liu, Y. D., and Zuo, L. (2008). Tensile behaviors of IF steel with different cold-rolling reductions, *Materials Science and Engineering A,* **496**; 383.
14. Das, G., Sabita Ghosh, T. and Sahay, S. K. (2005). Use of ball indentation technique to determine the change of tensile properties of SS316L steel due to cold rolling, *Materials Letters,* **59**; 2246.
15. Ganesh Narayanan, R., Gopal, M. and Rajadurai, A. (2008). Influence of friction in simple upsetting and prediction of hardness distribution in a cold forged product, *Journal of Testing and Evaluation,* **36** (4); 1.

16. Kim, H., Lee, S. M. and Altan, T. (1996). Prediction of Hardness Distribution in Cold Backward Extruded Cups, *Journal of Materials Processing Technology*, **59**; 113.

17. Onuh, S. O., Ekoja, M. and Adeyemi, M. B. (2003). Effects of die geometry and extrusion speed on the cold extrusion of aluminium and lead alloys, *Journal of Materials Processing Technology*, **132**; 274.

18. Ajiboye, J. S. and Adeyemi, M. B. (2006). Effects of die land on the cold extrusion of lead alloy, *Journal of Materials Processing Technology*, **171**; 428.

19. Ajiboye, J. S. and Adeyemi, M. B. (2007). Upper bound analysis for extrusion at various die land lengths and shaped profiles, *International Journal of Mechanical Sciences*, **49**; 335.

20. Kudo, H. (1960). Some analytical and experimental studies of axi-symmetric cold forging and extrusion–I, *International Journal of Mechanical Sciences*, **2**; 102.

21. Choi, H. J., Choi, J. H. and Hwang, B. B. (2001). The forming characteristics of radial-backward extrusion, *Journal of Materials Processing Technology*, **113**; 141.

22. Chitkara, N. R. and Aleem, A. (2001). Extrusion of axi-symmetric tubes from hollow and solid circular billets: a generalised slab method of analysis and some experiments, *International Journal of Mechanical Sciences*, **43**; 1661.

23. Ebrahimi, R., Reihanian, M., Kanaani, M., and Moshksar, M. M. (2008). An upper-bound analysis of the tube extrusion process, *Journal of Materials Processing Technology*, **199**; 214.

24. Berski, S., Dyja, H., Maranda, A., Nowaczewski, and J., Banaszek, G. (2006). Analysis of quality of bimetallic rod after extrusion process, *Journal of Materials Processing Technology*, 177; 582.

25. Avitzur, B., Wu, R., Talbert, S., and Chou, Y. T. (1986). Analysis of core fracture in drawing of bimetal rods and wires, *Journal of Engineering for Industry*, 108 (2); 133.

26. Avitzur, B., Wu, R., Talbert, S. and Chou, Y. T. (1985). An analytical approach to the problem of core fracture during extrusion of bimetal rods, *Journal of Engineering for Industry*, 107 (3); 247.

27. Irena, N. and Andrzej, S. (2009). Influence of die geometric parameters on plastic flow of layer composites during extrusion process, *Journal of Materials Processing Technology*, **209**; 1943.

28. Schikorra, M. and Kleiner, M. (2007). Simulation based analysis of composite extrusion processes, *Annals of the CIRP*, **56** (1); 317.

29. Hwang, Y. M. and Hwang, T. F. (2002). An investigation into the plastic deformation behavior within a conical die during composite rod extrusion, *Journal of Materials Processing Technology*, **121**; 226.

30. Dodeja L. C. and Johnson, W. (1957). The cold extrusion of circular rods through square multiple hole dies, *Journal of the Mechanics and Physics of Solids*, **5**; 281.

31. Keife, H. (1993). Extrusion through two die openings: a 2D upper–bound analysis checked by plasticine experiments, *Journal of Material Processing Technology*, **37**; 189.

32. Ulysse, P. and Johnson, R. E. (1998). A study of the effect of the process variables in unsymmetrical single-hole and multi-hole extrusion processes, *Journal of Material Processing Technology*, **73**; 213.

33. Sinha, M. K., Deb, S. and Dixit, U. S. (2009). Design of a multi-hole extrusion process, *Materials and Design*, **30**; 330.

34. Sinha, M. K., Deb, S., Das, R., and Dixit, U. S. (2009). Theoretical and experimental investigations on multi-hole extrusion process, *Materials and Design*, **30**; 2386.

35. Rubio, E. M., Camacho, A. M., Sevilla, L., and Sebastian, M. A. (2005). Calculation of the forward tension in drawing processes, *Journal of Materials Processing Technology*, **162–163**; 551.

36. Dixit, U. S. and Dixit, P. M. (1995). An analysis of the steady-state wire drawing of strain-hardening materials, *Journal of Materials Processing Technology*, **47**; 201.

37. Vega, G., Haddi, A. and Imad, A. (2009). Investigation of process parameters effect on the copper-wire drawing, *Materials and Design*, **30**; 3308.

38. Avitzur, B. (ed.) (1968). *Metal Forming: Processes and Analysis*, McGraw-Hill, New York; also, (1979). Krieger, Huntington, NY.

39. Shinohara, T. and Yoshida, K. (2005). Deformation analysis of surface flaws in stainless steel wire drawing, *Journal of Materials Processing Technology*, **162–163**; 579.

40. Ko, D. C. and Kim, B. M. (2000). The prediction of central burst defects in extrusion and wire drawing, *Journal of Materials Processing Technology*, **102**; 19.

41. Budiansky, H. and Wang, N. M. (1966). On the swift cup test, *Journal of Mechanics and Physics of Solids*, **14**; 357.

42. Woo, D. M. (1968). On the complete solution of the deep-drawing problem, *International Journal of Mechanical Sciences*, **10**; 83.

43. Kinsey, B., Liu, Z. and Cao, J. (2000). A novel forming technology for tailor welded blanks, *Journal of Materials Processing Technology*, **99**; 145.

44. Padmanabhan, R., Oliveira, M. C., Alves, J. L., and Menezes, L. F. (2008). Numerical simulation and analysis on the deep drawing of LPG bottles, *Journal of Materials Processing Technology*, **200**; 416.

45. Hosford, W. F. and Caddell, R. M. (eds.) (1993). *Metal forming: Mechanics and Metallurgy*, Prentice-Hall, New York, ISBN 0-13-588526-4.

46. Zhao, L., Sowerby, R. and Sklad, M. P. (1996). A theoretical and experimental investigation of limit strains in sheet metal forming, *International Journal of Mechanical Sciences*, **38** (12); 1307.

47. Sowerby, R. and Duncan, J. L. (1971). Failure in sheet metal in biaxial tension, *International Journal of Mechanical Sciences*, **13**; 217.

48. Marciniak, Z., Duncan, J. L. and Hu, S. J. (eds.) (2002). *Mechanics of Sheet Metal Forming*, Elsevier, New Delhi.

49. Mullan, H. B (2004). Improved prediction of springback on final formed components, *Journal of Materials Processing Technology*, **153–154**; 464.

50. Gau, J. T. and Kinzel, G. L. (2001). A new model for springback prediction in which the Bauschinger effect is considered, *International Journal of Mechanical Sciences*, **43**; 1813.

51. Li, K. P., Carden, W. P. and Wagoner, R. H. (2002). Simulation of springback, *International Journal of Mechanical Sciences*, **44**; 103.

52. Wagoner, R. H. and Li, M. (2007). Simulation of springback: Through-thickness integration, *International Journal of Plasticity*, **23**; 345.

53. Bahloul, R., Ben-Elechi, S. and Potiron, A. (2006). Optimisation of springback predicted by experimental and numerical approach by using response surface methodology, *Journal of Materials Processing Technology*, 173; 101.

54. Karafillis, A. P. and Boyce, M. C. (1992). Tooling design accommodating springback errors, *Journal of Materials Processing Technology*, **32**; 499.

55. Gan, W. and Wagoner, R. H. (2004). Die design method for sheet springback, *International Journal of Mechanical Sciences*, **46**; 1097.

56. Wong, C. C., Dean, T. A. and Lin, J. (2003). A review of spinning, shear forming and flow forming processes, *International Journal of Machine Tools and Manufacture*, **43**; 1419.

57. Music, O., Allwood, J. M. and Kawai, K. (2010). A review of the mechanics of metal spinning, *Journal of Materials Processing Technology*, **210**; 3.

58. Nagasekhar, A. V., Uday Chakkingal and Venugopal, P. (2006). Equal Channel Angular Extrusion of tubular aluminum alloy specimens analysis of extrusion pressures and mechanical properties, *Journal of Manufacturing Processes*, **8** (2); 112.

59. Engel, U. and Eckstein, R. (2002). Microforming— from basic research to its realization, *Journal of Materials Processing Technology*, **125–126**; 35.

60. Saotome, Y., Yasuda, K. and Kaga, H. (2001). Microdeep drawability of very thin sheets, *Journal of Materials Processing Technology*, **113**; 641.

61. Emmens, W. C., Sebastiani, G. and van den Boogaard, A. H. (2010). The technology of Incremental Sheet Forming— A brief review of the history, *Journal of Materials Processing Technology*, **210**; 981.

62. Male, A. T. and Cockcroft, M. G. (1993). A method for the determination of the coefficient of friction of metals under condition of bulk plastic deformation, *Journal of Institute of Metals*, **64–65**; 38.

63. Yamada, R., Suzuki, M. and Harayama, Y. (1977). Application of finite element method to ring compression test, *Nuclear Engineering and Design*, **44**; 75.

64. Carter, W. T. Jr. and Lee, D. (1985). A finite element analysis of cylinder and ring compression and its experimental verification, *Computers & Structures*, **21** (1–2); 1.

65. Robinson, T., Ou, H. and Armstrong, C. G. (2004). Study on ring compression test using physical modeling and FE simulation, *Journal of Materials Processing Technology*, **153–154**; 54.

66. Tabata, T. and Masaki, S. (1978). Determination of the coefficient of friction in forging of porous metals from ring compression, *International Journal of Mechanical Sciences*, **20**; 505.

67. Rao, K. P. and Sivaram, K. (1993). A review of ring compression testing and applicability of the calibration curves, *Journal of Materials Processing Technology,* **37**; 295.

68. Ghobrial, M. I., Lee, J. Y. and Altan, T. (1993). Factors Affecting the Double Cup Extrusion Test for Evaluation of Friction in Cold and Warm Forging, *Annals of the CIRP,* **42** (1); 347.

69. Barcellona, A. and Cannizzaro L. (1996). Validation of frictional studies by double cup extrusion tests in cold forming, *Annals of the CIRP,* **45** (1); 211

70. Schrader, T., Shirgaokar, M. and Altan, T. (2007). A critical evaluation of the double cup extrusion test for selection of cold forging lubricants, *Journal of Materials Processing Technology,* 189; 36.

71. Buschhausen, A., Weinmann, K., Lee, J. Y., and Altan, T. (1992). Evaluation of lubrication and friction in cold forging using a double backward extrusion process, *Journal of Materials Processing Technology,* **33**; 95.

72. Xu, X. L. and Rao. K. P. (1997). Analysis of the deformation characteristics of spike forging process through FE simulations and experiments, *Journal of Materials Processing Technology,* **70**; 122.

73. Im, Y. T., Vardan, O., Shen, G., and Altan, T. (1988). Investigation of metal flow in non-isothermal forging using ring and spike tests, *Annals of the CIRP,* **37** (1); 225.

74. Rao, K. P., Doraivelu, S. M. and Sivaram, K. (1991). Physical modeling studies using spike forging to verify analytical predictions, *Journal of Materials Processing Technology,* **28**; 295.

75. Shen, G., Vedhanayagam, A., Kropp, E., and Altan, T. (1992). A method for evaluating friction using a backward extrusion type forging, *Journal of Materials Processing Technology,* **33**; 109.

76. Zhang, Q., Felder and E., Bruschi, S. (2009). Evaluation of friction condition in cold forging by using T-shape compression test, *Journal of Materials Processing Technology,* **209**; 5720.

77. Ebrahimi, R., and Najafizadeh, A. (2004), A new method for evaluation of friction in bulk metal forming, *Journal of Materials Processing Technology,* 152; 136.

CHAPTER 2 Fundamentals of Plasticity

Arbind Kumar Singh
Department of Civil Engineering, IIT Guwahati, India

2.1 INTRODUCTION

The subject of study of motion and deformation of a body under the action of forces acting on it is called mechanics. Solid mechanics primarily deals with motion and deformation of deformable solid under action of forces. There are two approaches to understand these physical behaviors (motion and deformation) of solid body:

(a) Physical approach
(b) Phenomenological approach

(a) **Physical approach:** In this approach, body is assumed to be made of atoms, molecules and crystals. The behavior of the body is studied based on interaction of these atoms and molecules. This is the subject matter of solid state physics. It is very difficult (almost impossible) to apply this approach to engineering structures such as aircrafts, machine tools or buildings. But the approach gives a very good physical insight into the deformation of the body.

(b) **Phenomenological approach:** Assumptions are made based on macroscopically observed behavior of the body and the basic principles of physics are applied. The obtained response is checked with the experimental results and hence theory is developed. Solid mechanics dealt in this chapter is based on phenomenological approach.

2.2 UNIAXIAL TENSION TEST

This is the simplest test conducted on metals. The results of the test provide basic information which has been widely used in the design of structures. In this test a cylindrical circular rod is tested by applying continuously increasing load and measuring the elongation of the specimen (gauge length, specified portion) simultaneously. Also, diameter of the specimen is measured. The specimen is tested in a machine (simplest case) called universal testing machine. The rate of loading is kept constant during the experiment and adjusted in such a way to avoid creep or dynamic effect. (The rate of loading should not be high such that inertia effect comes and also it should not be low such that the time dependent deformation comes into existence. To keep these points a standard has been developed for the test.) The tension test is conducted at constant temperature.

Although the uniaxial tension test is simple and easy to conduct, but the interpretation of its data is not so easy as it looks at the first glance. The complexity of the test can be judged by the following statement of E. Orowan [1].

"The extension of a piece of metal (is) in a sense more complicated than the working of a pocket watch and to hope to derive information about its mechanism from two or three data derived from measurement during the tensile test is perhaps as optimistic as would be an attempt to learn about the working of a pocket watch by determining its compressive strength."

In spite of this disheartening statement, tensile test has been a basic test to understand the primary mechanical behavior of the material. In most of the design procedures, data from the tensile test is widely employed. A typical specimen generally used in the tension test is shown in Fig. 2.1. We shall define the following terms to understand the results of the tension test.

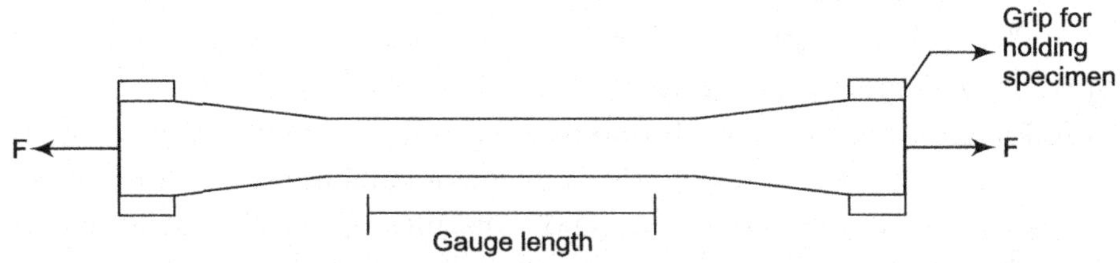

Fig. 2.1 A typical specimen used in the tension test

The initial distance between two points is called initial gauge length or gauge length. L_0 is the initial gauge length and ΔL is the increase in the gauge length. P is the applied force and A_0 is the initial cross section area of the gauge length portion and A is the area at any instant of time when length is L. We shall define the following stress and strain.

(a) **Nominal stress or engineering stress:** This is ratio of force applied on the specimen divided by initial area of the specimen [2, 3, 4].

$$\sigma_e = \frac{P}{A_0} \tag{2.1}$$

(b) **Nominal strain or engineering strain:** This measures the change in length of the specimen per unit of initial length of the specimen.

$$\varepsilon_e = \frac{\Delta L}{L} \tag{2.2}$$

One can note that this is not the one dimensional equivalent of Green Lagrange strain tensor although it is a measure of strain with respect to initial configuration. In the case of small deformation this strain is approximated with Green Lagrange strain tensor.

(c) **True stress or Cauchy stress:** This is the ratio of force per unit of current area of the specimen.

$$\sigma = \frac{P}{A} \tag{2.3}$$

This is one dimensional equivalent to Cauchy stress tensor.

(d) **Logarithmic strain or true strain:** Let us consider the specimen at current time. Let dl be the infinitesimal change in length when length was l, then infinitesimal strain introduced is

$$\mathrm{d}\varepsilon_L = \frac{\mathrm{d}l}{l} \tag{2.4}$$

Total strain accumulated during stretching of the specimen from length L to l is called logarithmic strain defined as

$$\varepsilon_L = \int_L^l \frac{\mathrm{d}l}{l} = \ln\left(\frac{l}{L}\right) \tag{2.5}$$

Engineering strain and engineering stress measure the deformation of a body with respect to original configuration whereas Cauchy stress and logarithmic strain deal with current deformed state.

The graphical relation between one dimensional stress and strain is called stress–strain diagram. A typical stress–strain diagram for the mild steel under uniaxial case is shown in Fig. 2.2. The stress and strain used are engineering stress and engineering strain. Following observation can be made from the curve.

The relationship between σ_e and ε_e is linear up to point A. The point "A" is called proportionality limit beyond which stress-strain relations are not linear. The ratio of the stress to strain is called Young's modulus. During this stage of loading, ratio of lateral strain to the longitudinal strain remains constant and this ratio is called Poisson's ratio, i.e., $v = \dfrac{|\varepsilon_{\text{lat}}|}{|\varepsilon_{\text{long}}|}$. From point A to B, loading–unloading curve is same but stress–strain curve is non linear.

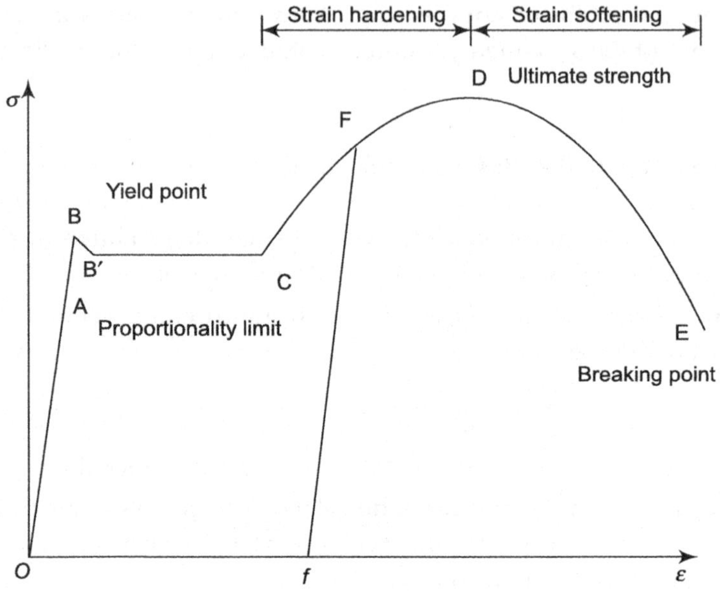

Fig. 2.2 Typical stress strain curve for mild steel

If we load beyond B, there is a drop in stress at almost constant strain and stress–strain curve drops to B′. Then stress remains constant with increase in strain from point B′ up to point C. If we unload the specimen after loading beyond B′, the unloading path is different than the loading curve. The point B is called *upper yield point* and B′ is called *lower yield point*. The behavior during B′–C is called perfectly plastic behavior. Generally, lower yield point is considered as the elastic limit. If we unload the specimen after loading beyond lower yield point, the unloading curve is parallel to the initial elastic curve. A permanent deformation remains on the specimen even if we completely remove the load. This permanent deformation is called plastic deformation. If we reload the specimen it follows the unloading curve as shown in Fig. 2.2. Hence we can observe that the unloading is elastic and elastic properties are not affected by plastic deformation. If we load the specimen beyond C, stress increases with strain and reaches a maximum value at D and stress starts decreasing after D and continues till it breaks at E. The stress level at D is called *ultimate stress* and E is called *breaking point*.

The phenomenon of increase in stress level during C–D is called strain hardening. The decrease in yield stress is called strain softening which is shown in Fig. 2.2 by curve segment D–E.

Sress-strain diagram is assumed to be independent of the size of the specimen and its gauge length. Stress–strain diagram for different materials differ widely. Even for the same material the stress–strain curves depend upon rate of loading, ambient temperature and many other variables during loading. For the material like aluminum, yield stress discussed above is not clearly obtained. For such materials, we define a stress called proof stress which limits the elastic domain. Proof stress is the offset stress corresponding to the 0.2 percent of plastic strain, *i.e.*, if the material is loaded up to proof stress and we unload along the line parallel to the initial elastic line, 0.2 percent plastic strain is obtained as shown in Fig. 2.3.

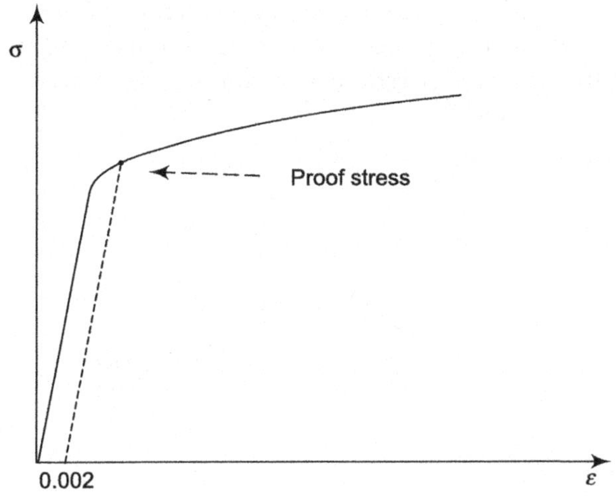

Fig. 2.3 Proof stress for a material like aluminum

It has been observed that during plastic deformation beyond ultimate load, cross section starts decreasing drastically at a particular section of the specimen as compared to other section and specimen breaks with the development of necking as shown in Fig. 2.4. Reduction of cross section area occurs in the localized portion of the specimen and it does not result in the entire portion of the specimen. The development of necking causes instability to the testing. At the breaking point specimen fractures. Some materials such as steel and aluminum exhibit a typical "cup and cone" fracture.

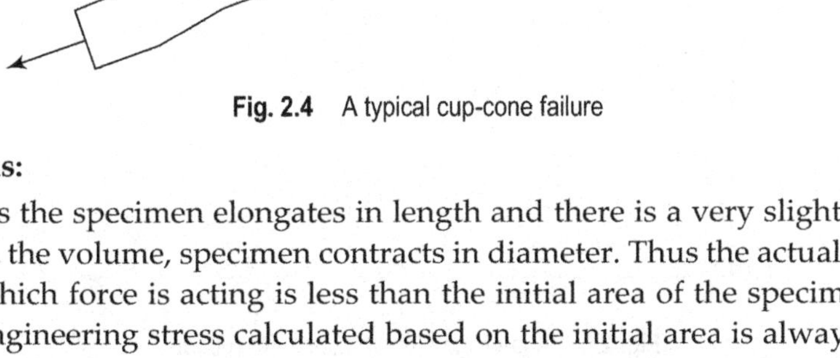

Fig. 2.4 A typical cup-cone failure

Remarks:

1. As the specimen elongates in length and there is a very slight change in the volume, specimen contracts in diameter. Thus the actual area on which force is acting is less than the initial area of the specimen. The engineering stress calculated based on the initial area is always lower

than the true stress. But for small deformation, the difference between these two stresses are negligibly small and the stress–strain diagram will be the same for both the stresses during initial stage of loading and the true stress curve diverges from the engineering stress curve. Also, it is very difficult to measure the lateral contraction. Hence for all practical purposes, engineering stress is used for engineering design. We can say that there is very insignificant difference between σ_e and σ_{true} up to yield point (see Fig. 2.5).

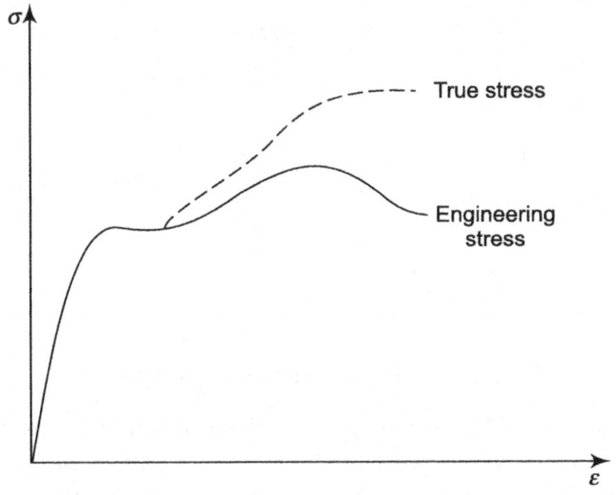

Fig. 2.5 Stress-strain curve for engineering stress and true stress (Not to the scale)

2. The initial straight line in stress–strain curve is not followed by all the materials. For material like rubber, the stress–strain curve is nonlinear even at the initial stage but it has been found that the loading–unloading curves are the same as shown in Fig. 2.6.

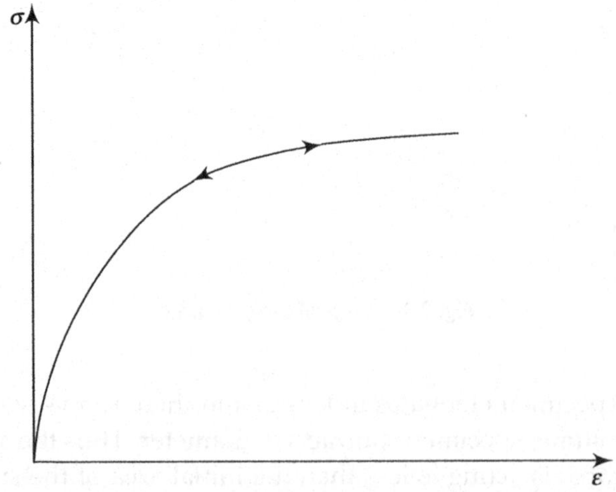

Fig. 2.6 A loading unloading curve for rubber

3. **Hardening:** We can see from Fig. 2.2 that the loading beyond C increases the yield stress of the material. If we unload the specimen and load it, it follows almost the same unloading curve up to the hardening curve and then it follows the original path. Suppose we load the material up to point F and unload it, we shall get permanent strain. If we load the material again, it behaves elastic up to point F and then follows exactly the same behavior as virgin curve. This process of increasing the yield stress by giving permanent deformation is called strain hardening. Although stress increases during hardening, ductility of the material decreases. The increased yield stress is called subsequent yield stress.

Energy during deformation A stress–strain curve is shown in Fig. 2.7 for linear elastic material. During loading, energy is supplied to the material which is stored as the energy of deformation. The stored energy is called strain energy. The energy can be equated with the shaded area in Fig. 2.7.

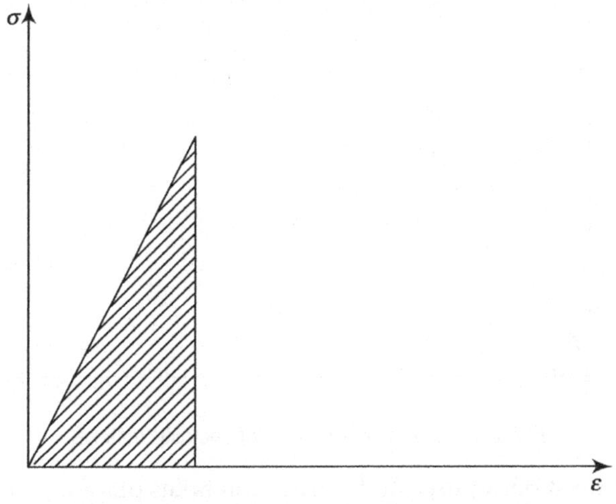

Fig. 2.7 Energy stored during elastic deformation

During elastic deformation, strain energy density is defined for linear behavior as

$$\Phi = \frac{1}{2}\sigma\varepsilon \tag{2.6}$$

where **Φ** is the strain energy in the material per unit volume called strain energy density. Total energy is given by

$$U = \int \Phi dV = \frac{1}{2}\int \sigma\varepsilon dV \tag{2.7}$$

For general nonlinear case, energy density of deformation is

$$\Phi = \int \sigma d\varepsilon \tag{2.8}$$

which is the area under stress–strain curve. Total energy of deformation is

$$U = \int_V \Phi dV = \int_V \int \sigma d\varepsilon dV \tag{2.9}$$

Elastic deformation During complete cycle of loading–unloading in elastic deformation, there is no dissipation of energy. Even if the material returns to its original shape but there is a dissipation of energy during deformation, then material behavior is not elastic.

Anelastic deformation There are materials which return to original configuration but through different path as shown in Fig. 2.8 and hence we get dissipation of energy. This type of material be-havior is called *Anelastic*. We can see that there is a loop formation in the stress–strain curve which shows the dissipated energy during the loading–unloading process.

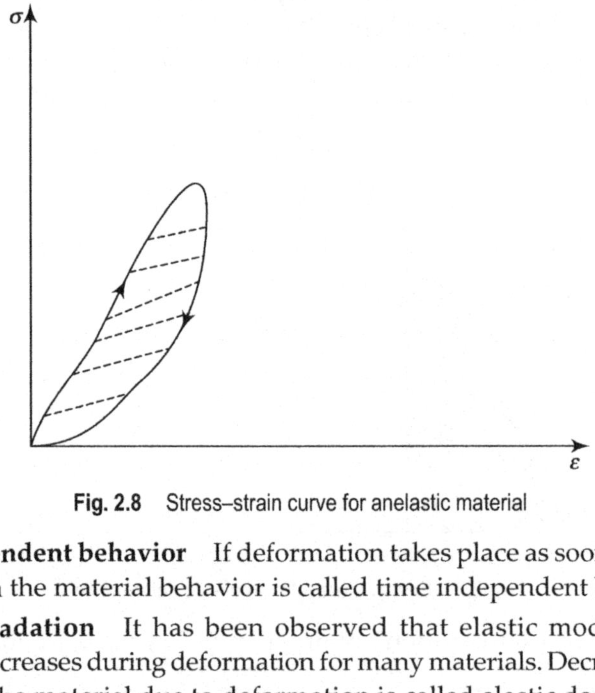

Fig. 2.8 Stress–strain curve for anelastic material

Time independent behavior If deformation takes place as soon as we apply the load then the material behavior is called time independent behavior.

Elastic degradation It has been observed that elastic modulus during unloading decreases during deformation for many materials. Decrease in elastic property of the material due to deformation is called elastic degradation.

2.3 YIELD CRITERIA

Basic aim in the design of a structure is to provide adequate stiffness and strength under loading and environment. One of the objectives is to analyze how strong a structure is and how much load can be applied on it before it fails or collapses. Mode of failure of a material/structure can be classified into two categories

(i) Ductile failure
(ii) Brittle failure

A failure is said to be *ductile* if it undergoes perceptible deformation before it fails whereas *brittle* failure is characterized by sudden failure without perceptible deformation. Ductile material shows elastic behavior up to certain point and loading beyond this point, material no more remains in elastic state. We can define yield point as the stress limit beyond which material does not remain elastic on loading. In other word, yield point for uniaxial loading is the stress limit beyond which path independent characteristics is no more valid. Under complex loading above criteria is termed *yield criteria* which provide the limit of elastic behavior.

Yield criterion is characteristics of ductile materials. After yielding, ductile material does not loose its strength. But brittle material fails suddenly. For granular or brittle materials, we refer failure criteria as failure strength because these materials cannot take load beyond the above limit whereas ductile material can take load even beyond the yield stress.

The failure analysis of a structure up to collapse is necessary to assess all safety aspect of the structure and find its deformation characteristics. This type of analysis is particularly desirable for certain structures such as reactor vessel, nuclear containment structures, parts of offshore structures, aircrafts structures, *etc.* Experimental studies on these structural systems are very expensive and empirical studies are not sufficient for the adequate evaluation of safety with respect to limit state. At least number of experiments can be reduced if we successfully predict the failure to certain level of accuracy. Failure criterion is one of the basic requirement for analysis of structure such as non linear finite element analysis.

In the present chapter, overview of failure criteria/yield criteria has been presented. In a general framework, yield criteria for ideal plastic isotropic material have been discussed and its geometric representation has been explained. Some commonly used yield criteria for isotropic, homogeneous material have been described first and their limitations have been presented. Brief review of failure criteria for anisotropic materials, high performance composite materials and granular materials like concrete have been presented.

2.4 PLASTICITY AND YIELD CRITERIA

2.4.1 Ideal Plastic Material

A material is said to be *ideal plastic* [5] if it has following properties.

(a) The theory developed is valid only at temperature for which recovery, creep and thermal phenomena are neglected. Ideal plastic material should obey the above properties.

(b) Non-uniformity on microscopic level is neglected, *i.e.*, the Bauschinger effect and hysteresis loop in loading and reloading are neglected. Material is assumed to be isotropic.

(c) Size effects are neglected.
(d) Yield strength in compression and tension are assumed to be the same.
(e) Yielding is independent of hydrostatic stress.
(f) Time plays no role during loading and reloading.

We shall first provide the mathematical and geometric aspect of yield criteria of isotropic materials.

2.4.2 Yield Criteria of Ideal Plastic Material

(i) *Existence of the yield criteria* The yield criterion limits the elastic region. It is a mathematical expression to define the combination of components of stress such that when it reaches material no more behaves elastically. Yield criterion gives the onset of plastic deformation. In other words, if a state of stress satisfies yield criterion, we can say that the plastification may start. It is assumed that initial yielding is dependent only on the state of stress and not on how the stress is reached. We can assume that there exists a function $f(\sigma_{ij})$ called *yield function* such that

Material is elastic

$$\text{if } f(\sigma_{ij}) < 0 \tag{2.10}$$

or

$$\text{if } f(\sigma_{ij}) = 0 \text{ and } \dot{f}(\sigma_{ij}) < 0 \tag{2.11}$$

where $f(\sigma_{ij}) = 0$ defines the yield surface in stress space and $\dot{f}(\sigma_{ij}) = 0$ indicates unloading. The latter combination tells us that onset of plastification has taken place, but unloading is going to take place elastically. As the yield criterion does not depend on the path of loading, it does not tell anything about deformation. If state of stress is already satisfied $f(\sigma_{ij}) = 0$, it tells us only that plastification has just started or taken place. But it does not say whether plastic deformation has taken place or not. But the yield function gives the information about loading.

Material behavior is plastic

$$\text{if } f(\sigma_{ij}) = 0 \quad \text{and} \quad \dot{f}(\sigma_{ij}) \geq 0 \tag{2.12}$$

Mathematically yield function is the closure of open set formed by elastic region. Also, one can observe that yield function is non-positive function. In general, function f maps nine dimensional stress space to non-positive real number $f: \Re^9 \to \Re^-$ where $\Re^-$ is the set of non-positive real number. For non polar media, stress is symmetric. Yield function is function of six components of stress, i.e., $f: \Re^9 \to \Re^-$ which can be written in expanded form as

$$f = f(\sigma_{xx}, \sigma_{yy}, \sigma_{zz}, \sigma_{yz}, \sigma_{zx}, \sigma_{xy}) \tag{2.13}$$

State of stress can be determined by specifying principal stress and orientation of principal axes. The three principal stress and three angle constitute the six dimensional space. We can write yield function in terms of principal space. Here functional form of yield function will change, but we keep the same function *f*. The yield function can be written as

$$f = f(\sigma_1, \sigma_2, \sigma_3, \alpha_1, \alpha_2, \alpha_3) \tag{2.14}$$

where σ_1, σ_2, σ_3, are principal stresses and α_1, α_2, α_3, are orientations of principal axes.

(ii) *Material is isotropic* For isotropic material, yield function must be independent of orientation of principal axes, i.e., yield function should be independent of α_1, α_2, α_3. Yield function *f* is function of σ_1, σ_2, σ_3 only which can be written as

$$f = f(\sigma_1, \sigma_2, \sigma_3) \tag{2.15}$$

Also, for isotropic material, if we interchange the direction of σ_1, σ_2 and σ_3, yield function does not change. Function *f* should be independent of sequence of principal stress, i.e., yield function is symmetric with respect to principal stress. Thus, we can write

$$f(\sigma_1, \sigma_2, \sigma_3) = f(\sigma_1, \sigma_3, \sigma_2) = f(\sigma_3, \sigma_2, \sigma_1) \text{ etc.} \tag{2.16}$$

We can represent the yield function by taking any three independent symmetric function of principal stress. We know that stress invariants are symmetric function of principal stresses. The above function can be written also in terms of stress tensor invariants as

$$f = f(I_1, I_2, I_3) \tag{2.17}$$

where I_1, I_2, I_3 are symmetric function of principal stress. The yield function can also be written as in terms of ρ, ξ, θ or σ_{oct}, τ_{oct}, θ as

$$f = f(\rho, \xi, \theta), \quad f = f(\sigma_{oct}, \tau_{oct}, \theta) \tag{2.18}$$

These combinations are used in developing yield function for geomaterials such as soil, rock concrete.

(iii) *Yield function is independent of the hydrostatic stress* We know that the state of stress at a point can be split into hydrostatic stress/volumetric stress and deviatoric stress. It has been observed (Bridgman) that yield function does not depend upon hydrostatic stress for metals and alloys. In these cases, we can write yield function as

$$f = f(s_{ij}) \tag{2.19}$$

where s_{ij} is the components of deviatoric stress tensor. The function *f* depends only on deviatoric stresses, we can write for isotropic materials

$$f = f(s_1, s_2, s_3) \tag{2.20}$$

where s_1, s_2, s_3 are the principal stresses of deviatoric stress tensor, defined as

$$s_{ij} = \sigma_{ij} - \delta_{ij}\,\sigma_m \tag{2.21}$$

where $$\sigma_m = \frac{1}{3}\sigma_{ii}$$

$$\delta_{ij} = 1 \text{ when } i = j \text{ and}$$
$$= 0 \text{ when } i \neq j$$

We can write Eq. (2.20) in terms of stress invariants

$$f = f(J_2, J_3) \tag{2.22}$$

where J_2, J_3 are stress invariants of deviatoric stress tensor. In terms of ρ, ξ, θ and σ_{oct}, τ_{oct}, θ, we can write yield function as

$$f = f(\xi, \theta)$$

$$f = f(\tau_{oct}, \theta)$$

(iv) *Identical tensile and compressive yield strength* It has been observed that yield stress in tension and compression is same for many of the metallic materials. Thus, the function form for tension–compression independent yield function is even function. Physically, the assumption requires that yield criteria should be independent of direction of loading. It should satisfy following condition

$$f(\sigma_{ij}) = f(-\sigma_{ij}) \tag{2.23}$$

For isotropic material, we have expressed yield function in terms of stress invariant. If we take material with hydrostatic mean independence and tension–compression independent, yield function will be even function of J_3. J_2 is always positive number.

(a) Geometric representation

There exists three principal stresses and their orientation for given state of stress at a point. Hence one can represent state of stress at a point in three dimensional vector space whose bases are principal stresses for isotropic material where orientation of axes is not important. For isotropic materials, $f(\sigma_1, \sigma_2, \sigma_3) = 0$ represent a surface in the principal stress space. A yield function represents a surface in three dimensional vector space.

If material is acted upon by only deviatoric stress, then yield surface can be represented as

$$f(\sigma_1, \sigma_2, \sigma_3) = 0 \tag{2.24}$$

subjected to

$$\sigma_1 + \sigma_2 + \sigma_3 = 0 \tag{2.25}$$

Here, $\sigma_1 + \sigma_2 + \sigma_3 = 0$ is a plane passing through origin and equally inclined to the principal axes. This plane is called π plane. The above equation represent a curve C on $\sigma_1 + \sigma_2 + \sigma_3 = 0$ plane (see Fig. 2.9). If the yield function is independent of hydrostatic stress then yield function is right cylinder with generator perpendicular to π plane and whose cross section from hydrostatic line is same as curve C.

In order to discuss possible form of curve, let us consider state of stress where hydrostatic components is zero. The locus on π plane may be convex or concave to origin but obviously not such that radius cut it twice. Due to isotropy, we have $f(\sigma_1, \sigma_2, \sigma_3) = f(\sigma_1, \sigma_3, \sigma_2)$. Hence, we can say that curve is symmetric about AA′ axis. Similarly it is symmetric about BB′ and CC′ axes. Hence the curve can be split of into six mutual symmetric parts.

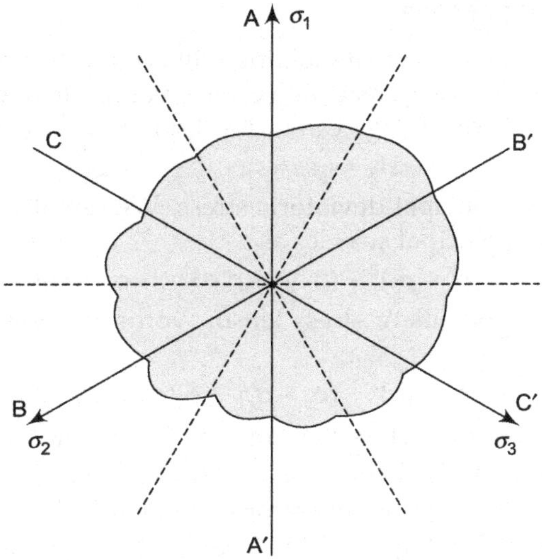

Fig. 2.9 Intersection of yield surface with π plane

If yield stress in tension and compression is same then locus is not only symmetric about the principal axes diameter but also about the three diameters orthogonal to them. Hence we can say that yield curve is made of 12 symmetric 30° segment on π plane.

There are other ways of representing yield function or stress in terms of (ρ, ξ, θ) and $(\sigma_{oct}, \tau_{oct}, \theta)$.

2.4.3 Some Commonly used Yield Criteria

The yield criteria of materials limit the elastic domain during loading whereas failure criteria give the maximum stress that can be applied. Traditionally, we use the term yield criteria for metals, alloys and failure criteria for geomaterials such as soil concrete. We shall discuss some of the yield criteria or failure criteria which have been widely used in design of structures and forming process. Suitable selection of these criteria depends upon the experimental behavior of the materials. There are a number of yield criteria proposed for different materials. Most of them are only of historical importance. Those which are widely used in the design are also important from the point of view of development of new criteria. Some of the commonly used yield criteria/failure criteria have been described which are as follows

(a) von Mises yield criteria,

(b) Tresca yield criteria,

(c) Mohr Coulomb yield criteria,

(d) Drucker–Prager yield criteria.

(a) von Mises yield criteria

von Mises (1913) suggested that yielding will occur when second invariants of deviatoric stress tensor, J_2 reaches a critical value. He does not take J_3 into account in the yield criteria. We can write at onset of yielding

$$2J_2 = s_{ij}s_{ij} = s_1^2 + s_2^2 + s_3^2 = 2k^2 \tag{2.26}$$

where s_1, s_2, s_3 are principal deviatoric stress. We can also write von Mises criteria in terms of principal stresses as

$$(\sigma_1 - \sigma_2)^2 + (\sigma_2 - \sigma_3)^2 + (\sigma_3 - \sigma_1)^2 = 6k^2 \tag{2.27}$$

In terms of components of stress tensor, von Mises yield criteria can be written as

$$(\sigma_x - \sigma_y)^2 + (\sigma_y - \sigma_z)^2 + (\sigma_z - \sigma_x)^2 + 6\,(\tau_{yz}^2 + \tau_{zx}^2 + \tau_{xy}^2) = 6k^2 \tag{2.28}$$

where k is a parameter which is constant for virgin material and depends on the amount of plastic strain. Geometrically, von Mises yield criterion represents right circular cylinder whose generator is equally inclined to the principal stress axes as shown in Fig. 2.10. Physical explanation of von Mises yield criteria was given by Hencky (1924) and Nadai (1937). Hencky interpretation of von Mises's law is that yielding begins when the elastic energy of distortion reaches a critical value. Nadai proposed that yielding begins when the shear stress acting on the octahedral plane reaches a critical value. This criterion sometimes also refers to as Hencky–Mises or the Huber–Mises criteria.

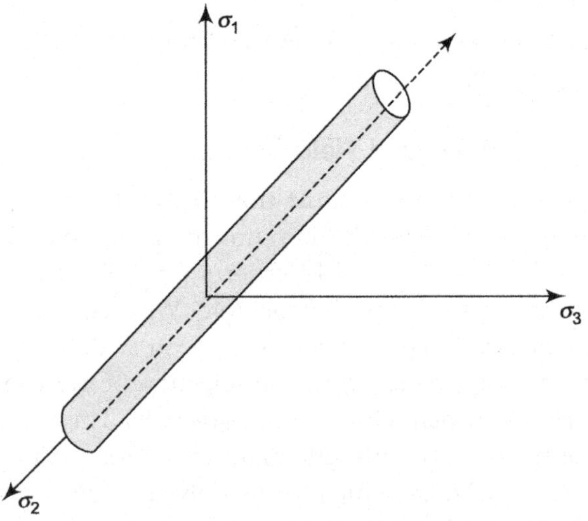

Fig. 2.10 von Mises yield criteria

Let us define effective stress σ_{eff} corresponding to stress tensor $\boldsymbol{\sigma}$ as

$$\sigma_{\text{eff}} = \sqrt{\frac{3}{2} s_{ij}s_{ij}} = \sqrt{\frac{3}{2} \mathbf{s} : \mathbf{s}}$$

where s_{ij} is the components of deviatoric stress tensor s. von Mises criteria can be written as

$$\sigma_{\text{eff}} - \sigma_y = 0$$

where σ_y is the yield stress of material in uniaxial tension/compression.

(b) Tresca yield criterion

According to the Tresca yield criterion (1864), yielding of the material will begin to occur when maximum shearing stress at a point reaches a critical value. If σ_1, σ_2, σ_3 are principal stresses arranged in descending order $\sigma_1 \geq \sigma_2 \geq \sigma_3$, we can write Tresca criterion as

$$\frac{1}{2}\left|\sigma_1 - \sigma_3\right| = k_T \tag{2.29}$$

where k_T is material dependent parameter determined experimentally. If σ_y be the yield stress, the maximum shear is $\dfrac{\sigma_y}{2}$. Tresca condition can be written as

$$\left|\sigma_1 - \sigma_3\right| = \sigma_y \tag{2.30}$$

or in terms of ρ and θ

$$\rho \sin\left(\theta + \frac{\pi}{3}\right) = \sqrt{2k} \tag{2.31}$$

The maximum shear stress at a point does not change when the state of stress at the point is changed hydrostatically. Tresca yield criterion represents a hexagonal cylinder in principal stress space.

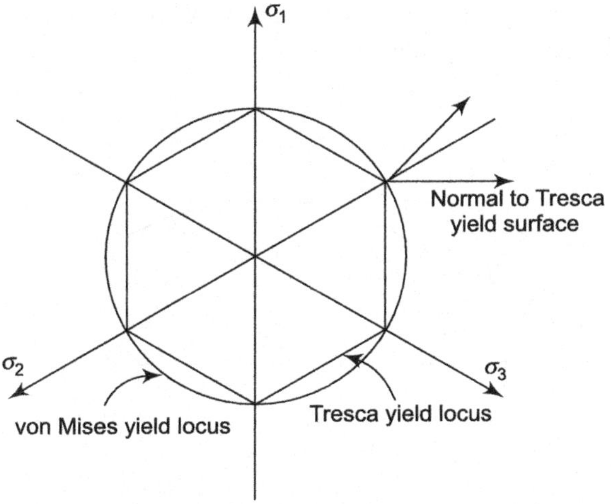

Fig. 2.11 Locus of Tresca and von Mises yield criteria on deviatoric plane

This cylinder is inside the von Mises right circular cylinder. One can see that Tresca yield surfaces have not unique normal at the corner (see Fig. 2.11) which poses problem in finite element formulation. But the expression is very simple. Due to simple expression of Tresca, it is preferred to von Mises for analytical solution although von mises criterion has been found experimentally closed for metal. But von Mises criteria has been used comprehensively for finite element implementation. This criteria is sometimes called Coulomb–Tresca or the Saint Venant–Tresca criteria.

Remarks : Both von Mises and Tresca criteria does not depend upon the hydrostatic mean stress. They are called pressure insensitive yield criteria. They model consequences of Bridgman experiments correctly.

2.5 BASICS FOR CONSTITUTIVE RELATIONS

Analysis of structure beyond yielding plays an important role in understanding the behavior of failure of the structure. It helps us to decide the factor of safety in design. Understanding the behavior beyond yield point of material is also needed in calculation of load to obtain desired shape in forming process. The relationship between stress and strain is not unique beyond yield point. Given a state of stress, there exists a unique state of strain for elastic material. Unique stress–strain relation does not exist if we load a material beyond yield point. If material has loading beyond yield point A and unloading/reloading, it is not possible to get unique stress–strain relations. Suppose if the stress level is σ_B, any strain along line BC can exists (see Fig. 2.12). This non-unique nature of stress–strain behavior poses problem in constitutive modeling. The strain at a point not only depends on the level of stress, but also on the history of loading. Thus, the load history plays an important role in describing the behavior of material beyond yield point. But we can talk about incremental relations for these materials. It is convenient to formulate governing equations of plastic deformation in rate form.

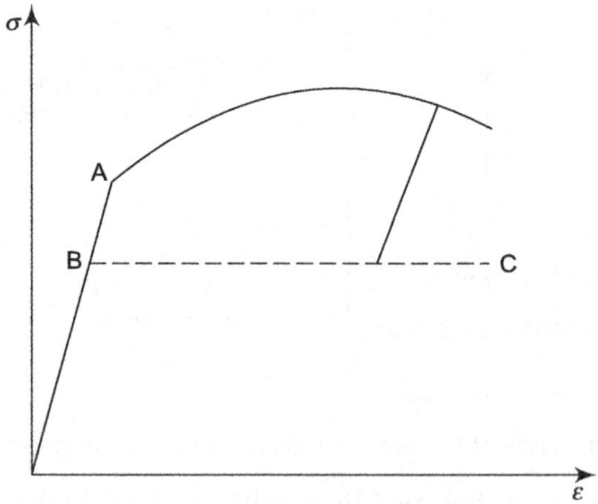

Fig. 2.12 A typical uniaxial stress strain relation

In subsequent sections, we shall discuss the basic equations of plasticity. We shall describe in general yield criteria, flow rule and hardening rule. We shall also talk about how to integrate these equations for given strain history. The treatment is general and brief. Our presentation is more toward development of basic equations and its computational aspects which can be applied in non-linear finite element analysis.

2.6 CONSTITUTIVE RELATIONS FOR ELASTO-PLASTIC MATERIALS

As we have discussed, the plastic behavior of material can be described in the rate form. For time independent deformation, these equations are homogeneous in derivatives with respect to time. Time is a pseudo parameter. Time plays a role in viscous material models such as viscoplasticity. During loading beyond yielding, the strain has two parts (a) recoverable parts (b) permanent deformation. These concepts will be utilized in the formulation of the theory. To describe the elasto-plastic behavior of material, we shall discuss the following points:

(a) Decomposition of strain into its elastic and plastic part,
(b) Elastic stress–strain relation,
(c) Initial yield criteria,
(d) Evolution of plastic strain (flow rule),
(e) Change in yield function or loading function during loading–unloading (hardening rule),
(f) Loading/unloading criteria.

These six topics form the backbone of plasticity theory. Our discussions are primarily restricted the infinitesimal deformation. Out of these six topics, following three are dependent on the nature of particular materials.

- Yield criterion
- Flow rule and
- Hardening Rule.

Yield criterion, flow rule, and hardening rules are material dependent. Various yield criteria, flow rule and hardening rules have been proposed for different material based on their observed experimental behavior. Our discussions here will be to develop the basic ingredients of the theory of plasticity.

2.6.1 Elastic and Plastic Strain

If we load a material beyond yield point and unload it, material does not regain its shape. But it has permanent deformation (see Fig. 2.13). We can see that at any level of loading, a portion of strain is recoverable whereas other part is permanent in nature. This separable nature of strain into recoverable and permanent part in tension test has lead to the assumption of decomposition of infinitesimal strain tensor into its elastic and plastic parts. If the strain is small, we can assume that the total strain is sum of elastic strain and plastic strain

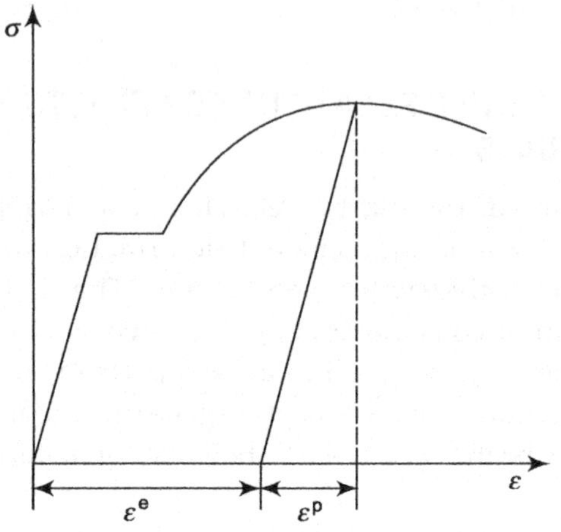

Fig. 2.13 Decomposition of strain into its elastic and plastic parts

$$\varepsilon_{ij} = \varepsilon_{ij}^{e} + \varepsilon_{ij}^{p} \tag{2.32}$$

where ε_{ij}^{e} is elastic part of strain and ε_{ij}^{p} is plastic part of strain. This intuitive separation of elastic and plastic strain based on one dimensional theory is valid only for small strain. If deformation is large, Eq. (2.32) does not hold good.

2.6.2 Elastic Stress–Strain Relations

We can observe from Fig. 2.13 that state of stress can be uniquely determined if the elastic strain is known at the load level. This observation is also assumed to hold in the case of general state of stress. This is based on the assumption that the elastic properties of material does not change during plastic deformation, *i.e.*, slope of loading/unloading curve is parallel to the initial loading curve. The elastic parameters do not change. We can write the stress–strain relations as

$$\sigma_{ij} = C_{ijkl}\,\varepsilon_{kl}^{e} \quad \text{or} \quad \boldsymbol{\sigma} = C : \boldsymbol{\varepsilon}^{e} \tag{2.33}$$

where C_{ijkl} is component of elastic stiffness tensor C of virgin materials. This assumption holds fairly well at moderate strain. Elastic degradation may takes place in the case of large deformation in metal and even at small deformation for brittle materials. An interested reader may consult book on damage mechanics [6, 7, 8]. We shall assume that C_{ijkl} does not depend upon the straining. For metal and alloy, it has been observed that the plastic deformation induces the change in only deviatoric part strain. Considering this, it is convenient to write stress–strain relation in its deviatoric and volumetric components as

$$p = K\varepsilon_{kk}^{e} \quad \text{and} \quad s = 2Ge^{e}$$

where $p = \frac{\sigma_{ii}}{3}$ and s is deviatoric part of stress tensor and e is deviatoric part of strain tensor. The elastic constants K and G are called bulk modulus and shear modulus and they are independent of strain in this case. We can write the constitutive relations in terms of K and G as

$$\sigma_{ij} = 2Ge^{e}_{ij} + K\varepsilon^{e}_{kk}\,\delta_{ij}$$

2.6.3 Initial Yield Criteria and Subsequent Yielding

The yield function defines the domain of elastic deformation. The yield function is non-positive function. A material is said to have onset of plastic deformation if

$$f(\sigma_{ij} - \beta_{ij}, \alpha_i) = 0 \tag{2.34}$$

where, f is a non-positive function called yield function and α_i, β_{ij} are hardening parameters. The hardening parameters are related to inelastic deformation of the material. For a given material, satisfaction of Eq. (2.34) represents the initial yield surface. With inelastic deformation, hardening parameter changes. Thus the yield surface changes. These yield surfaces are called subsequent yield surfaces.

In generalized isotropic hardening, yield surface changes its size whereas yield surface translates for kinematic hardening. Both isotropic as well as kinematic hardening can take place together called mixed hardening. Hardening rules are discussed in detail in Section 2.6.5. We shall illustrate initial and subsequent yield surfaces with the help of von Mises yield criterion in the following paragraphs.

von-Mises yield criterion can be written as

$$\sqrt{\frac{3}{2}s_{ij}s_{ij}} - R(\varepsilon^p) = 0 \quad \text{or} \quad \sigma_{\text{eff}} - R(\varepsilon^p) = 0 \tag{2.35}$$

where effective stress is defined as

$$\sigma_{\text{eff}} = \sqrt{\frac{3}{2}s_{ij}s_{ij}} \tag{2.36}$$

Initial yield function is given

$$\sigma_{\text{eff}} - \sigma_y = 0 \tag{2.37-a}$$

Subsequent yield function considering isotropic hardening can be written as

$$\sigma_{\text{eff}} - [\sigma_y + R(\varepsilon^p)] = 0 \tag{2.37-b}$$

where, $\varepsilon^p = \int \sqrt{\frac{2}{3}\dot{\varepsilon}^p_{ij}\dot{\varepsilon}^p_{ij}}\,dt$ is the effective plastic strain which is equal to uniaxial plastic strain in the case of incompressible deformation. Effective plastic strain has been taken as isotropic hardening parameter.

2.6.4 Flow Rule

Deformation can be split up into its elastic and plastic parts during infinitesimal elastoplastic deformation. We have already discussed that total strain is sum of elastic strain plus plastic strain. This statement is valid only when strain is infinitesimally small. For finite strain, splitting of elastic and plastic deformation has to be treated based on either deformation gradient tensor or deformation rate tensor. We shall restrict ourself to infinitesimal strain only. Elastic part of strain is governed by Eq. (2.33). Flow rule defines the evolution of plastic strain. It relates the evolution of plastic strain rate with current state of loading or deformation. Plastic strain rate is proportional to the gradient of a function of stress given as

$$\dot{\varepsilon}_{ij}^{p} = \dot{\lambda}\frac{\partial Q}{\partial \sigma_{ij}} \quad \text{or} \quad \dot{\boldsymbol{\varepsilon}}^{p} = \dot{\lambda}\frac{\partial Q}{\partial \boldsymbol{\sigma}} \tag{2.38}$$

where, $Q(\sigma_{ij})$ is function of stress called as plastic potential. The name plastic potential has been taken similar to elastic potential which defines the stress for hyper-elastic materials

If yield function is taken as plastic potential $Q \equiv f$, the flow rule is known as associated flow rule. Otherwise, it is called as non-associated flow rule. Associated flow rule has been employed for metals and alloys whereas non-associated flow rule is used for geomaterial. Geomaterial such as concrete and soil shows dilatancy which is taken care by non-associated flow rule. For associated flow rule

$$\dot{\varepsilon}_{ij}^{p} = \dot{\lambda}\frac{\partial f}{\partial \sigma_{ij}} \quad \text{or} \quad \dot{\boldsymbol{\varepsilon}}^{p} = \dot{\lambda}\frac{\partial f}{\partial \boldsymbol{\sigma}} \tag{2.39}$$

The associated flow rule is also called normality rule or normality criteria because gradient of a function is normal to the tangent plane of the surface at a point on the surface. The plastic straining takes place in the direction of normal to the yield surface in the case of associated flow rule. Please note that plastification takes place only when stress is on the yield surface, *i.e.*,

Equation (2.38) (Flow rule) should satisfy together with consistency condition $f(\sigma_{ij} - \beta_{ij}, \alpha_i) = 0$, but reverse is not true, *i.e.*, if $f(\sigma_{ij} - \beta_{ij}, \alpha_i) = 0$, $\dot{\boldsymbol{\varepsilon}}^{p}$ may not exist.

We shall discuss the particular case of the flow rule for metals and alloys considering von Mises yield criterion.

Using von Mises yield criterion

$$f(\sigma_{ij}, \varepsilon^{p}) = \sqrt{\frac{3}{2}S_{ij}S_{ij}} - R(\varepsilon^{p})$$

we can write flow rule

$$\dot{\varepsilon}_{ij}^p = \dot{\lambda}\frac{\partial f}{\partial \sigma_{ij}} = \frac{3}{2}\dot{\lambda}\frac{S_{ij}}{\sqrt{3/2 S_{ij}S_{ij}}} \tag{2.40}$$

Using $f(\sigma_{ij}, \varepsilon^p) = 0$, we can substitute $\sqrt{3/2 S_{ij}S_{ij}} = R$ in the above equation for flow rule as

$$\dot{\varepsilon}_{ij}^p = \frac{3}{2}\frac{\dot{\lambda}}{R}S_{ij} \tag{2.41}$$

We can see that plastic strain rate is proportional to the deviatoric stress. This flow rule was used by Prandtl (1927) and Reuss to derive the stress–strain relation for elastoplastic materials.

$$\dot{\varepsilon}^p = \sqrt{\frac{2}{3}\dot{\varepsilon}_{ij}\dot{\varepsilon}_{ij}} = \sqrt{\frac{2}{3}\frac{3}{2}\dot{\lambda}\frac{S_{ij}}{R}\frac{3}{2}\dot{\lambda}\frac{S_{ij}}{R}} \tag{2.42}$$

which can be written as

$$\dot{\varepsilon}^p = \dot{\lambda}\sqrt{\frac{\frac{3}{2}S_{ij}S_{ij}}{R^2}} \tag{2.43}$$

Using consistency of yield criterion, above equation reduces to

$$\dot{\varepsilon}^p = \dot{\lambda} \tag{2.44}$$

Substituting $\dot{\lambda}$ from above equation in Eq. (2.41), we can write flow rule as

$$\dot{\varepsilon}_{ij}^p = \frac{3}{2}\frac{\dot{\varepsilon}^p}{R}S_{ij} \tag{2.45}$$

substituting $R = \sigma_y$, we get

$$\dot{\varepsilon}_{ij}^p = \frac{3}{2}\frac{\dot{\varepsilon}^p}{\sigma_y}S_{ij} \tag{2.46}$$

We can note that $\dot{\varepsilon}_{kk}^p = 0$ in this case. Plastic deformation is isochoric and volumetric strain is purely elastic. This fact facilitates to write stress–elastic strain relation in volumetric and deviatoric part separately.

The above equation can be written explicitly in the incremental form as

$$d\varepsilon_{xx}{}^p = \frac{d\varepsilon^p}{\sigma_y}\left[\sigma_{xx} - \frac{1}{2}\left(\sigma_{yy} + \sigma_{zz}\right)\right] \tag{2.47}$$

$$d\varepsilon_{yy}{}^p = \frac{d\varepsilon^p}{\sigma_y}\left[\sigma_{yy} - \frac{1}{2}(\sigma_{zz} + \sigma_{xx})\right] \tag{2.48}$$

$$d\varepsilon_{zz}{}^p = \frac{d\varepsilon^p}{\sigma_y}\left[\sigma_{zz} - \frac{1}{2}(\sigma_{xx} + \sigma_{yy})\right] \tag{2.49}$$

$$d\varepsilon_{yz}^{p} = \frac{d\varepsilon^{p}}{\sigma_{y}}\sigma_{yz} \tag{2.50}$$

$$d\varepsilon_{zx}^{p} = \frac{d\varepsilon^{p}}{\sigma_{y}}\sigma_{zx} \tag{2.51}$$

$$d\varepsilon_{xy}^{p} = \frac{d\varepsilon^{p}}{\sigma_{y}}\sigma_{xx} \tag{2.52}$$

In the analysis of metal, elastic strain can be neglected as compared to plastic strain in many of the engineering analysis. Thus $\dot{\varepsilon}_{ij} \approx \dot{\varepsilon}_{ij}^{p}$ and we can write flow rule as

$$\dot{\varepsilon}_{ij} = \frac{3}{2}\frac{\dot{\varepsilon}^{p}}{\sigma_{y}}S_{ij} \tag{2.53}$$

This assumption is employed by Levy and Mise in their analysis. The corresponding plasticity constitutive equation thus derived is called Levy Mises equations.

2.6.5 Hardening Rule

Hardening rule tells us how the evolution of yield surfaces (which is also called loading surfaces) take place. The yield surfaces may expand or translate even distort during elasto-plastic deformation. In other words, hardening rules say how elastic domain changes during the deformation. For elastic-perfectly plastic material model, the domain of elastic deformation remains the same which gives rise to the fixity of yield surface. Depending upon the material under consideration, hardening rules are developed based on the observed behaviors of the material. Thus the experimental observation forms the basic ingredients to develop the hardening rules. There have been many theories to mathematically model the evolution of loading surfaces. The popular among them are

(a) Isotropic hardening
(b) Kinematic hardening

These are the simplest hardening rules to describe the most commonly observed behavior related to change in elastic domain.

(a) **Isotropic hardening** The expansion of loading surface is modeled by appropriately choosing hardening parameter (see Fig. 2.14). We shall consider only one isotropic hardening parameter although there has been multiple hardening parameter model for materials such as concrete. To model uniform expansion, yield function can be written by separating function of stress from the function of hardening parameters as

$$f(\sigma_{ij}, \alpha) = F(\sigma_{ji}) - R(\alpha) = 0 \tag{2.54}$$

where $F(\sigma_{ij})$ is only function of components of stress and $R(\alpha)$ is function of hardening parameter alpha. We can observe that with the increase of the value of $R(\alpha)$, the yield surface expands. This can be understood by taking the example of the equation $x^2 + y^2 + z^2 = a^2$. With the increase with a^2, the sphere expands. With the above discussion, it is clear that modeling of uniform expansion of yield surfaces reduces to two sub problems (a) selection of appropriate hardening parameter α and (b) selection of hardening function $R(\alpha)$. Isotropic hardening parameter α_i can be either effective plastic strain or plastic work in the yield surface equation $f(\sigma_{ij}, \alpha_i)$. If we choose effective plastic strain as hardening parameter, we can write

$$\alpha = \int_0^t \dot{\alpha}\, dt = \int_0^t \sqrt{\frac{2}{3}\dot{\varepsilon}_{ij}^p \dot{\varepsilon}_{ij}^p}\, dt = \int_0^t \sqrt{\frac{2}{3}\dot{\boldsymbol{\varepsilon}}^p : \dot{\boldsymbol{\varepsilon}}^p}\, dt$$

Substituting $\dot{\varepsilon}_{ij}^p$ in above equation, we can write

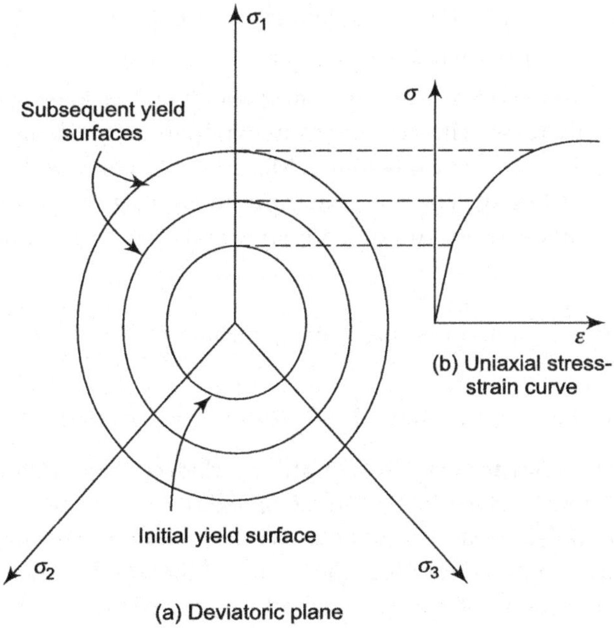

Fig. 2.14 Isotropic hardening

$$\alpha = \int_0^t \dot{\lambda} \sqrt{\frac{2}{3}\frac{\partial Q}{\partial \sigma_{ij}}\frac{\partial Q}{\partial \sigma_{ij}}}\, dt \quad \text{for non associated flow rule}$$

$$\alpha = \int_0^t \dot{\lambda} \sqrt{\frac{2}{3}\frac{\partial f}{\partial \sigma_{ij}}\frac{\partial f}{\partial \sigma_{ij}}}\, dt \quad \text{for associated flow rule}$$

If we choose plastic work as hardening parameter, we have

$$\alpha = \int \sigma_{ij}\dot{\varepsilon}_{ij}^{p}\,dt \quad \text{for associated flow rule}$$

Substituting $\dot{\varepsilon}_{ij}^{p}$ from flow rule in above equation, we can write

$$\alpha = \int \sigma_{ij}\,\dot{\lambda}\,\frac{\partial Q}{\partial \sigma_{ij}}\,dt \quad \text{non associated flow rule}$$

$$\alpha = \int \sigma_{ij}\,\dot{\lambda}\,\frac{\partial f}{\partial \sigma_{ij}}\,dt \quad \text{associated flow rule}$$

Many investigators have suggested different functions for hardening function $R(\alpha)$. The most popular among these hardening functions are power hardening law and exponential hardening law. In the power hardening law, $R(\alpha)$ is expressed by algebraic function as

$$R(\alpha) = \sigma_y + K\alpha^n \tag{2.55}$$

where K and n are material constants. For exponential hardening, $R(\alpha)$ is defined as exponential function of α as

$$R(\alpha) = R_\infty\,[1 - \exp(-K\alpha)] \tag{2.56}$$

where R_∞ and K are material constants.

Please note that in power hardening, yield surface keeps on expanding with the increase with α whereas in exponential hardening it increases exponentially and saturates at R_∞. Many of the materials show saturation during hardening. The material parameters K, n, R_∞, in the hardening expressions are obtained from uniaxial tests, primarily by curve fitting.

It can also be noted that hardening parameter $\alpha \neq \sqrt{\frac{2}{3}\boldsymbol{\varepsilon}^p : \boldsymbol{\varepsilon}^p}$, but it is obtained by integrating $\alpha = \int \dot{\alpha}\,dt$ where α is defined as $\dot{\alpha} = \sqrt{\frac{2}{3}\dot{\boldsymbol{\varepsilon}}^p : \dot{\boldsymbol{\varepsilon}}^p}$.

(b) **Kinematic hardening** Kinematic hardening models the movement of the yield surface in the stress space. It is assumed that the yield surface translates as a rigid body without changing its shape and size. In the kinematic hardening, the center of the yield surface is chosen as hardening parameter. It is assumed that the center of the yield surface changes with plastic deformation in the stress space (see Fig. 2.15). The center of the yield surface is called back stress. Kinematic hardening models the Bauschinger effect. Thus considering only kinematic hardening, we can write yield surface

$$f(\sigma_{ij}, \beta_{ij}) = F(\sigma_{ij} - \beta_{ij}) - R_o = 0 \tag{2.57}$$

where β_{ij} are components of the back stress tensor. The back stress tensor $\boldsymbol{\beta}$ is second order tensor. The hardening parameter β_{ij} is related to the plastic deformation. Thus our objective in the kinematic hardening modeling is to obtain the appropriate expression for the evolution of components of back stress β_{ij}. There have been various models for the evolution of back stress tensor. Some of them are

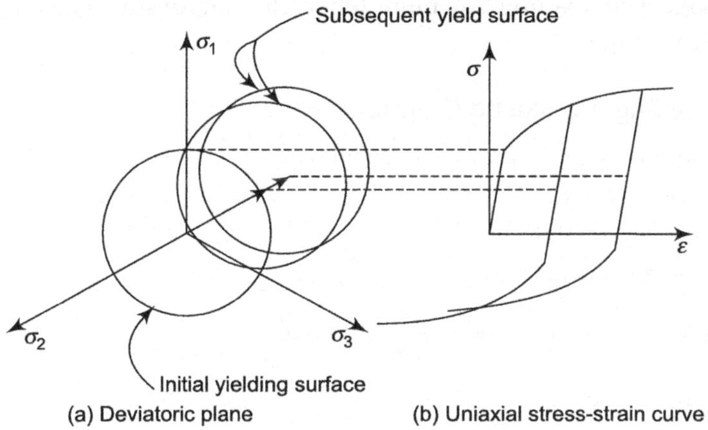

Fig. 2.15 Kinematic hardening

(a) **Prager's model:** Prager proposed that the rate of back stress tensor is proportional to rate of plastic strain tensor.

$$\dot{\beta}_{ij} = c\dot{\varepsilon}_{ij}^{p} \quad \text{or} \quad \dot{\boldsymbol{\beta}} = c\dot{\boldsymbol{\varepsilon}}^{p}$$

where c is the material parameter. Thus according to the Prager's kinematic hardening model, yield surface translates in the direction of plastic strain rate or along the normal to the yield surface in the case of associated flow rule. Substituting plastic strain rate $\dot{\varepsilon}_{ij}^{p}$ in the above equation, we can write

$$\dot{\beta}_{ij} = c\dot{\lambda}\frac{\partial Q}{\partial \sigma_{ij}} \quad \text{for non-associated flow rule}$$

$$\dot{\beta}_{ij} = c\dot{\lambda}\frac{\partial f}{\partial \sigma_{ij}} \quad \text{for associated flow rule}$$

Ziegler pointed out the deficiency in the hardening law proposed by Prager. According to Ziegler, Prager's kinematic hardening does not give consistent result in three dimensional, two dimensional and one dimensional cases. It can be proved that it introduces hardening in one direction and softening in other two directions in uniaxial case which has not been observed experimentally [9, 10].

(b) **Ziegler's model:** Ziegler proposed another kinematic hardening rule which states that the movement of yield surface takes place in the direction determined by the vector $\alpha - \beta$. Mathematically it can be stated that

$$\dot{\beta}_{ij} = \dot{\mu}(\sigma_{ij} - \beta_{ij})$$

where $\dot{\mu}$ is the proportionality constant determined by particular yield criteria.

Material behaves very complex manner under cyclic loading. There has been developed many approaches to model behavior of material under cyclic loading. An interested reader can consult book on plasticity [11, 12, 13].

2.6.6 Loading–Unloading Criteria

Loading-unloading criteria can be written as

(a) Elastic loading

$$f(\sigma_{ij} - \beta_{ij}, \alpha_i) < 0$$

(b) Elasto-plastic loading

$$f(\sigma_{ij} - \beta_{ij}, \alpha_i) = 0 \quad \text{and} \quad \frac{\partial f}{\partial \sigma_{ij}} d\sigma_{ij} > 0$$

(c) Neutral loading

$$f(\sigma_{ij} - \beta_{ij}, \alpha_i) = 0 \quad \text{and} \quad \frac{\partial f}{\partial \sigma_{ij}} d\sigma_{ij} = 0$$

(d) Elastic unloading

$$f(\sigma_{ij} - \beta_{ij}, \alpha_i) = 0 \quad \text{and} \quad \frac{\partial f}{\partial \sigma_{ij}} d\sigma_{ij} < 0$$

Loading-unloading criteria can be combined as Kuhn Tucker optimality criteria as

$$f \leq 0 \quad \dot{\lambda} \geq 0 \quad \dot{\lambda} f = 0$$

and consistency condition

$$\dot{\lambda} \dot{f} = 0$$

2.7 CONSTITUTIVE EQUATIONS FOR ELASTO-PLASTIC MATERIALS

Constitutive relations for elastoplastic deformation can be given as

$$\varepsilon_{ij} = \varepsilon_{ij}^e + \varepsilon_{ij}^p \tag{2.58-a}$$

$$\sigma_{ij} = C_{ijkl} \varepsilon_{kl}^e \tag{2.58-b}$$

$$\dot{\varepsilon}_{ij}^p = \dot{\lambda} \frac{\partial Q}{\partial \sigma_{ij}} \quad \text{or} \quad \dot{\varepsilon}_{ij}^p = \dot{\lambda} \frac{\partial f}{\partial \sigma_{ij}} \tag{2.58-c}$$

$$f(\sigma_{ij} - \beta_{ij}, \alpha) = 0 \tag{2.58-d}$$

$$\dot{\alpha} = \dot{\varepsilon}^p = \dot{\lambda} \sqrt{\frac{2}{3} \frac{\partial Q}{\partial \sigma_{ij}} \frac{\partial Q}{\partial \sigma_{ij}}} \quad \text{or} \quad \dot{\alpha} = \dot{\lambda} \sqrt{\frac{2}{3} \frac{\partial f}{\partial \sigma_{ij}} \frac{\partial f}{\partial \sigma_{ij}}} \tag{2.58-e}$$

$$\dot{\beta}_{ij} = c\dot{\lambda} \frac{\partial Q}{\partial \sigma_{ij}} \quad \text{or} \quad \dot{\beta}_{ij} = c\dot{\lambda} \frac{\partial f}{\partial \sigma_{ij}} \tag{2.58-f}$$

Equations (2.58-a), (2.58-b) and (2.58-d) are algebraic equation and Eqs (2.58-c), (2.58-e) and (2.58-f) are differential equation which is homogeneous in time. For a given load history, these equations are integrated to obtain the variables $\sigma_{ij}, \varepsilon_{ij}^p, \varepsilon_{ij}^e, \alpha, \beta_{ij}$. We can combine all the equations. One can write the relations between stress and strain rate as follows.

Differentiating Eq. (2.58-d), we obtain

$$\frac{\partial f}{\partial \sigma_{ij}}\dot{\sigma}_{ij} + \frac{\partial f}{\partial \alpha}\dot{\alpha} + \frac{\partial f}{\partial \beta_{ij}}\beta_{ij} = 0 \tag{2.59}$$

Substituting and from Eq. (2.58-c) and (2.58-f) in the above equation, we get

$$\frac{\partial f}{\partial \sigma_{ij}}\dot{\sigma}_{ij} + \frac{\partial f}{\partial \alpha}\dot{\lambda}\sqrt{\frac{2}{3}\frac{\partial Q}{\partial \sigma_{ij}}\frac{\partial Q}{\partial \sigma_{ij}}} + \frac{\partial f}{\partial \beta_{ij}}\dot{\lambda}c\frac{\partial f}{\partial \sigma_{ij}} = 0 \tag{2.60}$$

Differentiating Eq. (2.58-b) and obtaining with Eq. (2.58-a), we get

$$\dot{\sigma}_{ij} = C_{ijkl}(\dot{\varepsilon}_{kl} - \dot{\varepsilon}_{kl}^p) \tag{2.61}$$

Substituting $\dot{\varepsilon}_{ij}^p$ from Eq. (2.58-c) in the above equation, we can write

$$\dot{\sigma}_{ij} = C_{ijkl}\left(\dot{\varepsilon}_{kl} - \dot{\lambda}\frac{\partial Q}{\partial \sigma_{kl}}\right) \tag{2.62}$$

Subsituting from above equation in Eq. (2.60), we get

$$\frac{\partial f}{\partial \sigma_{ij}}C_{ijkl}\left(\dot{\varepsilon}_{kl} - \dot{\lambda}\frac{\partial Q}{\partial \sigma_{kl}}\right) + \frac{\partial f}{\partial \alpha}\dot{\lambda}\sqrt{\frac{2}{3}\frac{\partial Q}{\partial \sigma_{ij}}\frac{\partial Q}{\partial \sigma_{ij}}} + \frac{\partial f}{\partial \beta_{ij}}\dot{\lambda}c\frac{\partial f}{\partial \sigma_{ij}} = 0 \tag{2.63}$$

Solving for $\dot{\lambda}$, we can write

$$\dot{\lambda} = \frac{\frac{\partial f}{\partial \sigma_{ij}}C_{ijkl}\dot{\varepsilon}_{kl}}{\frac{\partial f}{\partial \sigma_{pq}}C_{pqrs}\frac{\partial Q}{\partial \sigma_{rs}} - \frac{\partial f}{\partial \alpha}\sqrt{\frac{2}{3}\frac{\partial Q}{\partial \sigma_{pq}}\frac{\partial Q}{\partial \sigma_{pq}}} + \frac{\partial f}{\partial \beta_{pq}}c\frac{\partial f}{\partial \sigma_{pq}}} \tag{2.64}$$

Subsituting $\dot{\lambda}$, from above equation in Eq. (2.62), we can write

$$\dot{\sigma}_{ij} = C_{ijkl}\left(\dot{\varepsilon}_{kl} - \frac{\frac{\partial f}{\partial \sigma_{ab}}C_{abcd}\dot{\varepsilon}_{cd}\frac{\partial Q}{\partial \sigma_{kl}}}{\frac{\partial f}{\partial \sigma_{pq}}C_{pqrs}\frac{\partial Q}{\partial \sigma_{rs}} - \frac{\partial f}{\partial \alpha}\sqrt{\frac{2}{3}\frac{\partial Q}{\partial \sigma_{pq}}\frac{\partial Q}{\partial \sigma_{pq}}} - \frac{\partial f}{\partial \beta_{pq}}c\frac{\partial f}{\partial \sigma_{pq}}}\right) \tag{2.65}$$

which can be simplified as

$$\dot{\sigma}_{ij} = \left[C_{ijkl} - \frac{C_{ijcd}\frac{\partial f}{\partial \sigma_{ab}}C_{abkl}\frac{\partial Q}{\partial \sigma_{cd}}}{\frac{\partial f}{\partial \sigma_{pq}}C_{pqrs}\frac{\partial Q}{\partial \sigma_{rs}} - \frac{\partial f}{\partial \alpha}\sqrt{\frac{2}{3}\frac{\partial Q}{\partial \sigma_{pq}}\frac{\partial Q}{\partial \sigma_{pq}}} - \frac{\partial f}{\partial \beta_{pq}}c\frac{\partial f}{\partial \sigma_{pq}}}\right]\dot{\varepsilon}_{kl} \tag{2.66}$$

which can be written as

$$\dot{\sigma}_{ij} = C^{ep}_{ijkl}\dot{\varepsilon}_{kl} \tag{2.67}$$

where C^{ep}_{ijkl} is the component of elasto-plastic continuum tangent modulus which is given as

$$C^{ep}_{ijkl} = C_{ijkl} - \frac{C_{ijcd}\dfrac{\partial f}{\partial \sigma_{ab}}C_{abkl}\dfrac{\partial Q}{\partial \sigma_{cd}}}{\dfrac{\partial f}{\partial \sigma_{pq}}C_{pqrs}\dfrac{\partial Q}{\partial \sigma_{rs}} - \dfrac{\partial f}{\partial \alpha}\sqrt{\dfrac{2}{3}\dfrac{\partial Q}{\partial \sigma_{pq}}\dfrac{\partial Q}{\partial \sigma_{pq}}} - \dfrac{\partial f}{\partial \beta_{pq}}c\dfrac{\partial f}{\partial \sigma_{pq}}} \tag{2.68}$$

which can be written in the vector notation as

$$C^{ep} = C - \frac{\dfrac{\partial f}{\partial \sigma}:C \otimes C:\dfrac{\partial Q}{\partial \sigma}}{\dfrac{\partial f}{\partial \sigma}:C:\dfrac{\partial Q}{\partial \sigma} - \dfrac{\partial f}{\partial \alpha}\sqrt{\dfrac{2}{3}\dfrac{\partial Q}{\partial \sigma}:\dfrac{\partial Q}{\partial \sigma}} - c\dfrac{\partial f}{\partial \sigma}:\dfrac{\partial f}{\partial \beta}} \tag{2.69}$$

Remarks: The above derived modulus is called continuum tangent modulus because it has been derived using continuum equation (in time) of elasto-plasticity. There is another modulus which is called consistent tangent modulus which has been derived from discretized equations (in time) of plasticity [14,15]. Equation (2.67) is equivalent to Eqs (2.58-a)–(2.58-f) for continuous time. Integrated equations represent different consistency discussed in the next section.

2.8 INTEGRATION OF CONSTITUTIVE RELATIONS

The boundary value problems in plasticity are highly nonlinear. They are generally solved by numerical methods. In the numerical methods, one of the objectives is to integrate the constitutive equations discussed earlier. In this section, we shall discuss the integration of the constitutive relations between time t_n and t_{n+t}. In the rate independent plasticity, time t is a pseudo-parameter which can be taken as loading parameter. Strain driven algorithm has been popularly used in the literature. We shall discuss the strain driven integration algorithms which has been widely employed in nonlinear finite element analysis.

2.8.1 Definition of the Problem

We shall assume that all the quantities (strain field ε_n, plastic strain field ε^p_n, elastic strain field ε^e_n, back stress β_n, isotropic hardening parameter α_n and stress field σ_n) is known at time t_n, i.e., $\{\varepsilon_n, \varepsilon^e_n, \varepsilon^p_n, \beta_n, \alpha_n, \sigma_n\}$ are known at t_n and strain field $\varepsilon_{n+1} = \varepsilon_n + \Delta\varepsilon_n$ is known at time t_{n+1}. Our objective is to obtain the field at time t_{n+1} taking into account constitutive relations given by Eqs (2.58-a)–(2.58-f). Except for few cases, these equations are so complex that they are integrated numerically. The problem of integration of constitutive relations can be thought at an initial value problem. But some equations are algebraic whereas others are differential in time. This poses the problem in the solution. Many investigators have carried out work to obtain the accurate and stable solution [15, 16, 17].

We shall briefly discuss some of the issues here

2.8.2 Integration Algorithm

Numerical integration of constitutive relations can be grouped as

(a) Explicit integration scheme

(b) Implicit integration scheme

Explicit integration scheme does not require the solution of nonlinear algebraic equations, but obtained stress field may drift from the yield surface. We need to apply correction for drift from yield surface or sub-incrementation to increase the accuracy and stability of solution. The method is conditionally stable. Implicit algorithms are unconditionally stable but requires solution of nonlinear equations. These resulting nonlinear equations are solved by Newton–Raphson method.

A large number of algorithms can be described by so called generalized mid point method or generalized trapezoidal method.

Equations (2.58-c), (2.58-e), and (2.58-f) are integrated to give

$$(\varepsilon_{ij}^{p})_{n+1} = (\varepsilon_{ij}^{p})_{n} + \Delta\lambda \left(\frac{\partial Q}{\partial \sigma_{ij}}\right)_{n+p} \tag{2.70-a}$$

$$\alpha_{n+1} = \alpha_{n} + \Delta\lambda \sqrt{\frac{2}{3}\left(\frac{\partial Q}{\partial \sigma_{ij}}\right)_{n+p}\left(\frac{\partial Q}{\partial \sigma_{ij}}\right)_{n+p}} \tag{2.70-b}$$

$$(\beta_{ij})_{n+1} = (\beta_{ij})_{n} + c\Delta\lambda \left(\frac{\partial Q}{\partial \sigma_{ij}}\right)_{n+p} \tag{2.70-c}$$

where $\Delta\lambda = \lambda_{n+1} - \lambda_{n}$ and $\left(\frac{\partial Q}{\partial \sigma_{ij}}\right)_{n+p}$ can be calculated from (a) *Generalized Trapezoidal rule* as

$$\left(\frac{\partial Q}{\partial \sigma_{ij}}\right)_{n+p} = (1-p)\left(\frac{\partial Q}{\partial \sigma_{ij}}\right)_{n} + p\left(\frac{\partial Q}{\partial \sigma_{ij}}\right)_{n+1}, \quad 0 \le p \le 1 \tag{2.71}$$

or from (b) *Generalized Mid-point rule* as

$$\left(\frac{\partial Q}{\partial \sigma_{ij}}\right)_{n+p} = \frac{\partial Q\left(\left(\sigma_{ij}\right)_{n+p}\right)}{\partial \sigma_{ij}}, \quad \text{where } (\sigma_{ij})_{n+p} = (1-p)(\sigma_{ij})_{n} + p(\sigma_{ij})_{n+1} \tag{2.72}$$

We need to satisfy the algebraic part of constitutive Eqs (2.58-a)–(2.58-f). These equations are satisfied at time t_{n+1} as

$$(\varepsilon_{ij})_{n+1} = (\varepsilon^{e}_{ij})_{n+1} + (\varepsilon^{e}_{ij})_{n+1} \quad (2.73\text{-a})$$

$$(\sigma_{ij})_{n+1} = C_{ijkl}\,(\varepsilon^{e}_{kl})_{n+1} \quad (2.73\text{-b})$$

$$f((\sigma_{ij})_{n+1} - (\beta_{ij})_{n+1}, \alpha_{n+1}) = 0 \quad (2.73\text{-c})$$

Thus we have obtained algebraic equations in field variables at time t_{n+1}. The strain field at time t_{n+1}, $(\varepsilon_{ij})_{n+1}$ is known. The unknowns in Eqs (2.70-a) –(2.70-c) and Eqs (2.73-a)–(2.73-c) are $(\varepsilon^{p})_{ij}$, $(\varepsilon^{e})_{ij}$, $(\sigma_{ij})_{n+1}$, $(\beta_{ij})_{n+1}$, λ_{n+1} and α_{n+1}. We also have four tensor equations and two scalar equations. The number of equations are equal to number of unknown. These equations are solved by Newton–Raphson method.

2.8.3 Remarks on Integration

1. Equation (2.73-c) is called consistency condition. In the equation, we have imposed the consistency at time t_{n+1}. It has been reported that we need to impose consistency at the time t_{n+p} to obtain the B-stability, *i.e.*, $f((\sigma_{ij})_{n+p} - (\beta_{ij})_{n+p}, \alpha_{n+p})$[18].
2. The algorithms presented above is explicit for $p = 0$ otherwise it is implicit. The algorithm has second order accuracy at $p = 0.5$.
3. It has been pointed that the tangent modulus should be obtained by linearization of integrated Eqs (2.70-a)–(2.73-c) to preserve the quadratic convergence of Newton–Raphson method used to solve discretized boundary value problem in FEM. Tangent modulus thus obtained is called consistent tangent modulus. The efficient calculation of tangent modulus and accurate as well as stable integration algorithm has been active area of research for past two decades and many issues are still to be solved.
4. The integration of Eq. (2.67) is not the same as the integration of Eqs (2.58-a) and (2.58-f) numerically to obtain stress field σ_{n+1}. The integrated Eq. (2.67) satisfy the consistency approximately whereas we impose the exact consistency in the later case in terms of approximately calculated field quantities.

2.9 CONCLUSION

Plasticity plays pivot role in failure analysis of structures and obtaining required load in forming process. We have discussed some of the points of plastic modeling of material. Flow rule, Hardening rule and summary of constitutive relations have been presented. Boundary value problems in plasticity are highly nonlinear. Integration of constitutive relations plays an important role in the solution of boundary value problem. We have also presented numerical methods to integrate constitutive relations which are at central to the implementation of plasticity in numerical methods such as finite element method.

REFERENCES

1. Orowan, E., (1944), *Proceedings of Institutions of Mechanical Engineers,* **151,** pp.133.
2. Jog, C.S., (2001), "Foundations and applications of mechanics", Volume-I, *Continuum Mechanics,* Narosa Publishing House, New Delhi.
3. Gurtin, M.E., (2005), *An introduction to continuum mechanics,* Academic Press, Indian print.
4. Chatterjee, R., (1999), *Mathematical theory of continuum mechanics,* Narosa Publishing House, New Delhi.
5. Hill R. (1998), *The mathematical theory of plasticity,* Oxford University Press.
6. Lemaitre, J., (1996), *A course on damage mechanics,* Springer.
7. Krajcinovic, D., (1996), "Damage mechanics", North Holland series in *Applied math and mechanics.*
8. Lemaitre, J. and Desmorat, R., (2005), *Engineering damage mechanics: ductile, creep, fatigue and brittle failure,* Springer.
9. Ziegler, H. (1959), A modification to Prager's hardening rule, *Quarterly of applied mathematics,* **17,** pp. 55–59.
10. Shield, R.T. and Ziegler, H., (1958). On Prager's hardening rule, Zeitschrift für Angewandte Mathematik und Physik (ZAMP), **99,** pp. 260–276.
11. Khan, A.S. and Huang, S., (1995), *Continuum theory of plasticity,* John Wiley and Sons.
12. Lubarda, V.A. (2002), *Elastoplasticity theory,* CRC Press.
13. Jirasek, M. and Bazant, Z.P. (2002), *Inelastic analysis of structures,* John Wiley and Sons.
14. Simo, J.C. and Taylor R.L. (1985), "Consistent tangent operators for rate independent elasto-plasticity", *Com. Meth. Appl. Mech. Engineering,* **48,** pp. 101–118.
15. Simo, J. and Hughes, T.J.R., (1998), *Computational inelasticity,* Springer.
16. Dunne, F. and Petrinic, N. (2005). *Introduction to computational plasticity,* Oxford University Press.
17. D.R.J. Owen and Hinton, E. (1980). *Finite element in plasticity: theory and practice,* Pineridge Press, Swansea.
18. Simo, J. C. and Govindjee, S. (1991). Nonlinear B-stability and symmetry preserving return mapping algorithms for plasticity and viscoplasticity, *International Journal Numerical Methods of Engineering,* **31,** pp. 151–176.

CHAPTER 3 Modelling of Metal Forming Process

Uday S. Dixit, R. Ganesh Narayanan
Department of Mechanical Engineering
Indian Institute of Technology Guwahati, Assam 781 039, India

3.1 INTRODUCTION

A comprehensive description about various metal forming processes is given in Chapter 1. There are various material and process parameters that affect the forming behavior in a synergistic fashion. It is advantageous to perform experiments in order to understand the metal forming processes. In order to reduce the cost associated with conducting experiments and dependence on resources, theoretical models were developed for predicting the metal forming behavior and process performance. Modeling methods like slab method, upper and lower bound methods, slip line field theory, finite element method, and recently soft computing methods like neural networks, fuzzy sets, genetic algorithm are used in an integrated manner to predict the forming behavior under varied working conditions. The main objective of the modeling exercise is to predict the metal forming behavior at the design stage itself so that only limited trial experiments can be planned. Later on to improve the accuracy of prediction, the modeling methods are fine tuned for day-to-day life usage.

The application of various modeling methods can be understood from the published literature. For example, Slab method has been used for analyzing the sheet and foil rolling process by Salimi and Kadkhodaei [1], Salimi and Sassani [2], Kumar and Dixit [3], etc. This method has been used by Chitkara and Aleem [4] for tube extrusion process analysis. Similarly upper bound method is used to-analyze the sheet hydroforming process [5], evaluate friction in bulk metal forming [6], analyze axi-symmetric extrusion process assisted by cyclic torsion [7], predict the influence of die angle on the relative extrusion pressure during tube extrusion process [8] and many other works. Many times the modeling methods are used in an integrated fashion for forming process analyses. For instance, slab method and finite element simulation together are used to investigate the ring forming operation by Tzou *et al.* [9]. The effects of frictional coefficient, rotating angular speed, reduction and aspect ratio on compression force, effective stress, and the effective strain are predicted and discussed in their work. The compression behavior of double layer clad sheet compression forming have been investigated using the upper bound and slab methods by Hsu and Tzou [10]. They compared the effect of friction factor ratio on the compression force using both the methods.

The soft computing methods like neural network, fuzzy sets, genetic algorithm are also used to predict the metal forming behavior. Lorenzo *et al.* [11] and Sivakrishna and Ganesh Narayanan [12] predicted the fracture in cold forming operations and forming limit of welded blanks, respectively, using neural networks trained with data from finite element simulations. Hans Raj *et al.* [13] predicted the forging load during hot upsetting and hot extrusion operations using back propagation neural networks. The neural network predictions are compared with simulations results. Chan *et al.* [14] predicted the forging load and maximum effective stress during a typical metal forming operation using ANN integrated with finite element methods. Kazan *et al.* [15] developed the prediction model of springback in wipe-bending process using ANN approach. Fuzzy sets are being used for metal forming equipment design and process analysis as described by Dixit *et al.* [16], Abburi and Dixit [17], Shashi Kumar [18], Hsiang and Lin [19] and many others in the field of rolling, machining, material deformation studies, etc. ANN-GA hybrid system is also used for modeling and prediction of various material forming operations like minimizing springback during sheet forming [20] and reducing flatness values of hot rolled strips [21].

In this chapter, a brief description about the various modeling methods are given with their applications in the field of metal (or material, in general) forming.

3.2 SLAB ANALYSES

This method, also called as force equilibrium approach, is based on equating forces acting at an elemental region in a deforming billet in one direction. This produces a differential equation that is solved along with the boundary conditions. Assumptions usually followed while using this method for modeling metal forming operations are as follows:

1. The deformation is considered homogeneous, i.e., a plane section remains plane throughout forming.
2. The principal axes are the same as that of loading axes and friction does not change the principal axes direction for the purpose of applying yield criterion.

The slab method is applied to analyze various metal forming operations like plane strain compression, axially symmetric compression, rolling, sheet drawing, etc., and some of them are briefly explained here.

3.2.1 Plane Strain Compression of Solid Billet

Consider the compression of a solid billet under plane strain condition (no deformation in z-direction) shown in Fig. 3.1. Assuming sliding friction at the interface with a constant friction coefficient of μ, force balance across the elemental region in x direction gives

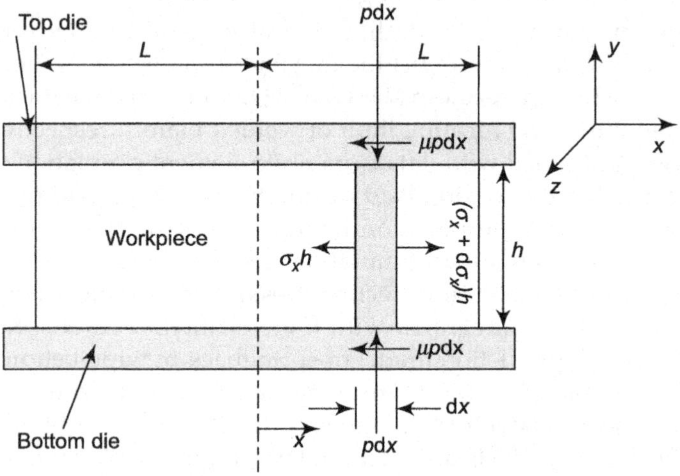

Fig. 3.1 Slab analysis of plane-strain compression of solid billet

$$\sigma_x h + 2\mu p\,dx = (\sigma_x + d\sigma_x)\,h \tag{3.1}$$

or,

$$2\mu p\,dx = h\,d\sigma_x \tag{3.2}$$

For plane-strain condition, by considering σ_x and σ_y (or $-p$) as principal stresses, and following von Mises yield criterion, we can obtain

$$\sigma_x - \sigma_y = 2k,$$

or,

$$\sigma_x + p = 2k = \frac{2}{\sqrt{3}}\sigma_0 \tag{3.3}$$

where σ_0 is the flow stress of the material in tension or compression. Differentiating Eq. (3.3), we get

$$d\sigma_x = -\,dp \tag{3.4}$$

Now Eq. (3.2) becomes,

$$2\mu p\,dx = -\,h\,dp$$

or,

$$\frac{dp}{p} = \frac{-2\mu}{h}\,dx \tag{3.5}$$

Integrating Eq. (3.5) by taking $\sigma_x = 0$, i.e., $p = 2k$ at $x = L$ provides

$$\ln\left(\frac{p}{2k}\right) = \frac{2\mu}{h}(L - x)$$

or

$$\frac{p}{2k} = \exp\left(\frac{2\mu}{h}(L - x)\right) \quad (3.6)$$

The maximum value for p can be obtained by putting $x = 0$ at the centerline. From Eq. (3.6),

$$\left(\frac{p}{2k}\right)_{\max} = \exp\left(\frac{2\mu L}{h}\right) \quad (3.7)$$

The graph between $\left(\frac{p}{2k}\right)$ and distance x is shown in Fig. 3.2 and is generally referred to as "friction hill".

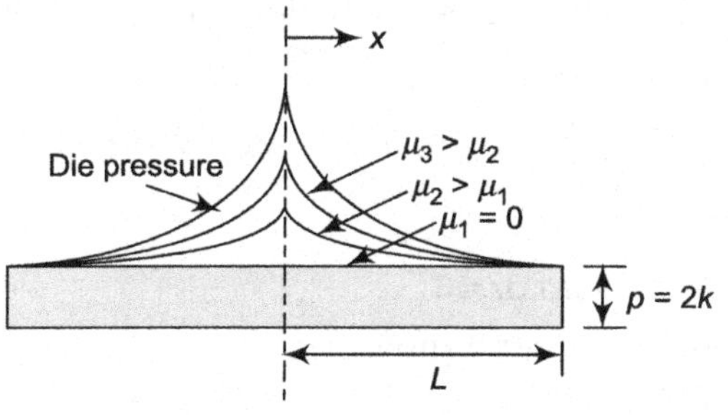

Fig. 3.2 Friction hill (die pressure variation) during plane-strain compression under sliding friction

The average pressure during plane strain compression (p_a) is given by

$$p_a = 2k\left(1 + \frac{\mu L}{h}\right) \text{ for smaller values of } \left(\frac{\mu L}{h}\right) \quad (3.8)$$

In sticking friction condition, the friction stress on the interface is equal to yield strength of the material in shear, k. In this case, Eq. (3.2) becomes,

$$2k\mathrm{d}x = h\mathrm{d}\sigma_x$$

and by following the same procedure as described above for sliding friction condition, Eq. (3.6) becomes,

$$\frac{p}{2k} = 1 + \frac{L - x}{h} \quad (3.9)$$

This shows the linear variation of $\left(\frac{p}{2k}\right)$ with distance x from the edge to the centerline (Fig. 3.3). The maximum value which occurs at the centerline ($x = 0$) is given by

$$\left(\frac{p}{2k}\right)_{max} = 1 + \frac{L}{h} \tag{3.10}$$

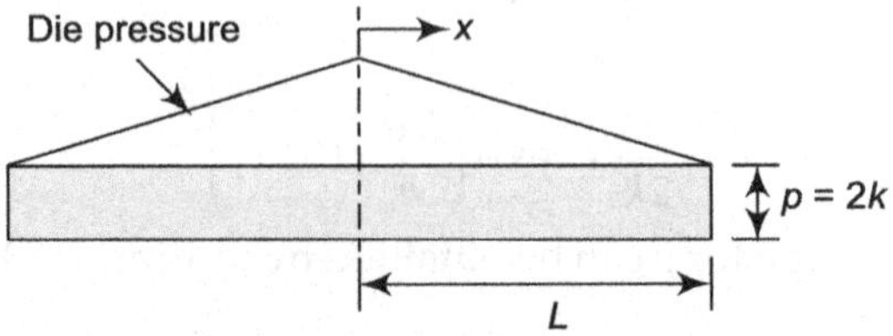

Fig. 3.3 Friction hill (die pressure variation) during plane-strain compression under sticking friction

3.2.2 Axially Symmetric Compression of Solid Billet

Similar to the analysis followed for plane strain-compression, assuming sliding friction exists at the interface with a constant friction coefficient of μ and force balance across the elemental region (Fig. 3.4) in the radial direction gives,

$$(\sigma_r + d\sigma_r)\, hd\theta\, (r + dr) = \sigma_r\, hrd\theta + 2\mu prdrd\theta + 2\sigma_\theta\, hdr\, \frac{d\theta}{2}$$

or

$$2\mu\, prdr + h\sigma_\theta dr - hrd\sigma_r - h\sigma_r dr = 0 \tag{3.11}$$

(by neglecting higher order terms).

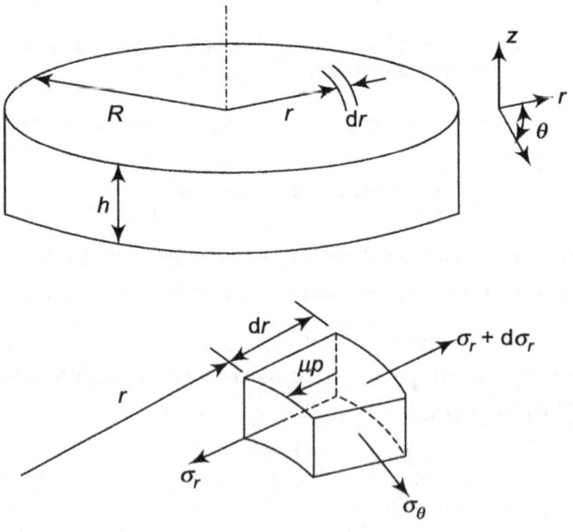

Fig. 3.4 Slab analysis of axi-symmetric compression of solid billet

Assuming that as a result of deformation R and r get increased by the same factor, $\varepsilon_r = \varepsilon_\theta$ and consequently $\sigma_r = \sigma_\theta$. At the onset of yielding, $\sigma_r + p = \sigma_0$ or $d\sigma_r = -dp$.

Substituting this in Eq. (3.11) gives

$$2\mu prdr = -hrdp$$

or,

$$\frac{dp}{p} = -\frac{2\mu dr}{h} \tag{3.12}$$

The solution of Eq. (3.12) can be obtained by taking $\sigma_r = 0$ at the edge of the disk ($r = R$). Hence with boundary condition $p = \sigma_0$ at $r = R$ provides the solution

$$p = \sigma_0 \exp\left(\frac{2\mu}{h}(R - r)\right) \tag{3.13}$$

By assuming sticking friction condition at the interface and following the above procedure, we get

$$p = \sigma_0 + \frac{2k}{h}(R - r) \tag{3.14}$$

Similarly slab analysis is applied for many other forming processes like deep drawing of sheet, strip drawing, extrusion, wire drawing, etc. [22, 23].

Salimi and Kadkhodaei [1] analyzed the asymmetrical sheet rolling using slab method for predicting the work roll pressure and mean pressure distributions, mean shear stress distribution, and variation of rolling force per unit width with initial thickness. They have also predicted the effect of friction coefficient and percentage reduction on the rolling force and torque per unit width. Some of the predicted results are compared with experiments. Salimi and Sassani [2] used a modified slab analysis method to predict the strip curvature during asymmetric sheet rolling operation. This modified method describes an enhancement to the rolling theory where friction becomes an integral part of deformation mechanics in the roll gap. The slab analyses predictions are compared with experimental results and a few other literature results. The slab method is also used for analyzing the extrusion of tubes from hollow and solid circular solid billets by Chitkara and Aleem [4]. The extrusion pressures predicted are agreeing well with the experiments. Kumar and Dixit [3] have used strain hardening law and a realistic friction model, Wanheim and Bay's friction model, during slab analysis of cold foil rolling process. A comparison between the roll torque and roll force predicted by Wanheim and Bay's friction model and Coulomb's friction model is shown in Fig. 3.5. Also the effect of strain hardening on the roll force and roll torque for different percentage reductions are discussed. It is shown that the model with strain hardening effect not only changes the magnitude of roll pressures, but also the distribution pattern.

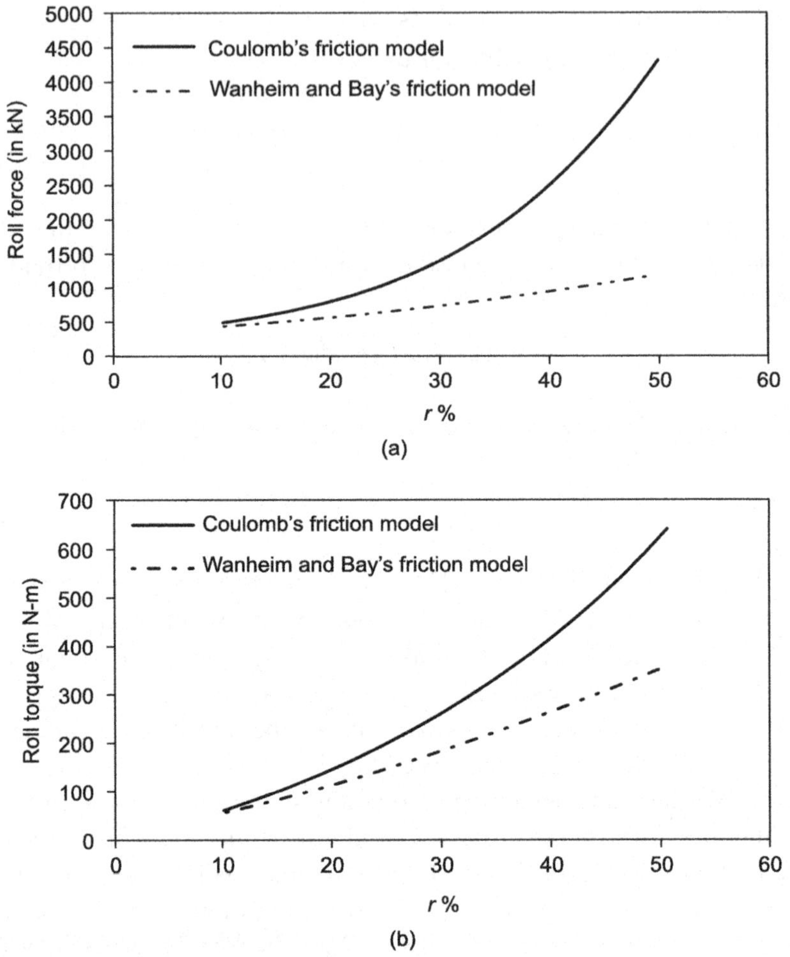

Fig. 3.5 Comparison between (a) roll force, and (b) roll torque predicted by Wanheim–Bay's and Coulomb's friction model [3]. With permission from Elsevier, Copyright [2005]

3.3 UPPER BOUND ANALYSES

Machine and tool designers are often interested to find the load required for plastic deformation in a metal forming process. The upper bound analyses predict the load requirement that is either equal to or greater than the actual load required for causing the plastic deformation. The upper bound theorem is stated as : "The upper bound of the load required for plastic deformation can be obtained by equating the internal rate of energy dissipation plus the energy spend in overcoming friction to the rate at which external forces produce work in some particular pattern of deformation." The load calculated in this manner will be greater than or equal to the actual forming load.

In upper bound analyses,

1. First, an internal flow field satisfying the incompressibility condition is assumed that should account for the required shape change. The flow field can also be assumed to consist of some rigid blocks that may have the tangential velocity discontinuities with the neighboring blocks. There is no velocity discontinuity in the normal direction on the surface of the block.
2. The power consumed internally in the deformation regions can be calculated. If the strain-rate is continuous, the power per unit volume is given by the scalar product of stress and strain-rate tensor. It is to be noted that actual stress tensor is not known. The stress tensor is calculated from the assumed strain-rate tensor using the constitutive relation for the material. At a surface of the velocity discontinuity, the power per unit area is equal to shear yield stress of the material multiplied by the magnitude of the velocity discontinuity. The total internal power is taken as the sum of the powers in continuous strain-rate zone and velocity discontinuity zone.
3. The internal power consumed plus the frictional power is equated to the external power for evaluating the external forces.

The energy dissipated on a plane of discrete shear or velocity discontinuity can be calculated as described elaborately in the book by Hosford and Caddell [24]. The same is briefly described here. Consider a rigid metal region *ABCD* moving at an initial velocity V_1 at an angle θ_1 and having unit width perpendicular to the paper [Fig. 3.6(a)]. The metal region after crossing the plane (or line) *OO′* becomes *A′B′C′D′* with different velocity V_2 at an angle θ_2. Thus, there is a change in velocity, shape, and direction of the region *ABCD* after crossing *OO″* plane. The velocity vector diagram shown in Fig. 3.6b depicts that the velocity discontinuity V_{12} occurs along the plane *OO′*. There will not be any velocity discontinuity normal to *OO′*. Now the rate of energy dissipation at *OO′* will be equal to the work done per volume times the volume per time crossing *OO′* This after few substitutions gives

$$\frac{dW}{dt} = kHV_{12} \tag{3.15}$$

where k is the shear strength, H is the length of the line crossing *AA′* (for unit width).

By considering many planes of discrete shear, we can write Eq. (3.15) as,

$$\frac{dW}{dt} = \sum_1^i kH_iV_i \tag{3.16}$$

where H_i and V_i belongs to that particular plane. In Section 3.3.1, the application of this approach is presented.

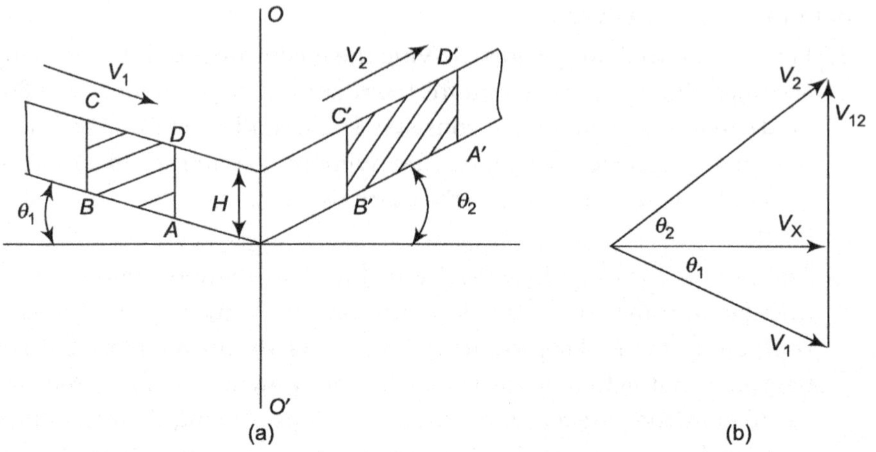

Fig. 3.6 (a) Plane of velocity discontinuity (b) Velocity vector diagram

3.3.1 Plane Strain Extrusion

Figure 3.7(a) describes the extrusion of billet through the die without considering friction between them. One half of the deformation (initial half thickness = H_0) is considered and the velocity discontinuities occur along *PQ* and *QR*. The velocity vector is constructed as shown in Fig. 3.7b in which the entry and exit velocities, V_0 and V_e, are drawn horizontally. As the material crosses line *PQ* the velocity is changed to, say V_1, and is parallel to *PR*. The velocity discontinuity along *PQ* is V_{PQ} which is drawn parallel to *PQ* starting at V_0. Now the velocity V_1 at plastic deformation zone will be the summation of V_0 and V_{PQ}. The material in this region then crosses line *QR*, wherein it

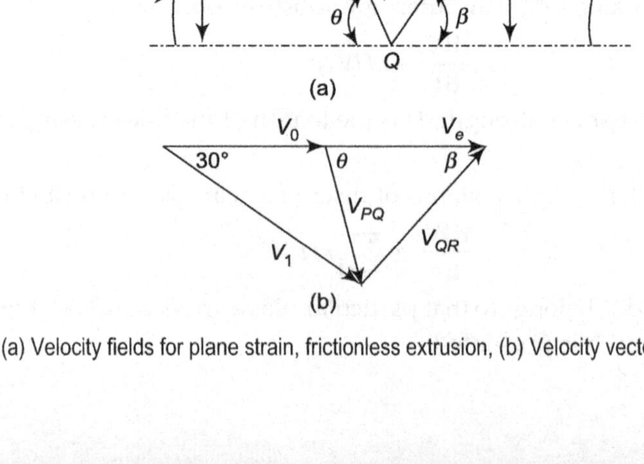

Fig. 3.7 (a) Velocity fields for plane strain, frictionless extrusion, (b) Velocity vector diagram

undergoes second velocity discontinuity, V_{QR}, which is represented by a vector V_{QR} parallel to QR, starting at V_{PQ}. Its magnitude is calculated by representing the exit velocity, V_e, parallel to V_0.

The work done per unit time (for unit width normal to the paper) is calculated as

$$\frac{dW}{dt} = P_e H_0 V_0 \tag{3.17}$$

where P_e is the extrusion pressure.

This initial energy is completely dissipated along the discontinuities PQ, QR and from Eq. (3.16), energy balancing gives,

$$P_e H_0 V_0 = k(V_{PQ}\overline{PQ} + V_{QR}\overline{QR})$$

$$\frac{P_e}{2k} = \frac{(V_{PQ}\overline{PQ} + V_{QR}\overline{QR})}{2H_0 V_0} \tag{3.18}$$

This equation can be solved graphically and numerically By considering 50% reduction (therefore, $H_0 = 1$; exit half thickness, $H_e = 0.5$), $\boldsymbol{\alpha} = 30°$, $\boldsymbol{\theta} = 90°$ (then $\boldsymbol{\beta} = 30°$), we have $\frac{P_e}{2k} = 0.8655$.

By varying $\boldsymbol{\theta}$ and hence $\boldsymbol{\beta}$, $\frac{P_e}{2k}$ will vary yielding a minimum value of 0.78 when $\boldsymbol{\theta} = 72°$ [24]. Similar values can be obtained from other methods like slip line field theory, slab analyses, lower bound analyses, etc.

By taking sticking friction condition at the interface, i.e., along PR, an additional term $\frac{V_1(\overline{PR})}{2H_0 V_0}$ will be included in Eq. (3.18).

$$\frac{P_e}{2k} = \frac{(V_{PQ}\overline{PQ} + V_{QR}\overline{QR} + V_1\overline{PR})}{2H_0 V_0} \tag{3.19}$$

For different values of $\boldsymbol{\theta}$, the minimum value of $\frac{P_e}{2k} = 1.432$ occurs at $\boldsymbol{\theta} = 83°$ [24]. It is also possible to propose more complex deformation field to obtain lower values of upper bound.

3.3.2 Other Forming Operations

An elaborate discussion about the application of upper bound analyses to other forming operations can be read from any standard textbook, viz. [24]. The following are the general upper bound solutions of different forming operations:

Plane-strain compression: $\frac{P}{2k} = \frac{1}{2}\left(\frac{h}{L} + \frac{L}{h}\right)$ (3.20)

Here h is the height and L is the width of the billet (Fig. 3.1).

Sheet drawing: $$\frac{\sigma_d}{2k} = \left(1 + \frac{m}{\sin 2\alpha}\right)\varepsilon + \frac{\tan \alpha}{2} \tag{3.21}$$

Here m is the friction factor, α is the semi-die angle, and $\varepsilon = \ln(t_o/t_f)$, where t_0 is the initial thickness and t_f is the final thickness.

Axi-symmetric rod drawing: $$\frac{\sigma_d}{2k} = \left(\frac{\sigma_0}{2k} + \frac{m}{\sin 2\alpha}\right)\varepsilon + \frac{2}{3}\tan \alpha \tag{3.22}$$

The upper bound analysis is used to understand and predict the forces involved in various metal forming operations. For instance, Wang and Rao [25] predicted the effect of friction during bulk forming using the upper bound analysis by considering velocity fields in single surface asperity. They derived expressions for normal pressures (p/σ_f) and studied the influence of some parameters on the friction coefficient, normal pressures, asperity deformation, etc. Wilson [26] established a general upper bound analysis of axi-symmetric metal forming problems and showed some examples for velocity discontinuities during axi-symmetric extrusion through square dies, conical dies, parabolic dies, ball indention, backward extrusion, and few other axi-symmetric problems. Choi *et al.* [27] analyzed the rotary forging of a cylindrical billet with the help of upper bound method. The total power dissipated during the operation is divided into three parts namely—power dissipated by internal deformation, power dissipated by velocity discontinuity, and power dissipated because of friction at the interfaces. In order to determine the forming force, the total power is minimized for representing the actual velocity distributions. The analysis is validated with experiments on carbon steel (AISI1020 and AISI1045) and Al 6061. The forming force predicted is compared with results presented in literature and good agreement is witnessed. The influence of spiral feed on the forming force is also studied. The velocity fields involved in various metal flows through converging dies was described by Avitzur and co-workers [28]. The metal flow involved in conically converging radial flow, cylindrically converging radial flow, parallel flow, and toroidal flow are described. The spherical, trapezoidal, triangular velocity fields are considered for their derivations.

3.4 SLIP LINE FIELD THEORY

Slip line field is a method of modeling plane-strain rigid-plastic metal forming problems, although there have been attempts to apply the method to strain hardening materials and for axi-symmetric problems as well. The method is based on the fact that through each point in the plane of plastic flow, one can consider a pair of orthogonal curves along which the shear stress has its maximum value. These curves are called slip lines or shear lines and material particles are assumed to flow along these lines.

In essence, the following equations are solved in the slip line method:

$$(\sigma_x - \sigma_y)^2 + 4\tau_{xy}^2 = 4k^2 \tag{3.23}$$

$$\frac{\partial \sigma_x}{\partial x} + \frac{\partial \tau_{xy}}{\partial y} = 0 \tag{3.24}$$

$$\frac{\partial \tau_{xy}}{\partial x} + \frac{\partial \sigma_y}{\partial y} = 0 \tag{3.25}$$

$$\frac{\partial v_x}{\partial x} + \frac{\partial v_y}{\partial y} = 0 \tag{3.26}$$

$$(\sigma_x - \sigma_y)\left(\frac{\partial v_x}{\partial y} + \frac{\partial v_y}{\partial x}\right) = 2\tau_{xy}\left(\frac{\partial v_x}{\partial x} - \frac{\partial v_y}{\partial y}\right) \tag{3.27}$$

where σ_x, σ_y and τ_{xy} are the stress components, v_x and v_y are the velocity components and k is the shear yield stress. Figure 3.8 shows a set of orthogonal slip lines. These are designated as α and β slip lines. By convention, the algebraically greatest principal stress direction bisects the angle between α and β direction taken as a right handed pair of curvilinear axes. Following equations and theorems are useful in analyzing and constructing the slip lines:

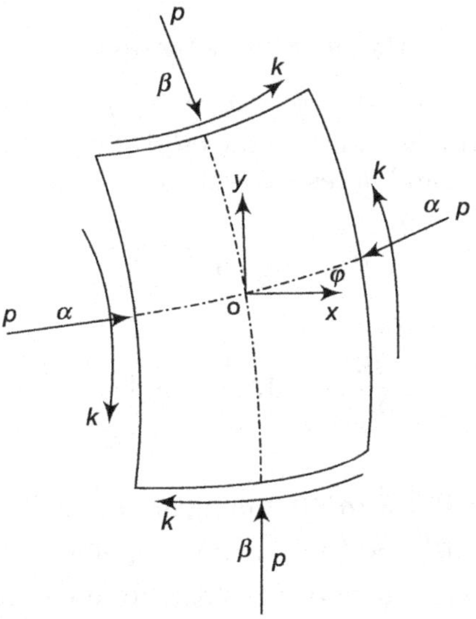

Fig. 3.8 A set of α and β slip lines

(i) Hencky equations:

$$p + 2k\phi = \text{constant along an } \alpha \text{ line} \tag{3.28}$$

$$p - 2k\phi = \text{constant along an } \beta \text{ line} \tag{3.29}$$

where ϕ is the counter clockwise angle of α line with the x-axis and p is the pressure (negative of hydrostatic stress).

(ii) Geiringer equation:

$$dv_x - v_y\, d\phi = 0 \quad \text{along an } \alpha \text{ line} \tag{3.30}$$

$$dv_y - u_x\, d\phi = 0 \quad \text{along an } \beta \text{ line} \tag{3.31}$$

(iii) Hencky's first theorem: The angle between the tangents to a pair of slip lines of one family at the points of interaction with a slip line of the other family is constant along their lengths. In other words, if we pass from one slip line to another of same family, the angle turned through and the change in hydrostatic pressure are the same along each intersecting slip lines. In Fig. 3.9:

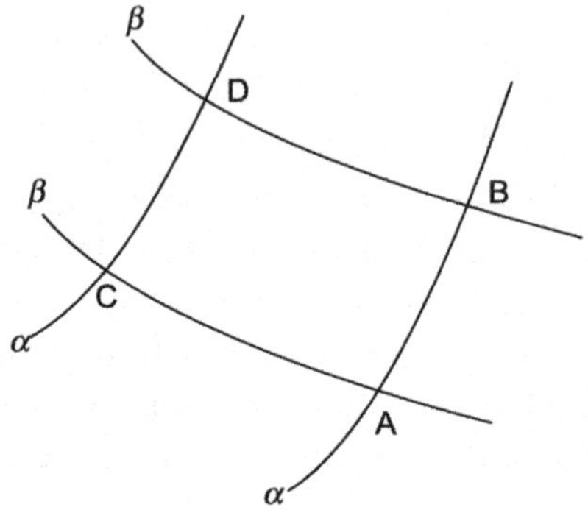

Fig. 3.9 Intersection of slip lines

$$\phi_C - \phi_D = \phi_A - \phi_B,\; p_C - p_D = p_A - p_B \tag{3.32}$$

(iv) Hencky's second theorem: Let the radii of curvature of the α and β lines be denoted by R and S, respectively, i.e.,

$$\frac{1}{R} = \frac{\partial \phi}{\partial s_\alpha},\; \frac{1}{S} = -\frac{\partial \phi}{\partial s_\beta} \tag{3.33}$$

It can be shown that

$$\frac{\partial R}{\partial s_\beta} = -1,\; \frac{\partial S}{\partial s_\alpha} = -1 \tag{3.34}$$

Alternatively,

$$dS + R d\phi = 0 \quad \text{along an } \alpha \text{ line} \tag{3.35}$$

$$dR - S d\phi = 0 \quad \text{along an } \beta \text{ line} \tag{3.36}$$

Usually, slip line field satisfying the stress boundary conditions is assumed. Then the solution is tested to ensure that the stress boundary conditions are also satisfied. In general, one may have to use trial and error, although the procedure may be computerized. Applicability to only plane strain case and the difficulty in the construction of slip line are two main causes that inhibit the use of slip line field in metal forming.

Just to give some idea, a very simple problem of flat punch indentation using slip line method is illustrated. Figure 3.10 shows the proposed slip line field for indentation of a semi-infinite medium by a flat punch. Punch is considered perfectly smooth; hence the shear stress on the punch face is zero. Therefore, the maximum principle stress is along the punch travel. It is clear that slip lines that are the lines of maximum shear stress must meet the punch face at 45°. Similarly, the slip lines are meeting at free surface (which is a principal plane) at 45°. In the triangular regions ABD, ACE and EFG the stresses are constant. These regions are joined by centered fan regions.

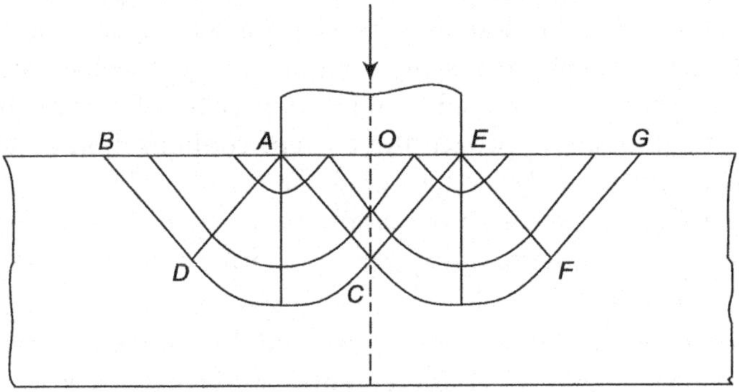

Fig. 3.10 Slip line field for the indentation by a flat punch

In the region ABD, one principal stress (along y-direction) is zero. Hence, the other principal stress has to be $(-2k)$. The Mohr circle is shown in Fig. 3.11. The hydrostatic which happens to be the mean of the two principal stresses in the plane strain case is $(-k)$. Now, by Hencky's equation, the mean stress in region ACE is $(-k - k\pi)$. A Mohr circle with this value as center is constructed. It is seen that the greatest (magnitude wise) principal stress is $(-2k - k\pi)$. Hence, the uniform pressure on the punch is given by

$$p_p = 2k\left(1 + \frac{\pi}{2}\right) \tag{3.37}$$

A hodograph showing the possible velocity field can also be constructed.

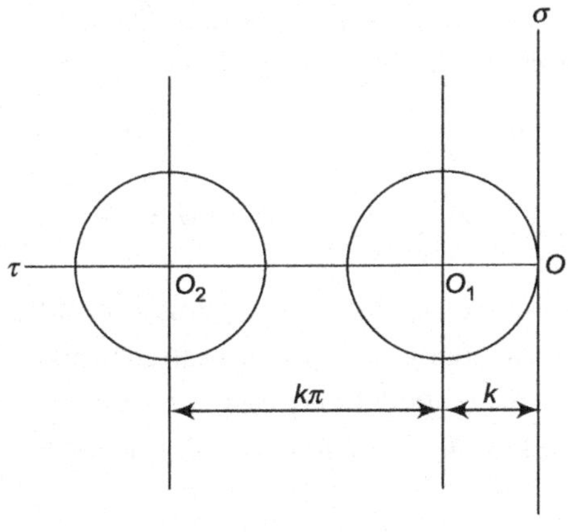

Fig. 3.11 Mohr circles for the slip line field of indentation by a flat punch

3.5 FINITE ELEMENT METHOD

Finite element method (FEM) is a numerical method of solving differential equations. As the behavior of metal forming processes can be expressed in the form of differential equations, FEM can be used to model metal forming processes. There are two prominent FEM approaches to model metal forming processes—updated Lagrangian approach and Eulerian approach. In updated Lagrangian approach the domain consists of a set of particles, which may keep changing their position in space. Lagrangian approach is a control mass approach that considers the initial configuration as the reference configuration. In updated Lagrangian approach the current configuration is taken as the reference configuration. Thus, the reference configuration keeps updating with the time. Here, the measure of deformation is an incremental strain tensor and the constitutive equation is expressed in terms of incremental stress and incremental strain tensor.

Three set of equations are required for modeling a metal forming process by updated Lagrangian FEM: incremental strain–displacement relations, incremental elastic–plastic stress–strain relations and incremental equations of motion [29]. These are briefly described below.

(i) Incremental strain–displacement relations: These are the set of six scalar equations. In index notations, they are represented by

$$d\varepsilon_{ij} = \frac{1}{2}(du_{i,j} + du_{j,i}) \tag{3.38}$$

where i and j vary from 1 to 3 for a three-dimensional case.

(ii) Incremental elastic–plastic stress–strain relations: These are six scalar equations. There are two types of equations: (a) equations applicable during loading with yielding and (b) equations applicable during loading without yielding and after unloading. During loading with yielding:

$$d^o\sigma_{ij} = C^{EP}_{ijkl}\, d\varepsilon_{kl} \tag{3.39}$$

where

$$C^{EP}_{ijkl} = 2G\left[\frac{\nu}{1-2\nu}\delta_{ij}\delta_{kl} + \delta_{ik}\delta_{jl} - \frac{9G}{2}\frac{S_{ij}S_{kl}}{(H' + 3G)\,\sigma_y^2}\right] \tag{3.40}$$

where G is the shear modulus, ν is the Poisson's ratio, S_{ij} is the component deviatoroic stress tensor, σ_y is the flow stress and H' is the rate of change of flow stress with respect to equivalent strain.

Following equation is applicable for loading without yielding and after yielding:

$$d^o\sigma_{ij} = C^{E}_{ijkl}\, d\varepsilon_{kl} \tag{3.41}$$

where

$$C^E_{ijkl} = \frac{2G\nu}{(1-2\nu)}\delta_{kl}\delta_{ij} + 2G\delta_{ik}\delta_{jl} \tag{3.42}$$

Here, the superscript 'o' on the stress increment indicates that it is the product of the Jaumann stress rate and the time incremen. Its relation with Cauchy stress (σ_{ij}) is as follows:

$$d^o\sigma_{ij} = d\sigma_{ij} - (d\omega_{il}\,\sigma_{lj} + \sigma_{il}\,d\omega_{jl}) \tag{3.43}$$

where

$$d\omega_{ij} = \frac{1}{2}(du_{i,j} - du_{j,i}) \tag{3.44}$$

(iii) Incremental equations of motion: These are three scalar equations given by

$$\rho\, da_i = \rho\, db_i + d\sigma_{ij,j} \tag{3.45}$$

where da_i is the component of acceleration vector, db_i is the component of incremental body force (per unit mass), and j in subscript denotes differentiation with respect to current coordinates x_j. Note that Einstein summation convention is employed here. As j is a dummy index:

$$d\sigma_{ij,j} = d\sigma_{i1,1} + d\sigma_{i2,2} + d\sigma_{i3,3} \tag{3.46}$$

The Eulerian formulation is convenient for the analysis of metal forming processes like rolling, drawing, extrusion, etc. In this formulation, a region fixed in space (called the control volume) is chosen as the domain for the analysis. The governing equations used in Eulerian formulation are as follows:

(i) Strain rate–velocity relations: These are six scalar equations given by

$$\dot{\varepsilon}_{ij} = \frac{1}{2}(v_{i,j} + v_{j,i}). \tag{3.47}$$

(ii) Elastic–plastic stress–strain rate relations: In plastic zone, the following equations are used:

$$\dot{\varepsilon}_{kk} = \frac{\sigma^o_{kk}}{3K},\ \dot{\varepsilon}'_{ij} = \frac{1}{2G}S^o_{ij} + \frac{3\dot{\varepsilon}^p_{eq}}{2\sigma_y}S_{ij} \tag{3.48}$$

where K is the bulk modulus of elasticity and $\dot{\varepsilon}'_{ij}$ denotes the deviatoric part of strain-rate tensor. In elastic zone, the following equations are used

$$\dot{\varepsilon}_{kk} = \frac{\sigma^o_{kk}}{3K},\ \dot{\varepsilon}'_{ij} = \frac{1}{2G}S^o_{ij} \tag{3.49}$$

Here, the superscript 'o' denotes that it is the *Jaumann stress rate.* The Jaumann stress rate is related to the Cauchy stress rate through spin tensor. Thus

$$\sigma^o_{kk} = \dot{\sigma}_{kk} - (\dot{\omega}_{kl}\sigma_{lk} + \sigma_{kl}\dot{\omega}_{kl}) \tag{3.50}$$

and

$$S^o_{ij} = \dot{S}_{ij} - (\dot{\omega}_{il}S_{lj} + S_{il}\dot{\omega}_{jl}) \tag{3.51}$$

where

$$\dot{\omega}_{ij} = \frac{1}{2}(v_{i,j} - v_{j,i}) \tag{3.52}$$

Note that the time derivative of the Cauchy stress has to be the material time derivative.

(iii) Equations of motion: Three scalar equations are

$$\rho\left(\frac{\partial v_i}{\partial t} + v_{i,j}v_j\right) = \rho b_i + \sigma_{ij,j}. \tag{3.53}$$

These equations can be modified for rigid-plastic case considering K to be infinite. Thus, we have incompressibility constraint:

$$\dot{\varepsilon}_{kk} = 0 \tag{3.54}$$

The equations can be solved by Galerkin FEM, in which the weighted residual of differential equation is minimized. In Galerkin formulation, the weight function is taken of the same form as that of approximating function.

3.6 SOFT COMPUTING METHOD

Soft computing methods are tolerant of imprecision, vagueness and incomplete information. That means that some inference can be drawn about the process in absence of proper input data by using soft computing tools. The most popular modeling constituents of soft computing are neural networks and fuzzy sets. The optimization can be carried out using genetic algorithm, an evolutionary algorithm considered to be part of soft computing. In the modeling by neural networks, a lot of training data in the form of input–output pairs is required. The trained network is tested by a number of testing data. A number of network architectures are tried and one which gives the minimum of training and testing errors is chosen. Two most common neural networks for function approximation are described in this section. Fuzzy set theory can be used for different purposes including input–output mapping. If the rule base is already available and the rules can be generated by experts, the fuzzy set based modeling does not require data. Automatic rule generation based on the data, will need a huge amount of data.

3.6.1 Multi Layer Preceptron (MLP) Neural Networks

This neural network is a feed forward neural network with one or more hidden layers. This consists of an 'input layer' to which input data are given, an 'output layer' corresponding to the number of outputs to be predicted and one or more 'hidden layers' that relates input to output layer. The presence of hidden layers is required as they provide complexity to the network architecture that is suitable for non-linear functional relationship. The input, output and hidden layers are arranged sequentially and they contain neurons that connect one layer to another layer. The output of neurons of the previous layer becomes input for neurons in the next layer.

An important step in neural network modeling is to train the selected neural architecture by providing data in the form of several input–output pairs. The main aim in the training process is to minimize the error between the predicted and desired outputs by adjusting the weights. A most common method to adjust the weights is the back propagation algorithm. In the forward pass, the input signals propagate from input layer to output layer through hidden layers. The output result is compared to the desired result, and in the backward pass, the error is propagated back towards the input layer where they are used to adjust the weights. Any optimization method can be used to minimize the error at the end of first iteration. During the backward pass, the weights are adjusted for minimizing the error, layer by layer. The weights in the output layer are adjusted first as the target value of output neurons are available for the weight adjustment. The weights of next hidden layer are adjusted after proper transformation of errors from the previous layer and are propagated through the network, layer after layer. This forward and backward pass is continued till the desired output is achieved or error is close to zero. Hence the algorithm is known as 'Back propagation algorithm'. After training the network, it is tested for efficiency by supplying the testing data. The schematic description of the steps involved in Feed forward back propagation (FFBP) neural network is shown in Fig. 3.12. This neural network consists of two inputs, two hidden layers containing four and three neurons, respectively, one output neuron as shown in the figure.The following are the steps involved in training a FFBP neural network.

Step 1: The learning rate $\eta > 0$ and desired maximum error are chosen. The initial weights (W_1 and W_2) are chosen at small random values.

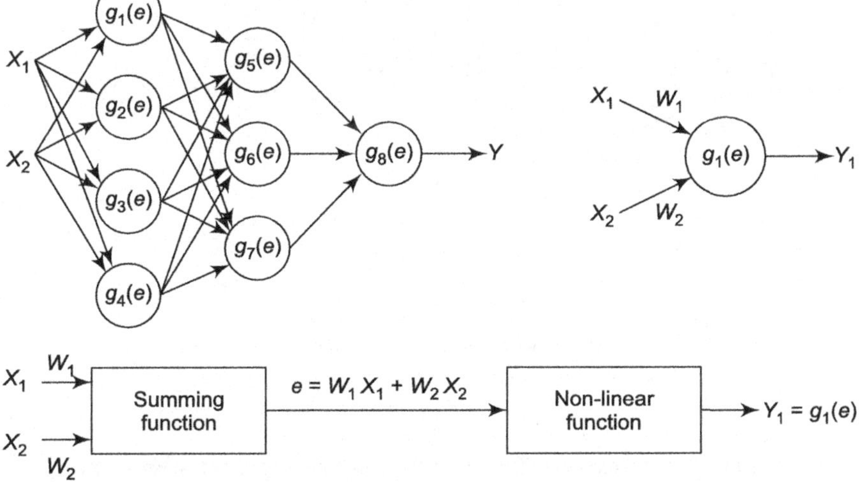

Fig. 3.12 Description of a neural network with two hidden layers

Step 2: Training of network starts here with the propagation of outputs from layer to layer till it reaches the output layer. For example, Y_1, Y_2, Y_3, …, Y_5 are evaluated and propagated till the output layer and the output Y is evaluated as shown in Fig. 3.13.

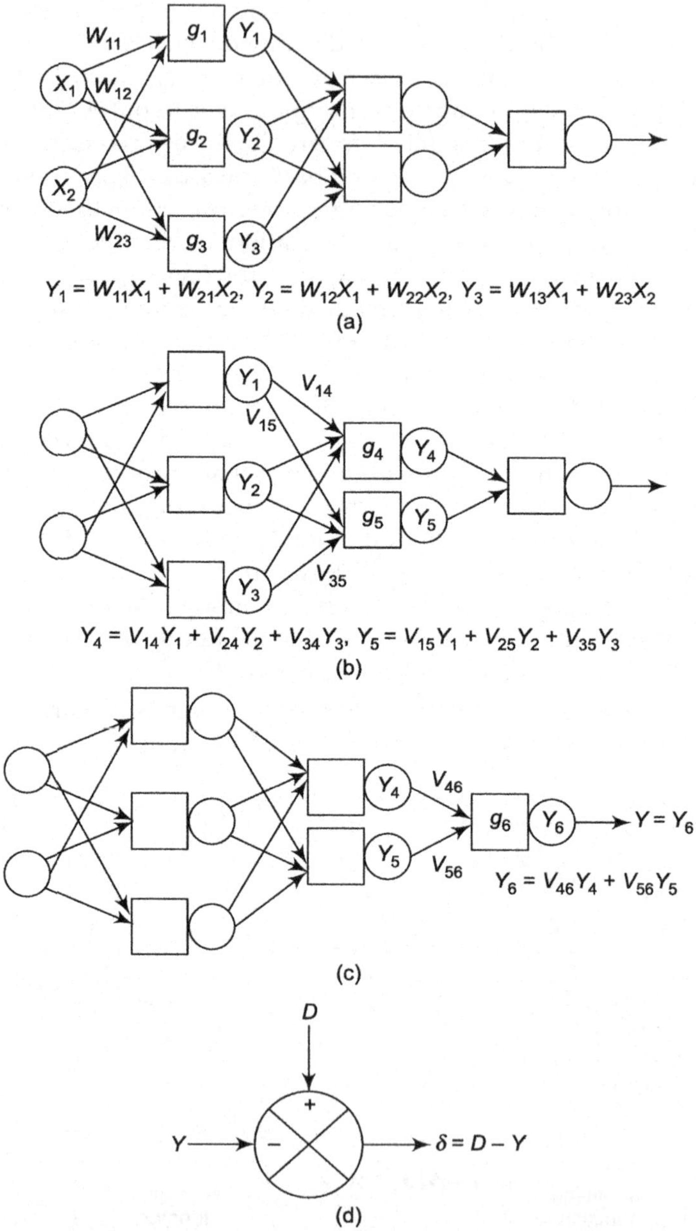

Fig. 3.13 Forward pass of FFBP algorithm: (a) signals sensed by first hidden layer, (b) signals sensed by second hidden layer, (c) signals sensed by output layer (d) error calculation

Step 3: The output Y is compared with the desired value D and the error δ is calculated. The main aim hereafter is to minimize the error.

Step 4: The error δ is propagated backwards as shown in Fig. 3.14 and the weights are modified.

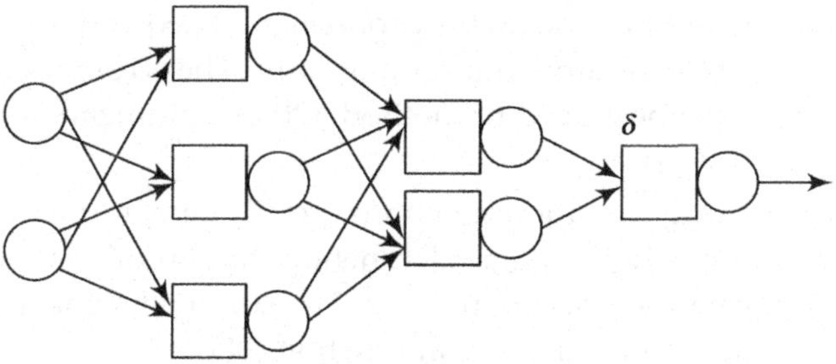

The error δ is used to adjust the weights between output and second hidden layer

(a)

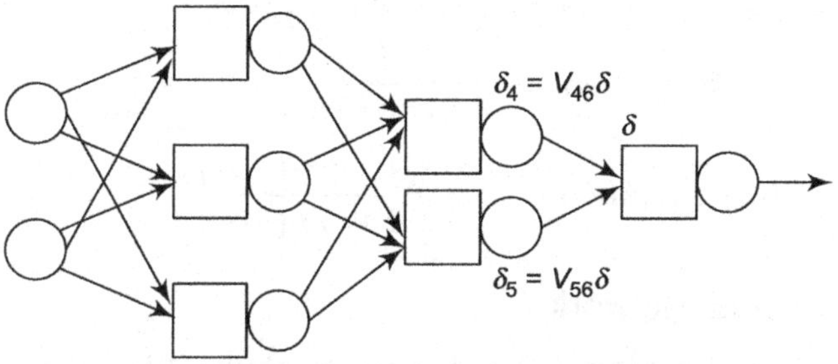

The back propagated errors δ_4 and δ_5 are used to adjust the weights between the first and second hidden layers

(b)

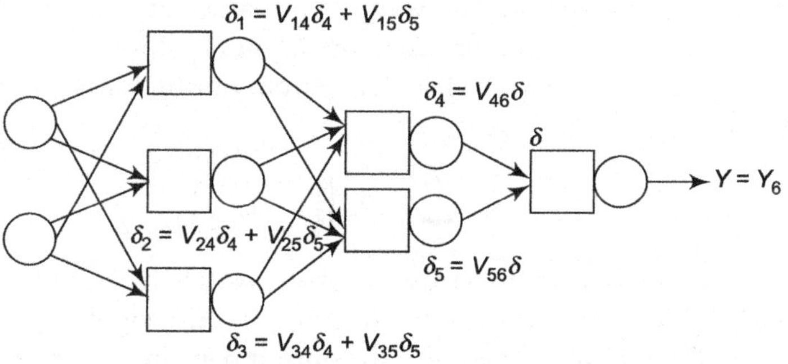

The back propagated errors δ_1, δ_2, δ_3 are used to adjust the weights between the first and second hidden layers

(c)

Fig. 3.14 ((a)–(e)) Propagation of error backwards for modification of I weights between (a) output layer and second hidden layer (b) first and second hidden layer (c) first hidden layer and input layer

Step 5: The output of each neuron is calculated with the modified weights and is propagated to the output layer. The output is compared to the desired output and error is calculated. The error is propagated backwards once again and initial weights are modified. This training process is continued till acceptable error is attained. The final weights are stored later.

The learning rate η decides the efficiency and convergence of the algorithm. However, the η values depend on the problem modeled and there is no single value for learning rate for different training cases. The error minimization can be carried out by steepest descent method. Other optimization methods are also used for the purpose.

There are standard activation functions $g(e)$ like linear function, threshold function (or step function), sigmoid function, hyperbolic tangent function that are used during the training process. The summed signal e is processed as below depending on the activation functions chosen:

Linear function: $g(e) = ke$ (3.55)

Threshold function: $g(e) = 1$ for $e > e_h$, $g(e) = 0$ (3.56)

otherwise, where e_h is the constant threshold value,

Sigmoidal function: $$g(e) = \frac{1}{1+\exp(-\beta e)} \tag{3.57}$$

Hyperbolic tangent function: $$g(e) = \frac{1-\exp(-\alpha e)}{1+\exp(-\alpha e)} \tag{3.58}$$

3.6.2 RBF Neural Network

The architecture of a typical radial basis function (RBF) network is shown in Fig. 3.15. It has one input layer, one output layer and one hidden layer. Associated with each neuron of the hidden layer is a center, which is a vector of same size as input vector. The weights are associated between hidden layer and output layer but not between the input and hidden layer. If there are n centers in the hidden layer, then the output of RBF for input vector $\boldsymbol{x}$ is calculated as

$$o = \sum_{j=1}^{n} w_j \phi_j \left(\left\| x - x_j^c \right\|_2\right) \tag{3.59}$$

where ϕ_j is a radial basis function from set of all positive real numbers to set of real numbers, $\|.\|_2$ denotes the Euclidean norm, w_j is the weight associated with the jth neuron of the hidden layer and the output neuron. The most common radial basis function is Gaussian function given by

$$\phi_j(x) = \exp\left\{\frac{-\left\| x - x_j^c \right\|_2^2}{2\sigma_j^2}\right\} \tag{3.60}$$

where σ_j^2 is called the variance, which controls the spread of the distribution about the center. Some other functions are as follows:

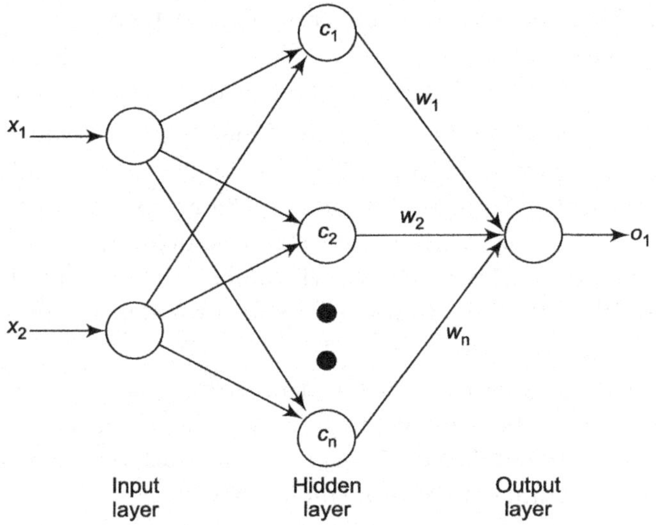

Fig. 3.15 A typical RBF neural network

Multiquadrics: $$\phi_j(x) = \left(\left\| x - x_j^c \right\|^2 + c_j^2 \right)^{1/2} \quad (3.61)$$

Inverse multiquadrics: $$\phi_j(x) = \left(\left\| x - x_j^c \right\|^2 + c_j^2 \right)^{-1/2} \quad (3.62)$$

Thin plate splines: $$\phi_j(x) = \left\| x - x_j^c \right\|^2 \ln\left(\left\| x - x_j^c \right\|^2 \right) \quad (3.63)$$

It is clear that for a fixed center, the output o is a linear combination of weights. The weights can be determined by the least square method.

The radial basis function neural networks can be trained at a faster speed compared to MLP neural networks. However, generally, they require relatively more training data. This is because the influence of a center reduces as the input data moves far away from it. For better input–output mapping, there should be lots of centers covering the entire input domain and enough training data around each center.

3.6.3 Fuzzy Set

A fuzzy set consists of elements whose membership grade can be any positive real number that is less than or equal to 1. A membership grade denotes the strength of the association of the element within the set with 1 indicating the perfect association. As an example, consider three types of steel: $s1$, $s2$, and $s3$. Suppose $s1$ is the best material for the work rolls of a rolling mill followed by $s2$ and $s3$. The suitability of the material to act as roll material may be graded in the scale of 0–1. If an expert grades them as 0.9, 0.7, and

0.5, respectively in the set S of suitable roll material, then the fuzzy set S may be represented as

$$S = \{0.9/s1, 0.7/s2, 0.5/s3\} \tag{3.64}$$

There is subjectivity in the allocation of membership grade and because of it the fuzzy set is often criticized. However, on the positive side, given the subjectivity present in day to day life, fuzzy set theory provides a mathematical frame work of tackling it. Many a times, only qualitative information in the form of language is available, which can be used in computation using the fuzzy set theory. Thus, fuzzy set theory provides a way to carry out computation with language.

There are various ways to assign the membership grades. The most popular among them seem to be using the expert's estimates. Various psychological methods may also be employed. One can also construct the membership grade from data using neural networks or similar tools.

Like for conventional crisp sets, various operations have been defined for fuzzy sets. For example, union of the fuzzy sets denoted by "$\cup$" associates the maximum membership grade with an element. In the previous example, assume that $s1$ is the costliest steel followed by $s2$ and $s3$. Assume that with the help of an expert, a fuzzy set C indicating the suitability from the point of view of cost has been formed and is given by

$$C = \{0.5/s1, 0.7/s2, 0.8/s3\} \tag{3.65}$$

The union of fuzzy set S and C will be a set which will have the maximum of its membership grades in C and S. Thus,

$$C \cup S = \{0.9/s1, 0.7/s2, 0.8/s3\} \tag{3.66}$$

In language form, '$C \cup S$' is interpreted as "C or S". Here, we have members that are good in either of the attributes. (Of course in fuzzy set we will have elements that are good in neither of the attributes, but their memberships will be low.) In the similar way, the intersection of fuzzy sets C and S can be constructed; in which each of its element will have the minimum of its membership grades in C and S. Thus,

$$C \cap S = \{0.5/s1, 0.7/s2, 0.5/s3\} \tag{3.67}$$

In language form, $C \cap S$ means "C and S" and is collection of members that are good in both the attributes.

Two prominent branches of fuzzy set theory are fuzzy arithmetic and fuzzy logic. Fuzzy arithmetic deals with mathematical operations on fuzzy numbers. Fuzzy numbers are generalization of interval numbers, where corresponding to each membership grade, there is an interval in which the fuzzy number lies. A fuzzy number is a convex set indicating that as the width of the interval reduces, membership grades keep on increasing. Figure 3.16 shows some examples of fuzzy numbers.

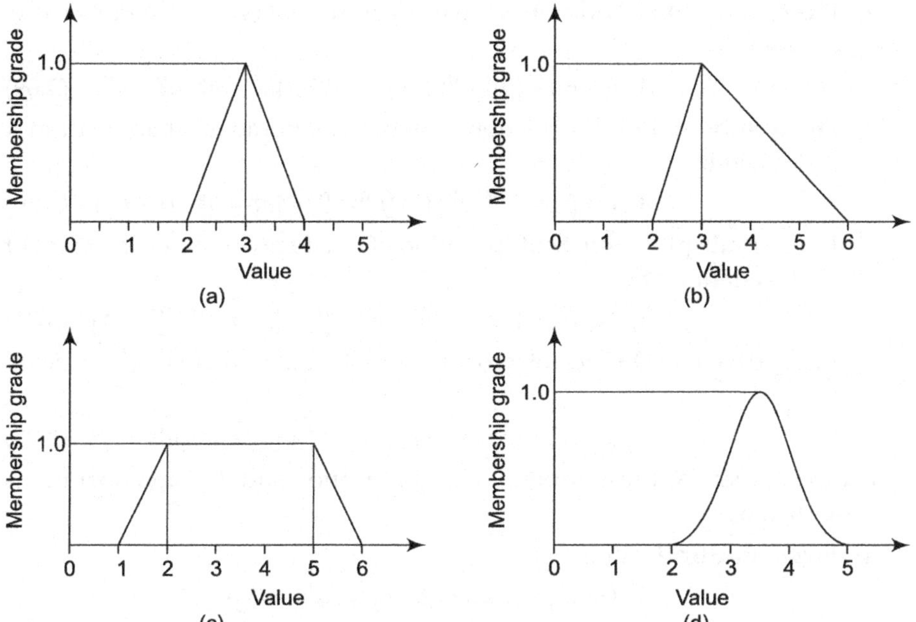

Fig. 3.16 Examples of fuzzy numbers: (a) symmetric linear triangular fuzzy number (b) non-symmetric linear triangular fuzzy number (c) trapezoidal fuzzy number (d) bell-shaped fuzzy number

An α-cut of a fuzzy number is the set containing members having membership grades greater than or equal to α. Figure 3.17 shows an α-cut. Note that all the values contained between A and B constitute an α-cut. At an α-cut, a typical fuzzy number may be represented by

$$A_\alpha = [a_1^\alpha, a_2^\alpha] \tag{3.68}$$

where A_α is the membership grade of α with a_1^α and a_2^α being the lower and upper bound.

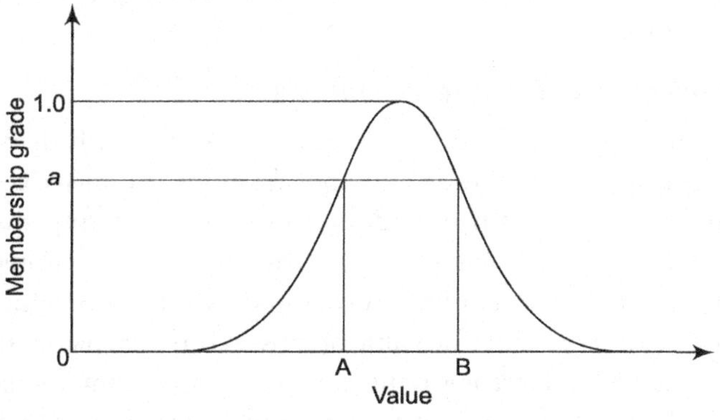

Fig. 3.17 An example of an α-cut

The four basic fuzzy operations are as follows:

(a) **Fuzzy addition:** Addition of two fuzzy numbers at an α-cut in R is defined by

$$A_\alpha (+) B_\alpha = [a_1^\alpha, a_2^\alpha] (+) [b_1^\alpha, b_2^\alpha] = [a_1^\alpha + b_1^\alpha, a_2^\alpha + b_2^\alpha] \quad (3.69)$$

(b) **Fuzzy subtraction:** Subtraction of two fuzzy numbers at an α-cut in R is defined by

$$A_\alpha (-) B_\alpha = [a_1^\alpha, a_2^\alpha] (-) [b_1^\alpha, b_2^\alpha] = [a_1^\alpha - b_2^\alpha, a_2^\alpha - b_1^\alpha] \quad (3.70)$$

(c) **Fuzzy multiplication:** Multiplication of two fuzzy numbers at an α-cut in R^+ is defined by

$$A_\alpha (\times) B_\alpha = [a_1^\alpha, a_2^\alpha] (\times) [b_1^\alpha, b_2^\alpha] = [a_1^\alpha \times b_1^\alpha, a_2^\alpha \times b_2^\alpha] \quad (3.71)$$

(d) **Fuzzy division:** Division of two fuzzy numbers at an α-cut in R^+ is defined by

$$A_\alpha (\div) B_\alpha = [a_1^\alpha, a_2^\alpha] (\div) [b_1^\alpha, b_2^\alpha] = [a_1^\alpha \div b_2^\alpha, a_2^\alpha \div b_1^\alpha] \quad (3.72)$$

For the domain R, the definition of multiplication and division operations are modified as

(a) **Fuzzy multiplication:**

$$A_\alpha (\times) B_\alpha = [\min (a_1b_1, a_1b_2, a_2b_1, a_2b_2), \max (a_1b_1, a_1b_2, a_2b_1, a_2b_2)] \quad (3.73)$$

(b) **Fuzzy division:**

$$A_\alpha (\div) B_\alpha = [\min (a_1/b_1, a_1/b_2, a_2/b_1, a_2/b_2), \max (a_1/b_1, a_1/b_2, a_2/b_1, a_2/b_2)] \quad (3.74)$$

Fuzzy logic is an approach of solving the problem by accepting the partial truth. A perfectly true statement has a truth value equal to 1 and a perfectly false statement has a truth value 0. In general, a statement can have the truth value between 0 and 1. In solving a problem, by fuzzy logic, inference is made with a rule base. The rule base consists of IF-THEN rules, in which either the antecedent (IF part) or the consequent (THEN part) or both contain fuzzy sets. In the following subsections the application of fuzzy set theory in metal forming is reviewed.

3.6.3.1 *Application of fuzzy set operations*

Dixit *et al.* [30] have discussed the design of a rolling mill using fuzzy set theory. In their approach, the design specifications are described by fuzzy membership functions, with membership grade indicating the customer satisfaction. Figure 3.18 shows the linear membership functions of desired strip velocity, reduction and power. A customer wants high value of possible reduction and strip speed but low value of power. But increasing the reduction and strip will lead to an increase in the power. Thus, customer's requirements can be satisfied simultaneously and a compromise solution is to be obtained. The overall membership of customer satisfaction can be expressed as a function of membership grade in different attributes. Thus,

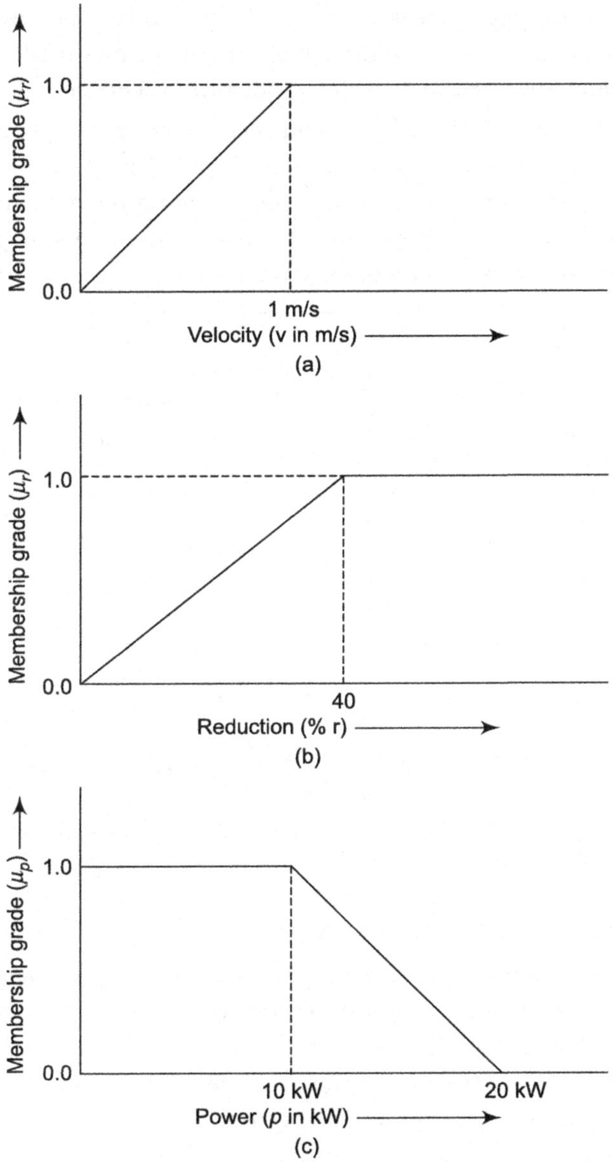

Fig. 3.18 Membership functions of desired (a) velocity, (b) reduction, and (c) power. From [30], copyright 2002, Elsevier

$$\mu_o = f(\mu_1, \mu_2, \mu_3) \tag{3.75}$$

where μ_1, μ_2, μ_3 are the membership grades for velocity, reduction and power, respectively, and μ_o is the overall membership grade. The function may mean either a fuzzy set theoretic operation similar to Eq. (3.67), where the overall membership grade is the minimum of the three membership grades, in which the overall membership grade is decided by the lowest membership grade or a function in which poor performance of one attribute can be compensated by the better performance of other attributes. Out of various available solutions, the solution maximizing the overall membership is chosen.

Figure 3.19 shows another application of choosing the proper roll diameter. Two criteria have been considered—minimization of power and defect prevention. Membership functions for these criteria have been plotted. The optimum diameter that will maximize the overall membership grade will be at the intersection of two membership functions.

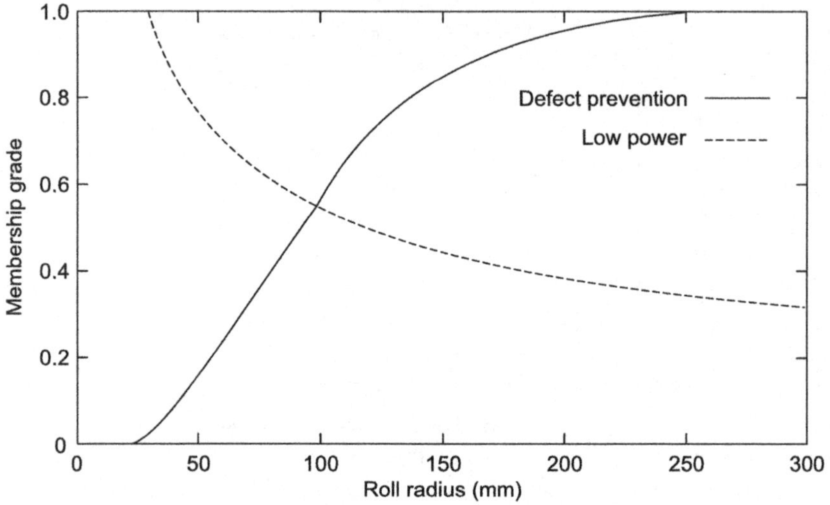

Fig. 3.19 Membership functions of roll diameter from the viewpoints of low power and defect-free rolling process. From [30], copyright 2002, Elsevier

3.6.3.2 Application of fuzzy arithmetic

In a typical metal forming process, material and process parameters are fuzzy. If the computation is carried out based on fuzzy input data, the prediction will also be in the fuzzy form, which is more realistic. A better inference can be made from a fuzzy prediction than from a crisp prediction without any knowledge about the uncertainly and accuracy of the solution.

Figure 3.20 shows the flow stress, hardening parameters, and friction as fuzzy numbers [31]. Figure 3.21 shows the torque required at 24% reduction based on this data. The predicted torque is a fuzzy number.

3.6.3.3 Application of fuzzy logic

Fuzzy logic can be used to make the inference system for prediction. A typical system will have a rule base. Depending on the input parameters, the rules are fired. Aggregate of different fired rules is obtained to obtain the fuzzy output. Finally, the output is defuzzified for making prediction.

Gudur and Dixit [32] used the first order Takagi–Sugeno (TS) fuzzy model for predicting the roll force and roll torque in a cold flat rolling process. Figure 3.22 shows the fuzzification of input variables for modeling the cold flat rolling process. The input variables are ratio of roll radius to inlet thickness (R/h_1), coefficient of friction (f), percentage reduction (r), hardening parameter (b) and hardening index (n). A huge amount of data was used for making a rule base. Generation of data by FEM takes a lot of computational time. Therefore, first the authors fitted a neural network (NN) model for which the training data was generated by an FEM code. Later NN was used for producing the required data for TS fuzzy model. Figure 3.23 compares the fuzzy model predicted roll force with NN predicted roll force. The validation data set consisted of the data that was not used in the development of fuzzy models. Figure 3.24

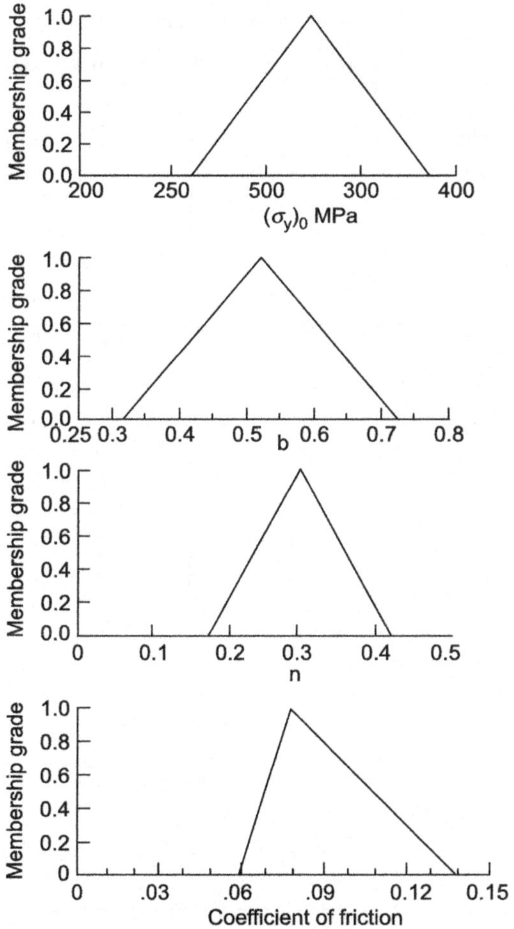

Fig. 3.20 Membership functions of (a) yield stress, $(\sigma_y)_o$ (b) hardening parameter b, (c) hardening index n, and (d) coefficient of friction. From [31] copyright [1996] Elsevier

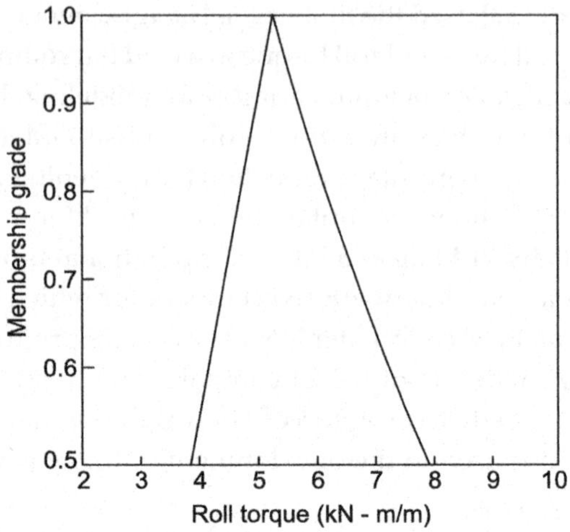

Fig. 3.21 Typical fuzzy output torque at 24% reduction. From [31] copyright [1996] Elsevier

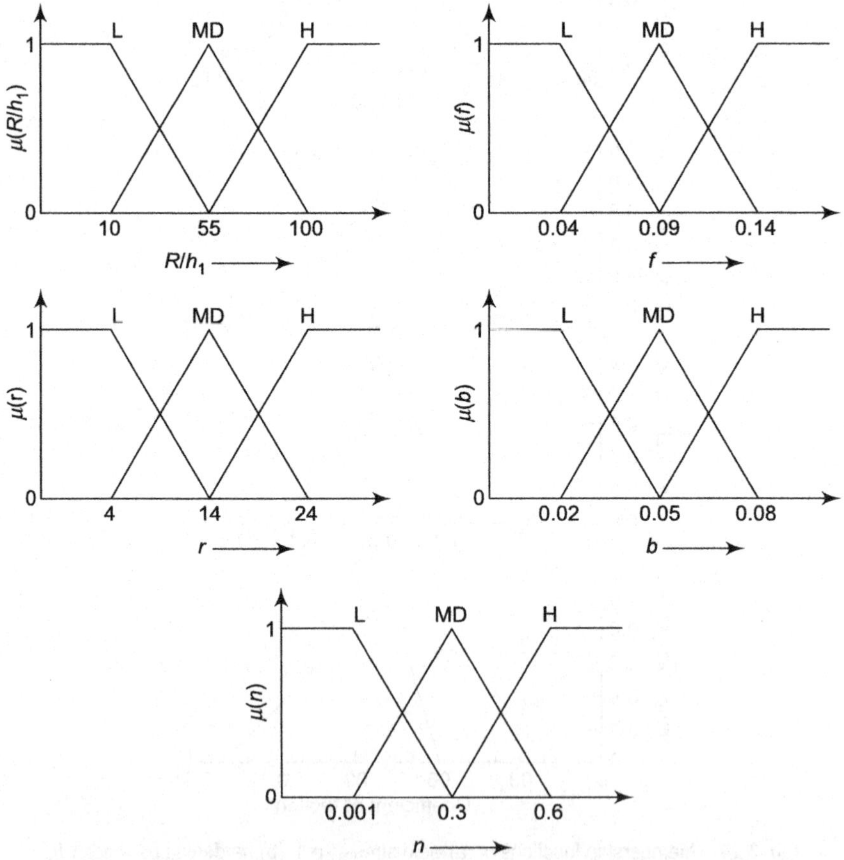

Fig. 3.22 Fuzzification of input variables for a typical cold rolling process. From [32] copyright [2009] Springer

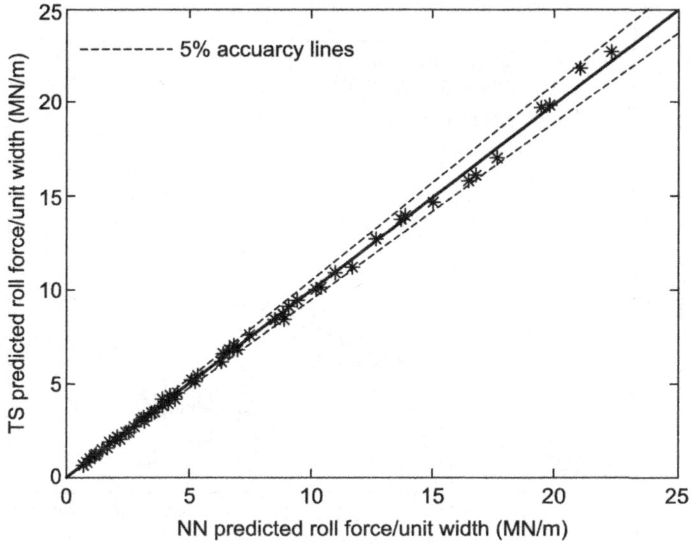

Fig. 3.23 TS fuzzy model predicted versus NN predicted roll force. From [32] copyright [2009] Springer

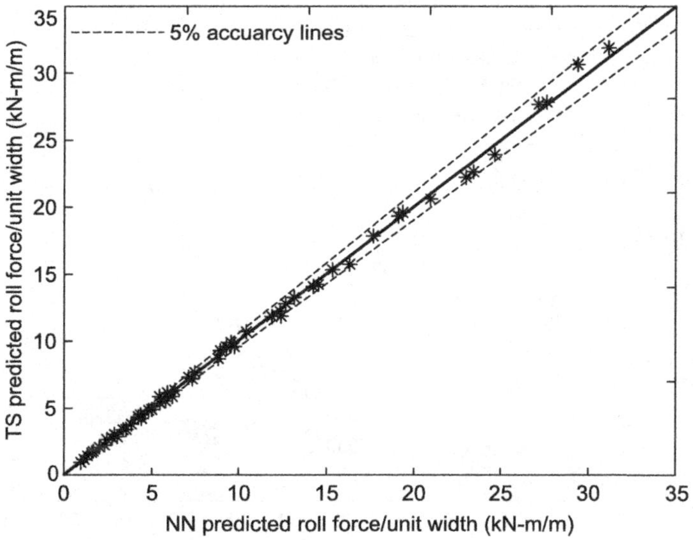

Fig. 3.24 TS fuzzy model predicted versus NN predicted roll torque. From [32] copyright [2009] Springer

compares the fuzzy model predicted roll torque with NN predicted roll torque. Both roll force and roll torque could be predicted within ±5% accuracy. The fuzzy models were also developed to predict the most likely, lower and upper estimates of the roll force and roll torque. Figures 3.25–3.26 show the validation results for roll force and roll torque, respectively. An algorithm to detect and suppress the outliers in the data was also proposed by the authors.

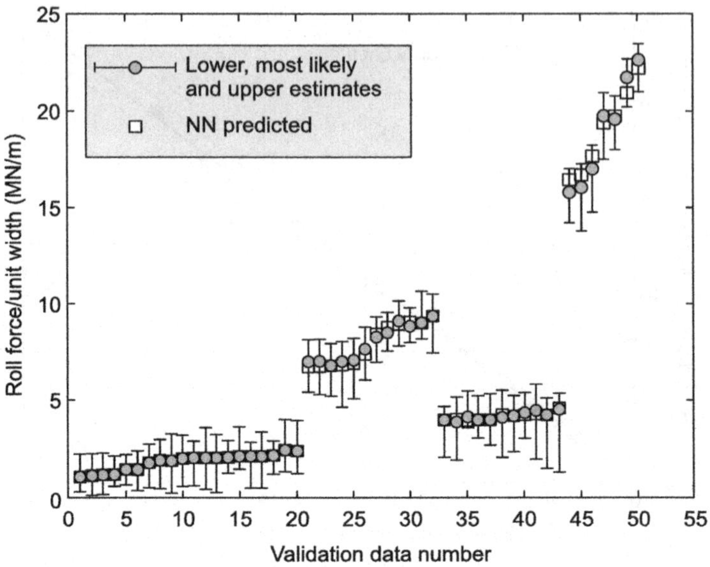

Fig. 3.25 Lower, most likely and upper estimates of roll force. From [32] copyright [2009] Springer

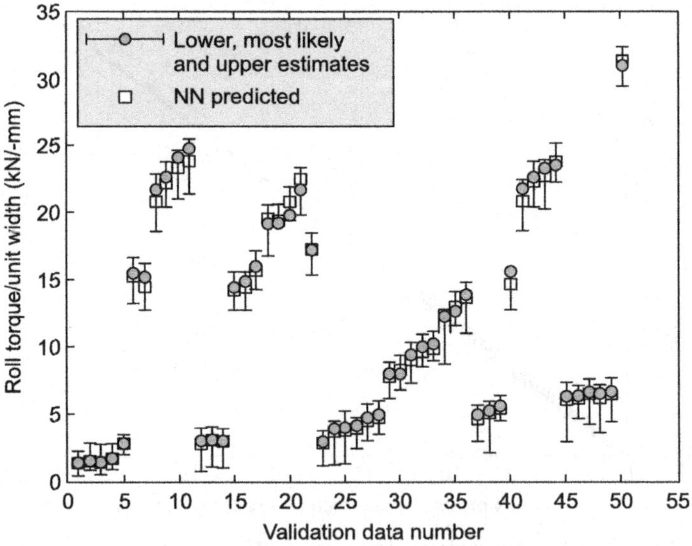

Fig. 3.26 Lower, most likely, and upper estimates of roll torque. From [32] copyright [2009] Springer

3.6.4 Hybrid Methods

Hybrid methods are the methods that use a combination of two are more methods. There are various ways in which two methods can be combined. The broad classification can be as follows: (i) Assisting a method by another method, (ii) Using two are more methods together and (iii) Developing a method that is inspired by more than one method. A brief discussion of these is provided below:

1. **Assisting a method by another method:** In this type of application, a well-known method is used for solving the problem, but the method is assisted by one or more methods. For example, the weights of a neural network can be optimized using genetic algorithm. For solving the problem, neural network plays the primary role and genetic algorithm plays an assistive role. The situation can be other way round also. The genetic algorithm may be used for optimizing a problem, where the functions are evaluated by neural networks. In that application, genetic algorithm plays the primary role and is assisted by neural networks.

   FEM and soft computing techniques can also be combined. Gudur and Dixit [33] have used neural network for assisting the FEM code. A general overview of their scheme is shown in Fig. 3.27. Here, a radial basis neural network is trained by the velocity field generated by an FEM model. Trained FEM model is used to provide a guess value of velocity field and location of neutral point. From this guess value, the FEM model is used to generate an accurate solution. A tenfold decrease in computational time is observed by this method. Figure 3.28 shows the comparison of roll force and roll torque by NN assisted FEM and simple FEM prediction. Figures 3.29–3.30 show the contours of strain rate.

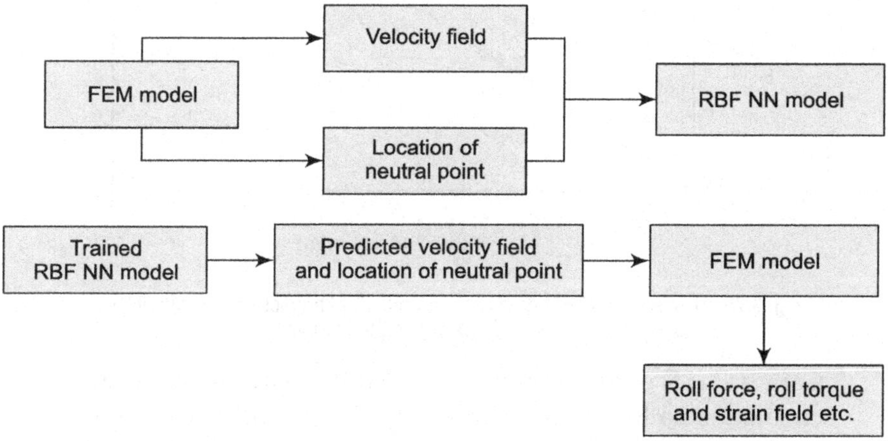

Fig. 3.27 A general overview of neural network assisted FEM model. (A label in the figure has been slightly modified). From [33] copyright [2005] Elsevier

2. **Using two are more methods together:** Two or more methods can be used together for modeling. For example, Abburi and Dixit [34] trained a neural network with experimental data for modeling the surface roughness in turning. The trained neural network was used to generate the rule base for fuzzy rule based system. The fuzzy rule based system

has transparency, ability to do inverse prediction and some extrapolation capability.

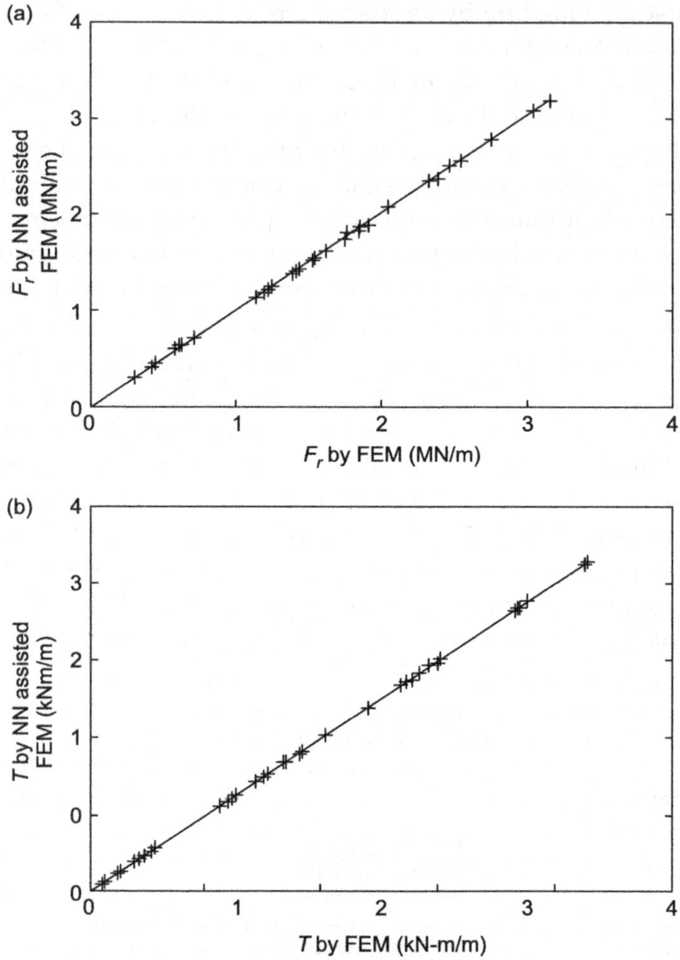

Fig. 3.28 NN assisted FEM predictions versus FEM predictions for (a) roll force, (b) roll torque. From [33] copyright [2005] Elsevier

Garg *et al.* [35] used a combination of MLP and RBF to model the natural circulation boiling water reactor. Initially, an MLP network was used to model the mass flow rate and void fraction with limited data. The trained network was used for providing the training and testing data for an RBF as it requires more data, but provides better prediction. Geerdes *et al.* [36] have used a combination of physics-based and artificial neural-network based model for the prediction of temperature in a hot strip mill. It was observed that combination of these two models provides better results than their individual performances.

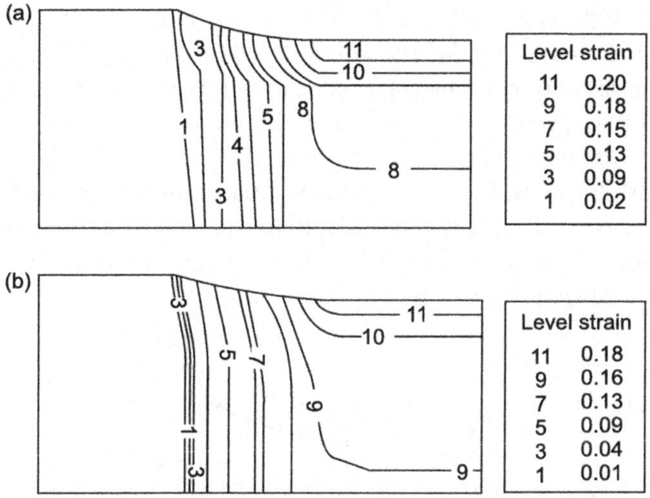

Fig. 3.29 Equivalent strain contours for high strain-hardening material ($b = 0.02$, $n = 0.6$) plotted by (a) FEM code and (b) NN-assisted FEM code. From [33] copyright [2005] Elsevier

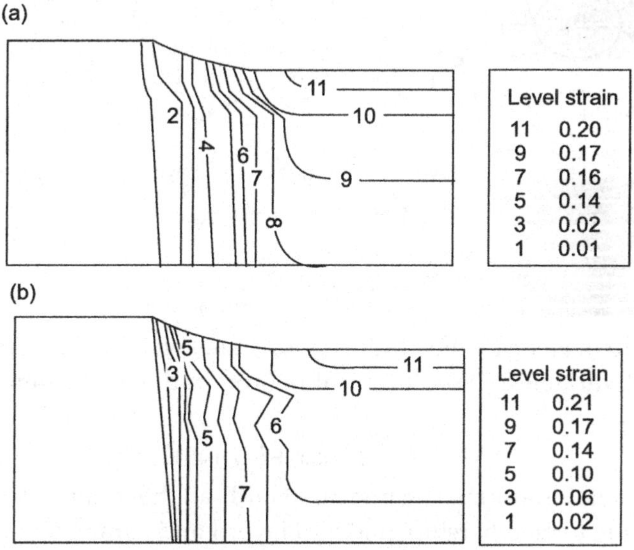

Fig. 3.30 Equivalent strain contours for high strain-hardening material ($b = 0.02$, $n = 0.6$) plotted by (a) FEM code and (b) NN-assisted FEM code. From [33] copyright [2005] Elsevier

3. **Developing a method that is inspired by more than one method:** In the neural network, the prediction is carried out by adjusting the weights. However, it acts as a black-box and requires a large number of data for training. The fuzzy set based system is transparent, but assignment of membership grades is arbitrary. An adaptive-network-based fuzzy inference system (ANFIS) [37] has been developed that combines the positive features of both the techniques.

A typical architecture of ANFIS is shown in Fig. 3.31. It consists of five layers. Layer 1 carries out fuzzification. A_1, A_2, B_1, B_2 are fuzzy subsets. The output of this layer is membership grades in different fuzzy subsets. In layer 2, the membership grades corresponding to IF part of a rule are multiplied. The output of this layer is the firing strength of various rules. As layer 1 fuzzifies the input variables x and y into two fuzzy subsets each, there are fourpossible rules. Layer 2 provides the firing strength of each rule. Layer 3 carries out the normalization. If the firing strength of the ith rule is denoted by w_i, then the normalized strength is given by

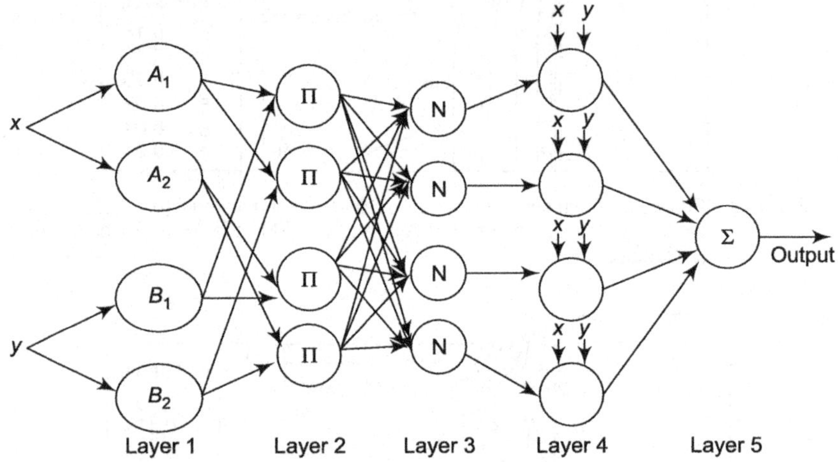

Fig. 3.31 ANFIS architecture

$$\bar{w}_i = \frac{w_i}{w_1 + w_2 + w_3 + w_4} \tag{3.76}$$

The output of layer 4 is the product normalized weight and a linear function of the input variables. Thus, a typical neuron in this layer emits the output o_i given by

$$o_i = \bar{w}_i\ (a_i x + b_i y + c_i) \tag{3.77}$$

Layer 5 contains a summing node that adds all the inputs. The parameters corresponding to membership grades of layer 1 and parameters a_i, b_i, and c_i in layer 4 are adjustable. These are adjusted to minimize the error based on the training data. The procedure for training is similar to the supervised training of neural network.

3.7 CONCLUSION

In this chapter, the various methods for modeling of metal forming processes are overviewed. The slab method, the upper bound method and the slip line field method are the popular methods for estimating the required load for deforming the material. The detailed information about the process can be obtained by FEM. However, it requires a large amount of computational time and needs the accurate data for material and friction. Soft computing techniques consisting of mainly the neural network and fuzzy set help to carry out modeling at faster computational speed and also with imprecise data. But these techniques are dependent on experimental data and/or expert's knowledge. A combination of FEM, soft computing and other suitable techniques can be a better strategy.

REFERENCES

1. Salimi, M. and Kadkhodaei, M. (2004). Slab analysis of asymmetrical sheet rolling, *Journal of Materials Processing Technology*, **150**; 215.
2. Salimi, M. and Sassani, F. (2002). Modified slab analysis of asymmetrical plate rolling, *International Journal of Mechanical Sciences*, **44**; 1999.
3. Kumar, D. and Dixit, U. S. (2006). A slab method study of strain hardening and friction effects in cold foil rolling process, *Journal of Materials Processing Technology*, **171**; 331.
4. Chitkara, N. R. and Aleem, A. (2001). Extrusion of axi-symmetric tubes from hollow and solid circular billets: A generalised slab method of analysis and some experiments, *International Journal of Mechanical Sciences*, **43**; 1661.
5. Assempour, A. and Emami, M. R. (2009). Pressure estimation in the hydroforming process of sheet metal pairs with the method of upper bound analysis, *Journal of Materials Processing Technology*, **209**; 2270.
6. Ebrahimi, R. and Najafizadeh, A. (2004). A new method for evaluation of friction in bulk metal forming, *Journal of Materials Processing Technology*, **152**; 136.
7. Maciejewski, J. and Mroz, Z. (2008). An upper-bound analysis of axi-symmetric extrusion assisted by cyclic torsion, *Journal of Materials Processing Technology*, **206**; 333.
8. Ebrahimi, R., Reihanian, M., Kanaani, M., and Moshksar, M. M. (2008). An upper-bound analysis of the tube extrusion process, *Journal of Materials Processing Technology*, **199**; 214.
9. Tzou, G. Y., Hsu, H. H. and Hsiao, Y. H. (2006). Investigation of a slab method analysis and FEM simulation on rotating compression forming of ring, *Journal of Materials Processing Technology*, **177**; 150.
10. Hsu, H. H., and Tzou G. Y. (2003). Two analytical models of double-layer clad sheet compression forming based on the upper bound and the slab methods, *Journal of Materials Processing Technology*, **140**; 604.
11. Lorenzo, R. D., Ingarao, G. and Micari, F. (2006). On the use of artificial intelligence tools for fracture forecast in cold forming operations, *Journal of Materials Processing Technology*, **177**; 315.
12. Siva Krishna, K., Ganesh Narayanan, R. and Saravanakumar, G. (2010). Tailor welded blank forming limit computation by neural network based expert system, *International Journal: Computer and Experimental Simulations in Engineering and Science (CESES)*, Issue No. 8,

13. Hans Raj, K., Sharma, R. S., Srivastava, S., and Patvardhan, C. (2000). Modeling of manufacturing processes with ANNs for intelligent manufacturing, *International Journal of Machine Tools & Manufacture*, **40**; 851.
14. Chan, W. L., Fu, M. W. and Lu, J. (2008). An integrated FEM and ANN methodology for metal-formed product design, *Engineering Applications of Artificial Intelligence*, **21**; 1170.
15. Kazan, R., Fırat, M. and Tiryaki, A. E. (2009). Prediction of springback in wipe-bending process of sheet metal using neural network, *Materials and Design*, **30**; 418.
16. Dixit, U. S., Robi, P. S. and Sarma, D. K. (2002). A systematic procedure for the design of the cold rolling mill, *Journal of Materials Processing Technology*, **121**; 69.
17. Abburi, N. R. and Dixit, U. S. (2006). A knowledge-based system for the prediction of surface roughness in turning process, *Robotics and Computer-Integrated Manufacturing*, **22**; 363.
18. Shashi Kumar, Sanjeev Kumar, Prakash, Ravi Shankar, and Tiwari, M. K. (2007). Shashi Bhushan Kumar, Prediction of flow stress for carbon steels using recurrent self-organizing neuro fuzzy networks, *Expert Systems with Applications*, **32**; 777.
19. Hsiang, S. H. and Lin, Y. W. (2008). Application of fuzzy theory to predict deformation behaviors of magnesium alloy sheets under hot extrusion, *Journal of Materials Processing Technology*, **201**; 138.
20. Liu, W., Liu, Q., Ruan, F., Liang, Z., and Qiu, H. (2007). Springback prediction for sheet metal forming based on GA-ANN technology, *Journal of Materials Processing Technology*, **187–188**; 227.
21. Shylu, J., Sikdara, S., Kumar Swamy, P., Das, S., and Maity, B. (2008). Hybrid neural–GA model to predict and minimize flatness value of hot rolled strips, *Journal of Materials Processing Technology*, **195**; 314.
22. Rubio, E. M., Camacho, A. M., Sevilla, and L., Sebastian, M. A. (2005). Calculation of the forward tension in drawing processes, *Journal of Materials Processing Technology*, **162–163**; 551.
23. Rojas, H. G. A, Calveta and J. V., Bubnovich, V. I. (2008). A new analytical solution for prediction of forward tension in the drawing process, *Journal of Materials Processing Technology*, **198**; 93.
24. Hosford, W. F. and Caddell, R. M. (2007), *Metal Forming Mechanics and Metallurgy*, 3rd edition, Cambridge University Press, USA, p. 1.
25. Wang, S. L. and Rao, K. P. (1997). An upper bound analysis of friction in metal forming, *International Journal of Mechanical Sciences*, **39** (2); 201.
26. Wilson, W. R. D. (1977). A simple upper bound method for axi-symmetric metal forming problems, *International Journal of Mechanical Sciences*, **19**; 103.
27. Choi, S., Na, K. H. and Kim, J. H. (1997). Upper-bound analysis of the rotary forging of a cylindrical billet, *Journal of Materials Processing Technology*, **67**; 78.

28. Avitzur, B., Narayan, C. and Chou, Y. T. (1982). Upper-bound solutions for flow through conical converging dies, *International Journal of Machine Tool and Manufacture,* **22** (3); 197.

29. Dixit, P.M. and Dixit, U.S. (eds.) (2008). *Modeling of Metal Forming and Machining Processes by Finite Element and Soft Computing Methods,* Springer–Verlag, London.

30. Dixit, U.S., Robi, P.S. and Sarma, D.K. (2002). A systematic procedure for the design of a cold rolling mill, *Journal of Materials Processing Technology,* **121**(1–14), 69.

31. Dixit, U.S. and Dixit, P.M. (1996). A finite element analysis of flat rolling and application of fuzzy set theory, *International Journal of Machine Tools and Manufacture,* **36** (8); 947–969.

32. Gudur, P.P. and Dixit, U.S. (2009). An application of fuzzy inference for studying the dependency of roll force and roll torque on process variables in cold flat rolling, *The International Journal of Advanced Manufacturing Technology,* **42**; 41.

33. Gudur, P.P. and Dixit, U.S. (2008). A neural network-assisted finite element analysis of cold flat rolling, *Engineering Applications of Artificial Intelligence,* **21**(1); 43.

34. Abburi, N.R. and Dixit, U.S. (2006). A knowledge-based system for the prediction of surface roughness in turning process, *Robotics and Computer-Integrated Manufacturing,* **22**(4); 363.

35. Garg, A., Sastry, P. S., Pandey, M., Dixit, U.S., and Gupta, S.K. (2007). Numerical simulation and artificial neural network modeling of natural circulation boiling water reactor, *Nuclear Engineering and Design,* **237**(3); 230.

36. Geerdes WM, Alvardo MAT, Cabrera-Ríos, M., and Cavazos, A. (2008). An application of physics-based and artificial neural network-based hybrid temperature prediction scheme in a hot strip mill. *J. Manuf. Sci. Eng.,* **130,** 014501.

37. Jang, J.-S.R. (1993). ANFIS: adaptive-network-based fuzzy inference system, *Systems, Man and Cybernetics, IEEE Transactions,* **23**(3); 665–685.

CHAPTER 4 Friction in Metal Forming

R.K. Uyyuru, Shimjith M., Satish V. Kailas
Department of Mechanical Engineering,
Indian Institute of Science, Bangalore 560 012, India

4.1 INTRODUCTION

From the understanding of basic physics, friction has been defined as the resistance encountered when a body is moved in contact with another [1]. In technical terms, friction force is the resisting force tangential to the common boundary between two bodies when, under the action of an external force, one body moves or tends to move relative to the surface of the other [2]. Friction is not a material property; it is property of a tribological system [3]that consists of at least two bodies in contact along with surrounding atmosphere and interface. It can be said that friction of a material is dependent on the mating material, surface preparation and operating conditions. The most obvious consequence of friction is that work has to be expended which would not otherwise be necessary. It is understandable that there is always a need for extra energy in addition to the energy needed to perform a physical task. And so, in general perception, friction is always detrimental! However, it is also true that in absence of friction this World would be in chaos. Thus there are many advantages of friction in daily life too. The usefulness of existence of friction along with its characterization and role in metal forming processes is explained in the following sections.

Since man discovered metal, he has had a need to shape it to make tools, machinery, weapons and jewelry. There are three basic methods to shape the metal: Casting—to give a shape by pouring in liquid state into a mold that holds the required shape, and solidifying the metal with or without external pressure; Forming—to give shape in solid state by applying pressure; and

Machining—in which material is removed in order to give it the required shape. Thus metal forming can simply mean the act or process of shaping things out of solid metal using usually very high external forces.

Metal forming—the act of shaping metal invariably involves relative movement between tool(s) and metal to be shaped (workpiece) under very high pressures that entail contact between two bodies—the tool and the workpiece. Thus friction, or in fact tribology, has an important role to play in metal forming. Friction and wear of metal forming tools (as a consequence of friction) have great influence on the economics of most metal forming processes. However, one of the last frontiers in the study of metal forming processes is a quantitative understanding of phenomenon called friction, i.e., the friction phenomenon in the metal forming, thus, is one of the most important aspects of the field with which the industry has not been able to fully understand or cope up yet!

The additional energy requirement due to friction arises mainly from three sources, namely: (1) friction loss along workpiece–tool interfacing surfaces of velocity discontinuity, (2) tangential shear along surfaces of velocity discontinuity within the workpiece, and (3) normal fracture losses across generated fracture surfaces within the workpiece. The extra energy spent during an operation is usually dispersed into the tribological system either in form of heat, sound, vibrations or due to plastic deformation across the interface. Heat generated at the interface might result in softening of the materials in contact, or may lead to phase changes; while an example for sound is squeal in brakes. Plastic deformation across the interface causes the softer material to tear away from its base either as wear debris or as a transfer film on to the harder surface.

Apart from its contribution to the additional energy requirement in a metal forming process, friction causes inhomogeneity in workpiece during the process. This inhomogeneity appears in two ways. The frictional stress causes a rotation of the directions of principal stresses, which in turn determines which of the possible crystallographic planes are most favorably oriented for atomic slip to occur. There is also a macroscopic inhomogeneity, observable using a grid scribed on a cross-section of the workpiece before working. The second inhomogeneity effect of friction is nothing but uneven material flow during a metal forming process, represented by unevenly distributed strain over a cross-section of the part formed.

Uneven material flow, i.e., inhomogeneous strain (e.g., bulging) results in a tri-axial state of stress. This, in turn, causes inhomogeneous distribution of mechanical properties, apart from controlling the incidence of failures due to cracking, scoring, wrinkling, tearing, and geometric variability. Thus it is important to have the knowledge of the friction effects on material flow to produce defect-free products. Since forces can be relatively easily measured, friction effects on load economics could readily be available. But the friction effects on material flow are not easily accessible for obvious reasons as no simple means of looking into friction interface is available till today. Hence a satisfactory solution for this problem must be found with the help of different modeling techniques and perhaps a closer look at the parameters that control the friction.

4.2 HISTORY

Metal forming processes are almost as old as mankind. Like many other branches of technology, scientific analysis and development have accompanied their evolution. The complexities of the theories involved, together with the amount of parameters, have, however, precluded what was achievable in any other cases, i.e., elegant, easy-to-use closed-form solutions, or reasonably accurate equations describing the phenomena and which can be immediately used in applications.

As it is for metal forming, the problem of friction between solid bodies is as old as the human race, yet the understanding of the basic mechanism of friction is still in its infancy. One of the earliest inventions of the human mind is the art of producing fire by friction between solid bodies. The cave man produced a spark by rubbing a piece of flint with a stone of pyrite. He aspired to make a fire at will. Later generations learned to make a fire by generating sufficient frictional heat with a spinning wooden rod to ignite wood shavings strewn around the bearing in which the rod was revolving. These two observations contain the essential features of all friction processes between solid bodies: Work must be done to move relatively to each other two solid bodies which are in frictional contact; Heat is always generated in the course of friction processes, i.e., mechanical energy is converted into heat; Sparks can be produced by friction between solid bodies.

Our modern civilization depends to a large extent on the control of frictional forces. It is generally appreciated that without friction there would be no walking, no driving, no riding and no railway traffic; it would be impossible to write with a piece of chalk on a blackboard, or to use a pencil or to paint with a brush or to turn the pages of this book; music could not be made by stroking of strings with a bow. On the other hand, with too much friction a prohibitively large motive power would be required, and the wear of moving parts would be excessive [4].

In the literature, a study of friction by Thermistius (350 BC) has been regarded as the oldest ever attempt to understand the friction. He found that the friction for sliding is greater than that for rolling. This finding should have led to the conclusion that the static friction coefficient is greater than the kinetic coefficient of friction. First noted by Leonardo da Vinci in the 1500's, later on re-discovered by Amontons in 1699, verified by Euler in 1750 and Coulomb in 1781 are the other empirical laws of friction stated as: friction is proportional to load and independent of the area of sliding surfaces. Thus the coefficient of friction is independent of load, and in the case of dry (un-lubricated) sliding, independent of velocity [5].

4.3 GENESIS OF FRICTION

Amontons, Coulomb and other early investigators proposed that metallic friction arises due to contact between two bodies in focus. Under applied loads, this contact may lead to mechanical interaction of asperities. As proposed by Coulomb in his theory, the action of the wedge shaped asperities causes the two surfaces to move apart as they slide from one position to another and then come close again. Potential energy stored, thus, will be recovered and only a small fraction of energy is dissipated. However, a realistic friction theory should include mechanisms of energy dissipation as friction is a dissipative process.

The widely accepted friction theory by Bowden and Tabor [1] states that during the contact, under external load, energy is required to move two bodies relatively. This is due to the work to be done against mechanical interlocking of asperities *or* to break micro-welds, where the junction may get fractured during relative motion or ploughing of softer material by asperities of harder material leading to physical changes (micro- and macro-scale deformation) of contacting surfaces. Thus it can be said that when two surfaces are brought into contact, the contact occurs at the tips of asperities and the load is supported by the deformation of contacting asperities.

Suh and Sin [6, 7] elaborated the genesis of friction in light of a new theory. Contrary to the postulates of the adhesion theory of friction, this theory postulates that the frictional force (and thus the friction coefficient μ) is affected by the sliding distance and the environment because of the changing contributions of three components of friction, i.e., that due to the deformation of surface asperities (denoted μ_d), that due to plowing by wear particles and hard asperities (denoted μ_p) and that due to the adhesion of the flat portions of the sliding surface (denoted μ_a). Of these three basic contributing factors, the plowing of the surfaces by hard and wear particles is found to be the most important in most sliding situations [8]. However, relative contribution from these factors depends up on surface preparation, i.e., surface roughness and texture.

Material loss because of sliding actions is termed as *wear*, which is certainly detrimental except in few cases where it is intended for, e.g., metallographic polishing, sharpening knife edges by grinding, run-in of newly built motors.

Thus friction and wear, in fact, are functions of one another, can be related with each other [9].The frictional behavior of materials is important in tribology not only because the frictional force between sliding surfaces is of interest but also because it generally affects the wear behavior [10–11]. To minimize, if not to avoid, the wear of surfaces, surfaces need to be protected. Surfaces can be protected mainly in two ways: separation of surfaces, or sacrifice by material coated on softer base surface. *Lubrication*, action of applying a lubricant, serves the purposes of separating the two surfaces. Lubricants are nothing but soft solid or fluid materials introduced at interfaces to minimize the friction. This article is intended to explain the phenomena of friction in metal forming.

4.4 BASIC UNDERSTANDING

As discussed earlier, in metalworking processes friction arises from sliding of the workpiece against the tool. Deformation of the workpiece usually proceeds with relative motion at the interface between workpiece and tool. This movement may originate in imposed velocity differences (as in wire drawing), or it may develop as a consequence of the deformation process itself (as in the upsetting of a cylinder), or it may result from a combination of both (as in the rolling of a strip).

In a metal forming process, the pressure and shear stresses transmitted through the tools affects plastic deformation of the workpiece. The frictional conditions prevailing at the tool–material interface greatly influence distribution of these pressures (or stresses), and thereby the material flow. The heavy loading of tools in metal forming processes makes it important to be able to estimate the local tool stresses accurately in order to design tools appropriately. As friction plays a central role in these calculations, due to its significant influence on the local pressures, *it is important to determine the friction as accurately as possible.* Although a tremendous amount of work and discussion has gone into this subject over the decades, especially during last few decades, friction in metal forming is still not very well understood.

In summarizing the effects of friction in material processing, magnitude of friction needs to be known for three important reasons:

- Pressures, forces, or energy requirements can be calculated only if interface conditions can be described by one of the variables like τ_i, μ, or m.
- Reduced friction is often the main criterion in choosing a lubricant that will be applied at the tool–workpiece interface. Lubricants are commonly used to minimize *or* prevent tool wear, thus increasing life of tools by reducing friction. However, to choose lubricant comparative values of friction coefficient are often sufficient. For example in case of steel friction coefficient value ranges from 0.3 to 0.5 and in case of aluminum

the friction coefficient value ranges from 0.4 to 0.8. The ultimate choice of the lubricant may have to be based on full-scale operating experience, which usually is disruptive and prohibitively expensive. Therefore, a preliminary evaluation is common in simulating tests or small-scale laboratory versions of actual metalworking processes.

- Interface friction has an important role to play in material flow during the forming operation.
- Stable and low friction conditions are attractive to produce products with desired qualities like good surface finish, internal microstructure, etc.

Friction between a workpiece and tools dominates the strain patterns and performance of many forming operations, and yet is often the least quantified of all phenomena involved in forming. This status is a result of two conditions: (1) friction laws depend sensitively on a host of variables that are not themselves usually well characterized, and are quite diverse for various forming operations, and (2) until recently, forming analysis was qualitative and there was little motivation to derive quantitative laws for friction under large-strain forming conditions [12].

4.5 FUNDAMENTAL CHARACTERIZATION

One may question as to why we know so little about friction processes? Stated simply by Tabor, the answer is *"we do not have a way of seeing what is actually taking place at the interface while sliding is taking place"*. Thus, one of the experimental difficulties in investigating friction is that friction occurs at a buried interface. The second difficulty is that friction events can take place at very rapid rate: less than nanoseconds for atomic events and microseconds for micrometer-sized asperity events. New approaches are therefore needed to investigate, in real time, the prompt loss processes buried in the interface of a sliding junction. The new ways of probing friction are Optical spectroscopy, Electromagnetic wave probes and Phonon spectroscopy. Atomic Force Microscopy and related tools are also being used to understand the fundamentals of friction.

Although three areas of tribological phenomena (thermal, chemical and mechanical) are represented (as in Fig. 4.1), the mechanical branch will be discussed in order to illustrate the hierarchy of levels of understanding. Moving away from the center, one first encounters a continuum elastic/plastic description with a single yield criterion (Y = constant). At the next level, there are polycrystalline grains which can produce inhomogeneous and anisotropic behavior in a material. Then, within each grain, deformation is controlled by dislocations and other defects. These defects are ultimately controlled by the inter-atomic interactions at the outermost perimeter of the branches. From this example it can be incurred that the study of friction phenomena is very complex.

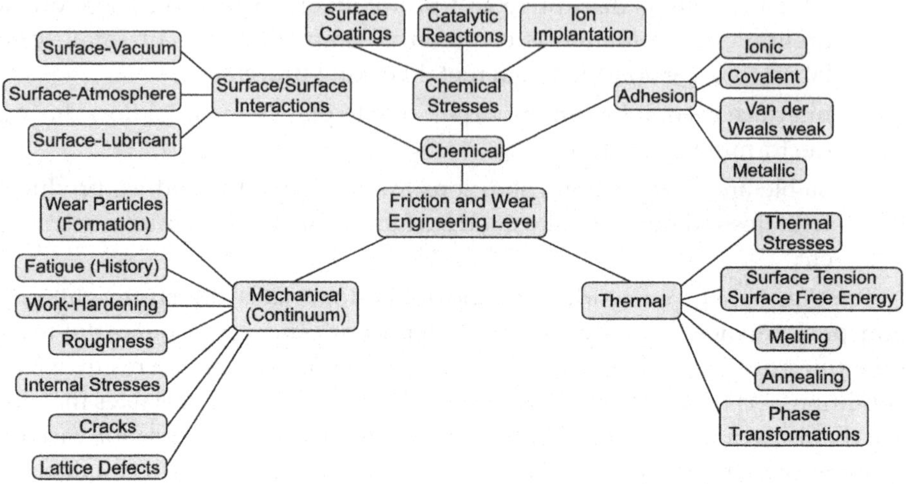

Fig. 4.1 Hierarchy of inputs required to calculate friction coefficient at an engineering level [13]

Metal forming is suitable for micro scale products because of high production rate, minimized material loss and high quality manufactured products. The main drawback in using the forming tool in this process is the die–work piece interface friction become very significant. It has been shown that the use of traditional friction coefficient can lead to erroneous results [13]. Furthermore, the drawback of most numerical simulation tools is that it is only possible to give global friction to the interface of die and work piece. In most of the metal forming simulations friction which has been traditionally used is Coulomb friction [14–17] and the constant friction law [18]. Variations in these have been developed and used [19–21].

In the case of sheet metal forming effective mathematical friction model is very important [22], especially in case where the surface to thickness ratio is large [23]. It does not yield the best results despite the well-developed material behavior models. This is due to the usage of simple friction model in which only a constant friction coefficient is considered. The most frequently used model is Coulomb friction model in which the ratio of friction force and normal load is called coefficient of friction, which is considered to be a constant [24] as used by Cheng *et al.* [25] and Huang [26]. However, friction depends on large number of parameters like micro geometry, sliding velocity temperature and normal load. If any of these parameter changes the coefficient of friction will also change [27].

Effective lubrication systems result in low friction levels which reduce the loads imposed on tooling and workpiece. Lower force levels also reduce tooling deflection and can improve the dimensional accuracy of the product.

In most of the metal forming processes, friction between the workpiece and tooling has an important effect on the mode of deformation and the required forces. In analyzing such processes it is important to have some method of characterizing the frictional stress at the interface. However, "dry metal forming", i.e., metal forming without lubrication has become possible in recent days, although it causes high friction at the tool–work piece interface. To achieve this it is necessary to understand the mechanism involved in dry friction. Dry metal forming is advantageous since the effect on the environment could effectively be reduced [28].

During the conceptual understanding of friction over the centuries, as explained in previous section, friction has been characterized by empirical laws. The laws state that (1) the force of static friction is proportional to the normal load between the surfaces and (2) Friction is independent of the areas in contact, and that (3) kinetic (or sliding) friction is less than the limiting static friction. The first two laws can be explained on the basis of the minute roughness present on all real surfaces resulting in plastic deformation and possibly local welding, and is therefore, dependent upon the load applied and the surface finish and not upon the superficial areas involved. The third law can be explained by considering that the heat generated during sliding reduces the yield strength of the contacting high spots.

The above empirical laws were summarized, and put forward by Coulomb in a mathematical form as:

$$F \propto P, \quad \text{and} \quad \mu = \frac{F}{P} = \frac{\tau}{p} \tag{4.1}$$

where F is the frictional force at interface, P is the normal force, μ is Coulomb coefficient of friction, τ is average frictional shear stress, and p is the normal pressure.

In the Eq. (4.1), τ and p are derived from F and P by dividing them with apparent area of contact, A_a. For a constant μ, the interface shear stress, τ, must increase at the same rate as the interface normal pressure, p. It is known to be valid for many physical contacts where the normal pressures are considerably less than the flow stress of either of the materials in contact.

Once the interface shear stress is increased, in proportion with increase of normal pressure, beyond the limit of shear flow stress, k, of either of materials in contact, material has a further option than sliding along the interface. As it tries to move relatively with the other, it will take less energy for the material to shear inside the body, while the surface remains immobile at the interface. This condition is termed as "sticking friction". Then Eq. (4.1) is modified in the following way to represent the condition of sticking friction:

$$\tau = \mu p > k \tag{4.2}$$

From theory of plasticity, shear flow stress of a material is constant and is related with flow stress, σ, as follows:

$$k = 0.577\ \sigma \tag{4.3}$$

according to von Mises yield criterion, and

$$k = 0.5\ \sigma \tag{4.4}$$

according to Tresca yield formulation. Thus μ assumes a maximum value of 0.577.

Let us assume a metal forming process where it is usual that the interface pressure reaches a multiple of flow stress. As k is constant for metals, with increasing pressure the calculated μ actually will have to drop. This is in contradiction to the friction law and can lead to misinterpretation of reported values of μ, and thus it can be concluded that definition of Coulomb coefficient of friction becomes meaningless when $\mu p > k$.

To overcome the difficulty with μ, in later developments interface friction is made a function of shear flow stress of the softer material. This is described by the following equation:

$$\tau = m\ k \tag{4.5}$$

where m is the frictional shear factor which has a value $m = 0$ for a frictionless interface, and $m = 1$ for sticking friction. This formulation of friction has great mathematical convenience as τ is defined with the aid of k, the value of which is known . In contrast, the use of μ can lead to complications because the value of p (which in itself is dependent on μ) has to be found.

As with the coefficient of friction, the friction factor is usually assumed to be a property of the workpiece, tooling materials, and lubricant, and independent of other parameters. However, the friction factor has an important advantage over the coefficient of friction as far as metal forming analyses are concerned. With the former characterization, the friction stress at a point is independent of the normal stress. This factor, coupled with the relative ease with which m may be measured, has led to increased use of the interface friction factor for describing friction in metal forming processes. It appears to be particularly suited for hot-working processes involving large deformations, such as forging and extrusion. In processes such as cold-rolling and wire-drawing the use of coefficient of friction appears to be well established. Some other simple characterizations of friction have been used in the literature. However, their advantages, if any, are usually limited to a narrow range of operation of a particular process and the coefficient of friction and friction factor approaches account for the vast majority of existing analyses.

The heavy loading of tools in bulk metal-forming processes makes it important to be able to estimate the local tool stresses accurately in order to design tools appropriately. Friction plays a central role in these calculations due to its large influence on the local normal stresses. The development and application of the finite-element method has allowed, when compared to the classical analytical methods, a more accurate calculation of process parameters including the tool stresses. However, increased accuracy in the prediction of the material flow at the tool–workpiece interface, and consequently in the calculation of the tool stresses, is unrealistic as long as inadequate friction models are applied.

When analyzing pressure distributions in bulk metal-forming processes, it is conventional to adopt either Amonton's friction law, $\tau = \mu p$ or the law of constant friction stress, $\tau = mk$. Neither of these two laws, however, is applicable universally as explained in earlier section. But commercially available finite element programs for the simulation of metal-forming processes usually model the frictional constraint acting at the interface between the workpiece and tools according to either of these friction laws. Unfortunately such description is often inadequate and does not represent the actual interface relation. In the simulation of metal-forming processes, the use of Amonton's law gives occasion for an overestimation of the friction stresses at the tool–workpiece interface, because the normal pressure often is considerably greater than the yield stress of the material. Consequently, the friction stress becomes greater than the yield stress of the material in pure shear.

The difficulties with Amonton's law prompted several researchers to include the law of constant friction in finite-element programs for bulk metal-forming. However, as the friction does not depend on the current state of stress at the tool–workpiece interface, but simply on a material property, the friction stress is overestimated at low normal pressures. Wanheim and Bay [29] developed a general model for expressing friction at the tool–workpiece interface. In practical terms the model assumes friction to be proportional to the normal stress at low normal pressure ($p/\sigma_o < 1.5$), but going towards a constant value at high normal pressure ($p/\sigma_o > 3$), (Fig. 4.2), the two ranges being combined via an intermediate transition region. It has been proved that the proposed general model by Wanheim and Bay is a major improvement in the ability to simulate processes where low tool–workpiece interface stresses may prevail. This is done through experimental and numerical investigations into the upsetting of a semi-tapered specimen between parallel dies [30].

4.6 BENEFICIAL EFFECTS OF FRICTION

As mentioned in earlier sections, it is not always desirable to reduce the friction to a minimum, but even in these instances pickup should be avoided. Some beneficial effects of friction are listed in the following [31–32]:

(a) *As a motivation force to supply required power*: In several metal forming processes including strip rolling, ironing, hydro-static extrusion with viscous drag. In rolling the energy required for the operation is transmitted to the strip through the friction or surface traction between the rolls and the strip. The improvement in roll bite by increasing friction

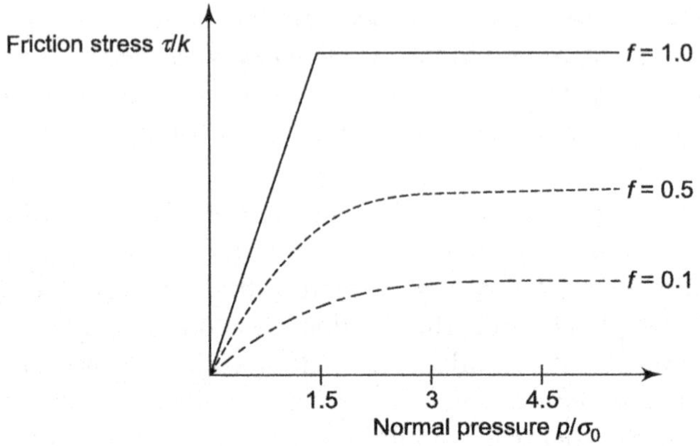

Fig. 4.2 Schematic representation of general friction model [15]

permits heavier drafts in cold-rolling of sheet and strip, i.e., it allows the metal to be drawn into the gap between the rolls, according to the relation (4.6) the roll bite is directly proportional to coefficient of friction and diameter of the roll.

$$\Delta h_{max} = \mu^2 R \tag{4.6}$$

where Δh_{max} is maximum roll bite, μ is interface coefficient of friction, and R is radius of rolls.

(b) *As a stabilizing factor*: In the process of strip rolling the position of the neutral point is automatically determined by the power required. Whenever the neutral point moves away from its stable position, the power balance is disturbed. For example, should the strip slow down temporarily, the neutral point will move toward the exit. The driving force increases, and the resisting friction decreases, causing the total power supplied to rise and restore the higher exit speed.

(c) *As a deterrent to defect formation*: In extrusion, higher friction values serve to deter formation of central burst. In drawing, however, higher friction values promote the same defect. Moreover, in extrusion, the back end of the billet tends to flow over the face of the pressure pad during the final stage when the extrusion pressure becomes too high. The rear section of the billet thus becomes entrained along the axis of an extruded bar, and even a hollow pipe may form. This feature is so common that it is usually referred to as the extrusion defect, without further qualification, the defective material has to be sawn off, or the extrusion must be terminated at an earlier stage, reducing the yield. The lower the friction at the face of the pressure pad, the easier it is for metal to flow inwards and the sooner the extrusion defect starts. This is particularly noticeable when a plain graphite dummy-block is used to obtain a complete, discard less extrusion. Considerable lengths of rear-end pipe may then be produced, so the friction is deliberately increased by placing asbestos pads between the surfaces.

(d) In punch-bench manufacture of tube on a mandrel, it is advantageous to maintain a reasonably high friction on the mandrel, to carry part of the drawing force. In this way, the tensile stress on the leading end of the finished tube is reduced, and heavier reductions of area can be made without fracture. A somewhat analogous situation is found in deep drawing or pressing. To prevent the onset of necking in the walls while deep-drawing a cup there should be high friction on the punch and low friction against the die.

(e) In open-die forging it prevents the metal escaping from between the tools and in closed-die forging it provides the back pressure in the flash to ensure filling of the die cavity.

(f) High friction is desirable in varieties of tensile grips used in metalworking.

The primary role of friction in metal forming processes is evident. Friction not only affects the forces required but also the outcome of an operation, whether it will end up successfully or fail, may depend on the value of friction. Many operations such as rolling depend entirely on friction for their implementation. It is, therefore, obvious that sophisticated and more accurate and dependable means of measuring friction will aid the metal forming industry.

4.7 MEASURING FRICTION IN METAL FORMING

Friction in metal forming processes comprises a very complex situation with many parameters interacting with it. Some of the important variables which affect the friction between die-material interface are relative velocity, temperature, lubrication condition, roughness, and mechanical properties of material or die [33]. Ideally, one would prefer to include all the variables affecting friction in a simple friction test. However, unfortunately, it is impossible because of the complex interaction between many of these variables [34]. This is also due to the great variety of process conditions. To put it in simple words, as mentioned elsewhere [35], *a universal test to measure friction in metal forming processes has not been and will never be found*. This lack of universal test to measure friction can also be attributed to the facts that: (1) the friction laws depend sensitively on a host of variables that are not themselves usually well characterized, and are quite diverse for various forming operations, and (2) until recently, forming analysis was qualitative, and there was little motivation to derive quantitative laws for friction under large-strain forming conditions.

There is at present no generally accepted method of measuring the value of the coefficient of friction for given surfaces and lubricant. Various factors can influence the result, and it is essential that test conditions should match the actual conditions of operation as closely as possible. This implies that the only strictly reliable friction data to use are those obtained from measurements during the operation considered.

Because of the difficulty in measuring the friction and its distribution, friction at the tool–material interface is usually measured globally, i.e., an average value for friction coefficient (*or* friction factor) will be used in calculations. Hydrodynamic lubrication, Coulomb friction, and constant shear factor are few assumptions to describe (global) friction phenomena [36]. A popular tribological guide to the material forming scientists is due to Schey [37].

An excellent overview of friction measuring methods (*or* evaluating the lubricant), imitating the particular working conditions, is due to Lange [35]. It also gives a brief explanation of different friction measuring methods and their pros and cons. On the basis of this overview, friction tests can be broadly classified according to means of measuring friction– by measuring loads *or* forces using sophisticated instrumentation, *or* just by measuring geometrical changes without resorting to instrumentation (e.g., ring compression). Again the first category of tests was classified, though not in very strict sense, into groups of tests depending on the—geometry of contact (e.g., twist compression), possibility of generating new surfaces with bulk deformation (e.g., wire drawing), etc. The basic principles guiding the selection of a friction test are simplicity (to separate frictional effects from others) and similitude (to assure that the many variables are similar to actual operation of interest) apart from tooling cost, specimen size, and so on [12].

Metal forming processes are basically categorized into two groups, bulk metal forming and sheet metal forming, for simplicity of analysis on the basis of physical changes the workpiece will undergo during a process. Thus it would be appropriate to discuss various friction measuring methods applicable to these two kinds of metal forming processes separately as explained in the following sections.

The earliest experimental work of measuring friction (published in the 1930s) is due to Ekelund [38] who investigated the influence of temperature on the coefficient of friction (inverse relation between them) through contact angle calculations in his hot rolling experiments with steel over the temperature range 700°–1100°C. An empirical relation between the coefficient of friction and rolling temperature resulting from his investigations is given by Eq. (4.7). Later, this relation has been modified for rolling-speed-effect by Geleji [40], Eq. (4.8).

$$\mu = 1.05 - 0.0005\ T \qquad (4.7)$$

$$\mu = 1.05 - 0.0005\ T - 0.056\ v \qquad (4.8)$$

where T is the temperature in Kelvin, v is the rolling speed in ms^{-1}.

After this early work on friction measurement in rolling, there was no further work for a couple of decades. This may be probably due to the dominance of rolling process in metal forming industry where the above equations might have served the purpose. Then in late 1950's and later, more work on friction measurement was documented as a result of search for alternative possibilities that can lead to a simplified test to carry out. The results of this work apply to a wider range of metal forming operations.

In early investigations, the technique of inserting feeler pins through the tool to the tool–workpiece interface was a popular method to measure interface loads and calculate friction values. Some short comings of this technique include the following: the method is expensive and delicate; measurements may not be reliable; method cannot be applied easily outside the laboratory, i.e., under production conditions. Such laboratory studies can indeed provide some much needed insight into the phenomenon and its effects on processing. It is well recognized today, however, the values existing in the plant would be better measured under production conditions. Furthermore, it is recognized that different friction values will be measured for identical lubricants under differing conditions of pressure, temperature, speed, and many other undefined variables. Friction values for extrusion may differ from those for drawing, forging, rolling, or other metal-forming processes.

4.8 FRICTION TESTS FOR BULK FORMING

Bulk metal forming processes basically involve high pressures ranging multiples of yield stress of workpiece material, larger strains than those usually found in sheet metal forming processes. Over the last few decades, numerous simulative bulk friction tests were developed such as twist compression test (TCT), upsetting, ring compression test (RCT), bucket test, open die backward extrusion test (ODBET), plane strain compression test (PSCT), rolling test, combined forward and backward extrusion test, double-backward extrusion test, and more. The choice of individual tests depends partly on the mechanisms judged important, and may be partly on individual experience and prejudices of the researcher.

4.8.1 Twist Compression Test

There are many variations of the test, and in these tests normal pressure is combined with continued sliding over the same surface. In one version of the test [Fig. 4.3(a)], two cylinders are pressed together on their end faces, and then one is rotated against the other. This has been used extensively in studies of adhesion and of lubricant properties at high normal pressure. In the twist compression, the pre-applied lubricant film is exposed to prolonged rotation under pressure, and its resistance to breakdown and potential for die pickup are evaluated. In modified version [Fig. 4.3(b)]—one cylinder will be hollow—when pressures are high, it shows remarkable correlation with lubricant performance in severe industrial operations such as cold extrusion. This is probably because the limited lubricant supply is exposed to continued sliding, and its durability and resistance to breakdown are evaluated. This test can be readily adapted to high temperatures. Pressures are, however, limited to the flow stress of the hollow specimen, and this limitation has been overcome by rotating a punch against a confined workpiece. In another version of the test [Fig. 4.3(c)], the principle of simultaneous twisting and compression is used where the interface pressure is increased by confining the workpiece in a container.

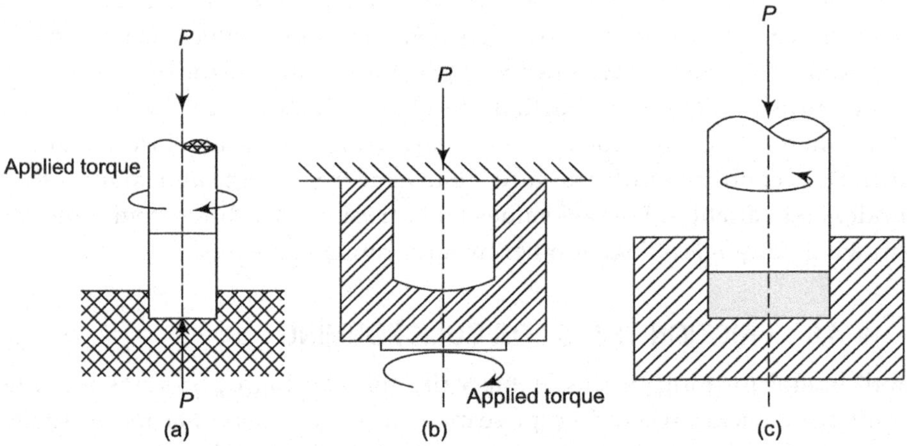

Fig. 4.3 Schematic presentation of geometries for different version of twist compression test

4.8.2 Upsetting

In ordinary upsetting of a billet to evaluate strength of a metallic material, the billet will be compressed between two flat dies. However, due to existence of friction at the interface between billet and dies, uneven deformation occurs across the longitudinal cross-section of the billet. This barreling phenomenon could be used to evaluate friction for open die forging operations. The method employs conical platens with a certain semi-cone angle to compress the billet as shown in Fig. 4.4. The cone angle which produces no barreling is equal to the friction angle, $\alpha = \tan^{-1}\mu$, from which friction parameters can be calculated. However, this method is tedious and may not be very accurate.

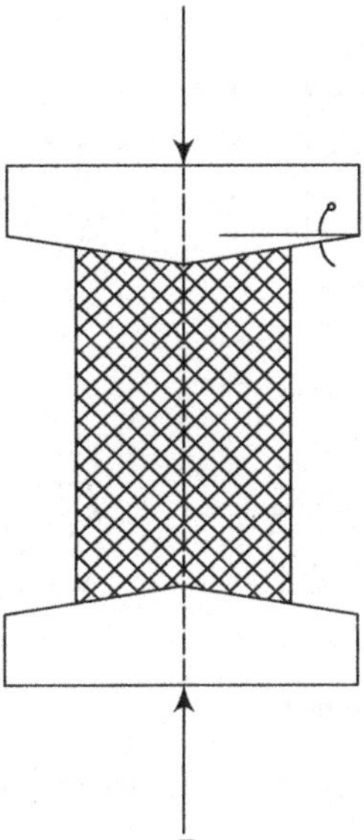

Fig. 4.4 Schematic view of upsetting using conical platens

4.8.3 Ring Compression Test

This test is now the most commonly used experimental method for quantitative evaluation of friction in bulk metal forming processes for a given tool, material and lubricant combination. This is probably because only simple interpretation of experimental results is enough to find friction values from calibration charts, and usefulness of these friction values in many metal forming processes, especially in forging operations, over a wide range of strain rates and temperatures. The popularity of the test can be attributed to its practical convenience including the fact that friction can be judged from deformation, i.e., geometrical changes alone.

RCT was first used by Kunogi [40] as a qualitative method of evaluating different lubricants in cold extrusion experiments with steels. The concept involved in these friction experiments is that an increase in the coefficient of friction causes an increase in the radius of the *no-slip circle* (R_n) as shown in Fig. 4.5 where R_i is inner radius of the ring, R_o is outer radius of the ring, and H is height of the ring. Later, Male and Cockcroft [41] (and others) developed the basic ring compression testing method of evaluating lubricants into today's more user-friendly technique.

The principle involved in RCT is as follows. When a short, hollow cylinder is compressed axially between flat, rigid platens, the diameter of the hole may instantaneously increase, decrease, or remain constant according to the amount of frictional constraint imposed by the platen–cylinder interface.

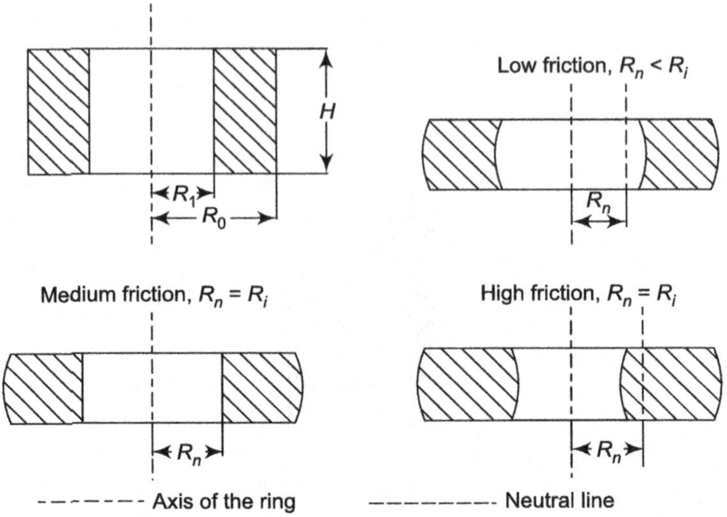

Fig. 4.5 Schematic representation of neutral line position for different friction conditions

After compressing a ring between two parallel platens, the changes in geometry are recorded in terms of percentage change in inner diameter as a function of deformation, measured as percentage reduction in height of the ring. Then the friction parameter can be obtained from calibration curves (Fig. 4.6). The calibration curves are developed theoretically, and relate the concerned variables—percentage change in inner diameter, percentage reduction in height, and the friction parameter. Mathematical analyses leading to the development of calibration curves are due to many researchers [e.g., 41–43].

The theoretical analysis of RCT was based on the following assumptions: (1) the deformation of the ring is uniform throughout the thickness, implying no bulging takes place on the background of uniformly distributed frictional constraint. Though this assumption is not valid in strict practical sense, barreling effects can be minimized by implementing appropriate geometrical conditions, (2) interface friction value is constant, which is not necessarily a valid assumption, (3) the ring material is rigid and perfectly plastic.

Based on theoretical studies [43–44], in the absence of data on best specimen geometry, the maximum accuracy can be obtained by using a ring of smaller height and large internal diameter as compared with the external diameter. Too large an internal diameter, however, makes deformation asymmetrical. Too low height of the cylinder makes the measurement of deformation inaccurate because of the presence of relatively thick lubricant related coatings. Thus it is important to choose a particular geometry ratio between outer diameter (*OD*), inner diameter (*ID*), and height of cylinder (*H*), for example—*OD*:*ID*:*H* = 6:3:2 or 6:3:1 or 6:3:0.5. For deformations under low friction conditions, i.e., in experiments with good lubrication, a somewhat greater thickness (e.g., *OD:ID:H*=6:3:2) may be used [45].

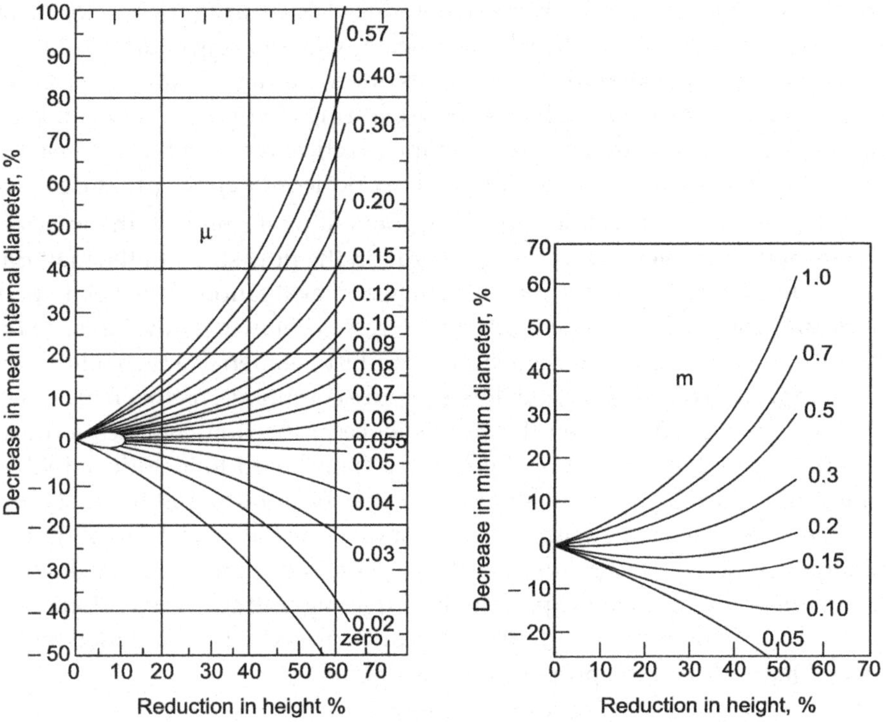

Fig. 4.6 Analytically derived calibration curves for ring compression test [37]

The major advantage of using the RCT for the evaluation of interface friction is that it involves only the physical measurement of shape changes. Thus, friction may conveniently be measured at the high temperatures and strain rates typical of hot-working processes. However, if deformation pressure is measured during compression of the ring, it is possible to calculate the flow stress in addition to the friction. Thus, the RCT can be used for two purposes: (1) to evaluate the friction parameter and (2) to evaluate the flow stress of a given material [46–47].

On the other hand, there exists certain drawbacks of RCT such as (1) the theoretical analysis from which the calibration curves are derived may not be valid in all cases (2) μ or m values will change depending on the choice of theory (3) the necessary interruptions may cause problems in boundary conditions.

There exists practical difficulty in conducting experiments where the unavoidable barreling *or* bulging of the cylindrical surface of rings due to friction causes inaccuracies in calculations of test-after geometry. The other difficulties involve the non-circularity of holes after compression due to anisotropy of the work material and 'tapering' [48]. However despite difficulties due to experimental nature and inaccuracies from material properties, RCT is still one of the most commonly used methods of measuring friction for metal forming processes. One of the recent critical reviews of ring compression test and applicability of calibration curves was due to Rao and Sivaram [49]. Other kinds of development of RCT involve use of rings with different cylindrical shapes [50–51], in an attempt to extend the practical range of the test to allow the evaluation of friction under low tool–workpiece interface pressures.

Due to the geometrical design of the conventional ring-test, the interface stresses will, however, always be greater than the flow stress of the material, thereby impeding quantification of friction, and evaluation of the behavior of lubricants, for processes where interface stresses below the flow stress of the material occur. A new complementary ring-test geometry, which allows the characterization of friction under low pressure conditions, has been proposed by Petersen *et al.* [50]. Given the convenience of the ring compression test, modification to the conventional geometry is proposed to include a double-conical outer profile (as shown in Fig. 4.7) has been proposed in an attempt to extend the practicable range for the test to allow the evaluation of friction under low tool–workpiece interface pressures ($p \leq \sigma$). Finite-element analysis in conjunction with metal experiments, applying both the conventional and the modified geometry for different lubricants, provides the validation for the general feasibility of the proposed test geometry.

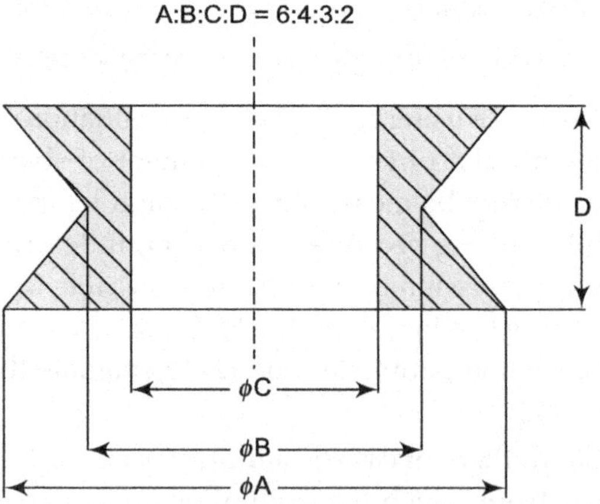

Fig. 4.7 Schematic presentation of complementary geometry for ring compression test [50]

Friction is difficult to measure even though there exist a large number of tests which have been developed, very few are suitable for hot metal forming applications. Friction is one of the most significant physical phenomena influencing the forming of metal, yet in comparison with metallurgy, heat transfer and mechanics it remains the least understood despite the large number of tests which have been developed over the decades. The importance of friction in hot metal deformation processes can hardly be over-emphasized as it contributes to the success or failure of the process to fulfill its objectives. For instance in hot rolling, friction is necessary at the interfaces of the roll bite in order to transmit deformation energy from the work rolls to the workpiece, though excessive friction tends to restrain the level of deformation which results in unacceptably high roll forces and spindle torques. Moreover, if the level of friction is too low, roll slippage or failure of the workpiece to enter the roll bite will occur. With the increasing use of finite elements to solve hot metal forming problems, a need for improved knowledge of how friction behaves under changing interface conditions, i.e., variations in strain rate and temperature, is required in order to improve the degree of accuracy of process models.

Few studies have been undertaken to understand friction in hot metal forming, especially when addressing the critical issue of whether changing magnitudes of strain rate and temperature at the interface influence the level of friction. The article by Rudkins *et al.* [52] describes an experimental investigation into friction under hot forming conditions using the RCT. In their work [52], the authors investigated the possibility of using the ring compression test to assess how friction is affected by variations in temperature. Additional observations were made with regard to the suitability of using the RCT for hot metal forming, especially their ability to reproduce process conditions in hot rolling. It was found that RCT is a good simulative friction test for hot metal forming and rolling; increasing the temperature which causes the level of friction to increase.

RCT is also used for different purposes such as to investigate the influence of scale on friction at the work–tool interface in hot metal working. Although it has been recognized that scale reduces interfacial friction, it has not previously been possible to separate the effects of scale thickness and of scale composition. In the work by Luong *et al.* [53], techniques were developed to obtain specimens in which the scale consisted exclusively of oxide. This enabled the authors to investigate the influence of both scale thickness and scale composition on interfacial friction. In the work by Luong *et al.* [53], specimens were heated in different atmospheres in order to obtain scales with different thicknesses and compositions. In a series of tests designed to study the influence of scale thickness on friction, specimens were heated in a CO_2 atmosphere to obtain scales which consisted essentially of wüstite. Heating times were varied to obtain different wüstite thicknesses. In another series of experiments, both the atmosphere and the heating time were varied to obtain specimens with scales of similar thicknesses but with different compositions. It was found that the friction at the work–tool interface during hot metal working is influenced by both the thickness of the scale and the bulk properties of the scale. The scale thickness appears to be more important in this respect than the scale composition. Scale reduces friction and, up to a certain limit, thicker scales give lower friction values.

In a ring test, both the forging load and the inner diameter are influenced by friction. However, the friction-calibration curve of RCT considers only the latter as a measure for interface friction. The reason is that the former, i.e., the forging load, is not sensitive to friction in ring tests because of the influence from the following two factors: (1) the forging pressure, i.e., the average value of the normal stress at the interface; (2) the contact area between the ring and the upper die. These two factors have a tendency to cancel each other at relatively low value of reduction in height. There is no "universal" ring-test friction-calibration curve. The friction-calibration curves must be defined for the specific ring material under specified ring- and die-temperatures and ram-speed conditions.

4.8.4 Bucket Test

In a ring test, metal flow and die geometry are simple, and the new surface generated, i.e., surface extension during deformation is small. Therefore, in a ring test the friction characteristics of a forging lubricant cannot be adequately evaluated. A new test, named as *bucket test* [54] (Figs. 4.8–4.9) has been formulated where surface extension is expected to be relatively high in addition to large amount of material flow.

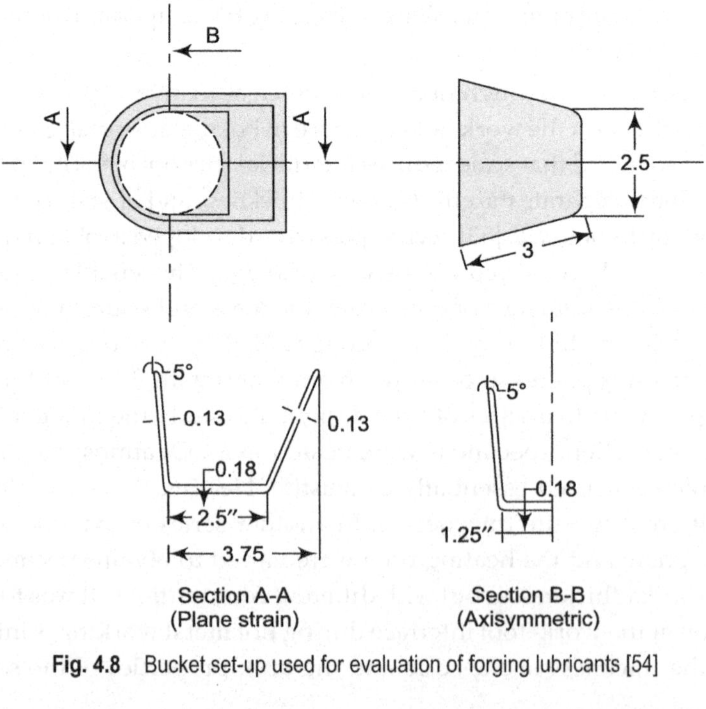

Fig. 4.8 Bucket set-up used for evaluation of forging lubricants [54]

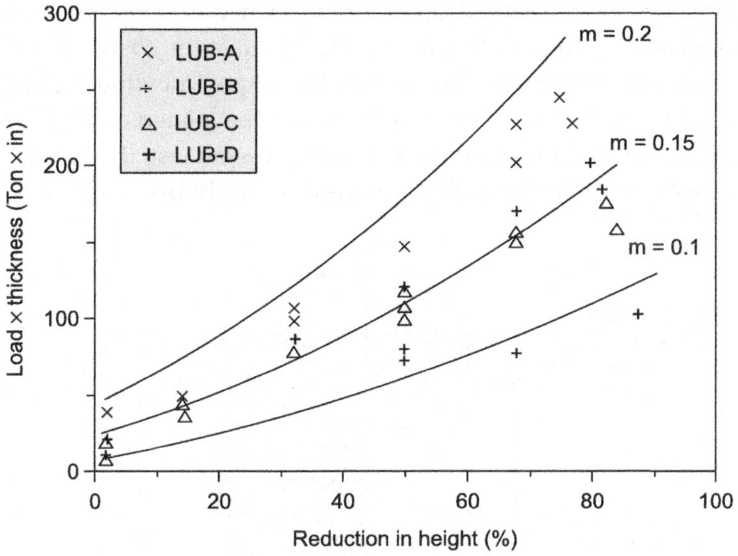

Fig. 4.9 Proposed method for estimating friction coefficient in bucket tests [54]

The quantitative determination of the friction factor, or coefficient, requires: (1) a friction sensitive parameter that can be used as a friction-calibration factor (for example, the internal diameter in the ring test), and (2) theoretically determined standard friction-calibration values. For the "bucket" test, the forging load and the bottom thickness together can be chosen as the friction-sensitive parameter because a good lubricant results in both lower load and thinner bottom thickness in this forging.

Very recent development of friction evaluation using backward cup extrusion is due to Uyyuru *et al.* [55]. The method is based on analysis of deformed physical grid pattern, embedded into the slug before cup extrusion. An ingenious physical pattern has been developed and used during the technique as it allows calculation of the extension of the surface all over the inner wall of the cup with ease. Degree of surface extension over the head was observed to vary along the cup length, which was compared with expansion curves obtained from numerical simulations. It was found that BCE with new physical pattern technique can be used for qualitative friction prediction, as well as quantitative evaluation of friction.

4.8.5 Open Die Backward Extrusion Test

The disagreement or inconsistency found while using RCT can be attributed to assumptions such as no barreling of the deforming specimen and a constant friction coefficient across the interface. Although the ring compression test is an effective method for determining the friction parameter for many deformation processes, the use of generalized friction calibration charts regardless of the material type and test conditions must be avoided.

The new technique, named open-die backward extrusion test (ODBET), was developed as an alternative method to evaluate the coefficient of friction in metal forming processes [56]. In this technique, a cylindrical billet was compressed between two rigid, upper and lower, platens (dies). Instead of making a hole in the billet as in the ring compression test, the hole is made in the top platen for materials to flow upward through this hole (Fig. 4.10).

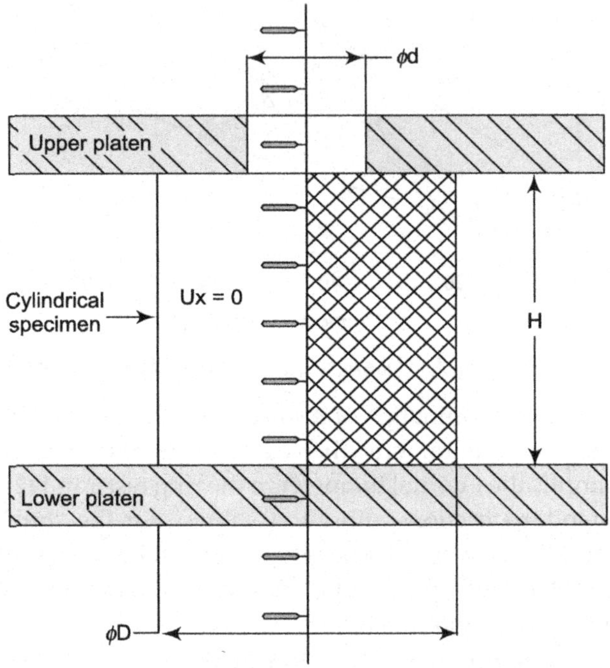

Fig. 4.10 The boundary condition used in ODBET [56]

As the top platen moves down to plastically deform the cylindrical billet, part of the material flows into the hole and generates the extruded height which is the main concern of this study while the other flows outward resulting in an increase in the diameter. Finite Element (FE) simulations of ODBET were performed for different aspect ratios (H/D) and different die geometries (d/D) by utilizing an elastic/plastic finite element code in order to obtain the friction calibration curves (FCCs). The results indicated that the extruded height is a function of the friction conditions at die–workpiece interface. The FCCs are, thus, generated by plotting the percentage increase in the extruded height versus the percentage reduction in the specimen height for numerical results.

Once the reduction in the height of billet and the increase in extruded height are known, the coefficient of friction, m, can be found from the appropriate plot. It was also shown that the extruded height was affected by the billet and the die geometry such as H/D and d/D ratios. When d/D is highest, the extruded height becomes largest and decreases as the ratio decreases for the same coefficient of friction. On the other hand, the extruded height increases as *H*/*D* decreases. It can therefore be concluded that the ODBET technique can be used for obtaining reliable data regarding the coefficient of friction encountered in large deformation processes as an alternative method to the ring compression test. However, the results should only be used in conjunction with calibration curves specifically derived for the material under investigation and the specific test conditions.

4.8.6 Plane Strain Compression Test

It is very interesting to note that the development of PSCT was related to solving the stress problems in rolling by taking advantage of plane-strain conditions. It was then used to evaluate yield properties of materials at high strain levels from 1950's. Though it was developed to solve the problems with stress and mechanical properties of materials, from early days of its development itself, PSCT has been also used in studies of tribological problems. A schematic picture of PSCT set-up is shown in Fig. 4.11.

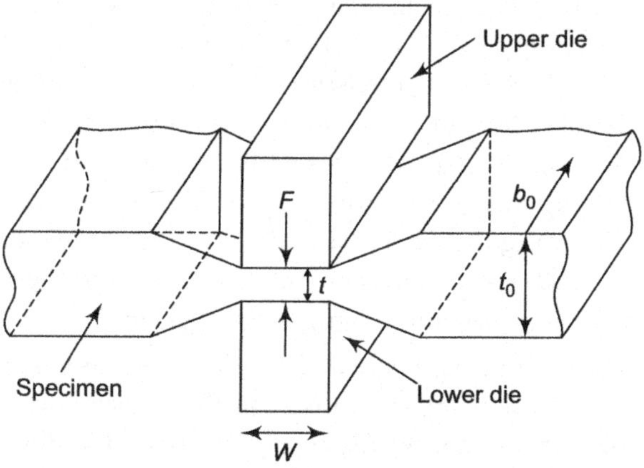

Fig. 4.11 Schematic 3D-presentation of PSCT [34]

Probably the earliest work in the direction of use of PSCT to quantify friction was from mid 1950's [57]. The first qualitative use of PSCT in order to distinguish between different lubricants, to find better lubricant for cold rolling processes, was reported at the end of the 1950's [58]. Alexander [57] suggested two methods to measure friction using plane strain conditions. First, the coefficient of friction was derived from geometric slip-line-field method for two integral ratios of (w/t) = 3–7. In the other experimental method, the coefficient was derived for all (w/t) ratios by comparing the mean pressures with the basic yield stress of the material using an approximate theory based on the simplifying assumption that plane sections remain plane.

Later on, Takahashi and Alexander [59] also showed experimentally that the PSCT could be used to derive considerable amount of information about the behavior of lubricants in metalworking processes. They were also successful in implementing the experimental method of comparing the mean die pressures with the basic yield stress of the material, and in obtaining quantitative values for the coefficient of friction. However, in that work, the basic yield curve of the material was derived at room temperature using graphite grease, which gives a very low coefficient of friction.

At elevated temperatures, problems may arise because of lack of suitable lubricants that can give very low friction. This problem was addressed by Bailey and Singer [60] in their work on PSCT. Here they adopted a different method called cylinder compression test to obtain basic yield curve developed by Cook and Larke [61]. In that method stress values were extrapolated to zero ordinate as they were decreased as the ratio (d_0/h_0) decreased to find the basic yield stress (d_0, initial diameter, h_0, initial height of the specimen). These methods of indirect measurement were tested to obtain good results though the process might be tedious.

One of the reported attempts to measure friction directly was due to Mizuno *et al.* [62]. In their attempt, they modified the common PSCT. Friction in this modified version of the test was measured directly by simultaneous sliding of one of the dies in the strip-width direction. Important experimental constraint was to make sure that V is much larger than either U or W, when V being the velocity of die in strip-width direction, U being the outflow velocity of workpiece, and W being the press ram velocity. Thus, the measured friction force component can be considered as true friction force. The procedure may not work at large thickness, and for higher reductions.

In early 1980's, the PSCT has been further developed for the direct evaluation of lubricants for metalworking purposes. In their research work, Lonn and Schey [63] calculated the magnitude of friction from the deformation of the edge in the compression zone instead of measuring compression forces. By doing that, they tried to put PSCT at par with ring compression test, which is used to measure friction directly from physical measurements instead of force measurements. They found that bulging of the specimen edge under the anvil zone is a reliable quantitative measure of friction. As in other attempts of using PSCT to measure friction, they had some constraints while standardizing the test procedure. One of them is to keep *initial* $(w/t) > 1$ to avoid non-homogeneous indentation in which the effect of in-homogeneity may overshadow friction effects. The other constraint is $(b/w) \geq 2$ to assure that the friction hill develops fully. Here they also found that larger (b/w) ratios do not offer any advantage. Thus, they were able to quantify friction coefficient by standardizing the test procedure up to a 75% reduction in case of hot and

87.5% reduction in cold working conditions. The principle involved in their calculations is that though the total spread is almost independent of friction, the shape of the edge zone under the anvil varies from almost straight for low friction to a very peaked profile for high friction. That shape has been quantified as *bulging* (defined as the differences in widths over the contact zone) as shown in Fig. 4.12.

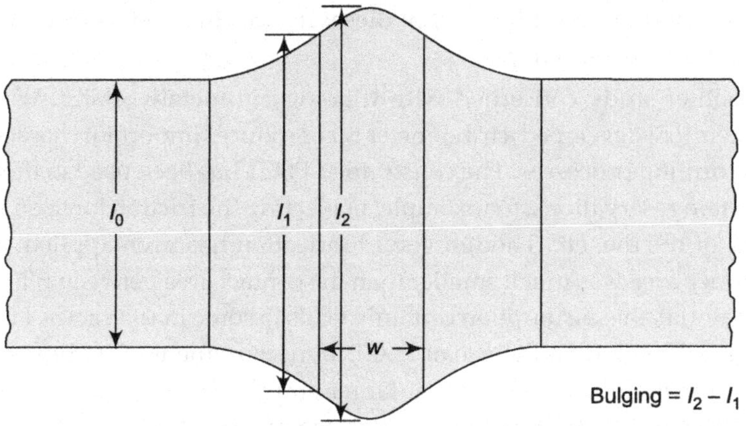

Fig. 4.12 Schematic representation of bulging

Another kind of PSCT called 'constrained PSCT' has also been used in some applications. In this test, it is always questionable if it is possible to achieve the real plane strain conditions during the test. As shown in Fig. 4.13, in this test the side surfaces of the work material are in contact with the grooved die walls during the whole test, thus frictional effects cause inhomogeneous material flow. However, the problem of sideways bulging is, of course, avoided.

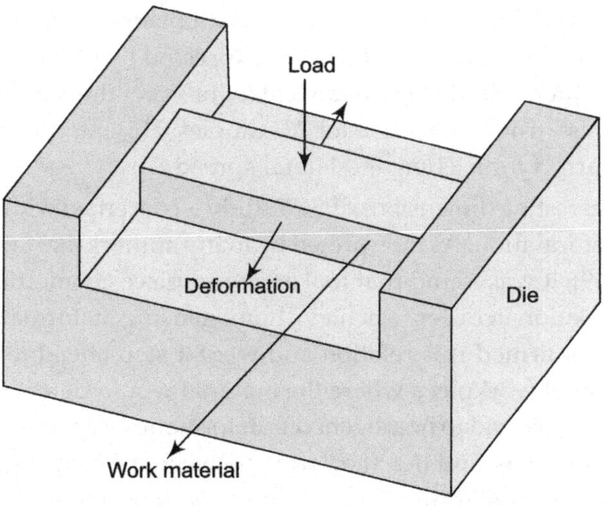

Fig. 4.13 Schematic presentation of constrained PSCT

A study concerned with use of PSCT for identification of flow stress is by Becker [64]. He tried changing the boundary conditions at edges of the tools according to plasto-mechanics theory, in order to obtain the flow properties of the workpiece material. In addition, he proposed a method that is based on an iterative procedure to calculate the interface friction. However, this method must be considered as inaccurate. His friction measuring method includes experiments with different dies with varying widths thus obtaining different force–stroke curves.

In another study concerned with tribology in metalworking, Wang and Ramaekers [65] developed a tribo-meter that produces important characteristics of bulk forming processes. The constrained PSCT has been used in their work with certain reservations, for example, neglecting the friction force on the side surfaces of the die, etc. Though good lubrication has been applied, and the side surface area was much smaller than the contact area between platens and work material, this assumption certainly leads to some inaccuracies. However, their study highlighted the use of PSCT in measurement of friction and flow stress of the material, especially at larger strains thus bringing in the PSCT as an alternative to RCT and/or the Rastegaev test whose specific specimen design render perfect lubrication at higher strains between the platens and workpiece.

Gurrero [66] applied FE-codes to analyze the hot plane strain compression testing, where she focused on friction and lateral spread. In this study, the author found that deformation patterns are sensitive to both initial geometry and friction, while they are insensitive (or dependency is inconsiderable) to the other major test variables like material type, strain rate, thermal effects, etc. It was suggested that a more complex relationship exists between friction and lateral spread in contrast to the view expressed by others [63, 67]. At the same time, an attempt has been made [68] to interpret the hot PSCT results to obtain true stress–true strain curve for Al-samples. This interpretation includes methods to correct for friction and lateral spread.

Another interesting dimension of PSCT study is concerned with homogeneity of material deformation. As interpreted by many authors like Lonn and Schey [63], Dieter [69], it was found that tool and workpiece geometries must agree to a specific relation in order to achieve homogeneous deformation. Recently Valberg [70] confirmed this relation and went a step ahead to find the best possible region of workpiece where the material would deform at a constant strain rate, i.e., subjected to homogeneous deformation. He found that location is on vertical mid-axis, and lies at halfway between the center and the surface of the workpiece. Similar investigation has been reported earlier by Gurrero [66] where it was found that strain distribution does not change significantly in the region, comprised center plane and its surroundings up to 20% away from it in the lateral direction. Metallurgists will be looking for this kind of information, which is useful in studies of micro-structural changes during forming operation, which in turn is important for understanding the whole process.

An interesting study of friction characterization during plane strain compression was due to Uyyuru *et al.* [71]. In that study, PSCTs were conducted on Al-alloy work-pieces embedded with internal physical grid patterns for the purpose of studying lubrication effects through deformed physical grid patterns. Experiments were complimented by numerical simulations. It was found that both the interface friction and die stroke levels influence the deformation of physical patterns in Plane Strain Compression. A new friction formulation was put forwarded for PSCT based on *Inverse Parameter Identification* approach, where deformed patterns in experiments are compared with corresponding patterns from FE-simulations to predict friction. By quantifying the deformed patterns in FE simulations, a set of friction calibration curves were developed, and were used with reasonable success.

4.8.7 Rolling Test

In rolling the friction conditions are unusual, since the direction of the frictional force reverses at the neutral line, and the interface is not subjected to pure slip. Frictional balance in the roll gap determines the position of the neutral point, where strip moves with the roll at input speed v_0. The strip emerges at a higher speed v_1 from the roll gap. The speed difference, called the forward slip $(v_1 - v_0)$, increases with increasing friction. A method which has been used in research is to apply a steadily increasing back tension until the neutral point coincides with the exit point, and the strip begins to skid. Under these conditions, the friction force acts in the same direction over the whole roll surface [Fig. 4.14(a)], and the coefficient of friction (μ) can be deduced as it will be simple function of the torque (T), the roll load (P), and roll radius (R) as given by Eq. (4.9).

$$\mu = T/PR \tag{4.9}$$

This method was first used by Whitton and Ford [72] for an evaluation of lubricants. If the rolling mill is not equipped with torque meters, the strip may be stopped by attaching it to a spring dynamometer [Fig. 4.14(b)] according to the technique of Pavlov [73]. If B is back tension, and α is the angle of bite, then,

$$\mu = \frac{B}{4P} + \tan\left(\frac{\alpha}{2}\right)$$

Even though these techniques are attractive because μ is derived without the aid of theory, interface conditions are not necessarily representative of those prevailing under normal rolling conditions.

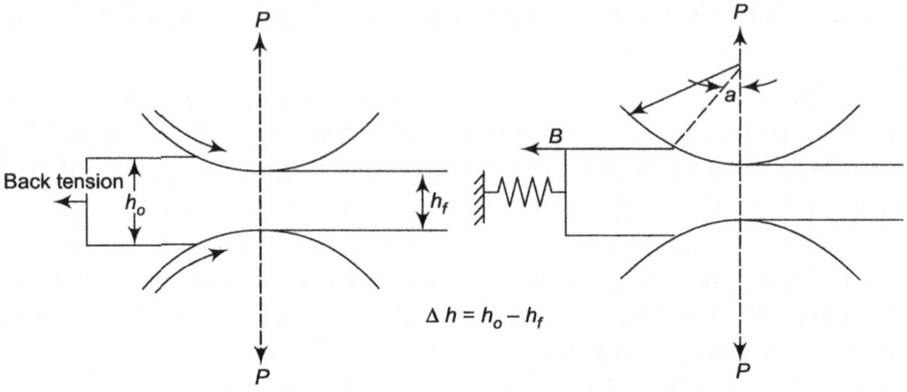

Fig. 4.14 Friction measuring techniques in rolling

On the other hand, the simplest practical assessment of friction is to determine the maximum draft which the rolls will accept, Eq. (4.10), thus

$$\mu = \sqrt{\frac{\Delta h}{R}} \tag{4.10}$$

where μ is coefficient of friction, Δh is maximum draft, and R is the roll radius.

4.8.8 Combined Forward and Backward Extrusion Test

A new test method for cold-forging lubrication method was proposed by Sanchez *et al.* [74]. This test is based on a combined forward rod and backward cup extrusion process. A global friction factor is estimated quantitatively by relating it to the length of the forward extruded bar for a specified displacement of the punch. The procedure involves comparison of the extrudate length with theoretically computed lengths keeping all other variables constant. The procedure was analyzed using an upper bound approach, the calibration curves for different friction factors were determined by analyzing the flow characteristics of the extruded material. Calibration curves, based on upper-bound calculations, deliver m-values between $m = 0$ and $m = 1$. This test was developed for the cold extrusion of aluminum. Corresponding experiments conducted by Popilek [75] showed, however, that the test is not very sensitive to small changes in the actual lubrication conditions. Figure 4.15 depicts the schematic view of the test; while a set of friction curves are shown in Fig. 4.16 [12]. In Fig. 4.15, α is die angle for forward extrusion, R_1 is radius of forward extrudate, R_2 is radius of punch, and R_o is billet radius/outer radius of backward extruded cup.

The advantages of the test are: it can closely simulate extrusion operation and no load measurements are required to evaluate friction. However, disadvantages of the test are that it is not very sensitive to small changes in lubricating conditions and accuracy of test results depends on assumptions used in the analysis.

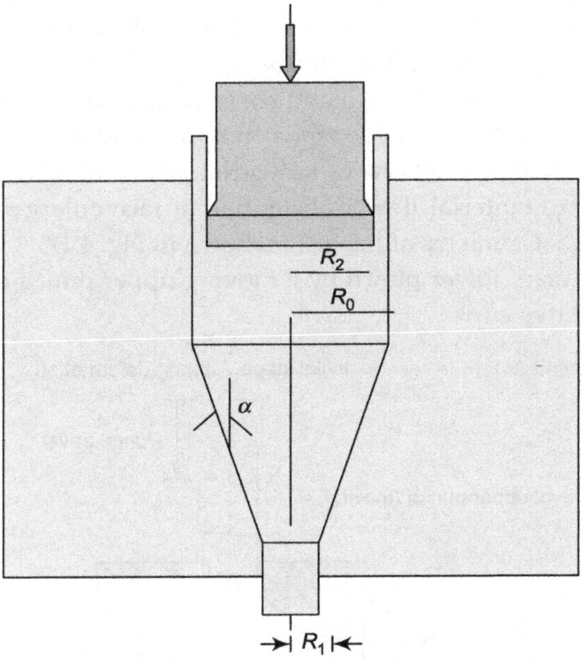

Fig. 4.15 Combined forward and backward extrusion test

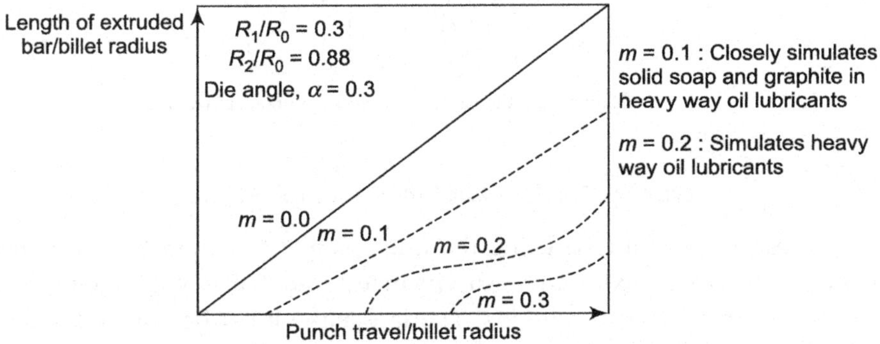

Fig. 4.16 Schematic of friction curves for a particular test geometry, along with lubricants closely simulating different friction conditions [12]

With the advent of FEM and better computational tools, a combination of computational methods and experiments has been used to get better understanding of many variables, including interface friction. The simultaneous usage of computational and experimental work makes it easier to find the process parameters more accurately as one method can be used to validate the other.

4.8.9 Double Backward Extrusion Test

In cold-forging operations, the material flow and the quality of forged parts are functions mainly of the tool geometry, the amount of deformation or strain, and the conditions at the workpiece–tool interface (e.g., friction, relative surface velocity, surface finish and heat transfer). While many of these parameters are well known and controllable, the influence of friction is often difficult to predict and depends on a variety of factors. A lubrication test that reflects the conditions of industrial production, i.e., high surface pressure, severe material flow, substantial surface enlargement, etc. has been developed. Geometry of the test is shown in Fig. 4.17. A billet is forced against a stationary lower punch by a moving upper punch in a stationary die to produce two cups.

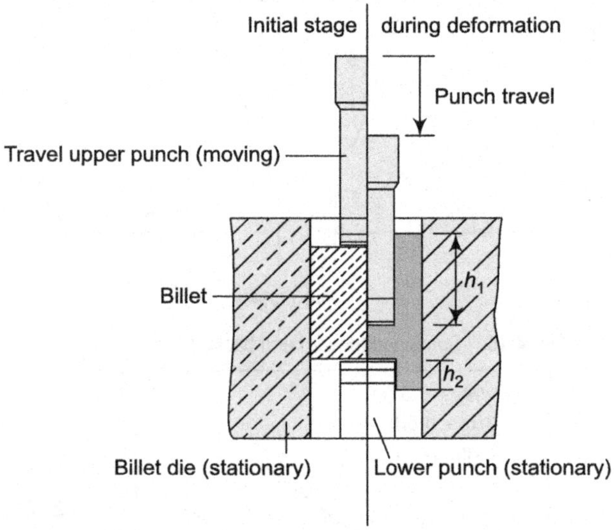

Fig. 4.17 Double backward extrusion test geometry [76]

This test was optimized using the finite element method to design test geometry such that maximum sensitivity in test results was obtained for small changes in lubricating conditions. The larger the billet height-to-diameter ratio, the bigger is the maximum difference in cup heights. The ratio of the two cup heights was used as a measure of the friction existing in the system. The difference in the cup heights arises due to different relative velocities resulting in different friction conditions and is a measure of lubrication. The height of the upper cup h_1 is bigger than the height of the lower cup h_2. To find calibration curves, an FEM analyses using program DEFORM^{TM} was carried out. The experimental value of the cup height ratio can then be compared to the predicted ratios of the friction coefficient. The advantages of this test are: geometry is very sensitive to small changes in lubricating conditions, and the test is simulative and does not require measurement of loads and strains to evaluate the friction. [76].

4.8.10 Forward Rod Extrusion Test

A simulative friction measuring test for extrusion processes was proposed by Tan *et al.* [77]. Friction stresses were obtained from measurements of slopes of extrusion pressure–punch travel curves at the steady state stage. Experimental investigations of measuring the friction stress at the container–workpiece interface were carried out. Extrusion pressure–punch travel curves before the steady state can be divided into four stages: elastic deformation, filling container deformation, filling die aperture deformation and exiting inhomogeneous deformation as shown in Fig. 4.18. It is found that the relationship between normal pressure and friction stress appears linear, and therefore, Coulomb's friction model fits the experimental data very well. It should be noted that the experimental data obtained here were under limited cold-forming conditions: at constant room temperature and constant strain rate. It is also to be noted that the temperature at the tool–workpiece interface, which might have a significant influence on friction, was not considered in friction stress calculations.

In Fig. 4.18 the following nomenclature is observed: **O** is initial point, **A** is point of the maximum elastic (linear) deformation, **B** is point of the maximum expanded deformation filling the container, **C** is point of the maximum extrusion pressure, **D** is start point of steady state, and **E** is final point of steady state.

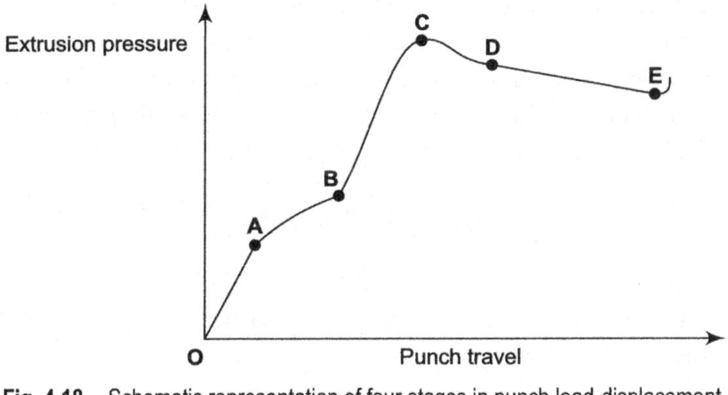

Fig. 4.18 Schematic representation of four stages in punch load-displacement curve for forward rod extrusion [17]

4.8.11 Barrel Compression Test

Analogous of RCT, where only measures of geometry of the specimen before and after compression would suffice to calculate friction parameter, a new method for evaluation of friction in bulk forming has been reported recently [78]. As the name suggests, in this test, a cylinder with specific aspect ratio was compressed under appropriate test conditions representing real forming process. Equation (4.11) based on upper-bound analyses, modified to include velocity discontinues [79], gives a minimum requirement of power for the actual velocity field:

$$J^* = \frac{2}{\sqrt{3}}\sigma_0 \int_V \sqrt{\frac{1}{2}\dot{\varepsilon}_{ij}\dot{\varepsilon}_{ij}}\, dV + \int_S K|\Delta V_i|\,dS + \int_{S_N} mK|\Delta V_i|\,dS - \int_{S_T} T_i V_i dS \quad (4.11)$$

where J^* is upper-bound power, σ_0 is flow stress of material, V is volume of deformation zone, $\dot{\varepsilon}_{ij}$ is components of strain rate tensor, S, discontinuity and friction surface, K, shear yield stress of material, m is constant friction factor, and T_i is the external traction.

After some simplifications, an equation for m can be written as follows:

$$m = \frac{(R/H)b}{(4/\sqrt{3}) - (2b/3\sqrt{3})} \quad (4.12)$$

where

$$b = 4\frac{\Delta R}{R}\frac{H}{\Delta H},\; R = R_0\sqrt{\frac{H_0}{H}} \quad (4.13)$$

and R is average radius of cylinder after deformation, H is height of cylinder after deformation, b is the barrel parameter, ΔR is the difference between maximum and top radius, ΔH is reduction of height of cylinder after deformation, R_0 is initial radius of cylinder, H_0 is initial height of cylinder.

After measuring top radius (R_T), maximum radius (R_M) and height of cylinder after deformation (H), Eqs (4.12)–(4.13) can be used to calculate the constant friction factor. However, the accurate measurement of the top radius of cylinder after deformation is difficult in practical conditions. Therefore, with approximation of the profile of the barreled specimens with an arc of a circle, top radius can be determined by Eq. (4.14):

$$R_T = \sqrt{3\frac{H_0}{H}R_0^2 - 2R_M^2} \quad (4.14)$$

Thus, using this method the constant friction factor, m, can be calculated by measuring only maximum radius, R_M, and height of the cylinder after deformation.

The advantages of the above method are as follows:

(i) It is a very simple method.

(ii) There is a high sensitivity of geometrical shape of cylinder to friction condition.

(iii) It does not need force measurement.

(iv) The constant friction factor can be determined for forming processes corresponding to actual processing condition.

4.8.12 Tip Test

A very specific test that is based on backward extrusion of a material is the tip test. The special experimental set-up of this method induces the formation of a sharp radial tip on the extruded end of the workpiece [80]. It has been found that the simple geometrical measurement. the radial distance from the radial tip to the side wall of the specimen, *b* in Fig. 4.19, increases with higher levels of friction, and thus can be used to effectively determine the friction condition. Both numerical simulations and physical experiments were employed during the study and found to correlate well with each other.

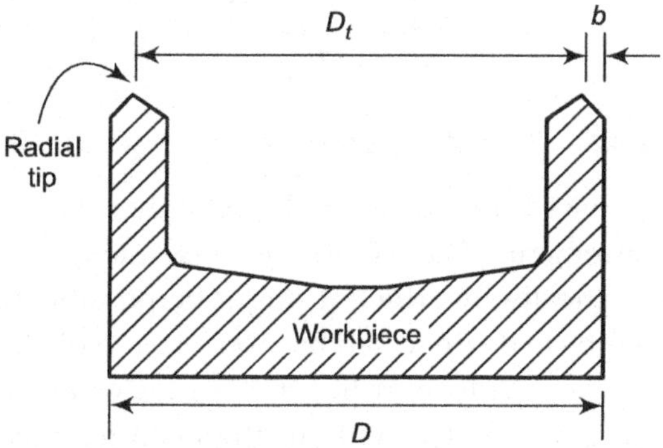

Fig. 4.19 Schematic view of tip position in backward extrusion process

4.8.13 Strip Drawing Test

This test was proposed by Le and Sutcliffe [81] in relation with friction model in cold rolling process. The test procedure involves pulling a strip through two stationary dies (with specific wedge angle, **Φ**) placed in close approximation by the application of indentation force, *N*. Force, *F*, required to pull the strip through the dies is measured in presence of different lubrication mediums, i.e., this test describes the measurements of friction in the bite during strip drawing under thin film conditions. The simplicity of the strip drawing process makes it attractive for this type of investigation. The tool geometry makes it easier to control tool roughness and the absence of a reversal in sliding direction, as compared with rolling, allows a more confident evaluation of friction. A schematic drawing of strip drawing test is shown in Fig. 4.20.

It was assumed that the friction coefficient, μ, is constant through the bite, the force, *F*, required to pull and indentation force, *N*, are used to calculate the constant friction coefficient according to the Eq. (4.15).

$$\frac{F}{N} = \frac{2(\mu + \tan\varphi)}{1 - \tan\varphi} \tag{4.15}$$

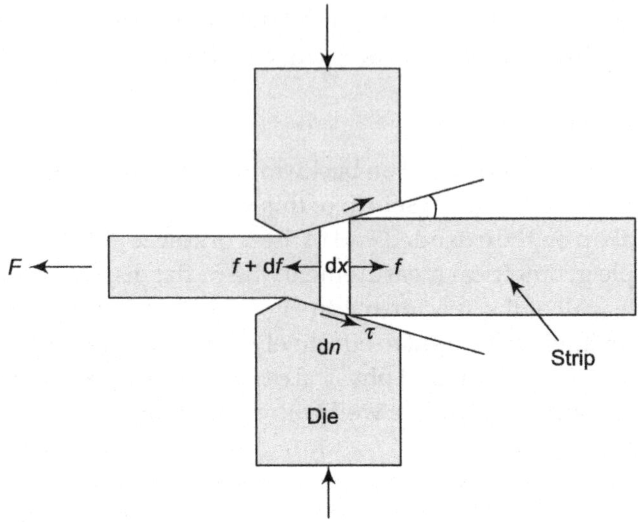

Fig. 4.20 Schematic representation of strip drawing test [81]

All the tests discussed until now are based either on force measurement or geometry measurements. There have been few attempts to predict friction parameter mathematically by which it might be possible to avoid errors in measurement and thus in the reported friction parameter. A method has been proposed by Lee *et al.* [82] to find the friction factor at the die–workpiece interface for the forging process without the need for measurement of the shape changes of the workpiece. It is based on the concept of the inverse analysis and requires only the definition of the problem inversely and does not need complicated mathematics to invert the problem. Using this method, it was shown that friction factor decreases when the die velocity is increased, and thus confirming the validity of the proposed method.

4.9 FRICTION TESTS FOR SHEET FORMING

In most of the sheet forming processes friction between the workpiece and tool surface has an important effect on the formability by affecting the strain distribution in the various regions of the sheets. In analyzing such processes, it is important to have some method of characterizing the frictional stress at the interface. Especially, in recent years, more computational programs based on finite element schemes have been developed and used to study plastic deformation in sheet forming processes. The accuracy of such analysis is limited by the knowledge of boundary conditions. The lack of knowledge of friction behavior during forming operations is a serious deficiency in the current state of the art. A good understanding of friction is crucial for improving accuracy of simulation of sheet forming processes.

The friction behavior is known to be a complex function of interacting variables that constantly change during the forming processes. Therefore, it is important to study friction behavior and estimate friction coefficients under the conditions that can closely represent those in real sheet forming processes. Such conditions include contact pressures, deformation magnitudes, tool geometry, forming speed, surface roughness, sliding velocities and lubrication conditions. Because it is experimentally difficult to isolate and identify all of the variables independently, an alternative method of studying friction behavior of sheet metal is needed for production condition. Although no single test can be used to simulate the wide range of practical sheet metal forming conditions, some areas have received a great deal of attention.

There are currently several different types of friction tests for sheet metal operations which have been used to evaluate the interfacial sliding behavior of materials. There are certain basic elements inherent in all friction tests, namely two surfaces in sliding contact with one another. There must be a force to create a relative sliding motion between the two contacting bodies, and a normal force to create a frictional condition at the interface. To calculate a friction coefficient these forces must be measurable.

4.9.1 Pinch-Type Tests (Sheet Friction Tests)

Several types of geometries for pinch-type friction tests of sheet metal are shown in Fig. 4.21. These tests may be one-sided, two-sided, and may have a flat, cylindrical, on inclined contact geometries. A back force may be applied to introduce a degree of tensile deformation that can be controlled independently from the normal force, f_N. These tests have the advantage of simple interpretation and friction coefficient between the sheet and die is found out using Eq. (4.16). However, the sticking case is complicated by the unknown contact area if curved dies are used. Tests are much more similar to conditions in rolling, drawing, or extrusion operations. Disadvantages of these tests are: geometry and deformation patterns induced are not similar to the tensile deformation and sliding in press dies, where through-thickness compression is nearly always absent.

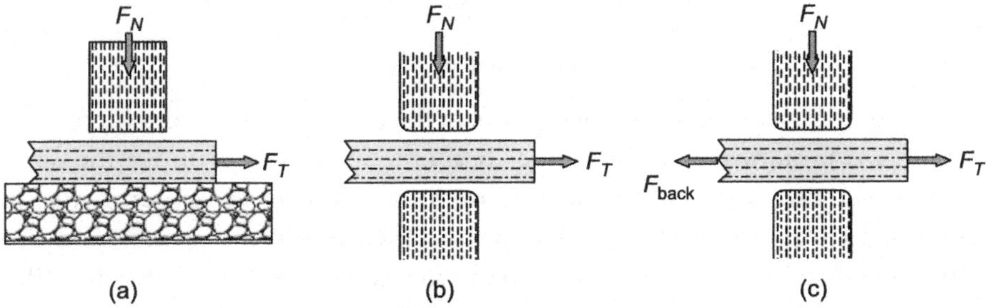

Fig. 4.21 Pinch-type friction tests for metal sheets

$$\mu = \frac{1}{2}\frac{f_T}{f_N} \quad \text{or} \quad m = \frac{1}{2}\frac{f_T}{\tau_{\max} A} \tag{4.16}$$

4.9.2 Simulative Tests for Deep Drawing and Stretch Forming

The effectiveness of boundary lubricants can be determined by finding the operative coefficient of friction between sheet metal and the tool. For this purpose, two different methods for the determination of coefficient of friction were developed by Kaftanoglu [83]. The first method is applicable to the radial drawing region (between sheet metal and die) in deep-drawing (Fig. 4.22). The second method is applicable by finding the operative coefficient of friction between sheet metal and the tool. From experimentally measured quantities such as loads and strain distributions, the coefficients of friction can be calculated using analytical equations. The results indicated that the coefficient of friction varied as a function of the plastic deformation. The coefficients of friction in stretch forming were found to be much higher than those in radial drawing. The experimental results indicated that plastic films could be more effective in reducing friction than grease type lubricants. The methods developed can be used in the evaluation of the lubricants in such processes. They can also be used to provide the necessary coefficients of friction to be used in metal forming theories.

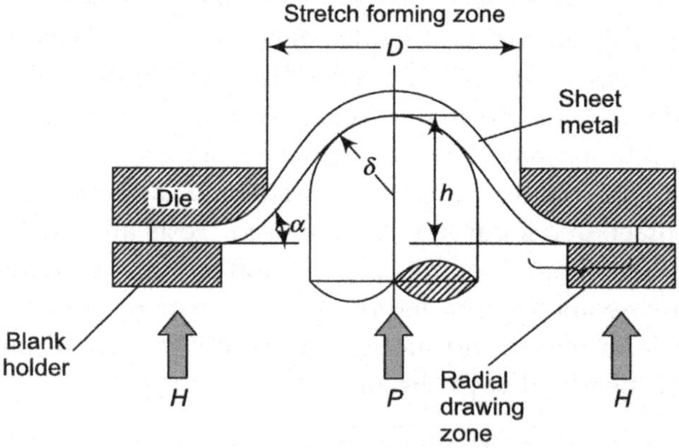

Fig. 4.22 Schematic view of deep drawing [83]

To determine the coefficient of friction, in the radial drawing zone, according to Eq. (4.17), blanks with the same diameter were tested under the same conditions using different blank-holding loads. During the experiment, the punch load was accurately measured. The increase in the punch load due to the increase in the blank-holder load was determined for different bulge depths. Then the coefficient of friction could be computed for different δH's, and μ versus δH curve could be plotted and extrapolated to zero δH to find the correct μ. The coefficients of friction for the stretch forming zone were found to be higher than those for the radial drawing region.

$$\mu = \frac{\delta P \cot \alpha}{2\delta H + \delta P} \tag{4.17}$$

Where μ is coefficient of friction, P is the punch load, and H is blank holding load.

The main drawback of the test is that while measuring large numbers which are close, small errors in measurement can lead to considerably larger errors in friction values.

4.9.3 Simulative Tests for Sheet Forming Operations

In sheet metal forming frictional forces are important boundary conditions and so must be known for accurate analytical and numerical process modeling. Accurate measurements of friction forces can provide valuable information for process model development and validation. The importance and unique features of friction in sheet metal forming have led to the development of friction measurement apparatus which simulate the workpiece deformation in forming operations.

The tensile strip test developed by Duncan *et al.* [84] is widely used. In this test, shown schematically in Fig. 4.23, a strip specimen of sheet metal is pulled over the cylindrical surfaces of pins to simulate stretching and drawing processes. The pulling force on one side of the pin is measured along with the strain in a section of the test specimen on the other side of the pin. The strip force on this second side of the pin is calculated from the measured strain using the stress–strain characteristics of the test material. Assuming that the coefficient of friction is constant over the pin surface, the strip tensions are used in the capstan friction model to calculate the coefficient of friction. There are questions that arise when the strip friction test, other process simulation tests, and experimental results are considered. Some of the results reported indicate a dependence of coefficient of friction on the test configuration,

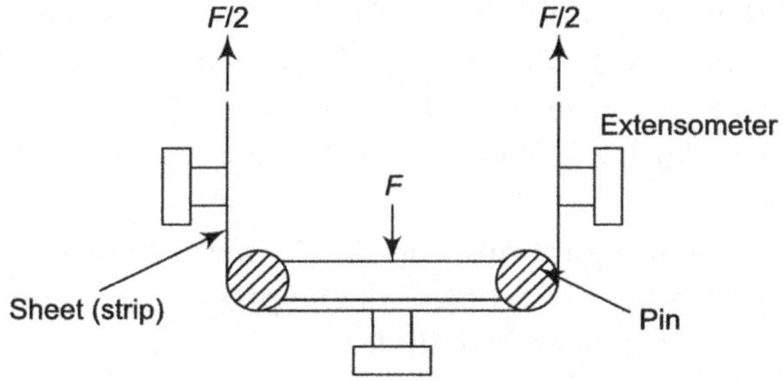

Fig. 4.23 Tensile strip test for measurement of coefficient of friction [84]

e.g., pin radius. In terms of the test apparatus itself, a drawback of the tensile strip test is that the load on one portion of the strip is estimated from measurement of strain.

In a successful effort to overcome the above mentioned difficulty, Hao *et al.* [85] proposed two different simulative tests for measuring friction in sheet forming operations. A major advantage of these test apparatus is that strain does not have to be measured to measure coefficient of friction, but by measuring the strip tensions of each side of the pin using capstan or belt friction model given by Eqs (4.18)–(4.20). A schematic representation of 'U' shape sheet forming friction test, is shown in Fig. 4.24. The strip specimen is pulled over two pins and around a surface which is mounted on a load cell. This apparatus is set up in a tensile test machine and the pulling force, F_1, is measured by the test machine load cell. The strip tension on the other side of the pins, F_2, is measured with the load cell in the apparatus. The radius and location of the pin can be changed to investigate the effects of the pin radius and wrap angle on friction.

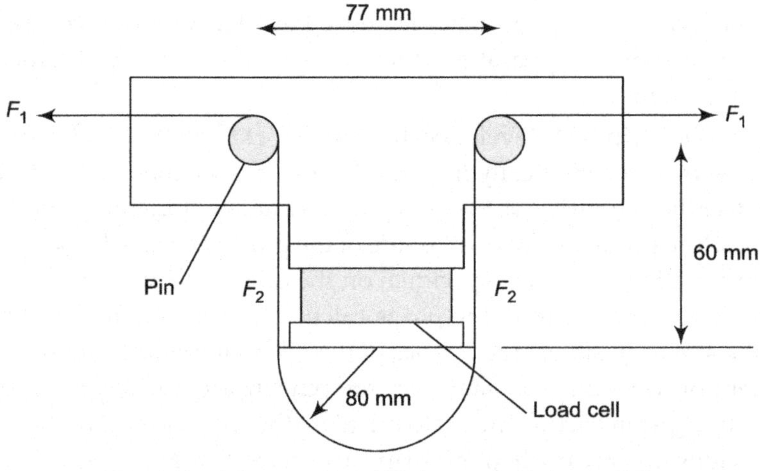

Fig. 4.24 Schematic view of 'U' shape strip friction test [85]

Figure 4.25 shows the setup for the 'L' shape sheet forming friction test. A test strip is held at one end in a grip supported by a load cell. The specimen is wrapped around a cylindrical pin and loaded in a tensile testing machine. The strip tensions and F_1 and F_2 are measured simultaneously during the test. An extensometer is used to monitor strain and to determine strain rate during the test.

Force and moment equilibrium on an element of the strip result in the following descriptions of the average contact pressure, p, the average friction stress, f, and the coefficient of friction, μ.

$$p = \frac{F_1 + F_2}{2WR} \tag{4.18}$$

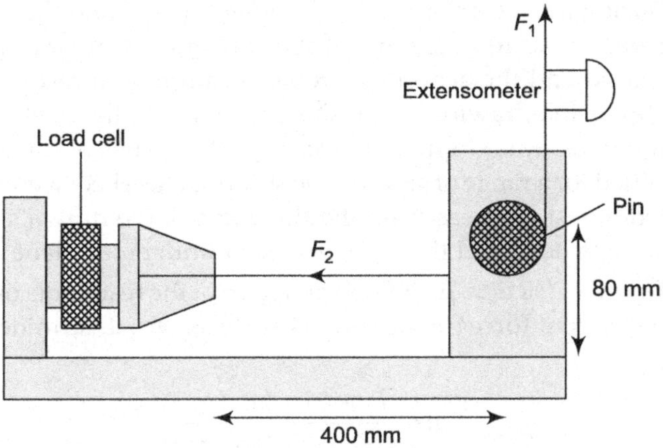

Fig. 4.25 Schematic view of 'L' shape strip friction test [85]

$$f = \frac{F_1 - F_2}{WR\Phi} \tag{4.19}$$

$$\mu = \frac{1}{\Phi} \ln \frac{F_1}{F_2} \tag{4.20}$$

where F_1 and F_2 are the strip tensions, W is the width of the strip, R is the radius of the pin and Φ is the wrap angle of the strip on the pin.

4.9.4 Sheet Drawing Test

A test device was constructed [86] which allow one to measure the normal force F_N and the tangential force F_T whilst drawing a strip of sheet between two cylindrical-surface (23.4 mm diameter) dies clamping the sheet. A schematic representation of the test is presented in Fig. 4.26.

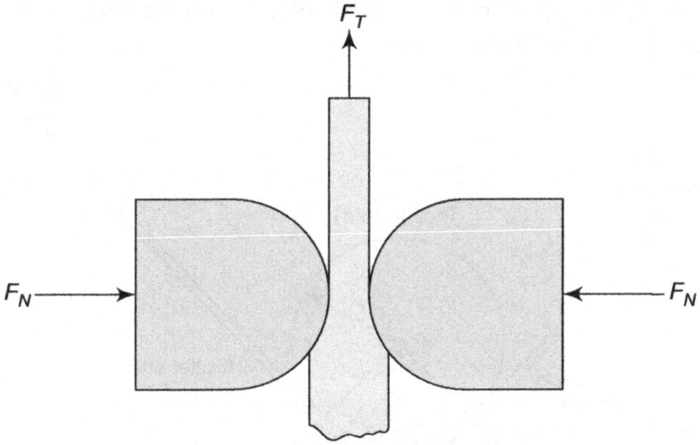

Fig. 4.26 Schematic view of sheet drawing test for friction evaluation [86]

The test was carried out in such a way that a strip of the sheet was clamped with force F_N so that the clearance between the dies was smaller than the sheet's thickness. During the recording of the drawing force F_T and the normal force F_N the sheet was drawn for a distance of about 10 mm. Then the clamping force F_N was increased and the action was repeated until, as a result of too great a clamping force, the drawing of the sheet turned into the stretching of it, at which point the test was stopped. In this way the values of forces F_N and F_T were determined for a range of strains. The strains of the sheet were determined after the friction test by measuring the thickness of the drawn sheet. As the specimen starts to be forced through, the sheet undergoes plane strain.

The coefficient of friction, μ, defined as a ratio of the resultants of the normal force and the tangent force on the contact surface, was determined using the relationship

$$\mu = \frac{F_T - 2F_N \tan(\alpha/2)}{2F_N + F_T \tan(\alpha/2)} \tag{4.21}$$

where α is the angle of the arc of the surface contact of the cylindrical dies with the sheet and $\tan(\alpha/2)$ is represented by

$$\tan(\alpha/2) = \sqrt{\frac{\Delta h}{4R - \Delta h}} \tag{4.22}$$

where R is the radius of the cylindrical surface of the dies and Δh is characterized by the difference between the thickness of the sheet before deformation and that after deformation in the friction test.

4.9.5 Sheet Metal Forming Simulator

A sheet metal forming simulator (SMFS) was developed by Wilson *et al.* [87], which stretches the sheet around a cylindrical pin as shown in Fig. 4.27. SMFS was used [88] to investigate the relationships between friction and process variables including sliding speed, angle of wrap and strain rate, in the boundary lubrication regime, with electro-galvanized steel sheet and commercially pure aluminum sheet.

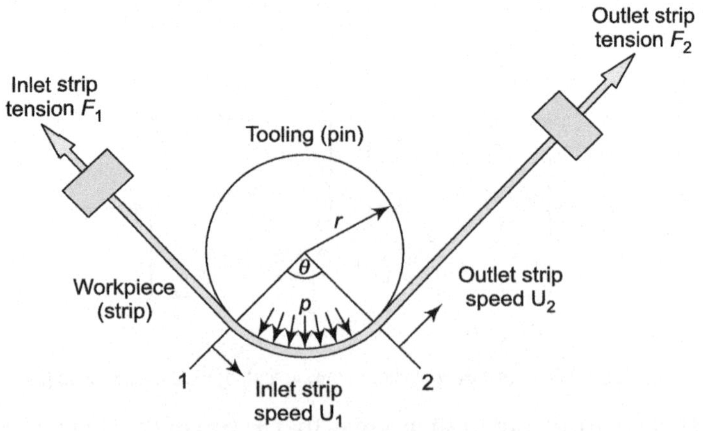

Fig. 4.27 Sheet metal forming simulator [88]

The average coefficient of friction, μ, can be calculated from Eq. (4.23).

$$\mu = \frac{2(F_2 - F_1 - F_b)}{\theta(F_1 + F_2)} \tag{4.23}$$

where F_1 and F_2 are the inlet and outlet tensions, respectively, F_b is the Swift bending tension, and θ is the angle of wrap. If the workpiece thickness is small compared to the pin radius, F_b may be neglected compared with the difference $(F_2 - F_1)$.

4.10 FRICTION TESTS FOR HYDRO FORMING

Hydro forming is comparatively a new class of sheet forming method. This class of forming technique has found many applications especially in automotive industry where there is a need for thin-walled hollow complex shaped parts. The main advantages of hydro formed parts are low weight and high stiffness. Tribology plays an important role in this class of forming methods too. The available tests to experimentally determine friction parameter in sheet forming are not satisfactory as they do not simulate the hydro forming process.

As shown in Fig. 4.28, a typical hydro forming process consists of two areas—feed area and forming area. The feed area is meant to feed material into forming zone, i.e., it acts like a reservoir for material which is pushed into the forming zone. Actual gross deformation takes place in deformation area where sheet attains its complex shape because of pressure induced by the fluid. The different classes of tests are developed to find friction parameters in these two areas.

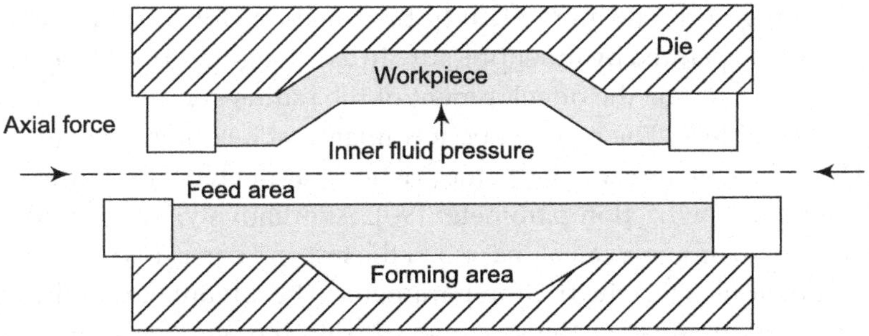

Fig. 4.28 Different zones in hydro forming

One class of tests, flow-through tests, is meant for measuring friction in feed area. In these tests a tube will be expanded by an internal pressure against the tool wall. Later on the tube will be forced out of the tool by die against the friction force at the interface between tube and the tool. This force is measured either as a difference of the punch forces or as a resulting force on the tool, measured in a horizontal or vertical test layout, as shown in Fig. 4.29. The coefficient of friction is calculated by dividing the friction force with the nominal area of contact.

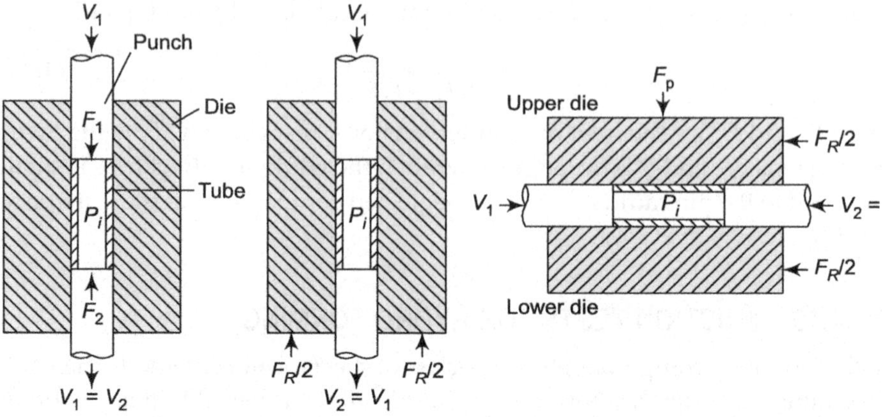

Fig. 4.29 Schematic presentation of feed area friction tests for hydro forming processes [89]

The first method is based on the direct measurement of the difference in punch force, while the second is based on the force measurement on the tool, i.e., only one force has to be determined, which minimizes the possible errors. In the case of the horizontal layout, the die is split parallel to the tube axis. The closing force F_p is measured and load cells between the two die halves measure the tool contact force. The difference of the two forces is taken as contact force between the tube and the die. The disadvantage of this method is that it is more complex than others.

Figure 4.30 shows the different tests for measuring friction parameter in hydro forming at the forming area. In first method (tube expansion test), a round tube is expanded by inner pressure in a rectangular tool. Depending on the friction conditions the development of the radius and the wall thickness in the edges differs. The test can be terminated at a certain pressure, using the edge radius as a measure for the friction parameter. The radius will be proportional to the friction parameter [90]. Alternatively, the wall thickness distribution can be analyzed at the end of the test and correlated, e.g., by using FEM calculations, with the friction parameter. The advantages of the test are the plastic deformation of the specimen during the whole test and that the forming process is not affected by the testing method itself, like it is the case in the direct measurement, explained in the following section. The disadvantage is that there is very little sliding of the specimen against the tool surface and that the contact pressure is not constant but increases during the test.

In direct measurement of friction for hydro-forming process, a ring segment is placed at the position where the dome of the part is formed, which will be shifted by the friction force against load cells. The disadvantages of the method are very typical, and arise due to the elasticity of the load cell and gap between the ring section and the remainder of the tool. This cause severe disturbances

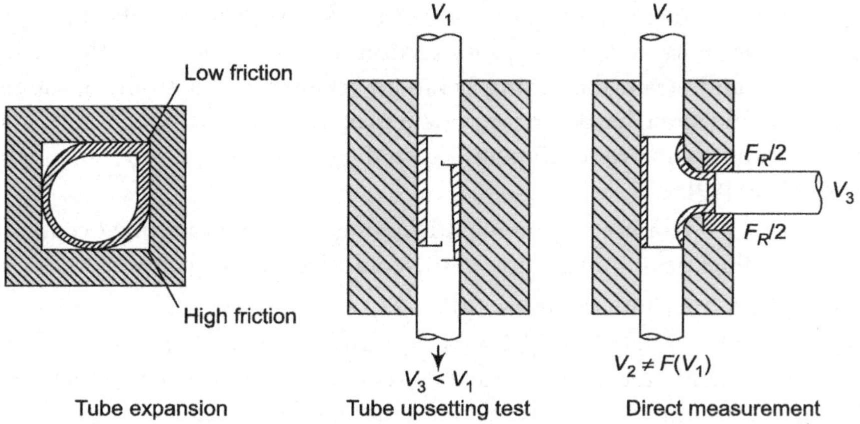

Fig. 4.30 Schematic presentation of forming area friction tests for hydro forming processes [89]

or even errors in the measurement. The dome must have a minimum length, as the material has to be formed so that it covers the tool radius. Before the radius is formed, it is not possible to measure any friction force in this test.

The third test, the tube upsetting test, was developed to overcome these difficulties. The principle of the test is similar to that of the ring compression test or the combined forward–backward extrusion test. The friction parameter is determined from the geometry of the test piece after forming, using a suitable diagram (Fig. 4.31). Such diagrams are again calculated from analytical solutions, or can be found from FEM calculations.

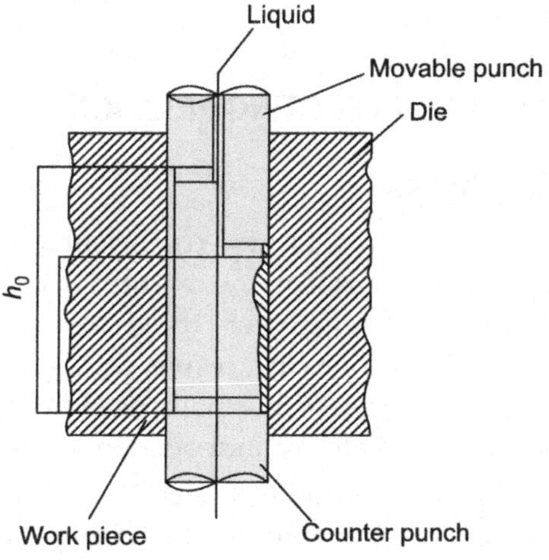

Fig. 4.31 Tube upsetting test

During the test, a tube is continuously upset in a closed die, while it is elastic/plastically expanded towards the tool wall by an internal pressure. Due to the friction forces the wall thickness increases non-homogeneously. The biggest increase of the wall thickness occurs near the punch with the higher velocity. The shape of the wall cross-section will be a function of the friction coefficient, and strain hardening behavior (characterized strain hardening behavior C and strain hardening exponent n) of the material under deformation and the compression ratio. Depending on these parameters, different shapes can develop [89].

(a) $\mu = 0$, $n = 0$: a homogeneous thickening of the whole tube occurs, the wall thickness is a function of the compression ratio only

(b) $\mu = 0$, $n > 0$: same as in case (a).

(c) $\mu > 0$, $n = 0$: the wall thickness increases linearly from one end to the other, while the largest thickness occurs at the tube end which is in contact with the punch having the higher velocity.

(d) $\mu > 0$, $n > 0$: the wall thickness is similar to that of the case (c), but the difference in thickness along the tube wall is smaller as the hardening coefficient increases.

Compared to the other test principles which are mentioned above, this method has the following advantages: the processing parameters (internal pressure, sliding velocity, deformation speed and upsetting grade) can be chosen freely during this test; the equipment used for this test is simple compared to some of the other tests and does not interfere with the process itself; by changing the length of the tube, the test can be adapted to different friction conditions. In the case of low friction, long tubes would be beneficial, as the friction stresses are accumulated to forces which significantly influence the thickening of the tube.

4.11 PARAMETERS AFFECTING FRICTION

4.11.1 Surface Roughness

Figure 4.32 depicts the variation of friction coefficient as a function of surface roughness. With very smooth surfaces, the real area of contact grows rapidly, and so does the friction, whereas with very rough surfaces the friction is high again because of the need to lift one surface over the asperities on the other. In the intermediate range of roughness (normally used in engineering practice), the friction is at minimum and almost independent of the roughness.

With sheet metal forming industrial backdrop, Bello and Walton [91] studied the combined effect of surface roughness and lubrication on friction at the tool–metal interface during sliding contact in presence of lubricant. In their experiments, strips of commercial pure aluminum were pulled through steel dies designed to give partial simulation of the conditions, which exists in the flange and die radius profile regions of the deep drawing process. They found that the conventional surface roughness parameters do not provide a

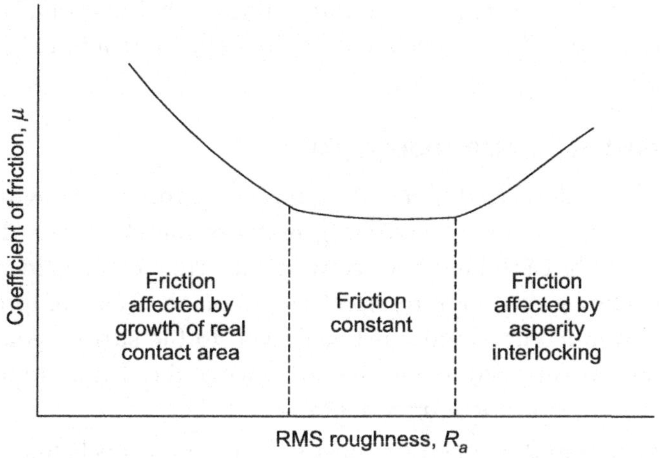

Fig. 4.32 Schematic depiction of variation of friction against surface roughness

satisfactory functional characterization of the surface roughness in the context of the friction developed in sliding contact. They coined a new roughness parameter, Bearing Length Product (BLP), that is equivalent to the product of average peak width (w_p) and RMS roughness parameter (R_a) for the purpose. It was pointed out that BLP was a better functional characteristic of the surface. However, the nature of the proportionality between BLP and friction parameter (μ) depends on the way lubricant behaves, for example as shown in Fig. 4.33

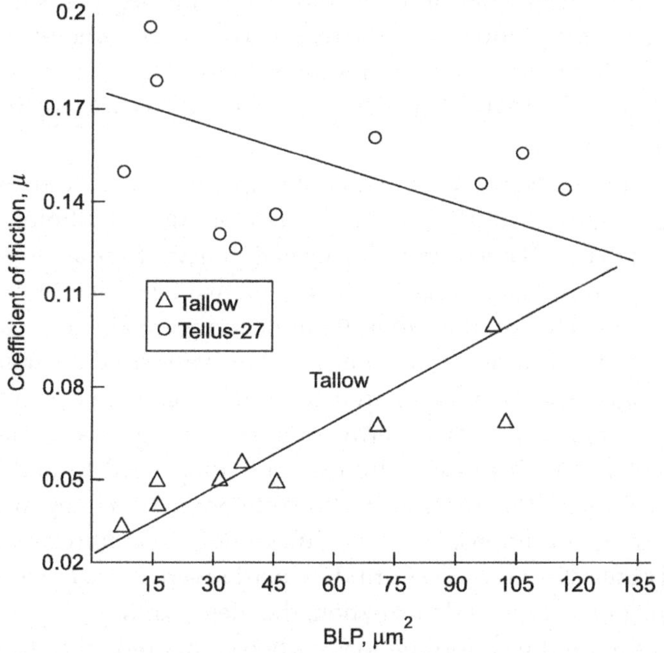

Fig. 4.33 Variation of friction parameter with BLP for different lubricants [91]

where lubricant Tallow acts as boundary lubricant because of adsorption on surface, and Tellus-27 was effective in hydrostatic lubrication regime due to its viscosity.

4.11.2 Effect of Surface Topography

Lubricants are widely used in metal forming operations for obvious reasons like for reducing friction by avoiding metal-to-metal contact thus easing material flow. Lubricants have been conveniently characterized in many ways, popular ones being ring compression test and plane strain compression test. However, not much attention has been given to the surface characteristics that lubricants are applied to and the influence of the surface topography of workpiece surfaces on friction reduction.

Early research in this direction mostly dealt with aspects like real area of contact and lubricant trapping. For example, in his theoretical work, using the plane-strain theory of plasticity for rigid-non-hardening materials, Kudo [92] proposed a methodology to estimate the effect of trapped lubricant in microscopic surface pockets of a work material by the tool surface on the apparent coefficient of friction. It was also pointed out that the effect was more pronounced for lubricant having a higher bulk modulus. Thus Kudo tried to highlight the significance of surface topography. Later-on, Nelleman *et al.* [93] investigated the effect of different surface topography geometries by varying asperity angles and concluded that normal pressure and bulk modulus have a great influence on the real area of contact, whereas the asperity slope and friction factor are of minor importance. The friction factor is also of minor importance to the real area of contact, but has a strong influence on the friction stress.

In one of the earliest experimental attempts to understand the effect of surface topography, Schey [94] demonstrated the effect of different machined surfaces though he did not examine in detail the effect of surface topography as such. More recently, Geiger *et al.* [95] identified two types of lubricant pockets, closed and connected, in their investigation and were able to determine the proportion of surface area at the interface they comprised by analysis of the surface topography. They compared the difference between two types of surface topography but not in terms of surface roughness. Lakshmipathy and Sagar [96] in their industrially relevant study tried to understand the influence of die grinding marks directionality on friction in open die forging under lubricated conditions. Two sets of dies, one with unidirectional grinding marks and other with criss-cross grinding marks were used. It was found that, for the same percentage of deformation, the dies with the criss-cross ground pattern required reduced forging loads when compared with the die having uni-directionally ground pattern (Fig. 4.34). Lower friction value was also observed for the forging process when the die with the criss-cross surface pattern was used. They concluded that the lubrication breakdown tendency is increased when pressing is done with uni-directionally ground die compared to that done with a criss-cross ground die.

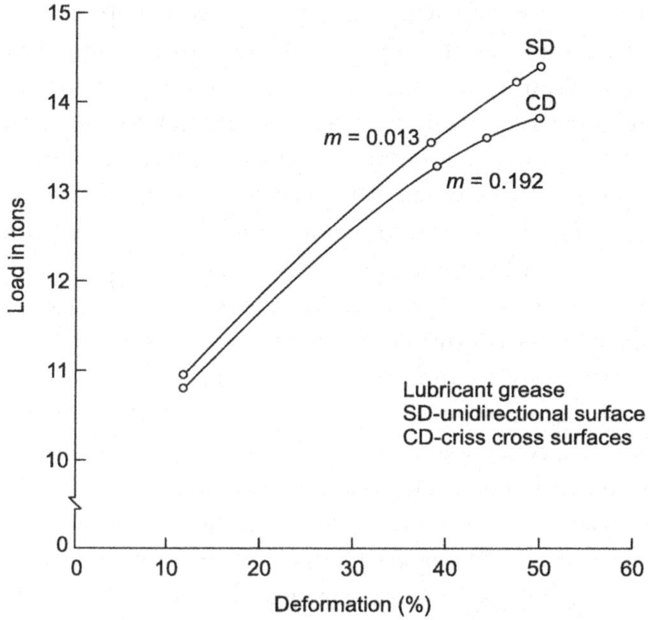

Fig. 4.34 Forging load versus deformation at different die surfaces [96]

Hu and Dean [97] studied the relationship between friction behavior and surface topography using ring upsetting test. The tests were carried out using either a liquid lubricant or under clean dry conditions. Two types of workpiece surfaces, random and directional, were prepared by either shot-blasting, or EDM or turning to obtain different levels of surface finish. It was found that, for random surfaces, smoother ones could retain more lubricant and decrease friction resistance. Rasp and Wichern [98] studied the effect of surface topography on frictional resistance using different kinds of surfaces. In their experiment, the five different specimen surfaces used were (i) as received, (ii) etched, (iii) coarse ground perpendicular to test direction, (iv) coarse ground parallel to test direction and (v) polished. They found that the arithmetic roughness value (R_a) and lubrication regime has greater influence than the directionality of the surface lay. It was also pointed out that the effects of surface lay on frictional behavior during upset without lubricant vary with upset velocity.

Experiments were conducted using an inclined scratch tester to study the influence of surface texture on friction [99, 100]. A special feature of the inclined scratch tester is that in a single experiment a range of loads can be studied. Various kinds of surface textures—namely unidirectional grinding marks, 8-ground (steel plate ground in 8 shaped motion), and random (polished)—were prepared using simple metallographic techniques. Roughness, represented by R_a, of surfaces was varied over a range as they were prepared using different grit emery papers and abrasive powders. Both surface texture and sliding angle (angle between unidirectional grinding marks and sliding direction), and their effects on friction and transfer layer formation were studied. It was observed that the plowing component varies significantly with both surface texture and sliding angle, whereas the adhesion component of friction does not vary with either. It was found that a surface texture that promotes plane strain conditions near the interface causes a higher plowing component and thus the higher coefficient of friction. On the other hand surface texture that promotes plane stress conditions at the interface results in lower value for plowing component of friction. It was found that the sliding perpendicular to unidirectional grinding marks gave maximum friction force contributed by higher plowing component, and at the other extreme random (polished) texture resulted in lower friction values. In addition it was observed that friction value varies with sliding angle, it was highest for 90° and lowest for 0° sliding angles. It was observed that the roughness as given by R_a within the range of the present tests does not significantly affect the friction values.

4.11.3 Effect of Crystal Structure

The effect of material crystal structure on friction depends on how easily material can undergo plastic deformation. The ease of plastic deformation of a material depends on the number of slip systems available. It is well known that a minimum of five independent slip systems are required for plastic deformation to take place by slip. Hexagonal metals have a limited number of slip systems, and thus the chance for forming higher true areas of contact is reduced due to lack of plastic deformation when compared with FCC. Hence, HCP metals experience low friction coefficient than FCC and BCC metals in which large number of slip systems are available for plastic deformation [101]. Crystal structure can affect friction behavior directly and indirectly through texture evolution. The factors that influence the friction behavior of metallic materials include evolution of crystallographic texture, plastic deformation and strain hardening. Preferred orientation, texture, can arise during fabrication or can be induced by deformation processes such as wear. Strain softening was also associated with texture development in many instances. The concept of strain softening with texture evolution was used by Kuhlmann–Wilsdorf [102] to explain the transition in the coefficient of friction with the sliding distance.

As early as 1940, Ernst and Merchant [103] recognized the influence of crystal structure on friction. Alison and Wilman [104] suggested that there is a fundamental difference in wear and friction behavior of hexagonal and cubic metals. As it could be expected from the above discussion, coefficient of friction for cubic metals is larger than that for hexagonal metals (Fig. 4.35). Buckley and Johnson [101], in their study of the influence of crystal structure on frictional behavior of HCP metals, found that crystal indeed have strong influence on friction by comparing two metals that are very close in nature except for crystal structure (Rhodium atomic number 45 with FCC crystal structure, and Ruthenium, atomic number 44 with HCP crystal structure). Since crystal structure affects friction behavior, it could be expected that allotropic metal will exhibit a similar influence on either side of its transition temperature of it. Experiments were performed using Cobalt which possesses HCP structure at room temperature and transforms to FCC structure at 417°C. A significant change in friction value was observed correspondingly on either side of the temperature, as shown in Fig. 4.36, which was attributed to the phase transformation. They also found that for HCP metals, the lattice parameters influence the number of operating slip systems, and thus ease of plastic deformation which in turn influences the friction. HCP metals with close to ideal atomic stacking ratio (c/a = 1.633), e.g., Co, Rh, Mg, primarily slip along basal planes (i.e., less number of slip systems), and metals which deviate considerably from ideal stacking such as titanium (c/a = 1.587) will exhibit primarily non-basal slip mechanisms (i.e., high number of slip systems). Variation of the friction with lattice parameter is shown in Fig. 4.37.

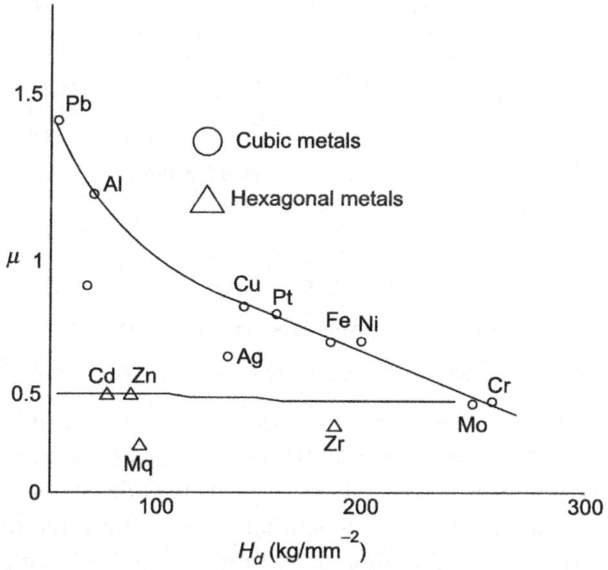

Fig. 4.35 Variation of coefficient of friction for cubic and hexagonal metals [104]

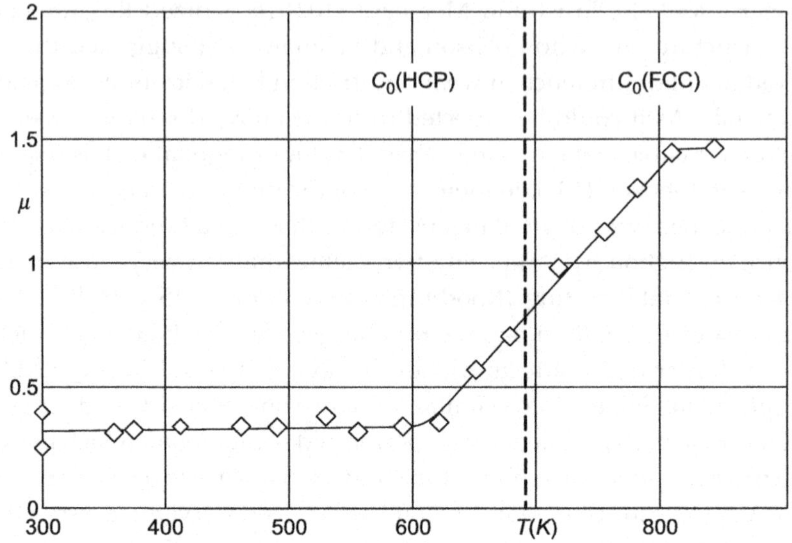

Fig. 4.36 Change in friction coefficient with allotropic transformation in cobalt [105]

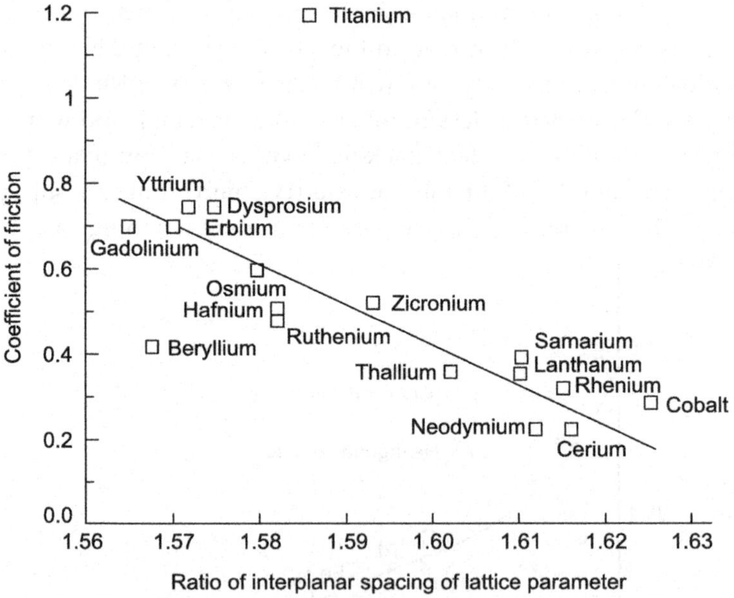

Fig. 4.37 Variation of friction with lattice parameters in HCP metals [101]

Texture development and its effect on friction were studied extensively over the decades. Wheeler and Buckley [106] found that during sliding for (1) FCC metals, a texture develops where (111) was tilted about 10° towards the sliding direction, (2) BCC metals, similar texture develops along (110) with [111] direction oriented in direction of sliding, and (3) HCP metals, no texture development was found at lower loads. However, basal texture evolution along (0001) for HCP metals was observed by many other researchers, for example [107]. It can be stated that texture evolution and thus the crystal structure of a metal plays an important role in its friction behavior. The lower coefficient of friction for hexagonal metals when compared with cubic metals can be attributed to the ease of metal shearing through higher number of slip systems.

4.11.4 Effect of Strain Hardening

It is well established fact that severe plastic deformation occurs in the surface regions (e.g., asperity contacts) of a metal sliding over a harder surface. This results in the surface metal being progressively work hardened and reaches a maximum hardness that depends on the method by which it has been deformed. The maximum hardness attainable by abrasion can differ appreciably from that resulting from trepanning or heavy compression [108].

There are very limited sources available in literature about the effect of material strain hardening capacity on friction. In one of these, Gaivoronskii *et al.* [109] formulated a methodology to relate the contact pressure with the friction stress and studied the hardness and strain hardening, and found that the degree of influence of these material bulk properties on friction and their interconnection with the surface layer properties can be used to determine friction stress.

On the other hand, Lin [110] studied the friction, wear, and micro-hardness of some metals as a function of degree of work hardening. It was found, as shown in Fig. 4.38, that there is a systematic relation between the friction coefficient and the surface micro-hardness as a function of increasing degree of work hardening for the face-centered cubic metals Al, Cu, Ag and Au. These loci are close to the locus for different fully work hardened cubic metals of a similar range of hardness [104], and they may be represented approximately in the range $40 < H_d < 140$ kg/mm^2 by an equation of the form

$$\mu = a - b\, H_d \tag{4. 24}$$

where a and b are constants.

Reid and Schey [111] performed experiments, and tried to correlate friction with bulk hardness, surface hardness and strain hardening rate. They found no correlation between bulk hardness and friction; however, they found satisfactory correlation between surface hardness and strain hardening. Other results indicate that a greater increase in surface hardness or a higher strain hardening rate do not imply lower friction. It has been concluded that hardness cannot be used as a criterion for predicting coefficient of friction in un-lubricated sliding. However, it was stated that metallurgical compatibility between materials plays an important role over their tribological properties.

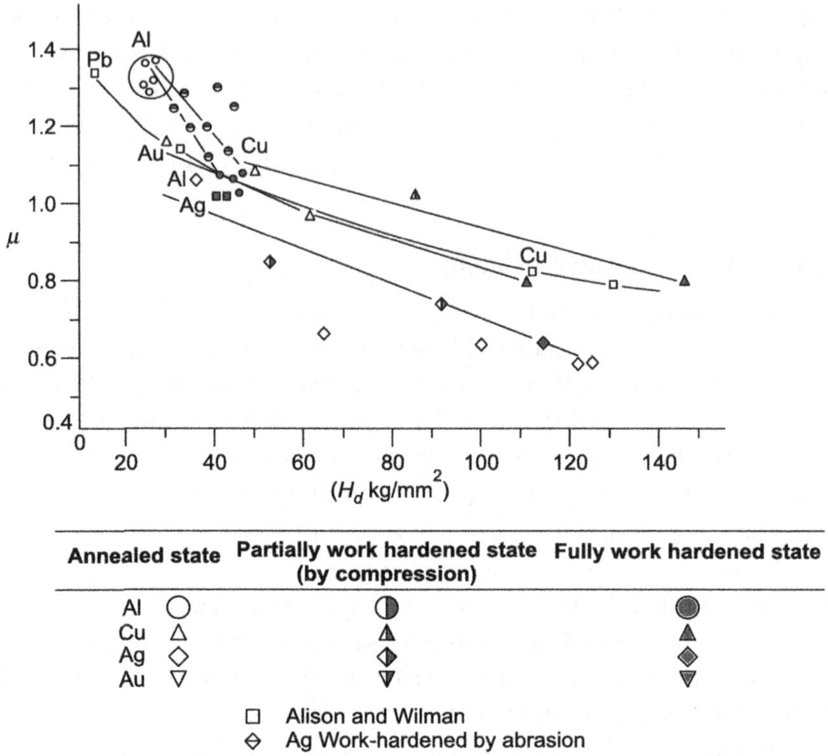

Fig. 4.38 Variation of coefficient of friction with degree of strain hardening [110]

Mokhtar [112] experimentally showed that hard metals have lower frictional resistance than softer metals owing to stronger inter-atomic bonds. He postulated that the atomic bonds in harder metals are strong and hence the resistance to adhesion is increased providing low frictional characteristics. Surfaces hardened by heat treatment were also characterized by low friction owing to phase transformation and increase in surface energy and the induction of internal stresses. Figure 4.39 depicts the marked decrease in coefficient of friction with increase in hardness for pure metals. Hardness has less but recognizable influence on friction in case of heat-treated alloys, as shown in Fig. 4.40. The influence of hardness on friction was attributed in two ways—firstly, strong inter-atomic bonds that minimize the plastic deformability of hard metals with a subsequent decrease in the ability of metals to adhere, and secondly higher surface energy associated with metals of higher hardness leading to formation of weak asperity contacts. This experimental evidence obeys theoretical prediction of coefficient of friction as an inverse function of hardness of metals. The effect of hardness on friction is attributed to the lack of plastic deformation of hard metals with subsequent decrease in the ability of metals to adhere resulting in low friction.

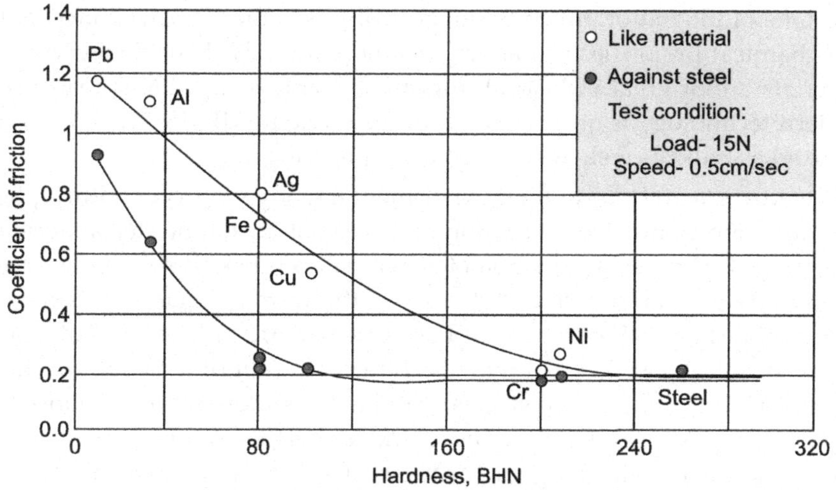

Fig. 4.39 Variation in friction coefficient with hardness for pure metals while sliding against themselves and against steel [112]

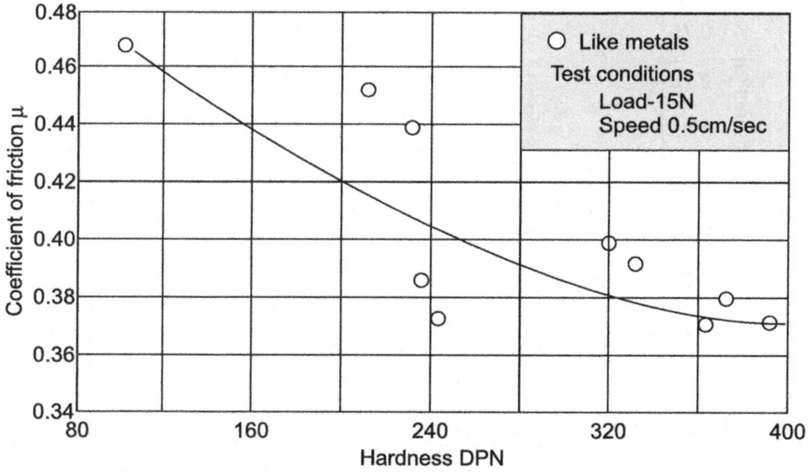

Fig. 4.40 Variation in friction coefficient with hardness for heat-treated beryllium bronze [112]

It was also claimed by Mohtkar [113] that the correlations between the coefficient of friction and the mechanical and physical properties of metals [114], which depend on inter-atomic bonds, and the periodic variation of all these phenomena with the atomic number, allow the coefficient of friction to be accepted as an intrinsic property of metals.

4.11.5 Effect of Grain Size

The effect of grain size on frictional characteristics of many materials has been studied in combination with their wear behavior. According to available results, it can be stated that friction coefficient decreases with decrease in grain size. One of the important consequences of this is the improved mechanical and chemical properties of nanocrystalline materials. Thus nanocrystalline materials are of great industrial importance, and their wide spread use in modern technology is quite evident, for example in MEMS. These materials are used as bulk, as well as coatings to engineering substrates.

Bregliozzi *et al.* [115] in their experimental study of effect of atmospheric humidity and grain size on friction and wear of austenitic steels measured friction coefficient using a tribometer in a ball-on-disk (BoD) configuration, whereby a steel ball (AISI 52100) slides over the austenitic stainless steel with a constant linear speed of 10 cm/s and under a load of 2 N. The effect of relative humidity and the grain size on the friction coefficient of AISI 304 are shown in Fig. 4.41. The fine grained steel is found to have more wear resistance than the large grained steel. The hardness of the steel is the predominant factor: the fine grained steel (with a higher hardness) shows a weight loss and a friction coefficient lower than the relatively softer large grain steel.

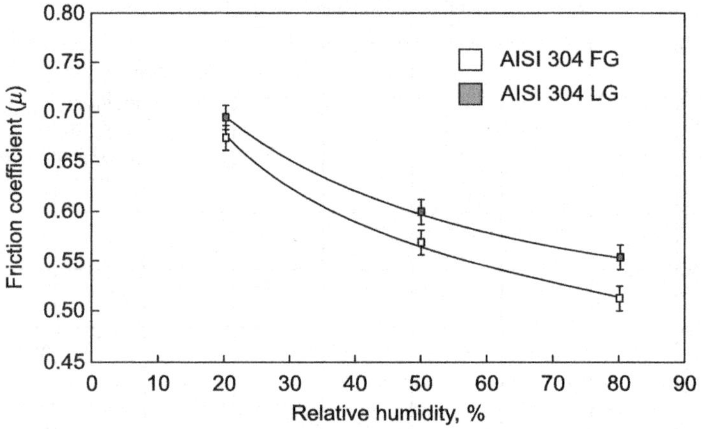

Fig. 4.41 Effect of humidity and grain size on friction coefficient of austenitic steels [115]

Wang *et al.* [116] created a thin nanocrystallized (*nc*) layer by surface mechanical attrition method on a low carbon steel plate, and studied the tribological characteristics of it. They found that the friction coefficient decreases and the wear resistance increases with the *nc* surface layer. The improvement in friction and wear properties were attributed to the harder *nc* surface layer, which reduces the degree of plowing and micro-cutting under the lower load, and alleviates plastic removal and surface fatigue fracture under the higher load.

Farhat *et al.* [117] produced nanocrystalline aluminium samples using radio frequency magnetron sputtering technique to study the wear characteristics as a function of grain size. They found that during early stages of sliding, the coefficient of friction increased to a peak value followed by a decrease to a steady state value. The peak coefficient of friction decreased from 1.4 for aluminum with a coarse grain size in order of 10^6 nm to 0.6 for nanocrrystalline aluminium with a grain size of 16.4 nm.

Similar results for nanocrystalline nickel over copper substrate were reported by Mishra *et al.* [118]. Nanocrystalline nickel was electrodeposited using Watt's bath ($NiSO_4 \cdot 6H_2O$ 240 gm/l, $NiCl_2 \cdot 6H_2O$ 30 gm/l, H_3BO_3 30 gm/l). The wear tests were performed at a normal load of 1N, an oscillation frequency of 8 Hz, tangential displacement amplitude of 100 μm and number of fretting cycles was 10,000 for all samples. All the wear tests were performed against steel ball (SAE 52100 grade, 8 mm diameter) without lubrication, at room temperature and in ambient air of approximately 40% relative humidity. The coefficient of friction (COF) was recorded continuously during the wear tests. The coefficient of friction was determined as a function of number of cycles. Figure 4.42 shows the variation of COF with number of cycles for different systems. In all the cases, it was noticed that the COF increased from a low value to a very high value within the running-in period (500 cycles) and thereafter, COF marginally decreased and attained a steady state value within the next 1000 cycles. By comparing the steady state COF values, it is clear that extremely low COF (down to 0.16 for 8 nm grain size Ni) is obtainable with electrodeposited Ni coatings. The coefficient of friction apparently decreases with decreasing grain size.

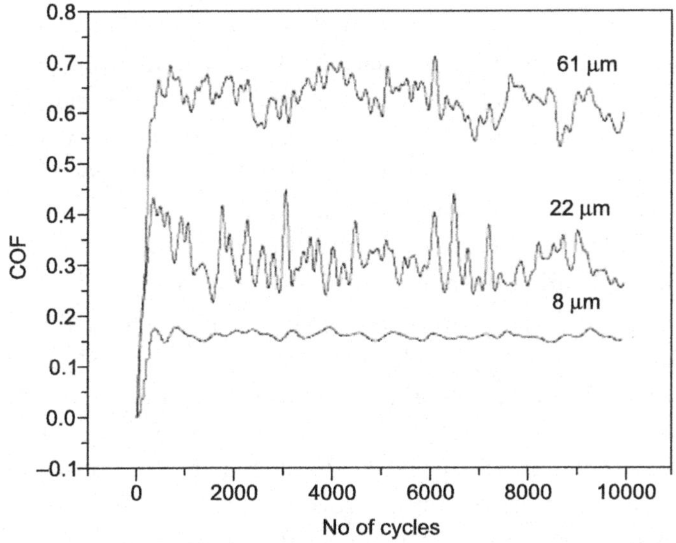

Fig. 4.42 Coefficient of friction for nanocrystalline Ni of grain sizes 8, 22 and 61 μm [118]

The grain size effect on friction, thus, is quite evident for metals and their alloys. Similar grain size effect on friction is observed in ceramic materials as in metals and their alloys. A weak relation between grain size and coefficient of friction was observed for alumina at elevated temperatures [119]. Zum Gahr *et al.* [120] studied the effect of the average grain size on friction and wear of self-mated Al_2O_3 and 3 mol%Y_2O_3-ZrO_2 in dry unidirectional and reciprocating sliding wear at ambient conditions. The stationary values of the friction coefficient were measured after a relatively short running-in period of less than about 1 km sliding distance. The average values of the friction coefficients were found to increase slightly with the average grain size.

4.12 CONCLUDING REMARKS

The difficulties in the determination of the friction parameter lie in the complexity of the friction phenomena itself, and in the inability to accurately measure shear stresses. Friction in metal forming processes comprises a very complex situation with many parameters interacting with it, for example, normal pressure, relative displacement (velocity), surface roughness, surface strain, lubricant, temperature, material (both tool- and work-material) properties, work hardening, surface coatings, surrounding atmosphere, etc., all play a part. It is impossible to formulate the friction in a universal way because of the complex interaction between many of these variables. Therefore, far-reaching approximations are used to describe friction behavior during metal forming. These approximations deal with apparent friction rather than with the fundamental phenomenon. One of the consequences of this approach is that friction must be measured separately for each forming process.

For example, for the last two decades it is evident from industrial and domestic applications that Al extrusions are very much in demand. Most of these Al extrusions are made in un-lubricated conditions. As expressed by Storen [121], 3D numerical analysis shall be able to give good prediction of material flow, and however, an important question left unanswered is: how can one model the friction condition and the formation of the surface of the extrudate in the bearing channel in un-lubricated extrusion?

The trend in material forming technology is characterized by increasing demand for complex shapes, close tolerances (near-net-shapes) and new materials such as powder metallurgy or refractory metals which have low ductility and demand a high forming load. The computer aided design, analysis, and simulation have proved to be of great utility in providing valuable information about metal forming processes. A trend for using numerical simulations in combination with physical experiments to analyze friction in metal forming processes is being used today. However, the accuracy of numerical simulation is strongly influenced by the tool–workpiece interfacial friction condition which is a major concern in metal forming design and analysis. On the other hand, a universal test to measure friction in metal forming processes has not been and will never be found [51]. There is a need for the use of different modeling techniques and/or a closer look at the parameters that control the friction. There is a growing trend in this direction, for example, analysis of surface roughness and texture effect on sliding friction [115–116].

One of the major concerns in manufacturing is the friction and wear of machines, tools, dies, and devices, for they control the accuracy, life, power, and size of machines as well as the quality of the final products. Yet the field of tribology has relied heavily on empiricism in solving technological problems, because until the recent decades scientific understanding has usually followed technological development. As assessed by Suh [122], we now have a fair scientific understanding of the first-order tribological effects, thanks to the significant scientific advances made during the past few decades. Nevertheless, major new scientific breakthroughs are still needed for creating nearly frictionless mechanisms for nanotechnologies, high-speed and high-stiffness spindles without ball bearings, and low-friction and non-wearing machines and dies. Authors stand-by Suh who believes that such scientific breakthroughs can be brought forth through research, and there are scientific indicators that support this optimism.

For example in the process of wire drawing, the independent parameters like reduction, die angle can be measured directly. Not so for friction. There is no instrument called *Friction-meter* to measure friction. Friction is not directly measurable, nor is it really an independent parameter. But, in many metal forming processes the effect of friction is as strong as that of reduction, die geometry, etc., and thus the importance of friction parameter and its elusive characteristics. *Friction,* it seems, *is the last frontier in the study of metal forming*! [123].

REFERENCES

1. Bowden, F. P. and Tabor, D. (eds.) (1950). *The friction and lubrication of solids,* Part-I, Clarendon Press, Oxford, UK.
2. (1969) *Glossary of terms and definitions in the field of Friction, Wear and Lubrication: Tribology,* OECD, Paris.
3. Bharat Bhushan (ed.) (2002). *Introduction to Tribology;* John Wiley & Sons, Inc.
4. Schnurmann, R. (1962). Friction and Wear, *Wear* **5;** 31–42.
5. Tichy, J. A. and Meyer, D. M. (2000). Review of solid mechanics in tribology, *International Journal of Solids and Structures,* **37**; 391–400.
6. Suh, N.P. and Sin, H.C. (1981). The genesis of friction, *Wear* 69 91–114.
7. Suh, N.P. and Sin, H.C. (1980). On the genesis of friction and its effect on wear, *ASME Applied Mechanics Division,* **39**; 167–183.
8. Kim, D.E. and Suh, N.P. (1991). On microscopic mechanisms of friction and wear, *International Conference on Wear of Materials,* **1**; 475–480.
9. Suh, N.P. and Sridharan, P. (1975). Relationship between the coefficient of friction and the wear rate of metals; *Wear* **34**; 291–299.

10. Tohkai, M. (1979). Microstructural aspects of friction; *M.S. Thesis*, Department of Mechanical Engineering, Massachusetts Institute of Technology, USA.
11. Rabinowicz, E. (1977). *Proc. Int. Conf. on Wear of Materials, St. Louis, MO,* American Society of Mechanical Engineers, New York, pp. 36 – 40.
12. Wagoner, R. H. and Chenot, J. L. (eds.) (1996). *Fundamentals of Metal Forming;* John Wiley & Sons Inc.
13. Geiger, M., Kleiner, M., Eckstein, R., Tiesler, N., and Engel, U. (2001). Microforming, *Annals CIRP* **50** (**2**); 445–462.
14. Tran, D.V., Lewis, R.W., Gethin, D.T., and Afriffin, A.K. (1993). Numerical modelling of powder compaction processes: displacement based finite element method. *Powder Metall,* **36**; 257–266.
15. Wanheim, T., Bay, N. (1978). A model for friction in metal forming processes, *Ann. CIRP* **27**; 189.
16. Joun, M.S. (1999). Iterative and non-iterative schemes for coulomb frictional law in metal forming simulation and their numerical characteristics, *Commun Numer Methods Eng;* **15** (7); 515–522.
17. Fedorciuc-Onisa, C. and Farrugia, D.C.J. (2003). Simulation of frictional conditions during long product hot rolling. *The 6th ESAFORM conference on material forming,* pp. 763–766.
18. Sivaprasad, P.V. and Davies, C.H.J. (2005). An assessment of the interface friction factor using the geometry of upset specimens, *Model Simul Mater Sci Eng,* **13**; 355–60.
19. Fedorciuc-Onisa, C. and Farrugia, D.C.J. (2006). Through process characterisation of frictional conditions under lubrication for long product hot rolling, *Steel rolling 2006—9th international and 4th European conferences,* Paris. June 19–21.
20. Fereshteh-Saniee, F. (2004). Friction modelling for the physical simulation of the bulk metal forming processes, *J Mater Proc Technol;* **153–154;**151–156.
21. Lebon, F. (2003). Contact problems with friction: models and simulations. *Simul Model Pract Theory,* **11**; 229–463.
22. Lee, B.H. and Keum, Y.T. (2002). Wagoner RH Modeling of the friction caused by lubrication and surface roughness in sheet metal forming. *J Mater Process Technol* **130–131**; 60–63.
23. Ramezani, M., Ripin, Z.M. and Ahmad, R. (2010). Modelling of kinetic friction in V-bending of ultra-high-strength steel sheets. *Int J Adv Manuf Technol* **46** (1–4); 101–110.
24. Ramezani, M., Ripin, Z.M. and Ahmad, R. (2009). A static friction model for tube bulge forming using a solid bulging medium, *Int J Adv Manuf Technol* **43** (3–4); 238–247.
25. Cheng, H.S., Cao, J. and Xia, Z.C. (2007). An accelerated springback compensation method, *Int J Mech Sci* **49**; 267–279.
26. Huang, Y.M. (2007). Finite element analysis on the V-die coining bend process of steel metal. *Int J Adv Manuf Technol* **34**; 287–294.
27. Matuszak, A. (2000). Factors influencing friction in steel sheet forming. *J Mater Process Technol* **106**; 250–253.

28. Leu, D.K. (2009). A simple dry friction model for metal forming process. *J Mater Process Technol* **209**; 2361–2368.

29. Bay, N. and Wanheim, T. (1976). Real area of contact and friction stress at high pressure sliding contact; *Wear*, **38**; 201–209.

30. Petersen, S.B., Martins, P.A.F. and Bay, N. (1997). Friction in bulk metal forming: a general friction model vs. the law of constant friction; *Journal of Materials Processing Technology*, **66**; 186–194.

31. Avitzur, B. (1980). *Metal forming—The application of limit analysis;* Marcel Dekker Inc.

32. Rowe, G.W. (1977). Principles of industrial metalworking processes; Edward Arnold Publishers Ltd.

33. Joun, M.S., Moon, H.G., Choi, I.S., Lee, M.C., and Jun, B.Y. 2009. Effects of friction laws on metal forming processes, *Tribology International*, **42** (2); 311–319.

34. Uyyuru, R.K. (2003). Engineering the friction and forming mechanics in cold Aluminium forming—Use of modeling techniques, Ph.D. dissertation, Dept. of Machine Design and Materials Technology, Norwegian University of Science and Technology, Trondheim (Norway), ISBN: 82-471-5584-2, ISSN: 0809-103x.

35. Lange, K. (1985). *Handbook of Metal Forming;* McGraw-Hill, New York.

36. Evans, W. and Avitzur, B. (1968); Measurement of friction in drawing, extrusion, and rolling; *Journal of Lubrication Technology/Transactions of ASME*, **90**; 72–80.

37. Schey, J. A. (1983). *Tribology in Metalworking: Friction, Lubrication, and Wear;* American Society for Metals, Metals Park, Ohio.

38. Ekelund, S. (1933). *Steel*, **93**; 38.

39. Geleji, A. (1960). Metal Forming in computation and Experiment (Bildsame Formung der Metalle in Rechnung und Versuch), *Akademie Varleg, Berlin* (in German).

40. Kunogi, M. (1956). New method of cold extrusion, *Journal of Scientific Research Institute (Tokyo)*, **50**; 215–246.

41. Male, T. and Cockcroft, M. G. (1964–65), A Method of the Determination of the Coefficient of Friction of Metals under Conditions of Bulk Plastic Deformation; *Journal of the Institute of Metals*, **93**; 38–46.

42. Avitzur, B. (ed.) (1968), *Metal Forming: Processes and Analysis*, McGraw-Hill, New York.

43. Kunogi, M. (1953). *Reports/Scientific Research Institute (Tokyo) (in Japanese)*, **30**; 63–93.

44. Kudo, H. (1955). An Analysis of Plastic Compressive Deformation of Lamella Between Rough Plates by Energy Method, *Proceedings of the 5th Japan National Congress for Applied Mechanics, Science Council of Japan /Tokyo*, **V**; 75–78.

45. Mielnik, E. M. (ed.) (1991). *Metalworking Science and Engineering*, McGraw-Hill.

46. Bramley, A. N. and Abdul, A. (1975). Stress Strain curves from ring test, *Proceedings of the 15th International Machine Tool Design and Research Conference*, Birmingham/Macmillan, London, 431–436.

47. Saul G., Male, A. T. and Depierre, V. (1971), Metal Forming: Interrelation between Theory and Practice, in Hoffmanner, A. L. (ed.), Plenum Press, New York.

48. Bhattacharya, D. and Brown, R. H. (1980). Sensitivity of Ring-Compression test, *Metalworking Lubrication/ASME*, 23–30.

49. Rao, K.P. and Sivaram, K. (1993). A review of ring-compression testing and applicability of the calibration curves, *Journal of Materials Processing Technology*, **37**; 295–318.

50. Petersen, S.B., Martins, P.A.F. and Bay, N. (1998). An alternative ring-test geometry for the evaluation of friction under low normal pressure, *Journal of Materials Processing Technology*, **79**; 14–24.

51. Tan, X., Martins, P.A.F., Bay, N., and Zhang, W. (1998). Friction studies at different normal pressures with alternative ring-compression tests, *Journal of Materials Processing technology*, **80**; 292–297.

52. Rudkins, N.T., Hartley, P., Pillinger, I. and Petty, D. (1996). Friction modelling and experimental observations in hot ring compression tests, *Journal of Materials Processing Technology*, **60**; 349–353.

53. Luong, L. H. S. and Heijkoo, P. T. (1981). The influence of scale on friction in hot metal working, *Wear*, **71**, 93 – 102.

54. Shen G., Vedhanayagam A., Kropp E. and Altan T. (1992). A method for evaluating friction using a backward extrusion-type forging, *Journal of Materials Processing Technology*, **33**; 109–123.

55. Uyyuru, R.K. and Valberg, H. (2006). Physical and numerical analysis of the metal flow over the punch head in Backward Cup Extrusion of Aluminium, *Journal of Materials Processing Technology*, **172**(2), 312–318.

56. Sofuoglu, H. and Gedikli, H. (2002). Determination of friction coefficient encountered in large deformation processes, *Tribology International*, **35**; 27–34.

57. Alexander, J.M. (1955). The effect of Coulomb friction in the Plane-Strain-Compression of a plastic-rigid material, *Journal of the Mechanics and Physics of Solids*, **3**, 233–245.

58. Guminski, R.D. and Willis, J. (1959–60). Development of Cold-Rolling Lubricants for Aluminium Alloys, *Journal of the Institute of Metals*, **88;** 481–492.

59. Takahashi, H. and Alexander, J.M. (1961–62). *Friction in Plane-Strain Compression test; Journal of the Institute of Metals*, **90**; 72–79.

60. Bailey, J.A., and Singer, A.R.E. (1963–64). The determination of the coefficient of friction at elevated temperatures using a Plane-Strain Compression test, *Journal of the Institute of Metals*, **92**; 378–380.

61. Cook, M., and Larke, E.C. (1945). Resistance of Copper and Copper alloys to homogeneous deformation in compression, *Journal of Institute of Metals*, **71**; 371–390.

62. Mizuno T., Kamiya, S. and Okamoto, M. (1978). Lubricant Behaviours in the Compression– Friction Test of Strip Metals, *Bulletin of the JSME*, **21**; 1424–1430.

63. Lonn, A.H., and Schey, J.A. (1974). Development of the plane-strain compression test, in *North American Metalworking Research Conference-II (NAMRC-II)*, Dearborn-Michigan pp. 165–178.

64. Becker, N. (1994). Further development of methods for accusation of flow curves in regions of large deformations, *PhD Thesis*, no. 122, University of Stuttgart/ Springer–Verlag (Germany), (in German).

65. Wang, S.L., and Ramaekers, J.A.H. (1996). Measurement of friction and material flow-stress by a plane strain compression tribometer, *Journal of Materials Processing Technology*, **57**; 345–350.

66. Gurrero Mata, M.P. (1996). Finite element modelling of hot plane strain compression testing, *PhD Thesis*, University of Sheffield / UK, .

67. Gelin, J.C., Ghouatio, O. and Shahani, R. (1994). Modelling the plane strain compression test to obtain constitutive equations of Aluminium alloys, *International Journal of Mechanical Sciences*, **36**; 773–796.

68. Slik, N.J. and van Der Winden, M.R. (1999). Interpretation of hot plane strain compression testing of Aluminium specimens, *Materials Science and Technology*, **15**; 295–300.

69. Dieter, G.E. (ed.) (1988). *Mechanical Metallurgy*, McGraw-Hill, Singapore.

70. Valberg, H. (2001). The non-homogeneous nature of plastic deformation in cylinder- and plane strain- compression investigated by FEM-analysis with experiment; *OPROTEH-2001* (TSTM7-Romanian Academy-2001), Bacau, Romania, pp. 334–352.

71. Uyyuru, R.K., and Valberg, H. (2005). Friction induced material flow to characterize lubricating conditions in Plane Strain Compression Test, *International Journal of Forming Processes*, **8**; 77–115.

72. Whitton, P.W. and Ford, H. (1955). Surface friction and lubrication in cold strip rolling *Proceedings of Institution of Mechanical Engineers*, **169**; 123–140.

73. Pavlov, I.M. (ed.) (1938). *Theory of Rolling and Fundamentals of Metalworking*, Gonti, Moscow, (Russia).

74. Sanchez, L.R., Jeinmann, K. and Story, J.M. (1985). *Proceedings of the.13th North American Manufacturing Research Conference*, Society of Manufacturing Engineers, 110–117.

75. Popilek, M. (1991). A friction test for extrusion using combined flow with emphasis on backward extrusion, *M.Sc. Thesis*, Department of Mechanical Engineering, Michigan Technological University, Houghton, MI,.

76. Buschhausen, A., Weinmann, K.L., Lee Y. and Altan T. (1992). Evaluation of lubrication and friction incold forging using a double backward-extrusion process, *Journal of Materials Processing Technology*, **33**; 95–108.

77. Tan, X., Bay, N. and Zhang, W. (2003). Friction measurement and modeling in forward rod extrusion tests, *Journal of Engineering Tribology (Proceedings of Institute of Mechanical Engineers)*, **217**; 71–82.

78. Ebrahimi, R. and Najafizadeh, A. (2004). A new method for evaluation of friction in bulk metal forming; *Journal of Materials Processing Technology*, **152**; 136–143.

79. Druker, D.C. and Providence R.I. (1954). Coulomb friction, *Journal of Applied Mechanics*, **21**; 71–74.

80. Im, Y.T., Cheon, J.S. and Kang, S.H. (2002). Determination of friction condition by geometrical measurement of backward extruded aluminium alloy specimen,

Journal of Manufacturing Science and Engineering (Transactions of ASME), **124**; 409–415.

81. Le, H.R. and Sutcliffe, M.P.F. (2002). Measurement of friction in strip drawing under thin film lubrication, *Tribology International*, **35**; 123–128.

82. Lee, C.D, Weng, C.I. and Chang, J.G. (2001). A prediction of the friction factor for the forging process; *Metallurgical and Materials transactions*, **32** B; 137–143.

83. Kaftanoglu, B. (1973). Determination of coefficient of friction under conditions of deep drawing and stretch forming; *Wear*, **25**; 177–188.

84. Duncan, J.L., Shabel, B.S., Filho, J.G. (1978). A tensile strip test for evaluating friction in sheet metal forming; *SAE Technical Paper*, No. 780391, 8.

85. Hao, S., Klamecki, B.E., and Ramalingam, S. (1999). Friction measurement apparatus for sheet metal forming, *Wear*, **224**; 1–7.

86. Matuszak, A. (2000); Determination of frictional properties of coated steel sheets, *Journal of Materials Processing Technology*, **106**; 107–111.

87. Wilson,W.R.D., Malkani, H.G. and Saha, P.K. (1991). Boundary friction measurements using a new sheet metal forming simulator, *Proc. NAMRC XIX*, **SME**; 37–42.

88. Saha, P.K., Wilson, W.R.D., Timsit, R.S. (1996). Influence of surface topography on the frictional characteristics of 3104 aluminium alloy sheet, *Wear*, **197**; 123–129.

89. Vollertsen, F. and Plancak, M. (2002). On possibilities for the determination of the coefficient of friction in hydroforming of tubes, *Journal of Materials Processing Technology*, **125–126**; 412–420.

90. Nielsen, M. (ed.) (2000); in *Proceedings of Symposium on New Coatings for Hydroforming Tooling, Next Generation Solutions*, Nashville, USA, 10.

91. Bello, D. O. and Walton, S. (1987). Surface Topography and Lubrication in Sheet-Metal Forming; *Tribology International*, **20**; 59–65.

92. Kudo, H. (1965). A note on the role of microscopically trapped lubricant at the tool-work interface; *International Journal of Mechanical Sciences*, **7**; 383–388.

93. Nelleman, T., Bay, N., Wanheim, T. (1977). Real area of contact and friction stress—The role of trapped lubricant, *Wear*,**43**; 45.

94. Schey, J.A., Mysliwy, R.E. The effect of die surface finish on lubrication in ring upsetting; *Euro Trib'77, 2nd European Tribology Congress*, Paper 67.

95. Geiger, M., Engel, U., Niederkom, S., Pfestorf, M. (1995). Experimental investigation of contact phenomena in cold forging, *ICFG plenary meeting*, Denmark.

96. Lakshmipathy, R. and Sagar, R. (1992). Effect of die surface topography on die-work interfacial friction in open die forging, *International Journal of Machine Tools and Manufacture*, **32**; 685–693.

97. Hu, Z. M. and Dean, T. A. (2000). A study of surface topography, friction and lubricants in metal forming, *International Journal of Machine Tools and Manufacture*, **40**; 1637–1649.

98. Rasp, W. and Wichern, C.M. (2002). Effects of surface-topography directionality and lubrication condition on frictional behaviour during plastic deformation; *Journal of Materials Processing Technology*,**125–126;** 379–386.

99. Menezes, P. L., Kishore ,and Kailas, S.V. (2006). Studies on friction and transfer layer using inclined scratch, *Tribology International*, **39(2)**; 175–183.

100. Menezes, P.L., Kishore and Kailas, S.V. (2005). Effect of directionality of grinding marks on friction at different surface roughness using inclined scratch test, to be presented at *World Tribology Congress III (WTC2005)*, Washington DC.

101. Buckley, D.H. and Johnson, R.L. (1968). The influence of crystal structure and some properties on hexagonal metals on friction and adhesion; *Wear*,**11**; 405–419.

102. Kuhlmann-Wilsdrof, D. (1980); Dislocation concepts in friction and wear, in Rigney D.A. (ed.), *Fundamentals of friction and wear of materials*, ASM, Metals Park, Ohio, 119.

103. Ernst, H. and Merchant, M.E. (eds.) (1940). *Chip formation, Friction and Finish*, Cincinnati Milling Machine Co.

104. Alison, P. J. and Wilman, H. (1964). The different behaviour of hexagonal and cubic metals in their friction, wear and work hardening during abrasion, *British Journal of Applied Physics*, **15**; 281–289.

105. Karapetian, S.S. and Korostelin, Y.I. (1983). The dependence of the friction coefficient on the crystal structure of solids: Physical principles, *Wear*, **85**; 133–141.

106. Wheeler, D.R. and Buckley, D.H. (1975). Texturing in metals as a result of sliding, *Wear*, 3365–74.

107. Farhat, Z.N. (2001). Contribution of crystallographic texturing to the sliding friction behaviour of fcc and hcp metals, *Wear*, **250**; 401–408.

108. Richardson, R.C.D. (1967). The maximum hardness of strained surfaces and the abrasive wear of metals and alloys, *Wear*, **10**; 353–382.

109. Gaivoronskii, A.T. (Acad. of Sciences of the USSR); Prokop'ev, G.A.; Pin'kovskii, V.P. (1988). Influence of material hardness and strain hardening on friction stress, *Soviet Journal of Friction and Wear (English translation of Trenie i Iznos)*, **9**; 109–111.

110. Lin, D.S. (1969). Effect of the degree of work hardening on the friction and wear of metals during abrasion, *Wear*, **13**; 91–97.

111. Reid, J.V. and Schey, J.A. (1987). The effect of surface hardness on friction, *Wear*, **118**; 113–125.

112. Mokhtar, M.O.A. (1982). The effect of hardness on the frictional behaviour of metals, *Wear*, **78**; 297–304.

113. Mokhtar, M.O.A. (1981). Friction: Is it intrinsic property of metals? *Wear*, **72**; 287–293.

114. Mokhtar, M.O.A., Zaki, M. and, Shawki, G.S.A. (1979). Effect of mechanical properties on frictional behaviour of metals, *Tribology International* , **12**; 265–268.

115. Bregliozzi, G., Di Schino, A., Kenny, J.M., and Haefke H. (2003). The influence of atmospheric humidity and grain size on the friction and wear of AISI 304 austenitic stainless steel, *Materials letters*, **57**; 4505–4508.

116. Wang, Z.B., Taao, N.R., Li, S., Wang, W., Liu, G., Lu, J., and Lu, K. (2003). Effect of surface nanocrystallization on friction and wear properties in low carbon steel, *Materials Science and Engineering* A352: 144–149.

117. Farhat, Z.N., Ding, Y., Northwood, D.O., and Alpas ,A.T. (1996). Effect of grain size on friction and wear of nanocrystalline aluminum, *Materials Science and Engineering*, **A206**; 302–313.

118. Mishra, R., Basu, B., and Balasubramaniam, R. (2004). Effect of grain size on the tribological behavior of nanocrystalline nickel, *Materials Science and Engineering*; **A373**; 370–373.

119. Senda, T., Yasuda, E., Kaji, M., and Bradt, R.C. (1999). Effect of grain size on the sliding wear and friction of alumina at elevated temperatures; *Journal of the American Ceramic Society* **82**;1505–1511.

120. Zum Gahr, K.-H., Bundschuh, W., and Zimmerlin, B. (1993). Effect of grain size on friction and sliding wear of oxide ceramics, *Wear*, **162–164**; 269–279.

121. Storen, S. (1993). The theory of extrusion—Advances and challenges, *International Journal of Mechanical sciences*, **35**; 1007–1020.

122. Suh, N.P. (1991). Tribology in manufacturing: A retrospective and prospective assessment, *ASME Production Engineering Division (Tribological Aspects in Manufacturing)*, **54**; 61–75.

123. Avitzur, B. (1990). A personal look on my involvement in the recent history of metal forming research, A keynote presentation at *Advanced Technology of Plasticity*, 4;1990.

CHAPTER 5

Metal Extrusion

Santosh Kumar
Department of Mechanical Engineering, IT-BHU Varanasi 221 005, India

5.1 INTRODUCTION

Extrusion is a near-net shape chipless manufacturing process, of forcing the work material to flow through a die opening of desired shape through a single-hole or a multiple-hole die to obtain the final shape of a component. Forward extrusion causes the product to be in the same direction as that of the ram/billet, whereas, in backward extrusion, the product comes out in opposite direction of the room/billet. Forward extrusion is very common for all kind of extrusions, while cold backward extrusion is popular to produce aluminum (compressed) gas cylinders. Dies of appropriate shape and sizes are required to impart the right shape and specified tolerance in their respective passes. A welding chamber die (Porthole, bridge, spider, etc.) is used under hot condition for common extrusion of hollow components. Square dies are widely used for multi-hole extrusion under hot condition. Continuous dies such as streamlined, conical, elliptical, hyperbolic, and cosine dies are becoming important and popular for manufacture of better mechanical and metallurgical property products at lower cost. Extrusion through multiple-hole dies, using a high capacity extrusion press is preferred for reasons of economy and high productivity. Some of the common solid and hollow extrudable metallic sections are shown in Fig. 5.1 [1].

Cold and hot extrusion may be defined in terms of working temperature (T) and metal melting point temperature (T_m). Cold Extrusion ($T/T_m < 0.3$) refers to extrusion that occurs usually, but not necessarily, at room temperature. For example, extruding lead at room temperature is a hot working process because the recrystallization temperature of lead is below room temperature (20°C). Hot Extrusion ($T/T_m > 0.6$) refers to extrusion carried out above the recrystallization temperature. Warm extrusion (T/T_m in range of 0.3–0.6) is carried out at intermediate temperatures. It is a compromise between cold and hot extrusion where T is the working temperature and T_m is the melting point of the metal, both on the absolute scale homologous temperature.

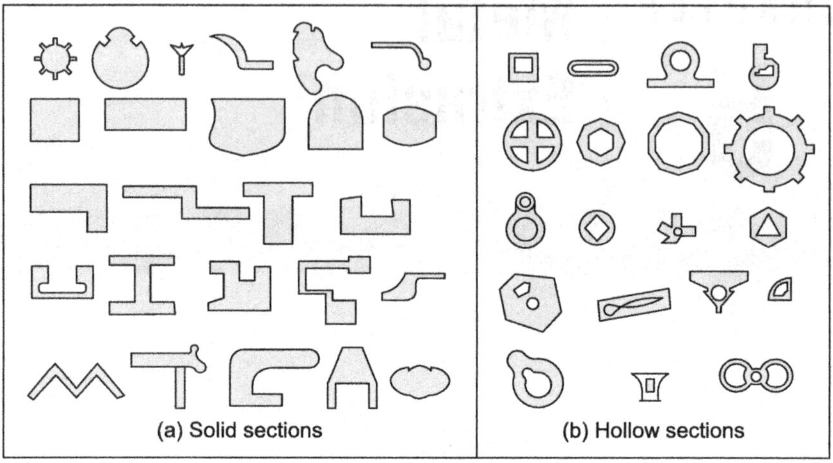

Fig. 5.1 Common extrudable sections. With permission from [1]

Extrudability (*W*) indicates the ease with which a metal can be deformed and the degree of deformation it can withstand. It is a ratio of cold ductility to flow stress in MPa (average of upper and lower field point). The two parameter of *W* are very difficult to measure in extrusion, and therefore, *W* is determined by laboratory studies, the results of which cannot replace extrusion trials but merely complement them. The values of *W* for Aluminum alloy may be classified as a practical guide: < 2 (poor), 2–4 (average), 4–15 (good) and >15 (very good) extrudability. The extrudability of many metals, including aluminum and magnesium alloys, nickel alloys and high-strength steels, falls between narrow limits and extrudability in individual cases is assessed according to various parameters like: extrusion temperature, permitted temperature range, load or specific pressure required, exit speed at a constant extrusion load, maximum extrusion speed (to the onset of hot shortness), and maximum extrusion ratio, etc.

Deformation using cold working results in higher stiffness and strength, but reduces malleability and ductility of the metal. In hot working, annealing takes place while the metal is worked and therefore is worked without getting work hardened. At high temperature, scaling and oxidation exists which generates undesirable surface finish. Most ferrous metals need to be cold worked after hot working in order to improve the surface finish. The amount of force needed to perform hot working is less than that for cold work. The mechanical properties of the material remain unchanged during hot working. The metal usually experiences a decrease in yield strength when hot worked. Therefore, it is possible to hot work the metal without causing any fracture. Quenching is the sudden immersion of a heated metal into cold water or oil. It is used to make the metal very hard. To reverse the effects of quenching, tempering is used (reheating of the metal for a period of time). Warm working depends on how close is the warm process to be to a cold or hot process. This means that the effects and characteristics get similar to the ones in a cold process when the warm process gets closer to 0.3. This also applies for hot processes. The effects and characteristics get similar to the ones in a hot process when the warm process gets closer to 0.5.

The extrusion process can produce simple as well as intricate components of different shapes and sizes with good accuracy, surface finish, mechanical as well as metallurgical properties at a lower unit cost. It is also recognized for its capability to produce even those components which are being fabricated by welding, casting, machining, etc. Recent technological developments have led the extrusion process to manufacture various types of domestic, consumer, aircraft, medical, engineering, and even food products.

Until recently both single and multiple-hole extrusion dies of complicated shapes are being designed on the basis of expertise gained through experience. Before designing such dies the designer has to identify the layout of component shapes that could be extruded simultaneously using the available press capacity. Non-symmetric component shapes both re-entry and non re-entry types, along with their locations are usually identified in such a way that the output speeds of each extruded component are the same. Even with a properly designed and manufactured die, several trial runs are necessary before the desired components, free from defects such as central burst, fir-tree cracking, roller mark, die spot, broken die, blisters, dross, speed cracks, water stains, die line, weld line, chatter mark, Orange peal, oil stains, graphite mark, joint inside crack, bulging and off flatness are obtained. Controlling all these defects in extrusion products is very difficult.

Trial and correction process is adopted in industries to control it completely. The process of trial and correction is tedious and expensive. These difficulties have motivated researchers and practitioners to look for an automated design and manufacturing practice using existing or new modeling techniques, optimization procedures, rules and algorithms, etc. with the help of computers. With the development of highly reliable and accurate FEM based plastic deformation techniques, it has now become possible commercially to model almost any metal forming process including extrusion. Availability of huge information in handbooks, online journals and material data websites, and CAD based large scale cheap computational facilities, it is now possible to design, optimize, and finalize the extrusion die design and process study for any complex profile with the help of CAE activity.

5.2 AN OVERVIEW OF ANALYTICAL MODELING TECHNIQUES FOR EXTRUSION

Modeling of extrusion process is required for designing the process parameters of extrusion and related tooling. It also helps to decide the tooling materials and process condition of extrusion process at desired output rate. Common modeling techniques are: Slab method, Slip line method, Upper bound method and Finite element method. Slab method has been used for axi-symmetric simple cases, whereas slip line method has been used for plane strain situations in 2D case. Extrusion of 3D cases has also been solved by slip line method but it becomes very tedious process. Upper bound method has been used in many ways for extrusion.

The slip-line field method is used to obtain the stress field in forming problem by constructing slip lines. Originally the method was proposed for plane strain analysis of rigid and perfectly plastic materials. However, lately there have been some attempts to extend the method to axisymmetric problems as well as to incorporate strain hardening.

Many researchers have analyzed the extrusion process by the slip-line field method for axisymmetric extrusion. Extension of slip-line method for non-axisymmetric problems is not possible mainly because the construction of slip lines in 3D space has not yet been established. It, therefore, appears that this method is limited only for axisymmetric extrusion.

5.2.1 Upper Bound Analysis for Solid Sections

The upper bound method is a technique for determining an estimate of the load which generally results into an over estimate, obtained from an assumed kinematically admissible velocity field that satisfy the incompressibility conditions and velocity boundary conditions. An extensive literature exists on upper bound analysis of metal forming processes and has been reviewed by many researchers. The plane strain and axisymmetric extrusion processes have been analyzed by many researchers. Many upper bound solutions are available for three-dimensional extrusion of solid sections as well as tubular hollow components. Lubricated cold extrusion of sections through continuous dies has been widely used for the manufacture of high strength products with good surface finish, minimum defects, and high productivity at lower unit cost.

Two kind of extrusion dies commonly used are continuous and square. Square dies are used in hot extrusion, whereas continuous dies are used for cold extrusion. Continuous dies may be divided into two categories. First category consists of streamlined continuous dies, e.g., third order, fourth order, fifth order polynomial dies and cosine dies, whereas second category encompass conical, hyperbolic, elliptical, Blazynasky's CRHS die (constancy of the ratio of successive generalized homogeneous strain increments), and modified CRHS dies etc.

CRHS dies are constructed in such a fashion that it minimizes the internal power of deformation (IPD) but there are power losses at the entry as well as exit sections. As a result, the total power consumption of CRHS die is almost as high as that for conical dies. In the modified CRHS dies, the discontinuities at the end sections are removed (i.e., it is made streamlined) while retaining the advantage of minimum IPD. Even then the total power consumption of the modified CRHS is more than the third, fourth order, etc., polynomial dies and cosine dies. Therefore, the concept of CRHS does not seem to be useful in minimizing the total power consumption.

It has been found that the first category dies require comparatively less amount of power for extrusion; and the extruded products, due to uniform deformation, have absence of intensive shear bands resulting in better microstructure and mechanical properties. However, second category dies are also preferred as a conventional practice. Despite the increased demand for 3D extrusion of generalized shaped sections through continuous dies, so far very little systematic work has been carried out for the determination of forming loads and flow characteristic of generalized shape, taken into consideration simultaneously, in order to design the extrusion process.

5.2.2 Constitutive Equations

The system of equations describing the behavior of a solid under various types of response is referred to as the constitutive equations. In steady-state forming processes such as extrusion, drawing, rolling, etc., the measure of deformation is the strain rate tensor $\dot{\epsilon}_{ij}$, which is expressed as

$$\dot{\epsilon}_{ij} = \frac{1}{2}\left(\frac{\partial V_i}{\partial x_j} + \frac{\partial V_j}{\partial x_i}\right) \quad \text{and} \quad \sigma_{ij} = -p\delta_{ij} + \sigma'_{ij} \quad \text{where} \quad p = -\frac{\sigma_{kk}}{3} \tag{5.1}$$

where V_i and V_j represent the velocity components along x_i and x_j direction, respectively, p is the hydrostatic part, δ_{ij} is Kronecker's delta and σ'_{ij} is the deviatoric part of the total stress.

In bulk metal forming processes such as extrusion, drawing, rolling, etc., the total strains are large as compared to the elastic strains. Therefore, elastic deformation can be considered as negligible compared to plastic/visco-plastic deformation. The constitutive law for rigid plastic–plastic material relating the deviatoric stress tensor σ'_{ij} and the strain rate $\dot{\epsilon}_{ij}$ is expressed as $\sigma'_{ij} = 2\mu\,\dot{\epsilon}_{ij}$. For a material yielding according to von-Mises criteria, the Levy-Mises coefficient μ is given by $\mu = \dfrac{\bar{\sigma}}{3\dot{\bar{\epsilon}}}$, where generalized yield stress $\bar{\sigma}$ and the generalized strain rate $\dot{\bar{\epsilon}}$, $\bar{\sigma} = \sqrt{\dfrac{3}{2}\sigma'_{ij}\,\sigma'_{ij}}$. The generalized strain $\bar{\epsilon}$ is therefore, defined as $\bar{\epsilon} = \int_0^t \dot{\bar{\epsilon}}\,dt$. The integration is to be carried along the particle path.

In general $\bar{\sigma}$ depends on $\bar{\epsilon}$, $\dot{\bar{\epsilon}}$ and temperature T. Since cold extrusion is carried out at room temperature, effect of temperature on $\bar{\sigma}$ can be neglected, i.e., $\bar{\sigma} = \Gamma(\bar{\epsilon}, \dot{\bar{\epsilon}})$. The specific functional form of Γ for the material under consideration is mentioned at respective places.

5.2.3 Upper Bound Formulations

Determination of extrusion power is very important for die design. Normally, the upper bound method provides an over estimate of extrusion power.

However, the upper bound model proposed here assumes a kinematically admissible velocity field which is quite close to actual velocity field. Another distinguishing characteristic of the proposed model is that the kinematically admissible field is expressed in terms of Cartesian coordinate system of streamlines which makes it easy to incorporate variable yield stress. The proposed model is appropriate for generalized non re-entry shape extrusion through a fairly general class of die profiles.

An upper bound theorem is required to satisfy only the kinematic conditions in terms of strain increments, strain rate and velocities in a plastically deforming medium and does not necessarily satisfy the stress equilibrium equations. An important concept involved is that of a kinematically admissible velocity field. Velocity fields that satisfy the constraint of volume constancy and the velocity boundary conditions are called kinematically admissible velocity fields. A kinematically admissible velocity field may have discontinuities in the tangential component along certain surfaces, but the normal component must be the same on both sides of such surfaces in order to satisfy the constraint of incompressibility. The unknown parameters in kinematically admissible velocity field are determined using the upper bound solution. The upper bound states that among all possible kinematically admissible velocity fields, the one which minimizes the total power field represents true velocity field.

5.2.4 Choice of Deformation Zone and Velocity Boundary Conditions

Since the upper bound analysis does not satisfy stress equilibrium, the prediction of complex shapes of plastic boundaries S_1 and S_2 as shown in Fig. 5.2 are never accurate. The upper bound solution obtained by earlier researchers [8,10] using straight and arbitrarily shaped plastic boundaries indicate that there is little effect of the shapes of surfaces S_1 and S_2 on the overall solution. Hence, in the present work, the deformation zone Ω is assumed to be bounded by straight plastic boundaries at the end sections of the die. Material is assumed to be rigid outside the entry and exit sections of the die. Therefore, the axial velocity at the entry and exit sections of the die should be uniform. These conditions are given by

$$\begin{aligned} V_z &= V_o \quad \text{on } S_1 \\ V_z &= V_{\text{ave}} \quad \text{on } S_2 \end{aligned} \tag{5.2}$$

At cross-section S_2, each point has different velocities. A point at the center has maximum velocity and a point on the periphery has minimum velocity. Corresponding to N different points, the common velocity of extrudate V_{ave} is defined as

$$V_{\text{ave}} = \left(\sum_{i=1}^{i=N} V_i \right) / N$$

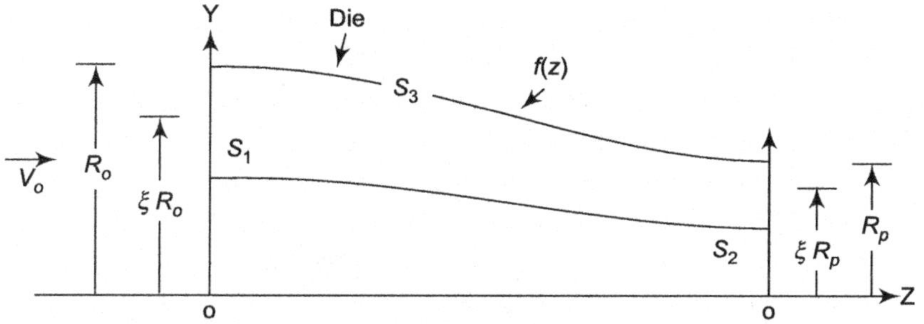

Fig. 5.2 Deformation zone and typical stream surfaces in extrusion. With permission from [3] publisher

There should not be any metal flow across boundary S_3 and the axis of symmetry. This condition on these boundaries can be expressed as

$$V_n = 0 \tag{5.3}$$

on die surface S_3 and the axis of symmetry.

5.2.5 Proposed Kinematically Admissible Velocity and Strain Rate Fields

Since deformation is under certain assumptions and streamlines are the basic flow path during the deformation process, it is necessary to build the proposed kinematically admissible velocity using the concept of streamlines. The geometry of die and stream lines in the steady flow process of extrusion is shown in Fig. 5.3. One power element OEGG′O′E′EO is chosen to demonstrate the analytical construction of velocity and strain rate fields. In Fig. 5.2 the billet with an initial radius (R_0) at entry, is extruded through a shaped die (continuous) constructed by a number of pre-defined streamlines representing the surface of die, to the final generalized given section at the exit. An arbitrary point E on the die surface at the entrance to the die can be combined with the corresponding point E′ on the die surface at the exit. It is assumed that the material of the billet passing through sector OEG at the entrance goes through the triangle O′E′G′ at exit, preserving the extrusion ratio. Then the surface defined by the points O, E, E′, and O′ becomes a three-dimensional stream surface.

The following assumptions are made throughout the analysis described in this work:

(i) The material is incompressible, rigid, perfectly-plastic and follows a particular strain hardening curve.

(ii) The deformation takes place under homogeneous and steady state conditions.

(iii) The deformation zone is assumed to be bounded by straight plastic boundaries at the end sections of the die.

(iv) The neutral line is the line joining CG of billet (i.e., point O) and C.G. of the extruded product cross-section (i.e., point O′).

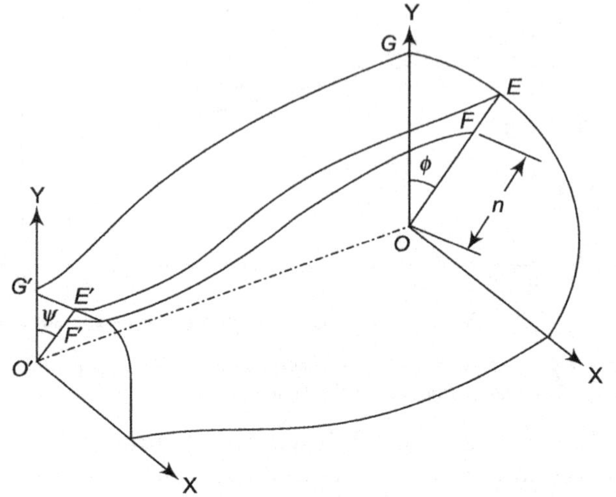

Fig. 5.3 Geometry of die and stream lines in an extrusion die. With permission from [3] publisher

(v) The stream surface OEE′O′ consists of a number of predefined streamlines, straight or curved.

(vi) The particle passing through an arbitrary point F (Fig. 5.3) on line OE goes towards the corresponding point F′ on line O′E′ maintaining the proportionality of position.

(vii) The elastic strain is small and can be neglected.

(viii) In the present analysis, friction factor between the die and workpiece material is assumed to be independent of slip.

Using assumption (*vi*), the extrusion ratio is constant for any element, i.e.,

$$\left(\frac{\text{Area of } \Delta\text{O}'\text{G}'\text{E}'}{\text{cross section area on product side}}\right) = \left(\frac{\text{Area of } \Delta\text{OEG}}{\text{cross section area on billet side}}\right) \tag{5.4}$$

Let length O′G′ = a_1 and O′E′ = a_2. If the coordinates of points G′ and E′ are (x_1, y_1) and (x_2, y_2), respectively, and A_a is the cross-sectional area of the extruded product shape, then $a_1 = \sqrt{(x_1^2 + y_1^2)}$ and $a_2 = \sqrt{(x_2^2 + y_2^2)}$. Angle ψ on the product side is found by $\cos\psi = b_1$, where $b_1 = \dfrac{b}{a_1 \times a_2}$, $b = (x_1x_2 + y_1y_2)$, and $\sin\psi = \dfrac{\phi \,.\, A_a}{\pi a_1 a_2}$.

Let streamlines on the stream surface be represented by a fourth order polynomial to satisfy the smooth entry and exit of the material flow. Any coordinate on the stream line is therefore, defined as

$$\left.\begin{aligned} x &= f_1(z) = b_1z^4 + b_2z^3 + b_3z^2 + b_4z + b_5 \\ y &= f_2(z) = c_1z^4 + c_2z^3 + c_3z^2 + c_4z + c_5 \\ z &= z \end{aligned}\right\} \tag{5.5}$$

where b_i and c_i (for $i = 1, 2, 3, 4$, and 5) are constants determined by the boundary conditions. Since the streamline does not produce any abrupt change in flow direction along the extrusion axis at entry and exit, the boundary conditions can be written as

$$\left.\begin{aligned} x &= n\sin\phi\,; \quad \left(\frac{\partial x}{\partial z}\right) = 0\,; \quad \frac{\partial^2 x}{\partial z^2} = 0 \\ y &= n\cos\phi\,; \quad \left(\frac{\partial y}{\partial z}\right) = 0\,; \quad \frac{\partial^2 y}{\partial z^2} = 0 \end{aligned}\right\} \text{at billet side } (z = 0) \tag{5.6}$$

$$\left.\begin{aligned} x &= \frac{n a_1 \sin y}{R_o}\,; \quad \left(\frac{\partial x}{\partial z}\right) = 0 \\ y &= \frac{n a_1}{R_o}\cos\psi\,; \quad \left(\frac{\partial y}{\partial z}\right) = 0 \end{aligned}\right\} \text{at product side } (z = L) \tag{5.7}$$

where R_o is the billet radius, n is the distance from the axis formed by joining the CG of the product and the billet shape (i.e., O-O′) to an arbitrary point F at the die entry, ϕ and ψ are the angles between the plane of symmetry and the stream surface at entry and exit of the die, respectively, and L is the length of the die. Defining Constant, C

$$C = \frac{A_a}{\pi . a_1 . a_2 . \cos\psi} \tag{5.8}$$

Substituting these boundary conditions from Eqs. (5.6) and (5.7) into Eq. (5.5) gives,

$$\left.\begin{aligned} x &= n\sin\phi + n\left(\phi\frac{a_1}{R_o}C - \sin\phi\right)\left[4\left(\frac{z}{L}\right)^3 - 3\left(\frac{z}{L}\right)^4\right] \\ y &= n\cos\phi + n\left(\frac{a_1}{R_o}\cos\psi - \cos\phi\right)\left[4\left(\frac{z}{L}\right)^3 - 3\left(\frac{z}{L}\right)^4\right] \\ z &= z \end{aligned}\right\} \tag{5.9}$$

Equation (5.9) can be rewritten in the following form

$$\left.\begin{aligned} x &= n\sin\phi + n\left(\phi\frac{a_1}{R_o}C - \sin\phi\right)f(z) \\ y &= n\cos\phi + n\left(\frac{a_1}{R_o}\cos\psi - \cos\phi\right)f(z) \\ z &= z \end{aligned}\right\} \tag{5.10}$$

where $f(z) = 0$ at $z = 0$ and $f(z) = 1$ at $z = L$. (5.11)

In the present analysis, function f is represented by the following fourth order curve.

$$f = f(z) = \left[4\left(\frac{z}{L}\right)^2 - 3\left(\frac{z}{L}\right)^3\right] \tag{5.12}$$

Equation (5.12) describes not only the coordinates inside the plastically deforming region but the relationship between the Cartesian and n, ϕ, z coordinate systems also. Although the present analysis employs a fourth order curve represented by Eq. (5.12) for the description of die profile and the assumed streamlines of particles, it is to be noted that function f in Eq. (5.12) can be any general function of z provided the function satisfies the given boundary conditions as given in Eq. (5.11).

The Jacobean of Eq. (5.10) can be found as

$$J = \begin{bmatrix} \frac{x}{n} & \left(y - \frac{A_a n}{\pi . R_o a_1} f \left\{\frac{\pi a_1 a_2}{A_a} \cos\psi - 1\right\}\right) & \frac{n}{L}\left(\frac{A_a \phi}{\pi R_o . a_1} - \sin\phi\right) \\ \frac{y}{n} & \left(\frac{A_a n}{\pi R_o a_1} \phi f - x\right) & \frac{n}{L}\left(\frac{a_1 \cos\psi}{R_o} - \cos\phi\right) \\ 0 & 0 & 1 \end{bmatrix} \tag{5.13}$$

Determinant of the Jacobean is therefore, written as det $J = -n . g(\phi, z)$, where

$$g(\phi, z) = \left[(1-f)^2 + \cos\left(\frac{a_1}{R_o} f\right)^2\right] + \left[(1+\cos)(1-f)\frac{a_1}{R_o} f\right] \cos\phi$$

$$+ \left[\cos(1-f)\frac{a_1}{R_o} f\right] \phi . \sin\phi$$

Assuming that the plastic zone is bounded by entry and exit shear surfaces, the velocity field components can be obtained. Because of volume continuity, the velocity component along z-direction in the Cartesian coordinate system (V_z) should be V_o at the entrance, or V_f at the exit of die. V_o and V_f are the speeds of the billet and the outgoing product, respectively. V_f can be described in terms of V_o as

$$V_f = \frac{\pi . R_o^2 \frac{\phi}{2\pi}}{\frac{1}{2} a_1 a_2 \sin\phi} V_o$$

These requirements are satisfied using Eqs (5.11) and (5.12). Therefore, other velocity components for incompressible material are determined as [1]:

$$V_x = \frac{n.\left(\frac{a_1}{R_o}\cos\phi - \sin\phi\right)f'(z)}{g(\phi, z)} V_o$$

$$V_y = \frac{n.\left(\frac{a_1}{R_o} - \cos\phi\right)f'(z)}{g(\phi, z)} V_o$$

$$V_z = \frac{1}{g(\phi, z)} V_o \tag{5.14}$$

Using the velocity fields given by the above equations, it is found that the velocity boundary conditions (Eq. 5.11) are satisfied. Different strain rate components have been found in Appendix [5]. It is found analytically that the proposed velocity and strain rate satisfy the incompressibility condition, i.e., $\dot{\epsilon} = (\dot{\epsilon}_{xx} + \dot{\epsilon}_{yy} + \dot{\epsilon}_{zz}) = 0$. Hence the proposed velocity field model fulfills the stringent requirement for the construction of a kinematically admissible condition.

5.2.6 Strain Hardening

The generalized yield stress $\bar{\sigma}$ at any point in the deformation zone is calculated using the value of generalized strain $\bar{\epsilon}$ at that point (section 5.2.2). The value of $\bar{\epsilon}$ in the deformation zone is obtained by integrating the generalized strain rate $\dot{\bar{\epsilon}}$ along the streamline. The generalized strain rate $\dot{\bar{\epsilon}}$ is obtained from the proposed strain rate field along any streamline, $dt = \frac{dz}{V_z}$, the generalized strain can be written as $\bar{\epsilon} = \int \frac{\dot{\bar{\epsilon}}}{V_z} dz$. Since the material enters the die as a rigid body, the generalized strain $\dot{\bar{\epsilon}}$ is taken as zero on the entrance boundary of the die. Integration of $\bar{\epsilon}$ is performed using ten point Gauss Quadrature rule to find the cumulative generalized strain along each streamline.

5.2.7 Estimation of Extrusion Power and Average Ram Pressure

From the proposed velocity and strain rate fields and the calculated $\bar{\sigma}$ field, the extrusion power and the average ram pressure are determined.

The total power consumed inside the die is the sum of total power consumed within different power elements. One such element shown in Fig. 5.3 is OEGG′O′E′EO. Total power consumed within the element is the sum of power losses due to plastic deformation (φ_i), the velocity discontinuities at entry (φ_e) and outlet (φ_o) of the die, and the friction power loss along the interface between the material and the die (φ_f). The predicted total power obtained through the present velocity field would be higher (if no other redundant power losses are present) than the power actually consumed. Each power in the power element is computed numerically using velocity, strain rate components, the generalized yield stress of the material and the given friction condition. The volume and surface integration are carried out numerically using ten point Gauss quadrature rule after extracting the necessary data from the geometry of die profile and the cross-section of the given component to be extruded.

Let $\bar{\epsilon}$ and $\bar{\sigma}$ be generalized yield strain and generalized yield stress for the given strain hardening material without consideration of the redundant work factor, and let σ_o be the yield stress of the given material without considering strain hardening effects. Then, various powers required for the calculation of total power (φ_T) are calculated as shown below.

(a) **Internal power of deformation :** By considering an element of volume dv in the deformation zone subjected to a stress system σ_{ij} which causes strain rates $\dot{\epsilon}_{ij}$, the incremental power of deformation ($d\varphi_i$) can be expressed as

$$d\varphi_i = \sigma'_{ij}\,\dot{\epsilon}_{ij}\,dv \tag{5.15}$$

The total power of deformation can be obtained by integrating $d\varphi_i$ over the entire volume of the deformation zone. Thus,

$$\varphi_i = \int_{\Omega} \sigma'_{ij}\,\dot{\epsilon}_{ij}\,dv = \int_{\Omega} \bar{\sigma}\,\epsilon_{ij}\,dv \tag{5.16}$$

The above equation in expanded form can be written as

$$\varphi_i = \frac{2}{\sqrt{3}}\,\bar{\sigma}\int_0^L\int_0^{\phi}\int_0^{R_o}\left[\frac{1}{2}\cdot\dot{\epsilon}_{ij}\,\dot{\epsilon}_{ij}\right]^{1/2}\left|\det J\right| n\,d\,nd\phi \tag{5.17}$$

(b) **Frictional power :** If the frictional resistance of the material along the total-work interface is τ and the slip velocity (or, the tangential velocity discontinuity) along the interface is ΔV_t, then the incremental frictional power at the interface can be expressed as

$$d\varphi_f = \tau\left|\Delta V_t\right| ds \tag{5.18}$$

where d_s is an elemental surface area. The total frictional power can be obtained by integrating the above equation along the total interface length. Thus,

$$\varphi_f = \int_s \tau\left|\Delta V_t\right| ds \tag{5.19}$$

Here s is the area of the die–workpiece interface.

In the present work, the frictional resistance on the interface is assumed to be a constant times the yield shear stress of the material ($\bar{\sigma}/\sqrt{3}$), i.e., ($m\bar{\sigma}/\sqrt{3}$), where m is the friction factor whose value is chosen on the basis of die–workpiece interface and the lubrication conditions. The friction factor varies from zero (for frictionless condition) to unity (for sticking friction condition). In the present analysis friction factor is assumed to be independent of slip.

Figure 5.4 shows the relationship between the die surface and the projected surface for frictional power calculations. A small unit square element s_1 on the die surface is selected. Let α be the maximum angle of inclination of the element (s_1) of the die surface with respect to the projected surface on the x-z plane. This angle (α) is obtained from the geometry of the die by knowing angle γ produced by the direction of a streamline (say EE′) and angle δ created by the position of this streamline. From Eq. (5.10), at the die surface,

$$E_1' = (x_1, y_1, L)\ E_2' = (x_2, y_2, L)$$

$$E_3' = (x_3, y_3, L)\ E_4' = (x_4, y_4, L)$$

$$E_5' = (x_5, y_5, L)\ E_6' = (x_6, y_6, L)$$

$$E_7' = (x_7, y_7, L)\ E_8' = (x_8, y_8, L)$$

$$(x_2 \,.\, y_2 \,.\, L) = (na_2 \sin \phi/R_o \,.\, na_2 \cos \phi/R_o\, L)$$

$$E = (R_o \sin \phi,\ na_2 \cos \phi/R_o \,.\, 0)$$

$$E_2 = (R_o \sin \phi,\ R_o \cos \phi,\ 0)$$

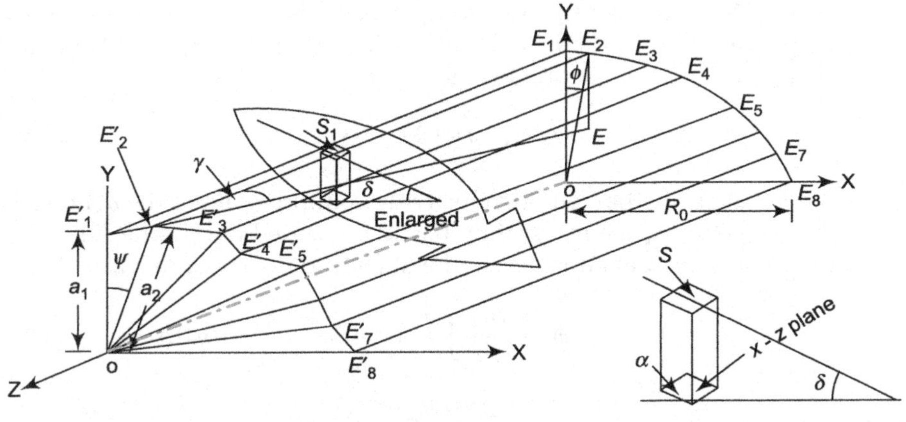

Fig. 5.4 Die surface and projected surface for frictional power calculations. With permission from [3] publisher

$$\frac{dy}{dz} = R_o\left(\frac{a_1}{R_o}\cos\psi - \cos\phi\right) f';\quad \frac{dy}{dz} = \tan\gamma \tag{5.20}$$

$$\frac{dy}{d\phi} = -R_o \sin\phi + R_o \sin\phi f \tag{5.21}$$

$$\frac{dx}{d\phi} = R_o \cos\phi + R_o\left(C\frac{a_1}{R_o} - \cos\phi\right) f \tag{5.22}$$

Therefore,

$$\frac{dy}{dx} = \frac{(dy/d\phi)}{(dx/d\phi)};\quad \frac{dy}{dx} = \tan\delta \tag{5.23}$$

The length of diagonal (*ll*) of the small unit square element on the die surface is found as $ll = \sqrt{(\cos\delta)^2 + (\cos\gamma)^2}$. Angles γ and δ can be found from Eqs (5.20) and (5.23). Therefore, angle α, which is subtended by *ll* with the *x*-*y* plane is found as

$$\cos\alpha = \frac{ll}{\sqrt{2}} \tag{5.24}$$

Based on Eq. (5.19), frictional power for the considered element is obtained from the above as

$$\cos\alpha = \frac{ll}{\sqrt{2}}$$

$$\varphi_f = \frac{m\sigma_o}{\sqrt{3}} \int_0^L \int_0^\phi \left[V_x^2 + V_y^2 + V_z^2\right]^{1/2}_{n=R_o} \frac{1}{\cos\alpha} \frac{\partial(x, z)}{\partial(\phi, z)} \, d\phi \, dz \tag{5.25}$$

(c) **Shear power:** In case the die profile has non-zero slope at the entry and exit sections, there will be shear losses due to velocity discontinuity. These planes of discontinuity can be called as the planes of sudden shear. The power dissipated along these surfaces can be calculated by employing the yield shear stress as the material resistance. Thus, the shear power can be expressed as

$$\varphi_s = \int_s \frac{\bar{\sigma}}{\sqrt{3}} |\Delta V_t| \, ds \tag{5.26}$$

where ΔV_t is the tangential velocity discontinuity along the d_s and s is the area over which integration has to be carried out. Using Eq. (5.26) the shear power at entry (φ_e) and exit (φ_o) side are obtained as:

$$\varphi_e = \frac{\sigma_o}{\sqrt{3}} \int_0^{\phi_m} \int_0^R \left[V_x^2 + V_y^2\right]^{1/2}_{z=0} n \, dn \, d\phi \tag{5.27}$$

$$\varphi_o = \frac{\sigma_o}{\sqrt{3}} \int_0^{\phi_m} \int_0^R \left[V_x^2 + V_y^2\right]^{1/2}_{z=L} nC \frac{a_1^2}{R_o^2} \, dn \, d\phi \tag{5.28}$$

(d) **Power due to interface shear**

$$\varphi_f = \int_0^L \int_0^\phi \frac{m\sigma_o}{\sqrt{3}} |v_t| \frac{1}{\cos\alpha} \frac{\partial(x, z)}{\partial(\phi, z)} \, d\phi \, dz \text{ where } v_t = \left[v_x^2 + v_y^2 + v_z^2\right]^{\frac{1}{2}}, \text{ when}$$

$$n = R_o \text{ and } \varphi_{\text{int}} = \frac{\sigma_o}{\sqrt{3}} \Delta v_i A_i$$

Here A_i is the area of the ith interface formed by the coordinate points (0, 0, *L*), (x_i, y_i, *L*), (R_o sin 0°, R_o cos 0°, 0) and (0, 0, 0). The velocity discontinuity (Δv_i) developed at the ith interface due to relative velocity of ith and $(i–1)^{\text{th}}$ elements is evaluated as $\Delta v_i = (v_{fi} - v_{f-1})$ where v_{fi} and v_{fi-1} are average tangential velocity of ith and $(i–1)^{\text{th}}$ elements. For i^{th}

element, v_{ft} is evaluated as $v_{ft} = \frac{\left[v_o + (v_f)_i\right]}{2}$. Therefore, for n power elements, the total interface power is ($\varphi_{\text{int-tot}}$) is $\varphi_{\text{int-tot}} = \sum_{i=1}^{n} \frac{\sigma_o}{\sqrt{3}} \Delta v_i \, A_i$.

(e) **Total power and average extrusion ram pressure :** The total power (φ_T) is the sum of its constituents including the redundant powers. Therefore, the total power (φ_T) is given as

$$\varphi_T = \varphi_i + \varphi_e + \varphi_o + \varphi_f + \varphi_{\text{int}} \tag{5.29}$$

Once the total power (φ_T) consumed during extrusion is obtained, the upper limit to the average ram pressure (P_{ave}) for extrusion is found as

$$P_{\text{ave}} = \frac{\varphi_T}{\pi R_o^2 V_o} \tag{5.30}$$

The average ram pressure (P_{ave}) can be converted to relative forming stress (R_s) as

$$R_s = \frac{P_{\text{ave}}}{\sigma_o} \tag{5.31}$$

where σ_o is the effective stress at zero hardening condition.

5.2.8 Upper Bound Anlaysis Applications

The above analysis is generalized in nature and can be used for any shape (re-entry and non re-entry) extrusion for calculation of load/power requirement and can be used for design of extrusion dies and processes under forward extrusion. These are explained below.

5.2.8.1 Application to non re-entry component shapes

Application of the proposed upper bound model is demonstrated for a general shape given in Fig. 5.4. The following steps are carried out for any given shape before going for the proposed upper bound solutions of power and other process parameters.

1. The Geometry of the component to be extruded is drawn using AutoCAD, Release-12. DXF file is created and processed to separate the data of line and arc entities [3].
2. The center of gravity (x_c, y_c) and cross-sectional area (A_a) of the given component is found using Advanced Modelling Extension (AME) of AutoCAD, Release-12.
3. Each arc and line entity is further divided into smaller lines to decide the total number of minimum triangular power elements, using the pseudo vertex concept. The total number of power elements corresponding to each line and arc entities are decided judicially such that the average error in incompressibility condition required in the upper bound

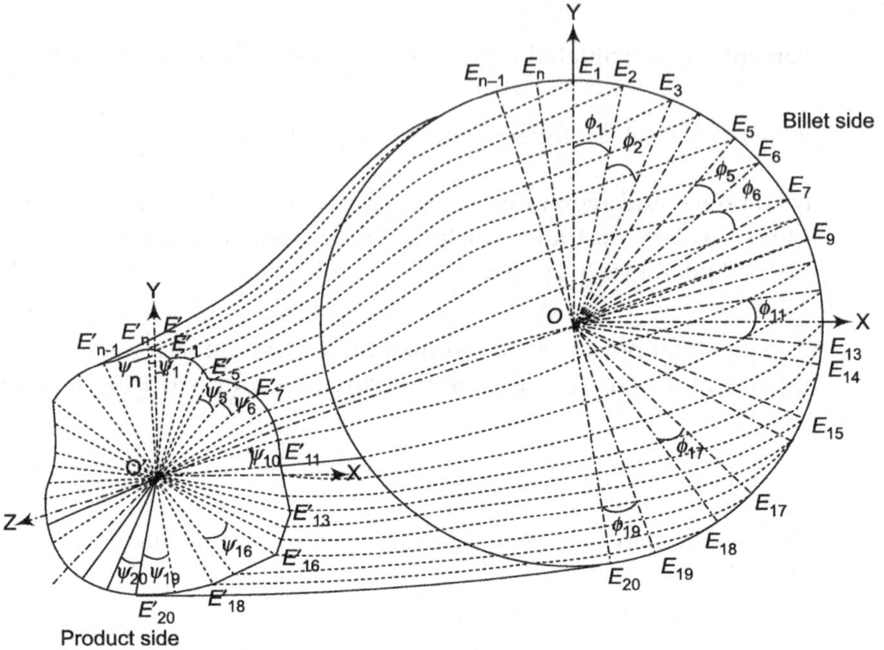

Fig. 5.5 Construction of power element for a general non re-entry shape. With permission from [3] publisher

solution does not exceed one percent for each power element and thus incompressibility condition ($\dot{\epsilon} = 0$) during extrusion is satisfied.

4. From step 3, coordinates of each triangular power elements are obtained. For the example given in Fig. 5.3, the coordinate points of power element one (1) are O, E_1, E_2 (not required to be known) on the billet side and O', E_1', E_2' on the extruded product side. Since coordinates of O', E_1', E_2' are now known from steps 2 and 3, angle ψ_1 for the power element is calculated using the relationships obtained from Eq. (5.4). Since percentage reduction (*red*) per pass is given, the billet radius R_o is calculated as $R_o = \sqrt{\dfrac{A_a}{\pi\,(1 - 0.01\,\text{red})}}$.

5. From the assumption that the extrusion ratio in each power element is maintained constant, i.e.,

$$\left(\frac{\text{Area of } \Delta O'E_1'E_2'}{A_a}\right) = \left(\frac{\text{Area of } \Delta OE_1E_2}{\pi R_o^2}\right) \quad \text{for power element 1,}$$

$$\left(\frac{\text{Area of } \Delta O'E_2'E_3'}{A_a}\right) = \left(\frac{\text{Area of } \Delta OE_2E_3}{\pi R_o^2}\right) \quad \text{for power element 2,}$$

. . . .

$$\left(\frac{\text{Area of } \Delta O'E'_n E'_1}{A_a}\right) = \left(\frac{\text{Area of } \Delta OE_n E_1}{\pi R_o^2}\right) \quad \text{for power element } n'.$$

6. The corresponding angle (ϕ_1) on billet side is also found using Eq. (5.4). Remaining ($n' - 1$) included angles, corresponding to various power elements on the billet side (i.e., ϕ_1, ϕ_2, ϕ_3, ϕ_n) and product side (ψ_1, ψ_2, ψ_3, ... ψ_n) can be found in the similar manner.
7. It is found numerically from step 6 that sum of angles (ψ) of all the power elements is very close to 2π radians (i.e., $\psi_1 + \psi_2 + \psi_3 + \ldots\ldots\ldots + \psi_n \cong 2\pi$). Then, $\phi + \phi_2 + \phi_3 + \ldots + \phi_n \cong 2\pi$ and average incompressibility, $\dot{\epsilon}_{ave} = \dfrac{\dot{\epsilon}_1 + \dot{\epsilon}_2 + \dot{\epsilon}_3 + \ldots\ldots\ldots\ldots + \dot{\epsilon}_{n'}}{n'} \cong \pm 0.02$ is also found to be within the tolerance limits. Here, $\dot{\epsilon}_1, \dot{\epsilon}_2, \dot{\epsilon}_3, \ldots \dot{\epsilon}_{n'}$, are individual incompressibility in various power elements within the deforming material.
8. The domain is divided into n' number of triangular power elements using step 7. For each power element (1 to n') the total power consumed (φ_T) can be found. Let φ_1, φ_2, φ_3, ... φ_n be the total power consumed in 1, 2, 3 ... n'^{th} power elements, then the overall power (φ_o) consumed in the extruding the given shape through the given fourth order die profile is found as the sum of individual powers in different power elements, i.e., $\varphi_o = \varphi_1 + \varphi_2 + \varphi_3 + \ldots\ldots\ldots + \varphi_{n'}$.
9. The minimum power ($\varphi_{\min}$) to φ_o is at an optimum die length that can be found using any single variable optimization technique. In this work Golden Section Unimodel Optimization Algorithm. Once the optimum power is found, the minimum average ram pressure (P_{ave}) and relative stress (R_s) on the die can be found easily.

5.2.8.2 *Application to re-entry components shapes*

Application of the proposed upper bound model is demonstrated for a general re-entry shape shown in Fig. 5.6. The following steps are carried out for the given re-entry to find the process parameters such as optimum power, average extrusion pressure and optimum die length.

1. The geometry of the re-entry component to be extruded is drawn using Auto CAD. The drawn component is then broken into minimum number of sub-components such that each of them represents a non re-entry component shape. The selection of entities to change the component drawing for this purpose should be made in such a way that each proposed non re-entry shape is separate and unambiguous. The designer should try to propose those non re-entry shapes for the re-entry shape that could form exactly two or more similar non re-entry shapes. Extra entities that are required to create the proposed non re-entry shapes should be recorded carefully. These entities are extra and do not require

frictional power calculation. Care should also be taken that frictional power is calculated only once for entities of the re-entry shape.

2. The center of gravity (x_c, y_c) and the cross-sectional area (A_a) for each proposed non re-entry shapes is found using Advanced Modelling Extension (AME) of AutoCAD.
3. Separate DXF file is created for each proposed non re-entry shape as obtained from step 1. Each of them is processed separately to extract line and arc entity data.

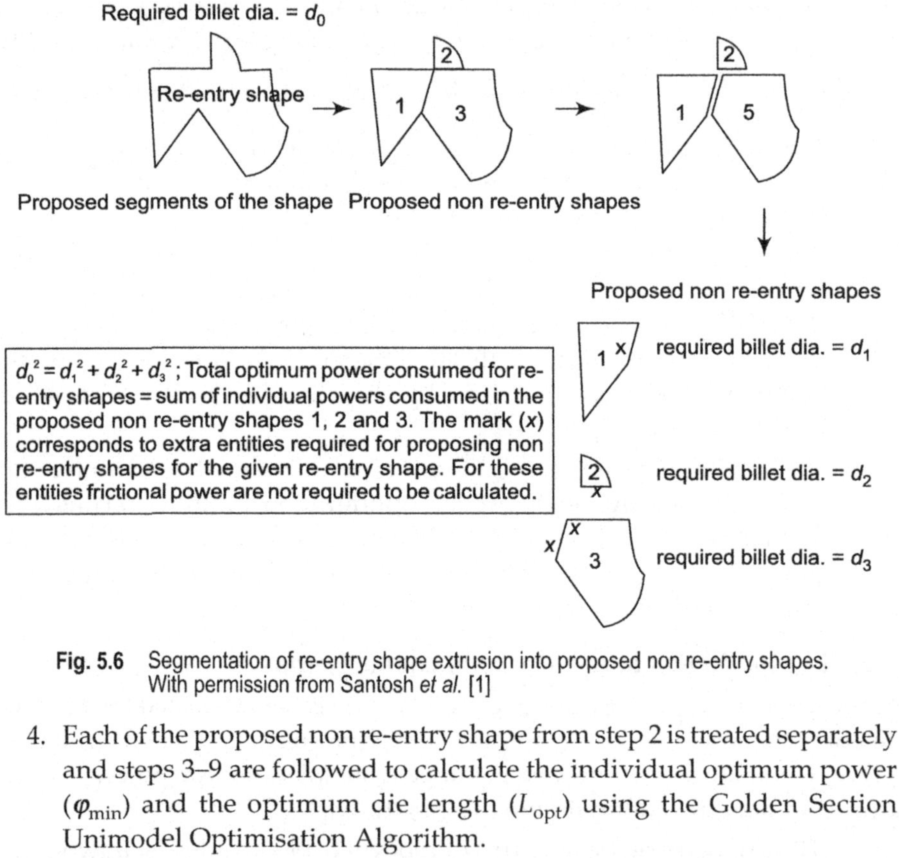

Fig. 5.6 Segmentation of re-entry shape extrusion into proposed non re-entry shapes. With permission from Santosh *et al.* [1]

4. Each of the proposed non re-entry shape from step 2 is treated separately and steps 3–9 are followed to calculate the individual optimum power (φ_{min}) and the optimum die length (L_{opt}) using the Golden Section Unimodel Optimisation Algorithm.
5. Let there be $\bar{n}$ number of non re-entry shapes proposed for a given re-entry shape, where $\bar{n} = 1, 2 \ldots \bar{n}$. CG and cross sectional area (A_a) for each proposed non re-entry shapes is known from step 2. For the required reduction (% red) separate billet for each of the proposed non re-entry shape could be obtained. The diameter of $\bar{n}^{\text{th}}$ billet is found as $d\bar{n} = \sqrt{\dfrac{(4/\pi)\, A_a^n}{(1 - 0.01\ \text{red})}}$, where $d\bar{n}$ and A_a^n are required billet diameter and the cross-sectional area of $\bar{n}^{th}$ proposed non re-entry shape. Let $\varphi_{\text{opt}}^{\bar{n}}$

and $L^{\bar{n}}_{opt}$ be the optimal power and optimal die length for $\bar{n}^{th}$ proposed non re-entry shape. With increase in friction conditions, optimal die length ($L^{\bar{n}}_{opt}$) reduces, whereas the corresponding optimal power ($\varphi^{\bar{n}}_{opt}$) increases.

The minimum overall power φ_o and the minimum billet diameter d_o for the re-entry shape (step 1) can be found as $\varphi_o = \varphi^1_{opt} + \varphi^2_{opt} + \varphi^3_{opt} + \ldots + \varphi^{\bar{n}}_{opt}$,

$$d_o = \sqrt{(d_1^2 + d_2^2 + d_3^2 + \ldots + d_{\bar{n}}^2)}$$

The average ram pressure $P_{ave} = \dfrac{\varphi_o}{\pi . R_o^2 . V_o}$ can be converted to relative extrusion pressure (R_s) as $R_s = \dfrac{P_{ave}}{\sigma_o}$.

6. The overall optimal die length (L_{opt}) for the re-entry shape is decided on the basis of power (φ_o) using Golden Section Optimisation technique for the proposed non re-entry shapes of the re-entry shape. The relative optimum die length is evaluated as $R_L = (L_{opt}/R_o)$. If all the proposed non re-entry shapes are of symmetrical in nature (similar profile and cross sectional area), individual die length (L) is same for all the proposed non re-entry shapes and is equal to L_{opt}. The values are different when the proposed non re-entry shapes are non-symmetrical in nature. For a non-symmetrical re-entry component, individual value of L for each non re-entry shape is taken as $L = (2\, L_{opt}/d_o)$. This kind of assignment of optimal die length for the proposed non-symmetrical non re-entry shape simplifies the solution without much loss in accuracy of the result. In case of symmetrical shapes, however, exact solution is obtained.

The above procedure for re-entry shape using the proposed upper bound solution methodology is general in nature and can be applied for any re-entry shape. This procedure, however, requires intelligent user involvement as different non re-entry shapes for a given re-entry shape can be proposed.

5.2.9 Examples

To test the validity of the proposed upper bound model for cold extrusions of re-entry and non re-entry shapes, several examples are selected and the results are compared with the experimental results obtained by earlier workers. These examples are explained below.

Example 1 (non re-entry)

Yang *et al.* [12] conducted experiments for axisymmetric extrusion of circular sections, details of material specifications and other process parameters taken from their paper are given below.

Materials: AISI 4041 Steel and Al 2024 (annealed).

Stress–strain relationships for these materials are

$$\bar{\sigma} = 418.9 + 699.0\,\bar{\epsilon}^{0.266} \quad \text{(MPa)} \quad \text{for AISI 4140 Steel}$$

$$\bar{\sigma} = 118.7 + 229.4\,\bar{\epsilon}^{0.247} \quad \text{(MPa)} \quad \text{for Al 2024 Aluminium alloy}$$

Billet radius (R_o) = 12.5 mm, Friction condition (m) = 0.12

Die profile: A fourth order die profile is used to validate the proposed model.

Let, ratio $\frac{L}{R_o}$ (initial die length to the billet radius be equal to 2.0, ram speed = 1.0 mm/s, temperature be the room temperature (20 °C).

Geometrical Construction of Shape

The final component geometry is circular with product radius R_p corresponding to different reductions are given in Table 5.1. These shapes are drawn accurately using AutoCAD. For the drawn shape, DXF file is created and preprocessed as given in the feature recognition methodology [1]. Minimum number of pseudo vertices (N_s) corresponding to each arc entity. The power elements found on this basis is minimum in number. This ensures that average error in incompressibility ($\dot{\epsilon}_{ave}$) condition does not exceed by ± 2 %, and total error in angles on billet side (ϕ) and on product side (Ψ) does not exceed by ± 1°. Minimum number of power elements generated on this basis are used for overall power (φ_o) calculations.

Results obtained at different reduction ratio (Table 5.1) are compared in Fig. 5.7 with the experimental results obtained by Yang *et al.* [12]. It can be seen from Fig. 5.7 that the results obtained from the proposed model are in very close agreement with the experimental results.

Table 5.1 Product radius (R_p) at different reductions

Billet radius (R_o) (mm)	Reductions (% red)	Output radius (R_p) (mm)	Area (A_a) (mm^2)
12.50	40	9.682	294.36
12.50	50	8.84	245.49
12.50	60	7.9056	196.56
12.50	70	6.846	147.42
12.50	80	5.5902	98.164

With permission from Santosh *et al.* [3] publisher.

Example 2 (non re-entry)

Yang *et al.* [7] have conducted experiments for non-axisymmetric products like rounded square Fig. 5.8 (a)], ellipse square [Fig. 5.8 (b)], clover [Fig. 5.8 (c)] and trocoidal gear [Fig. 5.8 (d)] sections. The details of material specifications and other process parameters of the experiment taken from Yang *et al.* [7] are given below.

Materials: Al2024 with stress–strain relationship as $\bar{\sigma} = 50.0 + 300.7\,\bar{\epsilon}^{0.144}$ (MPa).

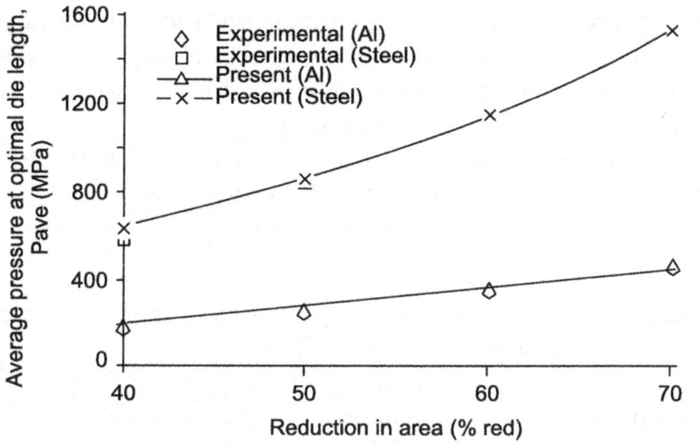

Fig. 5.7 Comparison of results with Yang *et al.* [12].
With permission from Yang *et al.* [12] and Santosh *et al.* [3] publishers

Billet radius (R_o) = 14.75 mm, die profile: Yang used a fourth order die profile. Equation (5.12) is taken as the die profile to validate the proposed upper bound model for the given non re-entry shapes.

Ratio $\frac{L}{R_o} = 2.0$, Ram speed = 1.0 mm/s and temperature is the room temperature (20 °C).

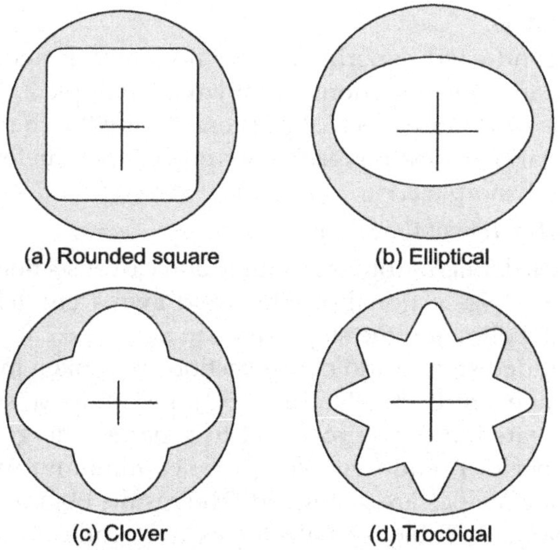

Fig. 5.8 Examples for validation of the model (non-re-entry).
With permission from Santosh *et al.* [3] publisher

Geometrical Construction of the Shape

The final geometrical details of rounded square and ellipse are taken from Yang *et al.* [8], and clover and trocoidal gear are taken from Yang *et al.* [7]. Their cross-sectional profiles are drawn accurately on true scale using AutoCAD. For these shapes, a separate DXF file is created and preprocessed to separate the geometrical data. Minimum number of pseudo vertices (N_s), i.e., minimum power elements for ellipse and square are generated, insuring that the average error in incompressibility not to exceed by ± 2% and total error in angles on billet (ϕ) and product side (Ψ) do not exceed by ± 1°. Minimum number of power elements generated on this basis are used for overall power calculations.

Table 5.2 Comparison of extrusion pressure with proposed model and experiment of Yang et al. [7]

S. No.	Section shape	% red	(γ^*)	Minimum number of power elements	Extrusion pressure (P_{ave}) at optimal, MPa Experimental	Proposed	Difference with experiment
1	Rounded square	60	1.0	210	317.3	376.63	+ 18 %
2	Ellipse	60	1.5	142	306.7	319.64	+ 4 %
3	Clover	60	1.5	288	349.9	398.70	+ 13 .9 %
4	Trocoidal gear	45.15	1.5	368	317.92	344.89	+ 8.0 %

The shape factor as proposed by Yang et al. [7] With Permission from Santosh et al. [3] publisher.

Experimental results obtained by Yang *et al.* [7] at various reductions are compared in Table 5.2. It can be seen from the table that the results obtained using the proposed model are in close agreement with the experiments.

Example 3 (re-entry)

Yang *et al.* [7] conducted experiments for non-axisymmetric extrusion of rounded square and clover sections as shown in Example 2. These examples are solved by the procedure (Sec. 5.2.8.2) considering the shape as a re-entry. Details of material specifications and other process parameters are taken the same as given in Example 2.

Geometrical Construction of the Shape

Initial geometrical details of rounded square and clover section are taken from Example 2. Their cross-sectional profiles are drawn accurately on true scale using AutoCAD. Following the steps given in Sec. 5.2.8.2 for re-entry shape procedures, rounded square and clover section are broken into two and five non re-entry shapes, respectively (Table 5.3). For all these shapes, separate DXF files are created and preprocessed to separate the geometrical data. Minimum number of pseudo vertices (N_s), i.e., minimum power elements for individual sub-shapes are generated. The results obtained for these two cases are compared in Table 5.4 with the experimental result of Yang *et al.* [12]. The results clearly show that Example 4 (Re-entry) extending the non re-entry solution methodology for re-entry shapes does not compromise on accuracy and there is no difference in splitting the re-entry shapes into two or more non re-entry shapes. Results obtained clearly show that the same results are obtained for all symmetric re-entry shapes. For non symmetric re-entry shapes, however, accuracy will get affected.

Kiuchi *et al.* [13] have demonstrated an analytical method to calculate power, extrusion pressure at optimum die length for extrusion of various kind of re entry (non-symmetric) cross-sections such as square, rectangle, hexagon, L, T, and H-sections, etc. Various results and trends obtained by him have been claimed to be of acceptable range. However, these results have not been validated with experiments. A re-entry shape extrusion of H-section (Fig. 5.9) is chosen to validate the trend obtained by Kiuchi *et al.* [13]. This shape is taken to study the relationships between relative die length and relative stress using a conical die. Details of material specifications and other process parameters for the study are given below. Since the billet material used by Kiuchi *et al.* [13] is not specified, AISI 4140 steel has been taken as the material. Because of this only trend in the result can be compared.

Table 5.3 Examples to validate re-entry shapes of extrusion

Sl. No.	Non re-entry shape	Proposed and re-entry shapes	Cross-sectional area, mm^2 of the shape (Required billet diameter, mm) and [% red] in brackets	Type (symmetric or not)
1		 1 - One no. 2 - One no.	273.4 (14.75) [60]	Symmetric
2		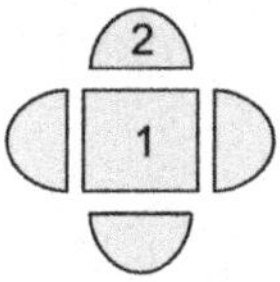 1 - One nos. 2 - Four nos.	273.397 (14.75) [60]	Symmetric

With permission from Santosh *et al.* [3] publisher.

Table 5.4 Proposed results from the model for re-entry shapes

Sl. No.	Non re-entry shape	Total no. of power elements	Proposed optimum		Proposed optimum die length, mm
			Extrusion power (watts)	Ram pressure (MPa)	
1		2 × 236 = 472	118.4 × 2 = 236.8	346.455	29.55
2		4 × 40 + 42 = 202	43.196 × 4 + 113.045 = 285.83	418.2	29.55

With permission from [3] publisher.

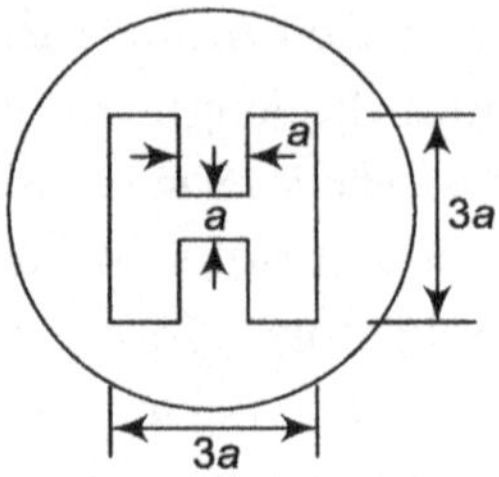

Fig. 5.9 An example (H-section).
With permission from Santosh *et al.* [3] publisher

Materials: Steel AISI 4140; $\bar{\sigma} = 418.9 + 699.0\ \bar{\epsilon}^{0.266}$ (MPa) for

Billet radius (R_o) = 40.8 mm, Die profile: Kiuchi *et al.* [13] used a conical die profile. Equation (5.12) is used as the die profile to validate the proposed upper bound model for the re-entry shape.

Ram speed = 1.0 mm/s, Temperature: Room temperature (20°C).

Geometrical Construction of the Shape

Initial geometrical details of H-section (Fig. 5.10) are taken from Kiuchi *et al.* [13]. Their cross-sectional profiles are drawn accurately on the true scale using AutoCAD. Following the steps given in Sec. 5.2.8.2 for re-entry shape procedures, the shape is broken into two sub-shapes (Fig. 5.10). For these shapes, separate DXF files are created and preprocessed. Minimum number of pseudo vertices (N_s), i.e., minimum power elements for individual sub-shapes are generated. Minimum number of power elements generated are used for overall power calculations.

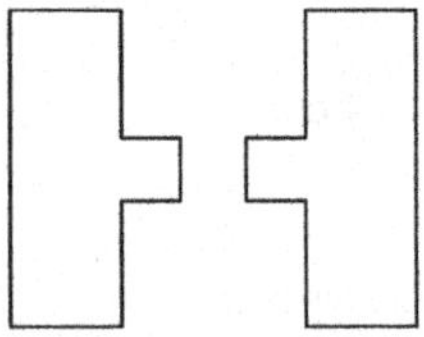

Fig. 5.10 Segmentation for example Fig. 5.9.
With permission from Santosh *et al.* [3] publisher

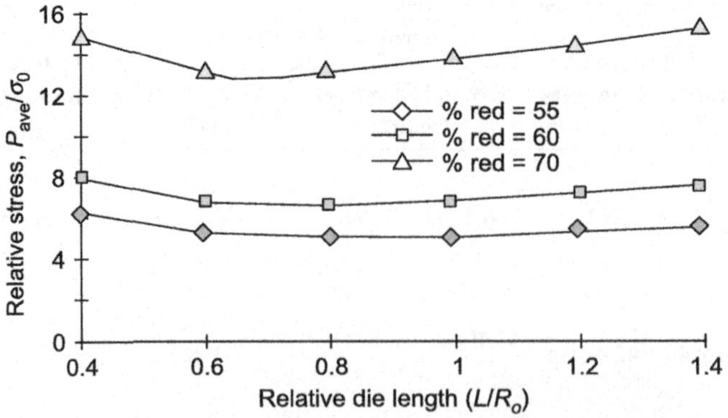

%red	a (mm)
55	17.98
60	16.94
70	14.68

Fig. 5.11 Variation of relative die length vs. relative stress in extrusion from round billet to H-type bar.
With permission from Santosh *et al.* [3] publisher

Results obtained at different reductions are compared in Fig. 5.11 with the analytical results obtained by Kiuchi *et al.* [13]. It can be seen that the trend obtained by the proposed methodology for re-entry shapes is similar to that obtained by Kiuchi *et al.* [13].

Example 5 (re-entry)

Table 5.5 lists eight commonly used re-entry shapes along with the details of their cross-sectional area and the initial billet diameter required for their extrusion. These shapes are chosen to demonstrate the result obtained by the proposed upper bound solution methodology for re-entry shapes. Geometrical details of the shapes have been taken from various sources. Details of material specifications and other process parameters for the analytical experiment are given below.

Materials: Steel, AISI 4041; $\sigma_o = 418.7 + 699\ \dot{\in}^{0.266}$ MPa

Billet radius (R_o) = 40 mm, Die profile : third order die profile given by Eq. (5.12).

Ram speed (V_o) = 1.0 mm/s, Friction condition: $m = 0.1$, Temperature: Room temperature (20°C).

5.3 UPPER BOUND ANALYSIS FOR HOLLOW EXTRUSION

To calculate the extrusion power in cold case, the upper bound model presented in the previous sections can be extended for hollow extrusion analysis. Any geometry, which is made by lines, polyline, arcs, circle, or ellipse, is the assembly of points. So geometry is divided into a number of points for geometry recognition and further processing of point co-ordinates of the geometry. A Bridge Die has been designed and manufactured for extrusion of commercially available Lead alloy (30% Sn + 70% Pb) as a working material for a circular profile. Results obtained by the experiments compared to theoretical results [7] shows a good agreement. Various parameters such as product profile (product shape factor) extrusion die and mandrel shapes, area reduction, friction at die, billet and mandrel interface, ram velocity, die length, etc., affect the extrusion power and average ram pressure have been taken for parametric study. A comparative study has been done by choosing two profiles (one is circular tube, second is a complex trocoidal shape tube with clover shape inner hole).

Area Division for Equal Velocity of Extruded Product

The data obtain in DATA.O and DATA.I is the large number of x and y coordinate of line at start and end position which after drawing in a continuous manner makes the actual product OD and ID profile. One needs the further division of these data using the following process:

(a) During initial division of geometry it is not necessary that points obtained for both OD and ID are started from same position.

(b) It is not necessary that data written in DATA.O is equal to data in DATA.I. It is also possible that both coordinates will not coincide at the line passing through the C.G. of final product shape and outer profile point.

(c) For the calculation of power, we divide whole geometry in large number of small elements having equal area of each small entity in a pseudo points on inner and outer periphery of product. So that all elements have same velocity during extrusion.

Generation of Equal Number of Data on Inner and Outer Profile

This is a necessary step for the generation of data file, which gives coordinates of elements in equal area. First of all C.G. of each profile is calculated by using equation

$$\bar{x} = \frac{x_1 l_1 + x_2 l_2 + x_3 l_3 + x_4 l_4 + \ldots + x_n l_n}{(l_1 + l_2 + l_3 + l_4 + \ldots + l_n)}$$

$$\bar{y} = \frac{y_1 l_1 + y_2 l_2 + y_3 l_3 + y_4 l_4 + \ldots + y_n l_n}{(l_1 + l_2 + l_3 + l_4 + \ldots + l_n)} \tag{5.32}$$

Due to above mentioned reasons; we need further division of area by considering points of OD and ID together.

$$\bar{x} = \frac{\sum_{i=1}^{n} x_i l_i}{\sum_{i=1}^{n} l_i}$$

or

$$\bar{y} = \frac{\sum_{i=1}^{n} y_i l_i}{\sum_{i=1}^{n} l_i} \tag{5.33}$$

where x_i and y_i is C.G. of small line for $i = 1$ to n, $\bar{x}$ and $\bar{y}$ are C.G. of product geometry, l_i is length of each line. By subtracting C.G. from each coordinate a new data file is generated with respect to C.G. of profile. Now the data are arranged in increasing tangent with respect to C.G. of each profile by using relation

$$\theta_l = \tan^{-1}\left(\frac{(y - \bar{y})}{(x - \bar{x})}\right) \tag{5.34}$$

Now take new arranged data file of both outer and inner profile together, and divide each profile by generating pseudo vertex in such a way that each profile has equal number of data, which represents points on the periphery of the profile. In addition, each vertex of both profiles coincide to the line joining C.G. of product and point at outer profile, i.e., each new generated point on each profile has same tangent. Tangent of line made by two points on periphery of any profile can be calculated by using relation and tangent of new vertex with respect to C.G. can be given by Eq. (5.34). So new point generated on either of profile will have to satisfy both Eqs (5.34) and (5.35) simultaneously. Figure 5.13 shows the generation of equal number of data on inner and outer profile by creating pseudo points using Eqs. (5.34) and (5.35).

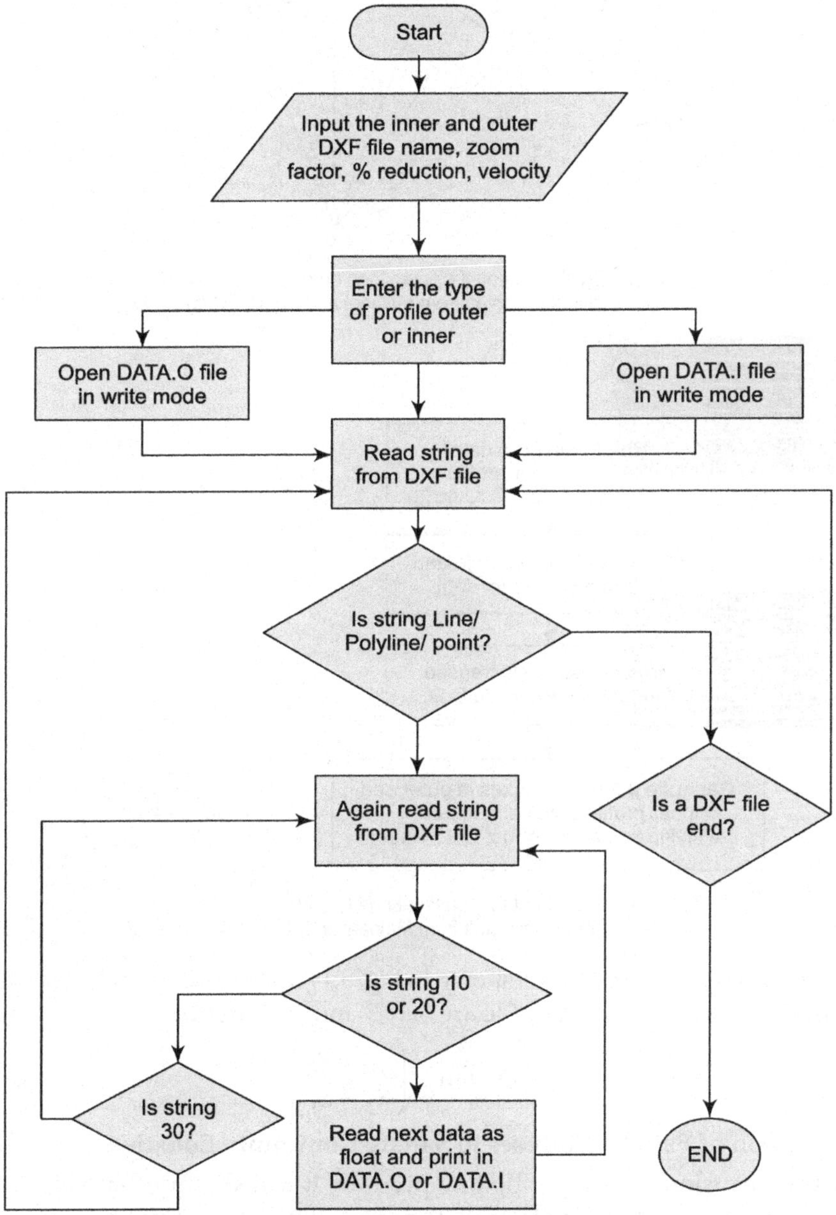

Fig. 5.12 Flow chart for reading DXF file.
With permission from Malpani and Kumar [4] publisher

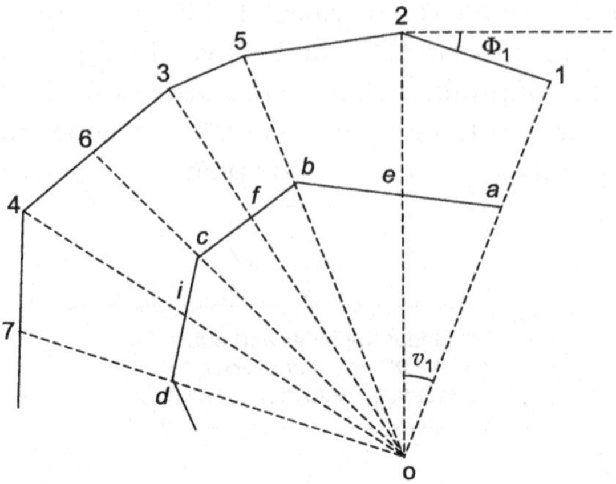

Fig. 5.13 Division of equal elements on outer and inner profile.
With permission from Malpani and Kumar [4] publisher

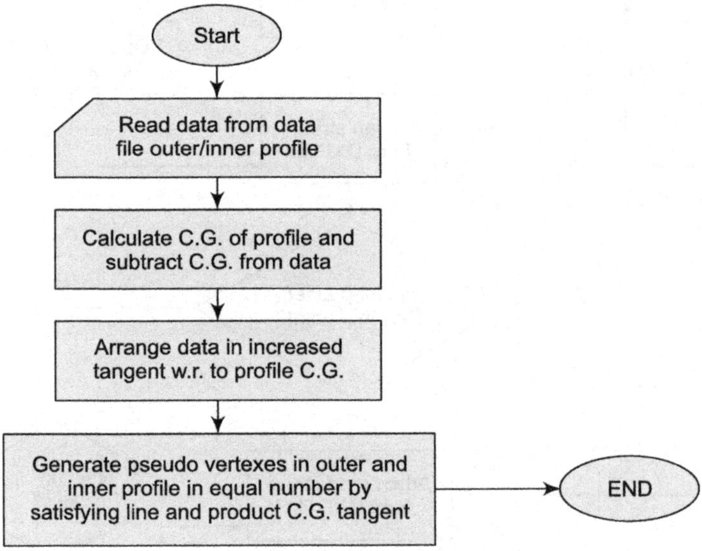

Fig. 5.14 Flow chart for equal data generation.
With permission from Malpani and Kumar [4] publisher

The flow chart of the process is shown in Fig. 5.14. In Fig. 5.13, newly generated pseudo vertexes on outer profile are 5, 6, 7 and on inner profile are e, f, g.

$$\phi_l = \tan^{-1}\left(\frac{y_2 - y_1}{x_2 - x_1}\right) \tag{5.35}$$

Generation of Pseudo Vertexes to Satisfy Continuity Equation

For the extrusion of straight hollow product, it is necessary that velocity of each extruded small quadrilateral element will be equal. This means area of each element is selected in such a way that it is equal so that by using law of continuity velocity of each element is also equal as shown by Fig. 5.15. Thus, if each small element is extruded with equal velocity, than there is no chance of twisting or bending of product. Figure 5.16 shows how data are further divided to get equal pseudo elemental extruded area. These are final data of vertexes of outer and inner profile used for calculating power in extrusion. Finally the whole procedure of data preprocessing and data generation shown in Fig. 5.16 can be summarized in following steps.

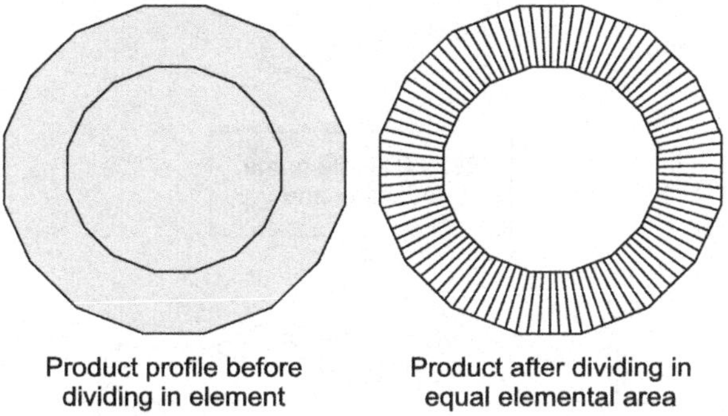

Fig. 5.15 Product profile before and after division in equal area elements. With permission from Malpani and Kumar [4] publisher

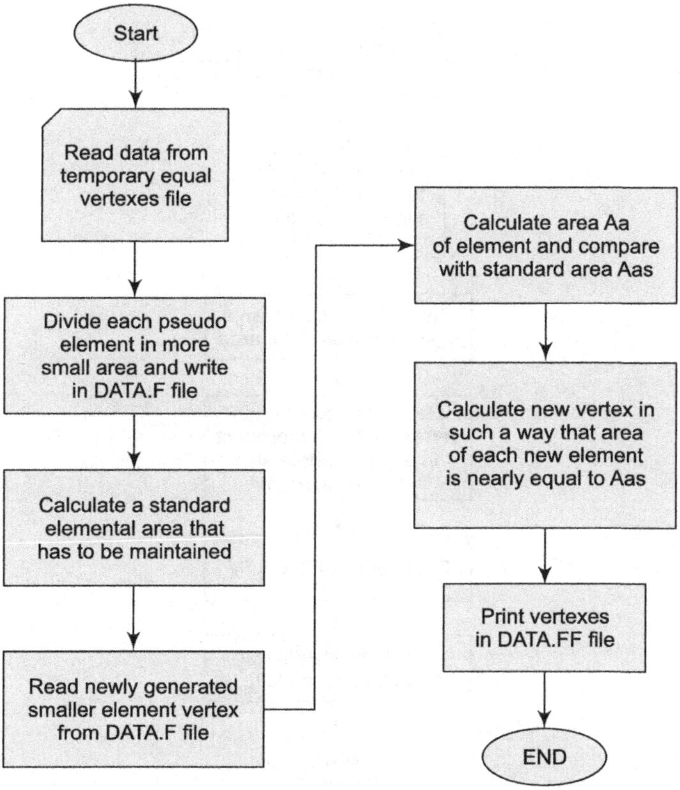

Fig. 5.16 Flow chart for generation of pseudo vertexes satisfying continuity equation. With permission from Malpani and Kumar [4] publisher

Step (1) Obtain the dimensions of the geometry of outer and inner profile of product from user and draw the profile in Auto CAD using suitable entities. Care should be taken that the inner profile area should not be greater than outer profile area.

Step (2) Divide the both profile in more then 100 point entities. Delete original geometry and save these points geometry of both profiles in separate

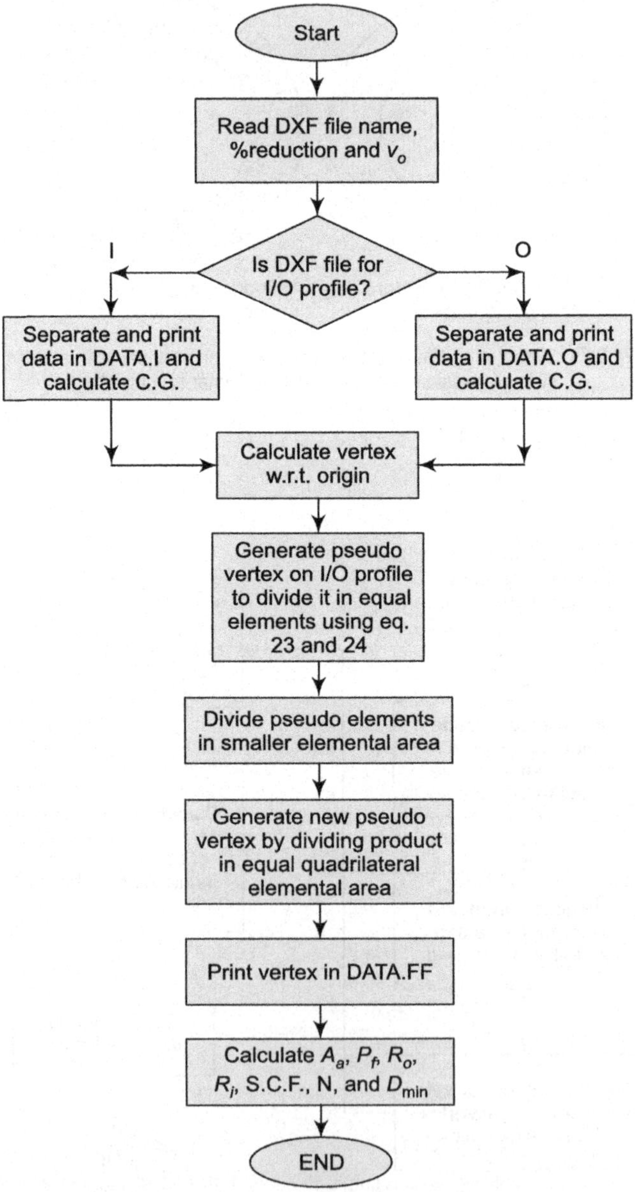

Fig. 5.17 Flow chart showing complete plans for data preprocessing and data generation. With permission from Malpani and Kumar [4] publisher

files in Auto Cad R12 DXF format. For higher accuracy and good results, we have to divide geometry in larger number of points.

Step (3) Read the coordinates from Entity Section of DXF file generate in step 2 for both profiles. Calculate their C.G. and subtract it from vertexes to get coordinates of each profile with respect to origin.

Step (4) Multiply coordinates to the zoom factor to get vertexes of original product. Divide the outer and inner profile in equal number of elementals by using the concept of pseudo coordinate generation.

Step (5) Use these pseudo vertexes and again divide the whole geometry in equal elemental area of hollow product by comparing it with required small elemental area and generating pseudo vertexes by satisfying Eqs (5.34) and (5.35). Save the data in DATA.FF file.

Step (6) Reopen the DATA.FF file and calculate product area, outer profile periphery, minimum inscribing circle diameter, shape complexity factor and total number of elements generated.

Step (7) Get the percentage reduction and velocity of extruded billet to get final product shape by user and calculate billet and mandrel radius.

Step (8) Print the data calculated in step 6 and step 7 in file ARPERI.F. This completes the process for generating coordinate having equal elemental area. These file are further used to calculate the power consumed during extrusion.

Upper bound model

The model is similar to the previous section. Figure 5.18 is taken to establish the model. The domain of deformation Ω is the plastic deformation zone, τ is the shear stress on velocity discontinuity surfaces S_i (Fig. 5.18).

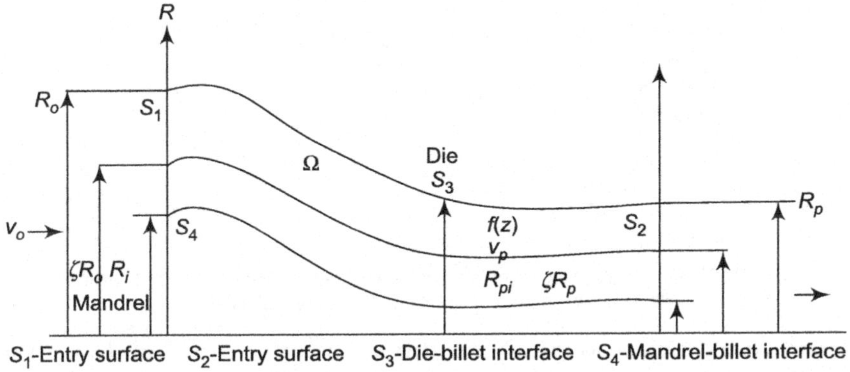

Fig. 5.18 Deformation zone typical streamline surfaces in extrusion. With permission from Malpani and Kumar [4] publisher

Under similar treatment of the analysis, with reference to Fig. 5.13, and in addition, at the product side coordinates are given by

$$x = (a_4 + n') \sin \phi \text{ and}$$
$$y = (a_4 + n') \cos \phi \tag{5.36}$$

where n' is given as

$$\frac{OE}{OD} = \frac{OE'}{OD'}; \text{ i.e., } \frac{R_i + n}{R_o} = \frac{a_4 + n'}{a_2}$$

$$n' = \left(\frac{R_i + n}{R_o}\right) a_2 - a_4$$

Putting values of n' we find

$$\left.\begin{aligned} x &= \frac{(R_i + n)\, a_2 \sin \psi}{R_o}; \quad \frac{\partial x}{\partial z} = 0 \\ y &= \frac{(R_i + n)\, a_2 \cos \psi}{R_o}; \quad \frac{\partial y}{\partial z} = 0 \end{aligned}\right\} \text{at product side } (z = L) \quad (5.37)$$

where R_o is the billet radius n_e is the distance from the axis formed by joining the C.G. of the product and the billet shape (i.e., O-O′) to an arbitrary point C at the die entry. ϕ and ψ are the angles between the plane of symmetry and the stream surface at entry and exit of the die, respectively, and L is the die length, R_i is the mandrel radius at entry.

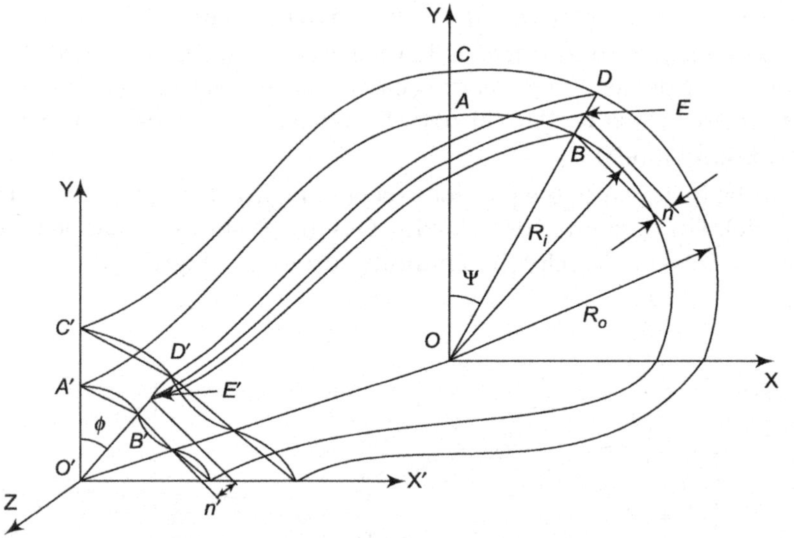

Fig. 5.19 Geometry of extrusion die and streamlines having generalized section. With permission from Malpani and Kumar [4] publisher

Substituting these boundary conditions from Eqs (5.9) and (5.10) into Eq. (5.8) gives

$$x = (R_i + n) \sin \phi + (R_i + n)\left(\frac{a_2 \sin \psi}{R_o} - \sin \phi\right)\left[4\left(\frac{z}{L}\right)^3 - 3\left(\frac{z}{L}\right)^4\right]$$

$$y = (R_i + n) \cos \phi + (R_i + n)\left(\frac{a_2 \cos \psi}{R_o} - \cos \phi\right)\left[4\left(\frac{z}{L}\right)^3 - 3\left(\frac{z}{L}\right)^4\right]$$

$$z = z \quad (5.38)$$

Equation (5.9) can be rewritten in the following form

$$x = n_e \sin\phi + n_e \left(\frac{a_2 \sin\psi}{R_o} - \sin\phi \right) f(z)$$

$$y = n_e \cos\phi + n_e \left(\frac{a_2 \cos\psi}{R_o} - \cos\phi \right) f(z) \quad (5.39)$$

$$z = z$$

where $f(z) = 0 \quad \text{at } z = 0$

and $f(z) = 1 \quad \text{at } z = L \quad (5.40)$

and $n_e = R_i + n$

In the present analysis, function f is represented by the following fourth order curve.

$$f = f(z) = \left[4\left(\frac{z}{L}\right)^3 - 3\left(\frac{z}{L}\right)^4 \right] \quad (5.41)$$

Equation (5.38) describes the coordinates inside the plastically deforming region but the relationship between the Cartesian and n, ϕ, z coordinate systems also. Although the present analysis employs a fourth order curve represented by Eq. (5.14) for the description of die and mandrel profile and the assumed streamlines of particles, it is to be noted that function f in Eq. (5.41) can be any general function of z provided the function satisfies the given boundary conditions as given in Eq. (5.40).

In most of cases mandrel is used as flat shaped for which function x and y coordinates are not the function of axial coordinate z. Hence Eq. (5.41) for flat mandrel or die profile is given as

$$\left.\begin{aligned} x &= f_1(z) = b_1 z^0 \\ y &= f_2(z) = c_1 z^0 \\ z &= z \end{aligned}\right\} \quad (5.42)$$

by putting the entry and exit boundary condition given by Eqs (5.9) and (5.10) into Eq. (5.39) gives

$$x = (R_i + n) \sin\phi$$

$$y = (R_i + n) \cos\phi \quad (5.43)$$

$$z = z$$

From Eq. (5.41) it is cleared that function $f(z) = 0$ for a flat shaped profile. So this equation can also be written as

$$x = n_e \sin\phi + n_e \left(\frac{a_2 \sin\psi}{R_o} - \sin\phi \right) f(z)$$

$$y = n_e \cos\phi + n_e \left(\frac{a_2 \cos\psi}{R_o} - \cos\phi \right) f(z) \quad (5.44)$$

$$z = z$$

where $f(z) = 0 \quad \text{at } z = 0$

and $f(z) = 0 \quad \text{at } z = L$ (5.45)

and $n_e = R_i + n$

In the present analysis, function f is represented by the following relation.

$$f = f(z) = \left[C_1 \left(\frac{z}{L} \right)^0 \right] \tag{5.46}$$

where C_1 is given as $C_1 = 0$.

The Jacobian of Eq. (5.38) can be found as

$$J = \begin{bmatrix} \frac{\partial x}{\partial n} & \frac{\partial y}{\partial n} & \frac{\partial z}{\partial n} \\ \frac{\partial x}{\partial \phi} & \frac{\partial y}{\partial \phi} & \frac{\partial z}{\partial \phi} \\ \frac{\partial x}{\partial z} & \frac{\partial y}{\partial z} & \frac{\partial z}{\partial z} \end{bmatrix} \tag{5.47}$$

Here

$$\begin{aligned} x &= n_e A + n_e Bf \\ y &= n_e C + n_e Df \\ z &= z \end{aligned} \tag{5.48}$$

where

$$\begin{aligned} n_e &= R_i + n \\ A &= \sin \phi \\ B &= \frac{a_2 \cos \phi}{R_o} - A; \cos = \frac{A}{\pi (a_1 a_2 - a_3 a_4)} \\ C &= \cos \phi \\ D &= \frac{a_2 \cos \psi}{R_o} - C \end{aligned} \tag{5.49}$$

Hence
$$J = \begin{bmatrix} A + Bf & C + Df & 0 \\ n_e A' + n_e B'f & n_e C' + n_e D'f & 0 \\ n_e Bf' & n_e Df' & 1 \end{bmatrix} \tag{5.50}$$

Determinant of the Jacobian is therefore written as

Det $J = n_e \, . \, g\,(\phi, z),$ (5.51)

where $g\,(\phi, z) = [(A + Bf)\,(C' + D'f) - (A' + B'f)\,(C + Df)]$ (5.52)

Hence,
$$J^{-1} = \frac{1}{\det J} \begin{bmatrix} K_{11} & K_{12} & K_{13} \\ K_{21} & K_{22} & K_{23} \\ K_{31} & K_{32} & K_{33} \end{bmatrix} \tag{5.53}$$

Strain rate components $\dot{\varepsilon}_{ij}$ are represented by,

$$\dot{\epsilon}_{ij} = \frac{1}{2}\left(\frac{\partial v_i}{\partial x_j} + \frac{\partial v_j}{\partial x_i}\right) \tag{5.54}$$

where v_i and v_j represent the velocity components along x_i and x_j axes in Cartesian coordinate system, respectively. The partial derivatives of the above equations are obtained with the aid of coordinate transformation, i.e.,

$$\frac{\partial v_i}{\partial x_j} = \sum_{k=1}^{3} \frac{\partial v_i}{\partial u_k}\frac{\partial u_k}{\partial x_j} \tag{5.55}$$

The expanded form of Eq. (5.55) can be seen in the work of Malpani and Kumar [4].

Subscripts of $g(\phi, z)_\phi$ and $g(\phi, z)_z$ indicate partial differentiation of function $g(\phi, z)$ with respect to ϕ or z, respectively. First and second primes on symbols, i.e., f', f'', etc., indicate the first and second derivatives of the symbol with respect to the given coordinate system. It has been found analytically from the above that $\epsilon_{xx} + \epsilon_{yy} + \epsilon_{zz} = 0$ and the proposed strain rate field satisfies the incompressibility condition.

Assuming that the plastic zone is bounded by entry and exit shear surfaces, the velocity field components can be obtained. Because of volume continuity, the velocity component along z-direction in the Cartesian coordinate system (v_z) should be v_o at the entrance and v_p at the exit of die. V_o and v_p are the speeds of the billet and the outgoing product, respectively. v_p can be described in terms of v_o as

$$v_p = \frac{\pi (R_o^2 - R_i^2)\dfrac{\phi}{2\pi}}{\dfrac{1}{2}(a_1 a_2 - a_3 a_4)\sin\psi} v_o \tag{5.56}$$

These requirements are satisfied using Eqs (5.40) and (5.41). Therefore, other velocity components for incompressible material are determined as

$$v_x = \frac{n_e B f'}{g(\phi, z)} v_o$$

$$v_y = \frac{n_e D f}{g(\phi, z)} v_o \tag{5.57}$$

$$v_z = \frac{1}{g(\phi, z)} v_o$$

Using the velocity fields given by the above equations it has been found that the velocity boundaries conditions are satisfied. Since the proposed velocity and strain rate satisfy the incompressibility condition, i.e., $\dot{\epsilon} = (\dot{\epsilon}_{xx} + \dot{\epsilon}_{yy} + \dot{\epsilon}_{zz}) = 0$. Hence, the proposed velocity field model fulfils the stringent requirement for the construction of a kinematically admissible condition.

Estimation of Extrusion Power and Average Ram Pressure

The total power consumed within the element is the sum of power losses due to plastic deformation (φ_i), the velocity discontinuities at entry (φ_e) and outlet (φ_o) of the die and mandrel, and the friction power loss along the interface between the material and the die and mandrel (φ_f). The above powers are consumed due to metal extrusion in die and to obtain total power consumed in extrusion process, container and die land frictional powers are added in it. The predicted total power obtained through the present velocity field would be higher (if no other redundant power losses are present) than the power actually consumed.

(a) **Internal Power of Deformation :** the incremental power of deformation ($d\varphi_i$) can be expressed as

$$\varphi_i = \int_0^L \int_0^\phi \int_{R_i}^{R_o} \bar{\sigma}\dot{\bar{\epsilon}} \left| \det J \right| dn_e \, d\phi \, dz \tag{5.58}$$

(b) **Frictional Power :** If the frictional resistance of the material along the total-work interface is τ and the slip velocity (or the tangential velocity discontinuity) along the interface is v_t, then the incremental frictional power at the interface can be expressed as

$$d\varphi_f = \tau \left| v_t \right| dS \tag{5.59}$$

Based on Eq. (5.59), frictional power for the considered element is obtained from the above as

$$\varphi_f = \int_0^L \int_0^\phi \frac{m\sigma_o}{\sqrt{3}} \left| v_t \right|_{n_e = R_o} \frac{1}{\cos\alpha} \frac{\partial(x, z)}{\partial(\phi, z)} d\phi \, dz + \int_0^L \int_0^\phi \frac{m\sigma_o}{\sqrt{3}} \left| v_t \right|_{n_e = R_i} \frac{1}{\cos\alpha} \frac{\partial(x, z)}{\partial(\phi, z)} d\phi \, dz \tag{5.60}$$

where $v_t = \left[v_x^2 + v_y^2 + v_z^2 \right]^{\frac{1}{2}}$, when $n_e = R_o$

(c) **Shear Power :** Thus, the shear power can be expressed as

$$\varphi_s = \int_S \frac{\bar{\sigma}}{\sqrt{3}} \left| v_t \right| dS \tag{5.61}$$

where v_t is the tangential velocity discontinuity along the dS and S is the area over which integration has to be carried out. The shear power at entry (φ_e) and exit (φ_o) side are obtained as:

$$\varphi_e = \int_0^{\phi_m} \int_{R_i}^{R_o} \frac{\sigma_o}{\sqrt{3}} [v_r]_{z=0} \, n_e \, d\, n_e d\phi + \int_0^{\phi_m} \int_0^{R_i} \frac{\sigma_o}{\sqrt{3}} [v_r]_{z=0} \, n_e \, dn_e \, d\phi \tag{5.62}$$

$$\varphi_o = \int_0^{\phi_m}\int_{R_i}^{R_o} \frac{\sigma_o}{\sqrt{3}} [v_r]_{z=L} \cos\frac{(a_2 - a_4)^2}{(R_o - R_i)^2} n_e\, d\, \underset{e}{r}d\phi$$

$$+ \int_0^{\phi_m}\int_0^{R_i} \frac{\sigma_o}{\sqrt{3}} [v_r]_{z=L} \cos\frac{(a_2 - a_4)^2}{(R_o - R_i)^2} n_e\, d\, \underset{e}{r}d\phi \qquad (5.63)$$

where $v_r = \left[v_x^2 + v_y^2\right]^{\frac{1}{2}}$

(d) **Container Wall Friction Power :** This includes the power consumed in friction between container and billet contact due to relative velocity between them.

$$\varphi_{c\,\mathrm{fri}} = (2\pi)\, m_c\, (\sigma_o/\sqrt{3})\, L_t\, R_o\, V_o \text{ Watts} \qquad (5.64)$$

where L_t=L_o in case of flat die, L_o is optimum billet length and for streamline die

$$L_t = L_o - V_o t;\ t = 0.5\, t_{pass} \text{ and } t_{pass} = 0.8(L_o/V_o)$$

t is the time taken as half of the 80% time of passing billet through die and m_c is friction factor between container and billet.

(e) **Die Land Friction Power :** The power consumed in die land friction is given by

$$\varphi_{dl\,\mathrm{fri}} = m\,(\bar{\sigma}/\sqrt{3})\, L_l\, P_f\, V_{dl} \qquad (5.65)$$

where L_l is length of die land which is taken as one seventh of optimum die length, P_f is outer profile periphery, V_{dl} is velocity of product inside die land and $\bar{\sigma}$ is the flow stress of material after strain hardening.

(f) **Total Power and Average Extrusion Ram Pressure :** The total power (φ_T) is the sum of its constituents including the redundant powers. Therefore, the total power (φ_T) is given as

$$\varphi_T = \varphi_i + \varphi_e + \varphi_o + \varphi_f + \varphi_{c\,\mathrm{fri}} + \varphi_{dl\,\mathrm{fri}} \qquad (5.66)$$

Once the total power (φ_T) consumed during extrusion is obtained, the upper limit to the average ram pressure (P_{avg}) for extrusion is found as

$$P_{avg} = \frac{\varphi_T}{\pi\,(R_o^2 - R_i^2)\, v_o} \qquad (5.67)$$

The average ram pressure (P_{avg}) can be converted to relative forming stress (R_s) as

$$R_s = \frac{P_{avg}}{\sigma_o} \qquad (5.68)$$

where σ_o is the effective stress at zero hardening condition. The relative die length is given as

$$R_l = \frac{L}{R_o} \qquad (5.69)$$

Case study: (I)

To perform the extrusion of a circular tube a conical die with flat mandrel profile is manufactured. The product shape is selected as circular for ease of manufacturing. The specifications of die and mandrel are given below.

Specifications of Die

- Profile - Conical profile with two cone angle 45° and 70.8°.
- Type - Welding chamber bridge die having four webs on the attached mandrel.
- Materials - Commercial grade die steel.
- Reduction - 80.94%
- Inlet diameter - 26.00 mm
- Outlet diameter - 13.00 mm
- Taper length - 7.00 mm and 6.00 mm, respectively.
- Bearing - 1.00 mm
- Total length - 13.00 mm
- Outer diameter - 38.00 mm

Mandrel Specifications

- Material - Commercial grade die steel.
- Length - 8.00 mm
- Web length - 7.00 mm
- Taper angle - 22.50° from vertical axis.

Experimental details

The extrusion of a circular tube is performed on pre fabricated extrusion set up using commercial grade Lead (70Pb, 30Sn) billet at 80°C temperature. Billets are machined to 25.10 mm diameter and 45.00 mm length. The extrusion is done vertically on a 2000 kN Hydraulic Press having constant strain rate facility. Lubricants used on ram and pressure pad is fine graphite power (10–20% by weight) mixed in Mobil oil. Attention is given on displacement and load, and is noted at fixed time interval. Figure 5.22 shows the extruded product.

Fig. 5.20 Assembled die and mandrel

Fig. 5.21 Assembled die and from inlet side mandrel from outlet side. With permission from Malpani and Kumar [4] publisher

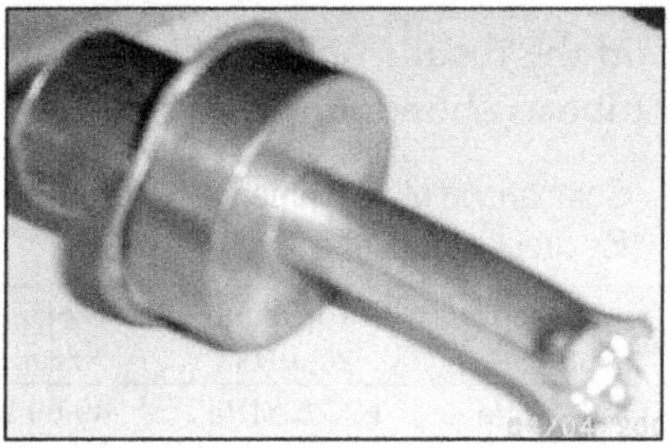

Fig. 5.22 Extruded circular tube attached with die assembly and left billet. With permission from Malpani and Kumar [4] publisher

As the lead is hot worked at room temperature and it is independent of strain hardening, the flow stress and other properties of the alloy is obtained experimentally using ring compression [9] test as:

$$\bar{\sigma} = 10.3223 \text{ (MPa) (no strain hardening)}$$

Melting temperature = 160 °C

Extrusion temperature = 80 °C

Lubricant is used as a mixture of graphite powder (10–20% weight) in mobil oil on the ram and pressure pad. Friction factor m for Lead is found to be 0.486 from the ring compression test at 80°C. Other details of experiment are given below.

Mandrel diameter = 7.00 mm

Product cross sectional area = 94.25 mm

Product outer diameter = 13.00 mm

Product inner hole diameter = 7.0 mm

Load vs. displacement and corresponding pressure obtain by experiment is given in Table 5.5.

The average velocity V_o is found to be 0.1239 mm/s and average pressure (P_{ave}) is found to be 64.83 MPa. A comparison has been done between

Table 5.5 Experimental results of lead tube extrusion

S. No.	Load in kN	Displacement in mm/10s	Average pressure (P_{ave}) in MPa
1	0	0	0
2	1	1.18	2.192
3	2	2.36	4.383
4	4	3.53	8.766
5	8	4.68	17.533
6	21	5.88	46.020
7	45	7.10	98.614
8	52	8.391	113.954
9	52	9.48	113.954
10	54	10.68	118.337
11	58	11.86	127.103
12	58	13.05	127.103

With permission from Malpani and Kumar [4] publisher.

experimental results and the results obtained by the proposed upper bound model for the circular tube as shown in Table 5.6.

Table 5.6 Comparison of experimental and theoretical results for circular lead tube

S. No.	Comparison properties	Experimental results	Proposed results	Optimum results	Difference (%) [(B/A) – 1]*100
1	Average pressure (P_{ave})	64.83 MPa	67.75 MPa	49.69 MPa	+4.51 %
2	Average power	3.67 KW	3.83 KW	2.808 KW	+4.36
3	Die length	13.05 mm	13.05 mm	17.75 mm	

Permission from Malpani and Kumar [4] publisher.

From Table 5.6, it is clear that theoretical results obtained by the proposed upper bound model shows a good agreement with experimental result for the circular tube of lead alloy. Figure 5.23 shows the variation of displacement rate with average pressure and load required for extrusion. It is clear from the graph that initially load increase at low rate and then after a certain displacement it increase sharply to reach at a maximum value and after that it become constant. It is because, as the extrusion start; upsetting of billet continues up to point (a) and billet accumulate itself in the container inner size, so load increases slowly. After point (a) piercing and shearing of billet starts and billet is sheared in four parts which is continued up to point (b) and after point (b), billet is sheared completely and material starts to flow over the mandrel hence again decreases in the load appears. After point (c) welding of the four sheared section takes place due to smaller cross sectional area available and hence more temperature rise takes place near the bearing which helps in welding to tubular shape. Finally material passes through bearing and deforms completely to obtain the final cross section with proper ID and OD of the circular shape extrusion. Thus extrusion is continuously done keeping the load constant after point (d).

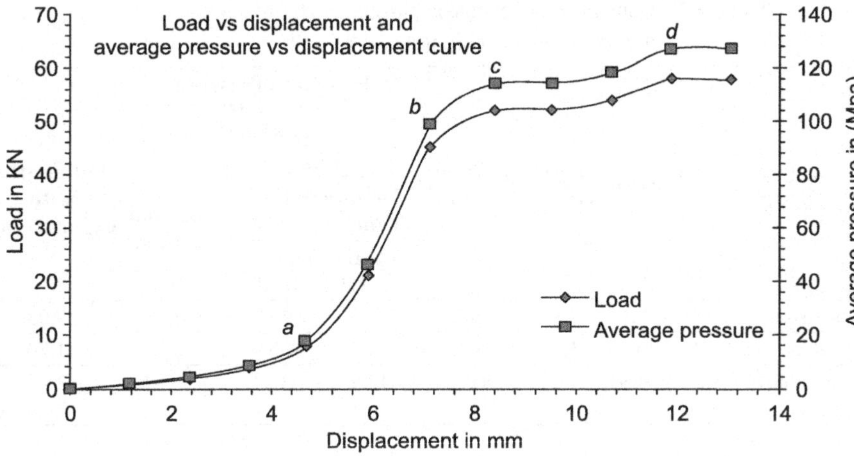

Fig. 5.23 Variation of average pressure and load with displacement rate for circular tube extrusion of Pb (V_o= 0.1239 mm/s, R_o=12.55 mm, %r = 80.94). With permission from Malpani and Kumar [4] publisher

II. Yang *et al.* [12] conducted experiment for three arbitrarily shaped tubular sections as: ellipse [Fig. 5.24(a)], clover [Fig. 5.24(b)], and trocoidal gear [Fig. 5.24(c)] with circular inner hole. Using Al2024 alloy as the working material and a forth order streamline die profile with a flat mandrel profile. The experimental conditions are taken as:

Material: Al2024 with Stress–Strain relationship as: $\bar{\sigma} = 441.5\,(\bar{\varepsilon})^{0.275}$ (MPa).

Billet diameter = 29.5 mm	Billet length = 80 mm
Friction factor m = 0.12	Ratio $L/R_o = 2.0$
Ram speed = 1.0 mm/s.	Temperature = Room temperature (20°C)

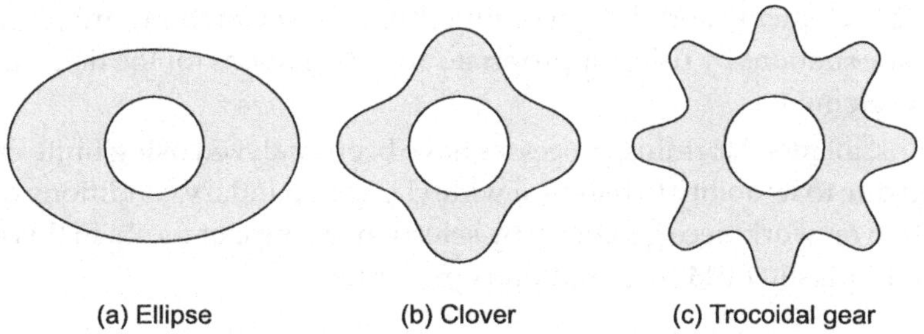

Fig. 5.24 Examples used for validation of the proposed model. With permission from Malpani and Kumar [4] publisher

The final geometrical details of elliptical, clover, and trocoidal gear are taken from Yang *et al.* [12]. The cross section of profile is drawn on true scale using AutoCAD Minimum number of power elements is generated for equal pseudo small elemental extruded area in such a way so that final geometry area calculated should be with in ±1% and total error in angles on billet (ϕ) and product side (ψ) do not exceed by ±1°.

Table 5.7 Comparison of extrusion pressure with experiment [12] and the proposed model (*r*: reduction)

S. No	Section shape	*r* (%)	Area	Shape factor	Element number	Extrusion Pressure at optimal (P_{ave}) in MPa			% Difference with experiments
						Results obtained by Yang *et al.* [12]		Proposed results	
						Experi-mental	Theore-tical		
1.	Ellipse	67.79	220.15	1.5	587	773.5	843.1	843.38	+9.03
2.	Clover	67.79	220.15	1.5	590	834.7	865.9	848.88	+1.70
3.	Trocoidal gear	51.01	334.84	1.5	825	617.9	632.4	639.49	+3.49

With permission from Malpani and Kumar [4] publisher.

Experimental results obtained by Yang *et al.* [12] at various reductions are compared in Table 5.7. It can be seen that the results obtained using the proposed model agrees well with experiment.

5.4 FINITE ELEMENT METHOD FOR EXTRUSION SIMULATION

Slip-line and upper bound method are useful in making good predictions for forming loads, the overall geometry changes with the deforming work piece, the qualitative modes of metal flow and the optimal process conditions. However, very accurate determination of various process parameters on metal flow became possible only after the finite element method was introduced. In the finite element method, the domain of interest is divided into a number of suitable elements and the governing differential equations are reduced to algebraic equations by using appropriate approximations for the field variables over the elements.

Almost all metal forming processes have been analyzed using finite element method due to its ability to handle a wide class of boundary conditions without restriction on workpiece geometry by selection of a proper mesh. In this section a 3D rigid plastic FEM formulation is presented.

5.4.1 A Rigid-Plastic FEM Formulation for 3D Extrusion

If the plastic strains are very large as compared to the elastic strains and can be considered to be insignificant as far as the deformation phenomena are concerned. Under the assumption of negligible elastic strains, the material can be idealized as rigid-plastic rigid-viscoplastic. Deformation of such materials can be conveniently analyzed by Eulerian or flow formulation for 3D extrusion that this section is concerned. The constitutive equations for rigid-plastic material used here are the same as described. Initially the domain (Fig. 5.5) is divided into many small slices (Figs 5.25 and 5.26) for simplification and easy solution.

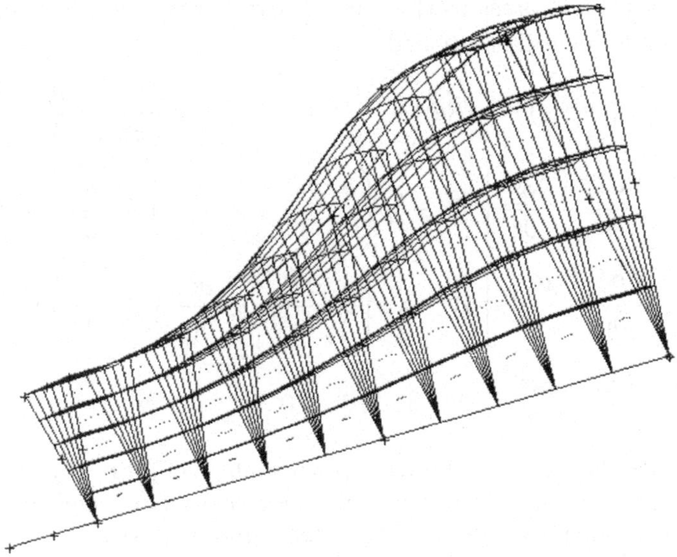

Fig. 5.25 Division of one slice into finite elements for the analysis. With permission from S. Kumar [5]

5.4.2 Governing Equations

In general, the material behavior of the deforming metal is governed by the equations of conservation of mass, conservation of momentum, and conservation of energy. Present formulation deals with the steady-state 3D cold extrusion process where the forming speed is low and therefore the temperature rise is small. Thus, the process is considered to be isothermal. Therefore, the equation of conservation of energy is not required for the analysis. The momentum equation for steady state extrusion process (neglecting the body forces) is given as:

$$\rho\left(v_x \frac{\partial v_x}{\partial x} + v_y \frac{\partial v_x}{\partial y} + v_z \frac{\partial v_x}{\partial z}\right) = -\frac{\partial p}{\partial x} + \frac{\partial \sigma'_{xx}}{\partial x} + \frac{\partial \sigma'_{xy}}{\partial y} + \frac{\partial \sigma'_{xz}}{\partial z}$$

$$\rho\left(v_x \frac{\partial v_y}{\partial x} + v_y \frac{\partial v_y}{\partial y} + v_z \frac{\partial v_y}{\partial z}\right) = -\frac{\partial p}{\partial y} + \frac{\partial \sigma'_{yx}}{\partial x} + \frac{\partial \sigma'_{yy}}{\partial y} + \frac{\partial \sigma'_{yz}}{\partial z} \quad (5.70)$$

$$\rho\left(v_x \frac{\partial v_z}{\partial x} + v_y \frac{\partial v_z}{\partial y} + v_z \frac{\partial v_z}{\partial z}\right) = -\frac{\partial p}{\partial z} + \frac{\partial \sigma'_{zx}}{\partial x} + \frac{\partial \sigma'_{zy}}{\partial y} + \frac{\partial \sigma'_{zz}}{\partial z}$$

Since volume remain constant in case of plastic deformation. The continuity equation is given as:

$$\dot{\varepsilon}_{xx} + \dot{\varepsilon}_{yy} + \dot{\varepsilon}_{zz} = 0. \quad (5.71)$$

5.4.2.1 Weak formulation

In the present work, mixed pressure–velocity formulation is used. Let v_x, v_y, v_z be velocities and p (pressure) be is the function that satisfy a weak solution if the following integral is satisfied:

$$-\int_V \Bigg[(\dot{\varepsilon}_{xx} + \dot{\varepsilon}_{yy} + \dot{\varepsilon}_{zz}) W_p + \left\{ \rho \left(v_x \frac{\partial v_x}{\partial x} + v_y \frac{\partial v_x}{\partial y} + v_z \frac{\partial v_x}{\partial z} \right) + \frac{\partial p}{\partial x} - \frac{\partial \sigma'_{xx}}{\partial x} - \frac{\partial \sigma'_{xy}}{\partial y} - \frac{\partial \sigma'_{xz}}{\partial z} \right\} W_x + \left\{ \rho \left(v_x \frac{\partial v_y}{\partial x} + v_y \frac{\partial v_y}{\partial y} + v_z \frac{\partial v_y}{\partial z} \right) + \frac{\partial p}{\partial y} - \frac{\partial \sigma'_{yx}}{\partial x} - \frac{\partial \sigma'_{yy}}{\partial y} - \frac{\partial \sigma'_{yz}}{\partial z} \right\} W_y + \left\{ \rho \left(v_z \frac{\partial v_z}{\partial x} + v_y \frac{\partial v_z}{\partial y} + v_z \frac{\partial v_z}{\partial z} \right) + \frac{\partial p}{\partial z} - \frac{\partial \sigma'_{zx}}{\partial x} - \frac{\partial \sigma'_{zy}}{\partial y} - \frac{\partial \sigma'_{zz}}{\partial z} \right\} W_z \Bigg] \mathrm{d}x\,\mathrm{d}y\,\mathrm{d}z = 0 \quad (5.72)$$

where W_p, W_x, W_y, and W_z be the appropriate weighting function. Since hydrostatic pressure is compressive and hence negative, weighting function W_p is also taken negative. Performing integration by parts:

$$\int_V (I_1 + I_2)\,\mathrm{d}x\,\mathrm{d}y\,\mathrm{d}z - \int_{A_1} I_3\,\mathrm{d}A_1 - \int_{A_2} I_4\,\mathrm{d}A_2 - \int_{A_3} I_5\,\mathrm{d}A_3 = 0 \quad (5.73)$$

Where

$$I_1 = -(\dot{\varepsilon}_{xx} + \dot{\varepsilon}_{xx} + \dot{\varepsilon}_{xx}) W_p$$

$$I_2 = \rho \left(v_x \frac{\partial v_x}{\partial x} + v_y \frac{\partial v_x}{\partial y} + v_z \frac{\partial v_x}{\partial z} \right) W_x + \rho \left(v_x \frac{\partial v_y}{\partial x} + v_y \frac{\partial v_y}{\partial y} + v_z \frac{\partial v_y}{\partial z} \right) W_y + \rho \left(v_x \frac{\partial v_z}{\partial x} + v_y \frac{\partial v_z}{\partial y} + v_z \frac{\partial v_z}{\partial z} \right) W_z - p\{\dot{\varepsilon}_{xx}(W) + \dot{\varepsilon}_{yy}(W) + \dot{\varepsilon}_{zz}(W)\} + 2\mu \left\{ \dot{\varepsilon}_{xx}\,\dot{\varepsilon}_{xx}(W) + \dot{\varepsilon}_{yy}\,\dot{\varepsilon}_{yy}(W) + \dot{\varepsilon}_{zz}\,\dot{\varepsilon}_{xx}(W) + 2\dot{\varepsilon}_{xy}\,\dot{\varepsilon}_{xy}(W) + 2\dot{\varepsilon}_{yz}\,\dot{\varepsilon}_{yz}(W) + 2\dot{\varepsilon}_{zx}\,\dot{\varepsilon}_{zx}(W) \right\} \quad (5.74)$$

$$I_3 = [((-p + \sigma'_{xx})\,\hat{n}_x + \sigma'_{xy}\hat{n}_y + \sigma'_{xz}\hat{n}_z)]\,W_x = t_x W_x$$

$$I_4 = [((-p + \sigma'_{yy})\,\hat{n}_y + \sigma'_{yx}\hat{n}_x + \sigma'_{yz}\hat{n}_z)]\,W_y = t_y W_y$$

$$I_5 = [((-p + \sigma'_{zz})\,\hat{n}_z + \sigma'_{zy}\hat{n}_y + \sigma'_{zx}\hat{n}_x)\,n_z]\,W_z = t_z W_z$$

$$\mathrm{d}A_1 = \mathrm{d}y\,\mathrm{d}z$$
$$\mathrm{d}A_2 = \mathrm{d}x\,\mathrm{d}z$$
$$\mathrm{d}A_3 = \mathrm{d}x\,\mathrm{d}y$$

$$\dot{\varepsilon}_{xx} = \frac{\partial v_x}{\partial x} \qquad \dot{\varepsilon}_{xx}(W) = \frac{\partial W_x}{\partial x}$$

$$\begin{aligned}
\dot{\varepsilon}_{yy} &= \frac{\partial v_y}{\partial y} & \dot{\varepsilon}_{yy}(W) &= \frac{\partial W_y}{\partial y} \\
\dot{\varepsilon}_{zz} &= \frac{\partial v_z}{\partial z} & \dot{\varepsilon}_{zz}(W) &= \frac{\partial W_z}{\partial z} \\
\dot{\varepsilon}_{xy} &= \frac{1}{2}\left(\frac{\partial v_y}{\partial x} + \frac{\partial v_x}{\partial y}\right) = \dot{\varepsilon}_{yx} & \dot{\varepsilon}_{xy}(W) &= \frac{1}{2}\left(\frac{\partial W_y}{\partial x} + \frac{\partial W_x}{\partial y}\right) = \dot{\varepsilon}_{yx}(W) \\
\dot{\varepsilon}_{yz} &= \frac{1}{2}\left(\frac{\partial v_z}{\partial y} + \frac{\partial v_y}{\partial z}\right) = \dot{\varepsilon}_{zy} & \dot{\varepsilon}_{yz}(W) &= \frac{1}{2}\left(\frac{\partial W_z}{\partial y} + \frac{\partial W_y}{\partial z}\right) = \dot{\varepsilon}_{zy}(W) \\
\dot{\varepsilon}_{xz} &= \frac{1}{2}\left(\frac{\partial v_z}{\partial x} + \frac{\partial v_x}{\partial z}\right) = \dot{\varepsilon}_{zx} & \dot{\varepsilon}_{xz}(W) &= \frac{1}{2}\left(\frac{\partial W_z}{\partial x} + \frac{\partial W_x}{\partial z}\right) = \dot{\varepsilon}_{zx}(W)
\end{aligned} \tag{5.75}$$

In finite element formulation, it is convenient to express integrals in matrix form as:

$$\int_V \left[-W_p^T \{m\}^T \{\dot{\varepsilon}\} + \rho \left[v_x [W_x\, W_y\, W_z] \begin{Bmatrix} \frac{\partial v_x}{\partial x} \\ \frac{\partial v_y}{\partial x} \\ \frac{\partial v_z}{\partial x} \end{Bmatrix} + v_y [W_x\, W_y\, W_z] \begin{Bmatrix} \frac{\partial v_x}{\partial y} \\ \frac{\partial v_y}{\partial y} \\ \frac{\partial v_z}{\partial y} \end{Bmatrix} \right.\right.$$

$$\left.\left. + v_z [W_x\, W_y\, W_z] \begin{Bmatrix} \frac{\partial v_x}{\partial z} \\ \frac{\partial v_y}{\partial z} \\ \frac{\partial v_z}{\partial z} \end{Bmatrix} \right] - \{\dot{\varepsilon}(W)\}^T \{m\}\, p + 2\mu \{\dot{\varepsilon}(W)\}^T \{\dot{\varepsilon}\} \right] dx\, dy\, dz$$

$$- \int t_x W_x\, dA_1 - \int t_y W_y\, dA_2 - \int t_z W_z\, dA_3 = 0 \tag{5.76}$$

where
$$\{m\} = \begin{Bmatrix} 1 \\ 1 \\ 1 \\ 0 \\ 0 \\ 0 \end{Bmatrix} \qquad \{\dot{\varepsilon}\} = \begin{Bmatrix} \dot{\varepsilon}_{xx} \\ \dot{\varepsilon}_{yy} \\ \dot{\varepsilon}_{zz} \\ \sqrt{2}\,\dot{\varepsilon}_{xy} \\ \sqrt{2}\,\dot{\varepsilon}_{yz} \\ \sqrt{2}\,\dot{\varepsilon}_{xz} \end{Bmatrix} \qquad \{\dot{\varepsilon}(W)\} = \begin{Bmatrix} \dot{\varepsilon}_{xx}(W) \\ \dot{\varepsilon}_{yy}(W) \\ \dot{\varepsilon}_{zz}(W) \\ \sqrt{2}\,\dot{\varepsilon}_{xy}(W) \\ \sqrt{2}\,\dot{\varepsilon}_{yz}(W) \\ \sqrt{2}\,\dot{\varepsilon}_{xz}(W) \end{Bmatrix}$$

5.4.2.2 *Finite element approximation*

In the present work a brick element is used for finite element approximation with 20 nodes in each corner. However, since the essential boundary condition is to satisfy the continuity equation, the primary variable velocity is taken at each node to be continuous inside the element. But pressure variable is taken constant inside the element. Another possibility may be taking the shape function for velocity as quadratic and the same for pressure linear at each corner nodes. For nodal velocity, 20 noded and for pressure, eight noded elements are selected.

Shape function

The square shape function as 20 noded elements with a corresponding master element is taken for velocity and eight noded (having corner nodes) for pressure is assumed constant over an element.

The natural coordinates for the 20 and eight noded elements as used are given as below:

20 noded:

1 (−1, −1, −1)	11 (1, −1, 0)
2 (−1, −1, 1)	12 (0, −1, −1)
3 (1, −1, 1)	13 (−1, 0, −1)
4 (1, −1, −1)	14 (−1, 0, 1)
5 (−1, 1, −1)	15 (1, 0, 1)
6 (−1, 1, 1)	16 (1, 0, −1)
7 (1, 1, 1)	17 (−1, 1, 0)
8 (1, 1, −1)	18 (0, 1, 1)
9 (−1, −1, 0)	19 (1, 1, 0)
10 (0, −1, 1)	20 (0, 1, −1)

where,

$$N_i = -\frac{1}{8}(1+\xi_i\,\xi)(1+\eta_i\,\eta)(1+\varsigma_1\,\varsigma)(2-\xi_1\,\xi-\eta_i\,\eta-\varsigma_i\,\varsigma) \text{ for } i = 1 \text{ to } 8.$$

$$N_j = \frac{1}{4}(1+\xi_1\,\xi)(1+\eta_i\,\eta)(1-\varsigma^2) \qquad \text{for } j = 9,11,19,17.$$

$$N_k = \frac{1}{4}(1-\xi^2)(1+\eta_i\,\eta)(1+\varsigma_i\,\varsigma) \qquad \text{for } k = 10,12,20,18. \tag{5.77}$$

$$N_l = \frac{1}{4}(1+\xi_i\,\xi)(1-\eta^2)(1+\varsigma_i\,\varsigma) \qquad \text{for } l = 13,14,15,16.$$

Eight noded:

1 (−1, −1, −1)	5 (−1, 1, −1)
2 (−1, −1, 1)	6 (−1, 1, 1)
3 (1, −1, 1)	7 (1, 1, 1)
4(1, −1, −1)	8 (1, 1, −1)

$$N_i = \frac{(1+\xi_i\,\xi)(1+\eta_i\,\eta)(1+\zeta_i\,\zeta)}{8}, \quad i = 1 \text{ to } 8$$

The shape function for die billet interface is chosen as eight noded as given below:

$$N_i = -\frac{1}{4}(1+\xi_1\,\xi)(1+\eta_i\,\eta)(1-\xi_1\,\xi-\eta_i\,\eta) \qquad \text{for } i = 1 \text{ to } 4 \tag{5.78}$$

$$N_j = \frac{1}{2}(1-\xi^2)(1+\eta_j\,\eta) \qquad \text{for } j = 5, 7$$

$$N_k = -\frac{1}{2}(1+\xi_1\,\xi)(1-\eta^2) \qquad \text{for } k = 6, 8 \tag{5.79}$$

where,

$$[J] = \begin{bmatrix} \frac{\partial x}{\partial \xi} & \frac{\partial y}{\partial \xi} & \frac{\partial z}{\partial \xi} \\ \frac{\partial x}{\partial \eta} & \frac{\partial y}{\partial \eta} & \frac{\partial z}{\partial \eta} \\ \frac{\partial x}{\partial \varsigma} & \frac{\partial y}{\partial \varsigma} & \frac{\partial z}{\partial \varsigma} \end{bmatrix} = \begin{bmatrix} J_{11} & J_{12} & J_{13} \\ J_{21} & J_{22} & J_{23} \\ J_{31} & J_{32} & J_{33} \end{bmatrix}; \begin{Bmatrix} \frac{\partial f}{\partial x} \\ \frac{\partial f}{\partial y} \\ \frac{\partial f}{\partial z} \end{Bmatrix} = [J]^{-1} \begin{Bmatrix} \frac{\partial f}{\partial \xi} \\ \frac{\partial f}{\partial \eta} \\ \frac{\partial f}{\partial \varsigma} \end{Bmatrix} \tag{5.80}$$

$$\det J = J_{11}(J_{22}J_{33} - J_{32}J_{23}) + J_{12}(J_{31}J_{23} - J_{21}J_{33}) + J_{13}(J_{21}J_{32} - J_{22}J_{31}) \tag{5.81}$$

$$[J]^{-1} = \frac{1}{\det J} \begin{bmatrix} (J_{22}J_{33} - J_{32}J_{23}) & (J_{32}J_{13} - J_{12}J_{33}) & (J_{12}J_{23} - J_{13}J_{22}) \\ (J_{31}J_{23} - J_{21}J_{33}) & (J_{11}J_{33} - J_{31}J_{13}) & (J_{21}J_{13} - J_{11}J_{23}) \\ (J_{21}J_{32} - J_{22}J_{31}) & (J_{31}J_{12} - J_{11}J_{32}) & (J_{11}J_{22} - J_{21}J_{12}) \end{bmatrix}$$

For boundary nodes (die billet interface):

$$A = \frac{\partial x}{\partial \xi}\frac{\partial y}{\partial \eta} - \frac{\partial x}{\partial \eta}\frac{\partial y}{\partial \xi}$$

$$\det J = \sqrt{A^2 + B^2 + C^2} \text{ with } B = \frac{\partial x}{\partial \eta}\frac{\partial z}{\partial \xi} - \frac{\partial x}{\partial \xi}\frac{\partial z}{\partial \eta} \tag{5.82}$$

$$C = \frac{\partial y}{\partial \xi}\frac{\partial z}{\partial \eta} - \frac{\partial y}{\partial \eta}\frac{\partial z}{\partial \xi}$$

5.4.2.3 FE formulation for $\dot{\varepsilon}$ and $\dot{\varepsilon}$ (W)

$$\text{Since} \quad \{\dot{\varepsilon}\} = \begin{Bmatrix} \dot{\varepsilon}_{xx} \\ \dot{\varepsilon}_{yy} \\ \dot{\varepsilon}_{zz} \\ \sqrt{2}\,\dot{\varepsilon}_{xy} \\ \sqrt{2}\,\dot{\varepsilon}_{yz} \\ \sqrt{2}\,\dot{\varepsilon}_{zx} \end{Bmatrix} = \begin{Bmatrix} \frac{\partial v_x}{\partial x} \\ \frac{\partial v_y}{\partial y} \\ \frac{\partial v_z}{\partial z} \\ \frac{1}{\sqrt{2}}\left(\frac{\partial v_x}{\partial y} + \frac{\partial v_y}{\partial x}\right) \\ \frac{1}{\sqrt{2}}\left(\frac{\partial v_y}{\partial z} + \frac{\partial v_z}{\partial y}\right) \\ \frac{1}{\sqrt{2}}\left(\frac{\partial v_z}{\partial x} + \frac{\partial v_x}{\partial z}\right) \end{Bmatrix} \tag{5.83}$$

The present formation an isoperimetric, representation is used, hence,

$$x = \sum_{i=1}^{20} N_i x_i = [N][x] \qquad v_x = \sum_{i=1}^{20} N_i x_i = [N]\{v_x\}$$

$$y = \sum_{i=1}^{20} N_i y_i = [N]\{y\} \quad (4.21) \qquad v_y = \sum_{i=1}^{20} N_i y_i = [N]\{v_y\} \quad (5.84)$$

$$z = \sum_{i=1}^{20} N_i z_i = [N]\{z\} \qquad v_z = \sum_{i=1}^{20} N_i z_i = [N]\{v_z\}$$

Now taking an arbitrary function *f* which is a function of *x*, *y* and *z*. so,

$$f(x, y, z) = f(x(\xi, \eta, \zeta), y(\xi, \eta, \zeta), z(\xi, \eta, \zeta)) \quad (5.85)$$

Taking differential

$$\begin{aligned} \frac{\partial f}{\partial \xi} &= \frac{\partial f}{\partial x}\frac{\partial x}{\partial \xi} + \frac{\partial f}{\partial y}\frac{\partial y}{\partial \xi} + \frac{\partial f}{\partial z}\frac{\partial z}{\partial \xi} \\ \frac{\partial f}{\partial \eta} &= \frac{\partial f}{\partial x}\frac{\partial x}{\partial \eta} + \frac{\partial f}{\partial y}\frac{\partial y}{\partial \eta} + \frac{\partial f}{\partial z}\frac{\partial z}{\partial \eta} \\ \frac{\partial f}{\partial \zeta} &= \frac{\partial f}{\partial x}\frac{\partial x}{\partial \zeta} + \frac{\partial f}{\partial y}\frac{\partial y}{\partial \zeta} + \frac{\partial f}{\partial z}\frac{\partial z}{\partial \zeta} \end{aligned} \quad (5.86)$$

or,

$$\begin{Bmatrix} \frac{\partial f}{\partial \xi} \\ \frac{\partial f}{\partial \eta} \\ \frac{\partial f}{\partial \zeta} \end{Bmatrix} = \begin{bmatrix} \frac{\partial x}{\partial \xi} & \frac{\partial y}{\partial \xi} & \frac{\partial z}{\partial \xi} \\ \frac{\partial x}{\partial \eta} & \frac{\partial y}{\partial \eta} & \frac{\partial z}{\partial \eta} \\ \frac{\partial x}{\partial \zeta} & \frac{\partial y}{\partial \zeta} & \frac{\partial z}{\partial \zeta} \end{bmatrix} \begin{Bmatrix} \frac{\partial f}{\partial x} \\ \frac{\partial f}{\partial y} \\ \frac{\partial f}{\partial z} \end{Bmatrix} = [J] \begin{Bmatrix} \frac{\partial f}{\partial x} \\ \frac{\partial f}{\partial y} \\ \frac{\partial f}{\partial z} \end{Bmatrix}$$

where [*J*] = Jacobian matrix

Hence

$$\begin{Bmatrix} \frac{\partial f}{\partial x} \\ \frac{\partial f}{\partial y} \\ \frac{\partial f}{\partial z} \end{Bmatrix} = [J]^{-1} \begin{Bmatrix} \frac{\partial f}{\partial \xi} \\ \frac{\partial f}{\partial \eta} \\ \frac{\partial f}{\partial \zeta} \end{Bmatrix} = \begin{bmatrix} JI_{11} & JI_{12} & JI_{13} \\ JI_{21} & JI_{22} & JI_{23} \\ JI_{31} & JI_{32} & JI_{33} \end{bmatrix} \begin{Bmatrix} \frac{\partial f}{\partial \xi} \\ \frac{\partial f}{\partial \eta} \\ \frac{\partial f}{\partial \zeta} \end{Bmatrix} \quad (5.87)$$

where *JI* is inverse of *J*. Now, if *f* is considered as v_x, v_y or v_z

Then,

$$\{\dot{\varepsilon}\} = \begin{bmatrix} JI_{11} & JI_{12} & JI_{13} & 0 & 0 & 0 & 0 & 0 & 0 \\ 0 & 0 & 0 & JI_{21} & JI_{22} & JI_{23} & 0 & 0 & 0 \\ 0 & 0 & 0 & 0 & 0 & 0 & JI_{23} & JI_{23} & JI_{33} \\ \frac{1}{\sqrt{2}} JI_{21} & \frac{1}{\sqrt{2}} JI_{22} & \frac{1}{\sqrt{2}} JI_{23} & \frac{1}{\sqrt{2}} JI_{11} & \frac{1}{\sqrt{2}} JI_{12} & \frac{1}{\sqrt{2}} JI_{13} & 0 & 0 & 0 \\ 0 & 0 & 0 & \frac{1}{\sqrt{2}} JI_{31} & \frac{1}{\sqrt{2}} JI_{32} & \frac{1}{\sqrt{2}} JI_{33} & \frac{1}{\sqrt{2}} JI_{21} & \frac{1}{\sqrt{2}} JI_{22} & \frac{1}{\sqrt{2}} JI_{23} \\ \frac{1}{\sqrt{2}} JI_{31} & \frac{1}{\sqrt{2}} JI_{32} & \frac{1}{\sqrt{2}} JI_{33} & 0 & 0 & 0 & \frac{1}{\sqrt{2}} JI_{11} & \frac{1}{\sqrt{2}} JI_{12} & \frac{1}{\sqrt{2}} JI_{33} \end{bmatrix}_{6\times 9} \begin{Bmatrix} \frac{\partial v_x}{\partial \xi} \\ \frac{\partial v_x}{\partial \eta} \\ \frac{\partial v_x}{\partial \zeta} \\ \frac{\partial v_y}{\partial \xi} \\ \frac{\partial v_y}{\partial \eta} \\ \frac{\partial v_y}{\partial \zeta} \\ \frac{\partial v_z}{\partial \xi} \\ \frac{\partial v_z}{\partial \eta} \\ \frac{\partial v_z}{\partial \zeta} \end{Bmatrix}_{9\times 1} \tag{5.88}$$

Since $$v_x = \sum_{i=1}^{20} N_i\, v_{xi}\,; \quad v_y = \sum_{i=1}^{20} N_i\, v_{yi} \quad \text{and} \quad v_z = \sum_{i=1}^{20} N_i\, v_{zi} \tag{5.89}$$

Hence, $$\frac{\partial v_x}{\partial \xi} = \sum_{i=1}^{20} \frac{\partial N_i}{\partial \xi} v_{xi} \text{ and } \frac{\partial N_i}{\partial \xi} = b_i, \frac{\partial N_i}{\partial \eta} = c_i, \frac{\partial N_i}{\partial \zeta} = d_i \tag{5.90}$$

$$\begin{Bmatrix} \dfrac{\partial v_x}{\partial \xi} \\ \dfrac{\partial v_x}{\partial \eta} \\ \dfrac{\partial v_x}{\partial \zeta} \\ \dfrac{\partial v_y}{\partial \xi} \\ \dfrac{\partial v_y}{\partial \eta} \\ \dfrac{\partial v_y}{\partial \zeta} \\ \dfrac{\partial v_z}{\partial \xi} \\ \dfrac{\partial v_z}{\partial \eta} \\ \dfrac{\partial v_z}{\partial \zeta} \end{Bmatrix} = \begin{bmatrix} b_1 & 0 & 0 & b_2 & 0 & 0 & -- & b_{20} & 0 & 0 \\ c_1 & 0 & 0 & c_2 & 0 & 0 & -- & c_{20} & 0 & 0 \\ d_1 & 0 & 0 & d_2 & 0 & 0 & -- & d_{20} & 0 & 0 \\ 0 & b_1 & 0 & 0 & b_2 & 0 & -- & 0 & b_{20} & 0 \\ 0 & c_1 & 0 & 0 & c_2 & 0 & -- & 0 & c_{20} & 0 - \\ 0 & d_1 & 0 & 0 & d_2 & 0 & -- & 0 & d_{20} & 0 \\ 0 & 0 & b_1 & 0 & 0 & b_2 & -- & 0 & 0 & b_{20} \\ 0 & 0 & c_1 & 0 & 0 & c_2 & -- & 0 & 0 & c_{20} \\ 0 & 0 & d_1 & 0 & 0 & d_2 & -- & 0 & 0 & d_{20} \end{bmatrix}_{9\times 60} \begin{Bmatrix} v_{x1} \\ v_{y1} \\ v_{z1} \\ - \\ - \\ - \\ v_{x20} \\ v_{y20} \\ v_{z20} \end{Bmatrix}_{60\times 1} \tag{5.91}$$

$$\{\dot{\varepsilon}\} = \begin{bmatrix} B_{11} & 0 & 0 & B_{14} & 0 & 0 & \dots\dots\dots & B_{1,58} & 0 & 0 & 0 \\ & B_{22} & 0 & 0 & B_{25} & 0 & \dots\dots\dots & 0 & B_{2,59} & 0 & 0 \\ B_{31} & 0 & 0 & B_{34} & 0 & 0 & \dots\dots\dots & 0 & 0 & B_{3,60} & 0 \\ B_{41} & B_{42} & B_{43} & B_{44} & B_{45} & 0 & \dots\dots\dots & B_{4,58} & B_{4,59} & 0 & 0 \\ 0 & B_{52} & B_{53} & 0 & B_{55} & B_{56} & \dots\dots\dots & 0 & B_{5,58} & B_{5,60} & B_{524} \\ B_{61} & 0 & B_{63} & B_{64} & 0 & B_{66} & \dots\dots\dots & B_{6,58} & 0 & B_{6,60} & B_{624} \end{bmatrix}_{6\times 60} \tag{5.92}$$

where, $B_{11} = (JI_{11}\, b_1 + JI_{12}\, c_1 + JI_{13}\, d_1)$

$B_{22} = (JI_{21}\, b_1 + JI_{22}\, c_1 + JI_{23}\, d_1)$

$B_{33} = (JI_{31}\, b_1 + JI_{32}\, c_1 + JI_{33}\, d_1)$

$B_{41} = \dfrac{1}{\sqrt{2}}(JI_{21}\, b_1 + JI_{22}\, c_1 + JI_{23}\, d_1) = \dfrac{1}{\sqrt{2}} B_{22}$, etc.

or,

$$\{\dot{\varepsilon}\} = [B]\,\{v^e\}$$

Similarly, $\{\dot{\varepsilon}(W)\} = [B]\{W_v^e\}$

Therefore,

$$\begin{Bmatrix} v_x \\ v_y \\ v_z \end{Bmatrix} = \begin{bmatrix} N_1 & 0 & 0 & N_2 & 0 & 0 & \ldots\ldots & N_{20} & 0 & 0 \\ 0 & N_1 & 0 & 0 & N_2 & 0 & \ldots\ldots & 0 & N_{20} & 0 \\ 0 & 0 & N_1 & 0 & 0 & N_2 & \ldots\ldots & 0 & 0 & N_{20} \end{bmatrix}_{3\times 60} \begin{Bmatrix} v_{x1} \\ v_{y1} \\ v_{z1} \\ v_{x2} \\ v_{y2} \\ v_{z2} \\ - \\ - \\ v_{x20} \\ v_{y20} \\ v_{z20} \end{Bmatrix}_{60\times 1}$$

$$= [N_v]\,\{v^e\}$$

$$\{W_v\} = \begin{Bmatrix} W_x \\ W_y \\ W_z \end{Bmatrix} = [N_v]\,\{W_v^e\}$$

$$p = [N_p]\,\{p^e\}$$

$$W_p = [N_p]\{W_p^e\} \qquad (5.93)$$

Secondary elements are approximated by taking an eight nodded quadrilateral element.

$$t_x = \sum_{i=1}^{8} N_{bi}\, t_{xi} = [N_b]\,\{t_x^b\}$$

$$t_y = \sum_{i=1}^{8} N_{bi}\, t_{yi} = [N_b]\,\{t_y^b\} \qquad (5.94)$$

$$t_z = \sum_{i=1}^{8} N_{bi}\, t_{zi} = [N_b]\,\{t_z^b\}$$

By substituting all these matrices along with approximation for boundary elements, the final finite element equation is given by

$$\left[\int_V \left[-([N_p]\{W_p^e\})^T \{m\}^T [B]\{v^e\} + \rho\left\{ v_x \{W_v\}^T \frac{\partial [N_v]}{\partial x}\{v^e\} + v_y \{W_v\}^T \frac{\partial [N_v]}{\partial y} \right.\right.\right.$$

$$\left.\{v^e\} + v_z \{W_v\}^T \frac{\partial [N_v]}{\partial z}\{v^e\}\right\} - ([B]\{W_v^e\})^T \{m\}[N_p]\{p^e\} + 2\mu\,([B]\{W_v^e\})^T$$

$$\left.([B]\{v^e\})\right]\mathrm{d}x\,\mathrm{d}y\,\mathrm{d}z$$

$$\sum_{b=1}^{NEB}\left[\int_{A_1}([N_b]\{W_x^b\})^T[N_b]\{t_x^b\}\,d\,{}_1A\int_{A_2}([N_b]\{W_y^b\})^T[N_b]\{t_y^b\}\,d\,{}_2A\right.$$
$$\left.+\int_{A_3}([N_b]\{W_z^b\})^T[N_b]\{t_z^b\}\,d\,{}_3A\right] \tag{5.95}$$

Hence,

$$\left[\{W_v^e\}^T\int(-[B]^T\{m\}[N_p]\{p^e\})\,\mathrm{d}x\,\mathrm{d}y\,\mathrm{d}z\right.$$
$$+\{W_p^e\}^T\int(-[N_p]^T\{m\}^T[B]\{v^e\})\,\mathrm{d}x\,\mathrm{d}y\,\mathrm{d}z+\{W_v^e\}^T\int\left(2\mu[B]^T[B]\right.$$
$$\left.\left.+\rho\left[v_x[N_v]^T\frac{\partial[N_v]}{\partial x}+v_y[N_v]^T\frac{\partial[N_v]}{\partial y}+v_z[N_v]^T\frac{\partial[N_v]}{\partial z}\right]\right)\{v^e\}\,\mathrm{d}x\,\mathrm{d}y\,\mathrm{d}z\right.$$
$$=\sum_{b=1}^{NEB}\left[\{W_x^b\}^T\int_{A_1}[N_b]^T[N_b]\{t_x^b\}\,dA_1+\{W_y^b\}^T\int_{A_2}[N_b]^T[N_b]\{t_y^b\}\,dA_2\right.$$
$$\left.+\{W_z^b\}^T\int_{A_3}[N_b]^T[N_b]\{t_z^b\}\,dA_3\right] \tag{5.96}$$

Denoting

$$[K_{11}]=0$$
$$[K_{12}]=\int\left(-[N_p]^T\{m\}^T[B]\right)\mathrm{d}x\,\mathrm{d}y\,\mathrm{d}z$$
$$[K_{21}]=\int\left(-[B]^T\{m\}[N_p]\right)\mathrm{d}x\,\mathrm{d}y\,\mathrm{d}z$$
$$[K_{22}]=\int\left(2\mu[B]^T[B]+\rho\left[v_x[N_v]^T\frac{\partial[N_v]}{\partial x}+v_y[N_v]^T\frac{\partial[N_v]}{\partial y}+v_z[N_v]^T\frac{\partial[N_v]}{\partial z}\right]\right)\mathrm{d}x\,\mathrm{d}y\,\mathrm{d}z$$

$$[W^e]=\begin{Bmatrix}\{W_p^e\}\\\{W_v^e\}\end{Bmatrix}$$
$$[X^e]=\begin{Bmatrix}\{p^e\}\\\{v^e\}\end{Bmatrix}$$
$$\{f_x^b\}=\int_{T_1}[N_b]^T[N_b]\{t_x^b\}\,\mathrm{d}A_1$$
$$\{f_y^b\}=\int_{T_2}[N_b]^T[N_b]\{t_y^b\}\,\mathrm{d}A_2$$

$$\{f_z^b\} = \int_{T_3} [N_b]^T [N_b] \{t_z^b\} \, dA_3$$

$$[K^e] = \begin{bmatrix} [K_{11}] & [K_{12}] \\ [K_{21}] & [K_{22}] \end{bmatrix}$$

where $$\{X^e\} = \begin{Bmatrix} \{p^e\} \\ \{v^e\} \end{Bmatrix}$$

Hence

$$\sum_{e=1}^{Ne} [\{W_v^e\}^T [K_{21}] \{p^e\} + \{W_p^e\}^T [K_{12}] \{v^e\} + [W_v^e]^T [K_{22}] \{v^e\}]$$

$$= \sum_{b=1}^{NEB} \Big[[W_x^b]^T \{f_x^b\} + [W_y^b]^T \{f_y^b\} + [W_z^b]^T \{f_z^b\} \Big]$$

or $$\sum_{e=1}^{Ne} [[W_v^e]^T [K_{21}] [W_p^e]^T [K_{12}] + [W_v^e]^T [K_{22}]] \begin{Bmatrix} \{p^e\} \\ \{v^e\} \end{Bmatrix} \tag{5.97}$$

$$= \sum_{b=1}^{NEB} [[W_x^b]^T \{f_x^b\} + [W_y^b]^T \{f_y^b\} + [W_z^b]^T \{f_z^b\}]$$

or $$\sum_{e=1}^{Ne} [[W_p^e]^T [W_v^e]^T] \begin{Bmatrix} \{p^e\} \\ \{v^e\} \end{Bmatrix} [K^e] = \begin{bmatrix} [K_{11}] & [K_{12}] \\ [K_{21}] & [K_{22}] \end{bmatrix}$$

Various boundry conditions on one slice are defined as given below.

1. Entry and Exit Boundary Conditions: On Entry face AEJ,

   $$v_x = v_y = 0,\ v_z = v_o \tag{5.98}$$

   on exit face DHM

   $$v_x = v_y = 0,\ t_x = t_y = 0 \tag{5.99}$$

2. Die billet interface B C L K,

   $$v_n = 0, \text{ Normal velocity} \tag{5.100}$$

   $$V_n = V_x \hat{n}_x + V_y \hat{n}_y + V_z \hat{n}_z = 0 \tag{5.101}$$

   In constant friction case, $$|\sigma_s| = \frac{m\bar{\sigma}}{\sqrt{3}} \tag{5.102}$$

3. Billet–container interface A B K J,

   $$v_n = 0,$$

   $$t_x = t_y = t_z = 0 \tag{5.103}$$

4. Free surface C D M L

   $$v_n = 0,$$

   $$t_x = t_y = t_z = 0 \tag{5.104}$$

5. Neutral axis E F G H,

$$v_x = v_y = 0 \quad \text{i.e.,} \quad v_n = 0$$

$$t_x = t_y = t_z = 0 \tag{5.105}$$

6. Contact surface, E H D A and E H M J: $v_n = 0$ (5.106)

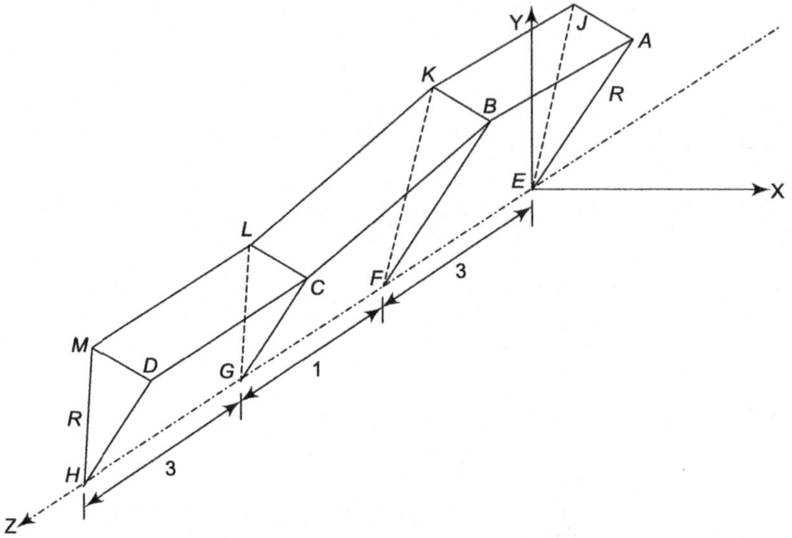

Fig. 5.26 One element from the power slices. With permission from S. Kumar [5]

5.4.2.4 Solution procedure

The above boundary conditions are imposed on the final equation at specified nodes. Resulting equation is solved iteratively by householder method. Initially for first iteration a suitable gauess value for μ is given. Then the result obtained is used as next iteration. Then iteration is conditioned until the nodal values of pressure and velocity converges within 0.1% in the successive iterations.

A program has been developed in FORTRAN and initially an axi-symmetric cold extrusion (consisting of 24 power elements) has been taken to validate [7].

5.4.3 Example

Yang *et al.* [7] conducted experiments for axi-symmetric extrusion of circular sections. Details of material specifications and other process parameters are given below.

Materials: Aluminum alloy Al 2024 (annealed). The Stress–strain relationship as:

$$\bar{\sigma} = 118.7 + 229.4\,\bar{\epsilon}^{0.247} \quad \text{(MPa) for Al 2024 Aluminium alloy}$$

Billet radius (R_o) = 12.5 mm.

Friction condition (m) = 0.12 (obtained from Ring Test with Molybdenum sulfite as the lubricant)

Die profile: A fourth order die profile has been used is taken as the die profile to validate the proposed model.

Ratio $\frac{L}{R_o}$ (initial die length to the billet radius) = 2.0

Ram speed = 1.0 mm/s

Temperature: Room temperature (20°C)

The final component geometry is with product radius R_p corresponding to different reductions are given in Table 5.8. The mesh corresponding to all these shapes are drawn accurately using/HyperMesh 5.0 and AutoCAD and Mesh data is created as input and the same is used with the FORTRAN program developed.

Table 5.8 Comparison of results with Yang *et al.* [7]

Billet radius (R_o), (mm)	Reductions (% red)	Output radius (R_p), (mm)	Result (work of Yang *et al.* [7]) Extrusion pressure in MPa	Result (proposed analysis) extrusion pressure in MPa
12.50	40	9.68	240	265
12.50	50	8.84	315	355
12.50	60	7.91	413	463
12.50	70	6.85	576	620

With permission from S. Kumar [5].

Results obtained at different reduction ratio (Table 5.9) are compared with the experimental results obtained by Yang *et al.* [7]. It can be seen from the table that the results obtained from the proposed model are satisfactory. Figure 5.27 shows variation of extrusion pressure with die length. This shows that

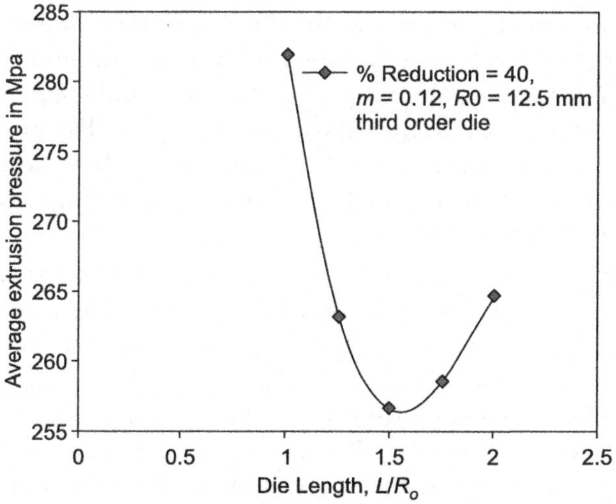

Fig. 5.27 Variation of extrusion average pressure with die length. With permission from S. Kumar [5]

an optimal die length should be selected for all practical cases. Other results have also been found satisfactory.

From the above 3D formulation, any general extrudable solid shape can be analyzed. The above FEM analysis would give the velocity and pressure field within the deformation region at various nodes. These velocity and pressure field may be taken as input for getting the temperature distribution within the deformation region and tooling in hot extrusion case.

5.5 CONTINUOUS EXTRUSION AND ANALYSIS

The process (Fig. 5.28) uses the concept of moving wheel, with friction between the billet and the wheel wall can create enough pressure in the billet that it is extruded. The equipment consists of a disc or pulley having a deep groove on its periphery. The groove may be rectangular, circular, trapezoidal, etc. The pulley is rotated by a motor through a gear box. A stationary shoe is mounted on the pulley. The shoe has a small tongue (abutment) which fits into the groove in the pulley. The groove and the tongue together make a hollow cavity whose three sides are moving and one side, i.e., tongue is stationary. On the forward side of the shoe, an extrusion die is fixed. The die completely blocks the groove except for a hole of desired shape in it. The billet which may be round or rectangular is only as wide as the groove and fits snugly into it.

The billet is pushed into the cavity its other end. The friction between the billet and the groove walls pulls the billet into the groove and presses it against the die. Since the tongue is stationary the frictional stresses on it is in the opposite direction. Nevertheless, there is net forward force because of three walls of the groove. The length of the shoe can be calculated to be sufficiently long so that pressure build up near the die is sufficient for extrusion to take place. The process has been developed to extrude aluminum, aluminum–magnesium–silicon alloy, aluminum–silicon alloys and copper. The process can be conveniently used to manufacture solder wire. The process requires high torque so hydraulic motors are best suited for it. The maximum possible reduction may be increased by making groove surface rough and the tongue surface smooth and polished.

The author and his team has developed a machine of this type based on the conceptual design process [6 and 11] and found that the total power (Pe) required for extrusion (by upper bound analysis) of a 8 mm aluminium feedstock is given as: $Pe = (P_D) + (Pc) + (P_B) + (P_b) + (P_f) + P_{turning}$, where P_D is die power which is obtained from the upper bound analysis as discussion earlier, (P_c) is container power, (P_B) is bearing power, (P_b) is biting power, (P_f) is deformation power or shear power at the shoe face and $P_{turning}$ is the turning power near the abutment. The powers may be calculated as: $P_c = 2\pi \mu (\sigma_0/\sqrt{3}) L_{opt} R_0 V_0$ watts, $P_B = \mu (\sigma'/\sqrt{3}) L_1 P_1 V_{end2}$ watts, $P_b = Y \mu x l_1 V_0$ watts,

(P_f) = Shear power = shear force* deformation velocity = $w\ l2K^*$ (peripheral velocity of wheel), and $P_{turning} = \sigma' 4/\sqrt{3} \times \mu \times L_1 \times P_1 \times V_0 \times \cos\theta$, where L_{opt} is optimum die length, R_0 *is* radius of feedstock, V_0 *is* peripheral velocity of wheel, σ_0 is the yield shear stress of feedstock material, L_1 is bearing length of feedstock, P_1 is perimeter of product, V_{end2} *is* velocity of product, and σ' is the flow stress of the feedstock material.

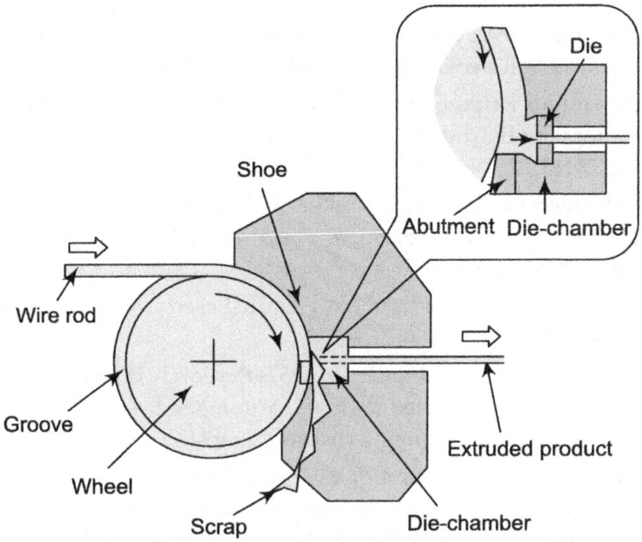

Fig. 5.28 Schematic elements of the continuous extrusion process. With permission from S. Kumar (2008)

Based on the above design a continuous extrusion set up (Fig. 5.29) has been developed for an extrusion wheel of diameter as 314 mm. The set up consists

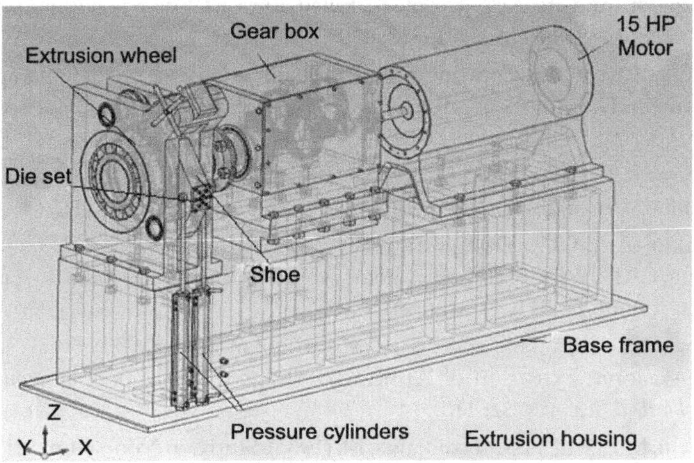

Fig. 5.29 Developed continuous extrusion machine. With permission from S. Kumar (2008)

of base frame, continuous extrusion housing, extrusion wheel, gear box, die set, and motor. The total power and total torque required for the 350 mm diameter wheel are calculated as 12.2 HP and 5710000 N-mm, respectively. Therefore, a motor of 15 HP has been selected.

REFERENCES

1. Santosh Kumar (1999). A Generative and Feature based Computer aided Process Planning system for cold Extrusion, *Ph.D. Thesis*, IIT-Kanpur, India.
2. Santosh Kumar, Kripa Shankar and Lal, G.K. (1999). A Feature Recognition Methodology for Extrudable Product Shapes, *International Journal of Production Research*, **37** (11); 2519–2544.
3. Santosh Kumar, Kripa Shankar and Lal, G.K. (2002). Analysis of Cold Extrusion of Non re-entry Product Shapes, *Trans. ASME J. of Manufacturing Sc. and Engineering*, **124** (1); 71–78.
4. Malpani M. and Kumar S. (2007). A Feature based Analysis of Tube Extrusion, *Int. J. of Mat. Proc. Tech.*, **190**; 363–374.
5. Santosh Kumar (2006). Report on DST Research Project [III.5 (109)/ 2001-SERC (ENGG)] on 'Knowledge Based Automatic System for Near-Net Shape Manufactured Products using Extrusion', 1–474.
6. Santosh Kumar (2008). Report on DST Research Project [SR/S3/MERC-07/2005 dated 12.05.06G] on 'Design, Development and Fabrication of a Continuous Extrusion Machine for Thin Sheets, Tubes and Sections', 1–185.
7. Yang, D.Y., Lee, C.M. and Cho, J.R. (1990). Analysis of Axisymmetric Extrusion of Rods by the Method of Weighted Residuals Using Body-Fitted Coordinate Transformation, *International Journal of Mechanical Sciences*, **32**; 101–114.
8. Yang, D.Y., Han, C.H. and Kiuchi, M. (1986). A New Formulation for Three Extrusion and its Application to Extrusion of Clover Sections, *International Journal of Mechanical Sciences*, **28;** 4201.
9. Gupta, A.K. (2004). Computer Aided Design and Drawing of Solid and Hollow Extrusion Dies for Generalized Shape, *Unpublished M. Tech. Thesis*, Mechanical Engineering Department, IT-BHU, Varanasi, India.
10. Chang, K.T. and Choi, J.C. (1972). Upper-bound solutions to tube extrusion problems through curved dies, *ASME J. Eng. Ind.* **94**; 1108–1115.
11. Shukla, A. (2007). Design and development of Con-Cast Extrusion Setup using CAE PROCESS, *Unpublished M. Tech. Thesis*, Mechanical Engineering Department, IT-BHU, Varanasi, India.
12. Yang, D.Y., Kim, H.S. and Lee, C.M., and Han, C.H. (1990). Analysis of Three Dimensional Extrusion of Arbitrarily Shaped Tubes, *International Journal of Mechanical Sciences*, **32;** 115.
13. Kiuchi, M., Kishi, H. and Ishikawa, M. (1981). Study on Non-symmetric Extmsion and Drawing, *Proceedings of 22nd International Machine Tool Design Research Conference*, 523.

CHAPTER 6 Forming Process Optimization and CAD/CAM/CAE of Dies

Jay S. Gunasekera
Department of Mechanical Engineering Ohio University, USA
Arjaan Buijk, Gary Huang, Gabriel Barton
Managing Director, Simufact-Americas LLC Plymouth, Michigan, USA

6.1 DIE MAKING [1]

Die making has been an art more than a science. The design and manufacture of dies have been passed down for decades from die and tool designers and tool makers to their apprentices and so on. This process has undergone some revolutionary changes with the advent of computers, CAD/CAM technology and powerful analysis such as the finite element method. Details of these will be presented in this chapter.

Dies for processes such as hot forging, extrusion, casting, and plastics molding were usually produced by manual machining, copy milling or by electric discharge machining (EDM). In the conventional EDM process, the electrode was frequently manufactured by copy milling from a suitable pattern. EDM is particularly useful for die manufacture as it can be employed for intricate shapes in hard-to-machine materials. Because of one of the author's experience, the dies considered in this chapter are mostly extrusion dies, and some forging dies.

6.2 EXTRUSION PROCESS AND DIE DESIGN [2]

Extrusion is a metal forming process in which a billet is forced to flow through a die to form a product of uniform cross-section along its length. The billet is usually of circular cross-section, whereas the desired product may have any shape for its cross-section. There are two common types of extrusion dies, viz.,

(i) the so-called "flat-faced" or "shear" die which is commonly used in the extrusion of aluminum; and

(ii) the "shaped" or "converging" die which has found applications in the lubricated extrusion of titanium, nickel and steel alloys (Fig. 6.1).

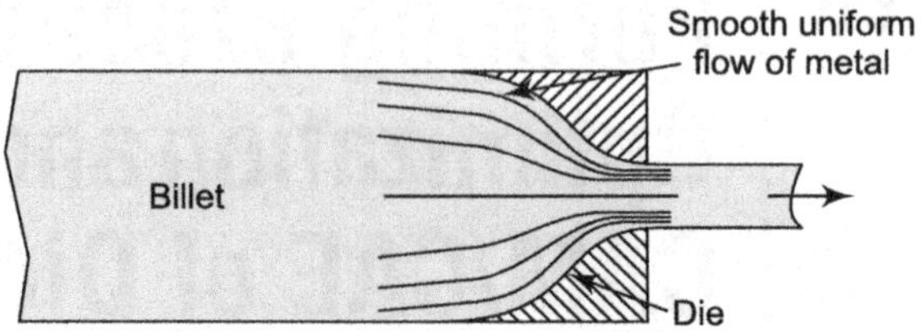

Fig. 6.1 Shaped extrusion die [1]

In conventional extrusion of soft alloys such as Aluminum using flat-faced dies, the material typically shears against itself and forms a dead, or stationary, zone at the die face (Fig. 6.2). The formation of the dead zone minimizes the overall rate of energy dissipation, but in general does not give, in extrusion of shapes, uniform metal flow at the die exit [2]. The non-uniform metal flow can result in twisting and bending of the emerging product. To prevent this, the flow rate is controlled through proper design of the die land and by proper positioning of the die cavity with respect to the billet center, as discussed by Nagpal and Altan [2] and Mockli and Locher [3].

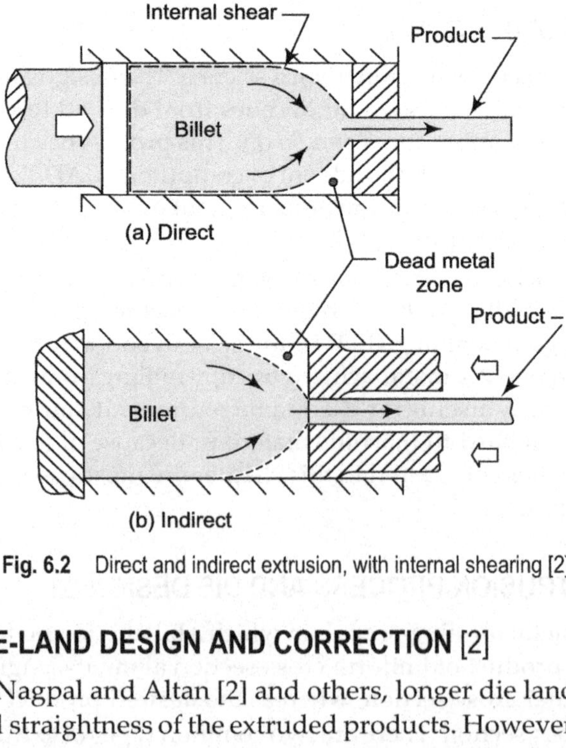

Fig. 6.2 Direct and indirect extrusion, with internal shearing [2]

6.3 DIE-LAND DESIGN AND CORRECTION [2]

According to Nagpal and Altan [2] and others, longer die lands improve the tolerances and straightness of the extruded products. However, the extrusion load increases with increasing length of the die land. Thus, the die land must be designed to give a uniformly strained product within desired tolerances and without excessive extrusion pressure.

Wilson [4] and Feldman [5] recommend certain land lengths for extrusion. Sieber [6] states that in axi-symmetric extrusion, there is an optimum land length which is given by the following equation:

$$L_e = 1.2d \text{ to } 0.8d \tag{6.1}$$

where L_e is the land length and d is the land diameter.

In shape extrusion, unlike in rod or tube extrusion, the die land length is altered to slow down or to speed up metal flow. According to Bello [7], the material flow can be enhanced by filing a relief bearing or can be slowed down by filing a choke surface on the die land, as seen in Fig. 6.3.

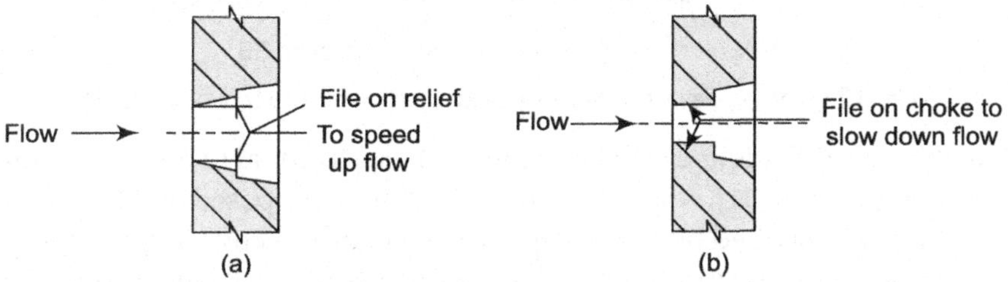

Fig. 6.3 Modified die land for control of extrusion flow [2, 7]

According to Nagpal and Altan [2], the shape of the extruded section can be modified by filing choke and relief on the die land, as shown in Fig. 6.4. In Fig. 6.4, the metal at the outside of the right leg flows faster than that inside. Therefore, with the die-land corrections indicated in Fig. 6.4, the right leg will tend to go toward the inside. A similar but reverse situation exists in Figs. 6.3(b) and 6.4.

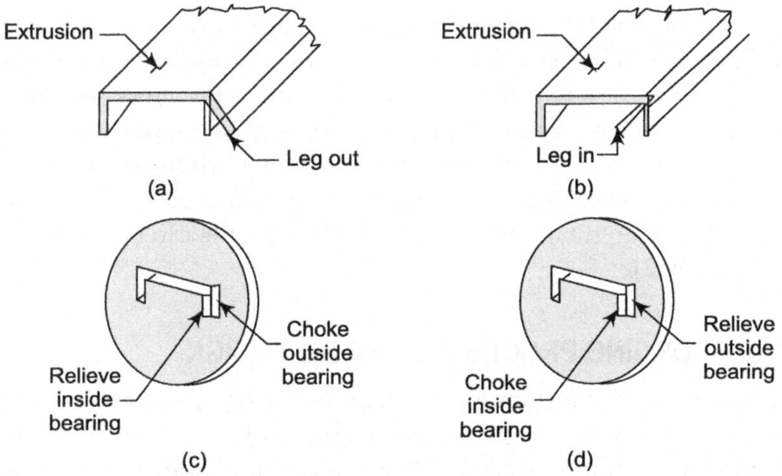

Fig. 6.4 Modified die land to control extruded product shape [2, 7]

In the practical design of the die land for extrusion of aluminum shapes, the land is varied in length according to section width, in order to obtain uniformity of flow. As described by Chadwick [8] and shown in Fig. 6.5, the thin section is provided with less land than the thicker section [2].

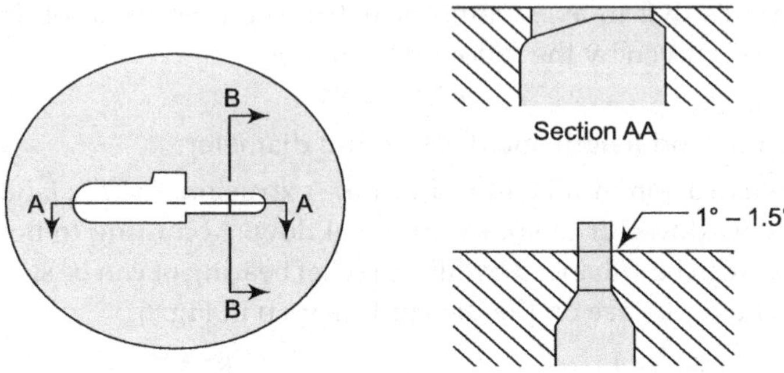

Fig. 6.5 Varying die land in accordance with section width to control flow [2, 8]

An empirical guideline is to keep the land length equal to one to two times the section thickness [1]. Another empirical relation, proposed by Matveev and Zhuravski as described by Perlin [9], is to make the effective length of the die land, at the various portions of the die opening profile, inversely proportional to the specific perimeters of these portions, as follows:

$$\frac{l_m}{l_n} = \frac{P_{sn}}{P_{sm}}$$
$$P_{sm} = \frac{P_m}{A_m} \tag{6.2}$$
$$P_{sn} = \frac{P_n}{A_n}$$

where l_m, P_m, A_m, and P_{sm} are effective land length, perimeter, cross-sectional area, and specific perimeter, respectively, of the portion '*m*' in the die profile, l_n, P_n, A_n, and P_{sn} are effective land length, perimeter, cross-sectional area, and specific perimeter, respectively, of portion '*n*' of the die profile. '*m*' and '*n*' are any two portions of the die profile which have different cross-sectional thicknesses. When the die land length is assigned to a specific portion of the profile, the land length at other portions of the profile can be determined by using Eq. (6.2) [2].

6.4 FORGING PROCESSES AND DIE DESIGN

Forging is one of the most important manufacturing operations. It is a plastic deformation process similar to extrusion but, unlike extrusion, it can be used to manufacture complex 3D parts. Forging can be classified into three main categories: open-die forging, impression-die forging and closed-die forging.

6.4.1 Open-Die Forging

The process is schematically illustrated in Fig. 6.6. At least one of the work-piece surfaces deform freely, and hence the open-die forging process produces parts of lesser accuracy and dimensional tolerance than impression-die or closed-die forging. However, the tooling is simple, relatively inexpensive and can be designed and manufactured with ease.

6.4.2 Impression-Die Forging

This forging process is schematically illustrated in Fig. 6.7. It can be used to produce complex 3D shapes having a greater accuracy than closed-die forging. The specially manufactured dies contain the negative of the forging to be produced. In one form of the process, the shape is obtained by filling the die cavity formed by the upper and lower dies. Excess material is allowed to escape into the flash. There may or may not be special provision for flash formation. These dies are illustrated in Figs. 6.8 and 6.9.

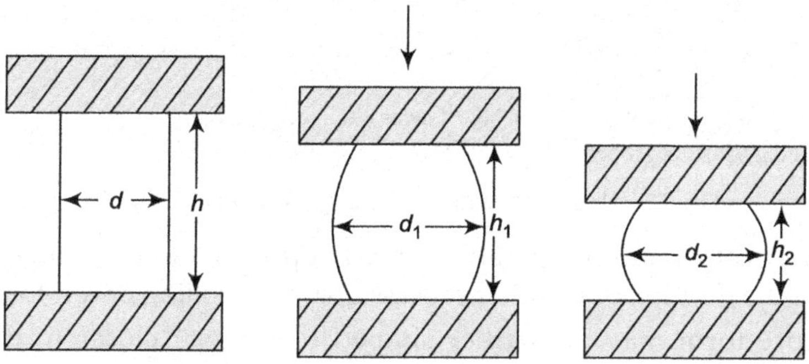

Fig. 6.6 Illustrating open-die forging [1]

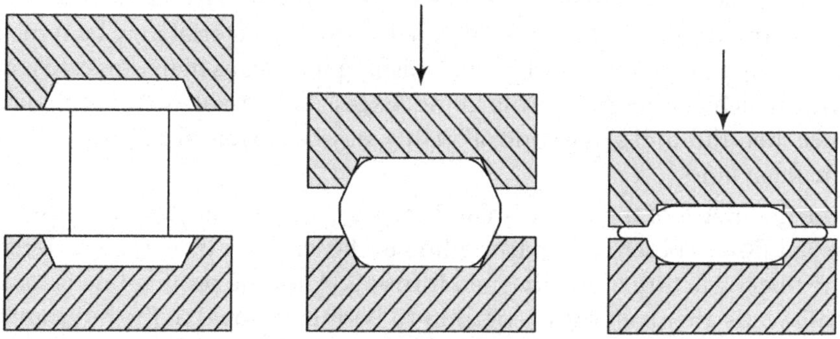

Fig. 6.7 Illustrating impression-die forging [1]

The die cavity must be filled without defects. A complex shape cannot be filled completely without defects in one operation starting with a rectangular or cylindrical shape. Some pre-forging steps are necessary to ensure complete filling of the dies without defect formation. The pre-forging or pre-forms can be

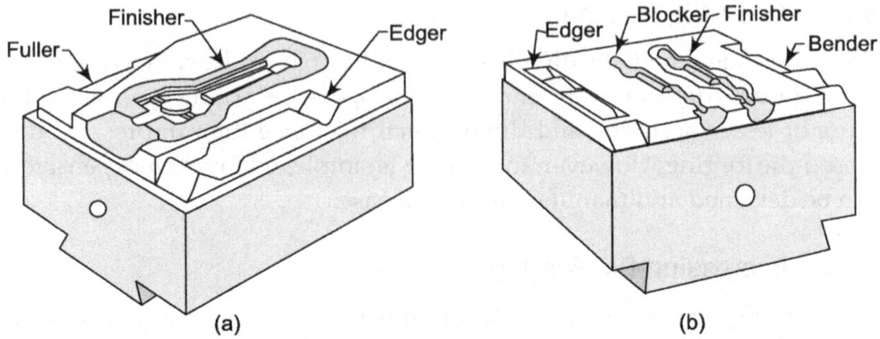

Fig. 6.8 Dies for impression-die forging (Courtesy of ASM) [10]

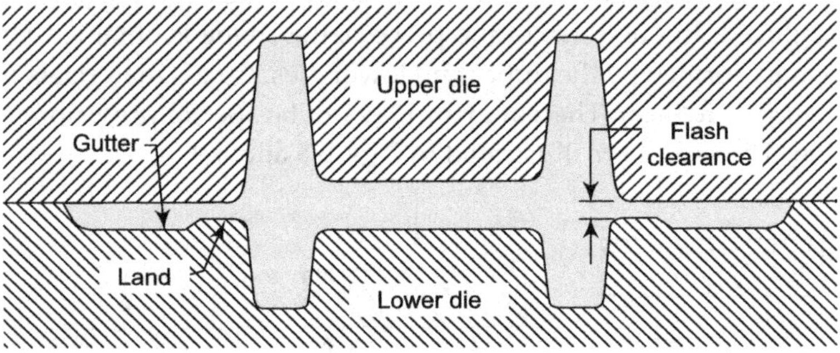

Fig. 6.9 Dies for impression-die forging (Courtesy of ASM) [10]

produced in other forging operations such as open-die or in the same die with different die cavities, or by other processes such as rolling or roll-forging.

The pre-form may be further shaped to bring it closer to the final configuration in a 'blocker die', which ensures proper distribution of material within the die. Excess material is allowed to run out between the dies into a flash. Before forging in the finishing die the excess material may be removed in a trimming die. A thick flash in the finishing die means high pressure within the die, which assures proper die filling. Excessive pressure may break dies or reduce their life, and some control on this may be exercised by proper control of the flash land.

The geometry of the pre-form and the forging dies must promote smooth material flow. Therefore, a parting line is chosen with proper consideration of the fiber structure of the finished forming. After the parting line is located the die walls are given sufficient draft to permit removal of the forging from the die cavities. Fillets and corners must be given appropriate radii to ensure smooth material flow and long die life.

6.4.3 Closed-Die Forging

In closed-die forging the work piece is completely trapped in the die and no flash is generated. Material utilization is very high, but the volume of the work-piece before and after forging is identical and hence control of incoming material volume becomes critical. Excess material can create large pressures, which are liable to cause die failure.

6.4.4 Die Manufacture

Conventional die manufacture is more of an art than a science. Several alternatives are available for impression-die manufacture. The common methods are conventional milling, copy milling, electric discharge machining (EDM) and electro-chemical machining (ECM). Dies may also be cast from special tool steels.

6.5 ADVANCED ANALYTICAL TECHNIQUES FOR COMPLEX EXTRUSION PROBLEMS [11]

Typical extrusion processes, such as with soft aluminum, are carried out using flat-faced dies or so-called square dies under hot conditions with or without lubricants. But, for certain important applications, such as with harder alloys like steel and titanium, it may not be possible to use flat-faced dies. The only solution is to resort to converging dies, with streamlined surfaces, because of their inherent advantages.

In the case of extrusion or drawing where the exit cross-section is different in shape from that at entry, the material flow does not remain on the same radial plane which contains the longitudinal axis, so that a 3D approach is necessary.

The upper bound method giving equations for the analysis of extrusion or the direct drawing of regular polygonal sections starting from a round bar stock, proposed by Gunasekara and Hoshino [11], is discussed here. The major advantage of this method is that it only requires the establishment of the generic streamline of a particle in a three-dimensional co-ordinate system.

6.5.1 Upper Bound Solutions for Extrusion and the Drawing

Upper bound solutions for the extrusion and the drawing of regular polygonal sections from cylindrical billets through straight converging dies, where the die surface comprises straight stream-lines, have been successfully demonstrated by Gunasekara and Hoshino [11]. These dies would produce changes of flow-direction at inlet and outlet of the dies. These were taken into account as velocity discontinuities. However, close observation of the dead metal zone reveals that the profile of the dead metal zone is, in general, a shaped envelope which minimizes the redundant shear, as shown schematically in Fig. 6.10. This profile might be considered as the best one under the given conditions where sticking friction occurs along the interface between the work-piece and the dead metal zone. For this reason, it becomes very important to consider curved dies as against straightly converging dies for the extrusion or drawing of sections.

In this section, previous concepts are directly extended to curved dies for the extrusion of regular polygonal and rectangular sections. The emphasis here is on the analysis of 'the streamlined die', though the method to be discussed is also applicable to other curved dies such as concave and convex shaped dies. The die surface of 'the streamlined die' is constructed by smooth curved stream-lines with zero gradient along the extruding direction at the entry and the exit of the die.

6.5.2 Streamlined Dies [12]

From a practical point of view, one of the major draw-backs of the use of streamlined dies for the drawing of cylindrical bars is that the diameter of the bar has to be controlled to a fine tolerance so as not to produce shearing of the material at the inlet of the die. Despite this drawback, the streamlined die, which might be considered as a derivative of 'the sigmoidal die', has such attractive features as to reduce the reforming stress and more importantly to improve the product quality.

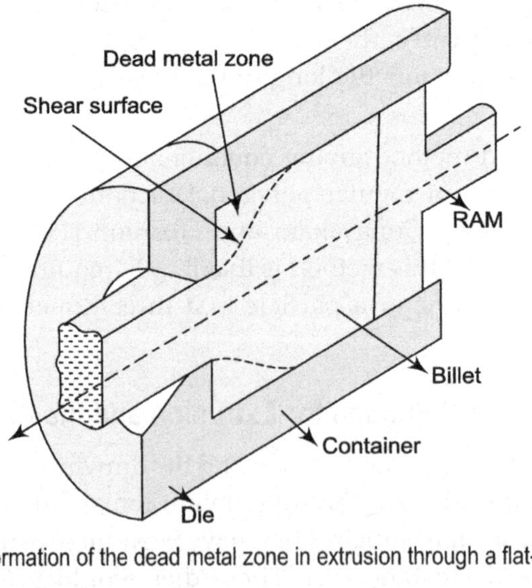

Fig. 6.10 Formation of the dead metal zone in extrusion through a flat-faced die [12]

Richmond and Devenpeck [13] first proposed the sigmoidal die as an ideal shape of die for the frictionless drawing of an ideal rigid-perfectly plastic strip. Theoretically, this die requires the minimal forming stress and produces homogeneous deformation. It has actually been shown by Devenpeck and Richmond [14] that the sigmoidal die can produce better mechanical properties of the product such as in fatigue life and ductility at least under well lubricated conditions, when compared with other straight, concave and convex shaped dies. However, the theoretically ideal die has a relatively large surface area, resulting in an increase of the frictional work done. Also, Hill [15] pointed out that many different streamlined profiles could be designed for a given reduction of area and in consequence the die profile cannot be uniquely determined. For these reasons Richmond and Morrison [16] proposed a drawing die of minimum length, assuming again a friction-free die surface. Devenpeck [17] has tested the efficiency of this die and concluded that a significant characteristic of die profile which minimizes structural damage and work is a zero-entrance angle. He also concluded that reducing the angle of the exit also appears desirable and thus consideration should be given to the ideal die which has zero exit angle as well as zero entry angle.

The above conclusions raised by Devenpeck have been incorporated by the present researchers in their analysis of the extrusion of non-circular sections. Other researchers such as Chen and Ling [18] and Avitzur [19] have developed upper-bound solutions to extrusion through straight and curved dies. Frisch and Mata-Pietri [20] and Mata-Pietri and Frisch [21] performed experiments using these dies and obtained good agreement with the theoretical results. However, the analyses were limited to axisymmetric extrusions. Nagpal and Altan [22] demonstrated the utility of the dual stream functions to obtain an upper bound solution for the extrusion of an elliptical cross-section from a cylindrical billet through streamlined dies. Their concept was later extended by Nagpal [23] to the deformation problem of rectangular sections from other rectangular bars. In contrast, Yang and Lee [24] proposed the conformal mapping approach to find a kinematically admissible velocity field for extrusion through concave and convex shaped dies where geometrical similarity is preserved throughout the deformation. This method eliminated any restriction of cross-sectional shape of sections. Yang *et al.* [25] also analyzed the extrusion of a helical section from cylindrical bars through straightly converging dies, and their method is in principle applicable to other curved dies. However, the methods discussed above still involve limitations on the geometry of the die and it is difficult to apply their methods to the die geometries analyzed here. The explanation of the present method follows in the next section.

6.5.3 Velocity Field [26]

The following assumptions are required to construct the kinematically admissible velocity field for the extrusion of regular polygonal sections from cylindrical billets.

1. The material of the billet passing through sector OEG (Fig. 6.11) at the die entry goes through triangle LFH at the die exit, preserving the extrusion (or the drawing ratio).
2. Stream surface OEFL consists of a number of curved stream-lines which start from a point (say E') at the entry and end at a corresponding point (say F') at the exit maintaining the proportionality of the position.

The above assumptions are illustrated in Fig. 6.11. As mentioned in the previous section, the present attempt is to produce the die configuration which requires no energy dissipation at the entry and the exit of die. Also, it has been reported, by Nagpal and Altan [22] that the extrusion force is not sensitive to higher order polynomial curves representing the die profile. Therefore, for simplicity of the present analysis, streamlines are represented by cubic curves satisfying the smooth entry and exit of the material flow [26].

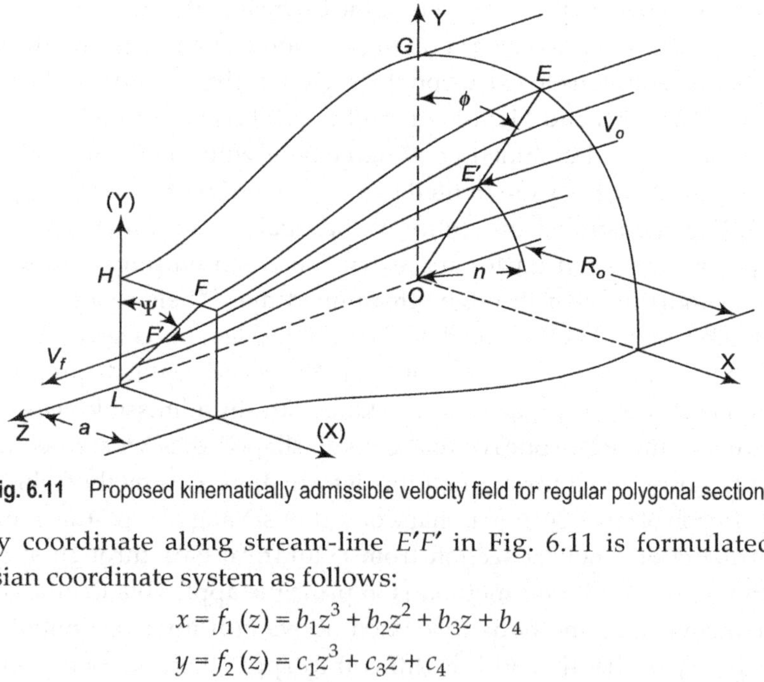

Fig. 6.11 Proposed kinematically admissible velocity field for regular polygonal sections

Any coordinate along stream-line $E'F'$ in Fig. 6.11 is formulated in a Cartesian coordinate system as follows:

$$x = f_1(z) = b_1 z^3 + b_2 z^2 + b_3 z + b_4 \qquad (6.3)$$

$$y = f_2(z) = c_1 z^3 + c_3 z + c_4$$

$$z = z$$

where b_i and c_i (i = 1, 2, 3, and 4) are constants determined by the boundary conditions. Consider that this stream-line does not produce any abrupt change of flow direction along the extrusion axis at the entry and the exit of die, the boundary conditions are given for Eq. (6.3) as:

$$x = n \sin \varphi \qquad \frac{\partial x}{\partial z} = 0 \text{ at } z = 0 \qquad (6.4a)$$

$$y = n \cos \varphi \qquad \frac{\partial y}{\partial z} = 0 \text{ at } z = 0 \tag{6.4b}$$

$$x = \frac{n}{R_0} a \tan \varphi \qquad \frac{\partial x}{\partial z} = 0 \text{ at } z = L \tag{6.4c}$$

$$y = \frac{n}{R_0} a \qquad \frac{\partial y}{\partial z} = 0 \text{ at } z = L \tag{6.4d}$$

where

$$\tan \varphi = \frac{N}{\pi} \tan\left[\frac{\pi}{N}\right]\varphi = c\varphi \tag{6.5}$$

R_o is the radius of billet, a is the half side length of product cross-section, N is the number of sides in regular polygon, L is the length of die, n is the distance from the axis to an arbitrary point E' at the die entry, and ϕ and φ are the angles between the plane of symmetry and the stream surface at entry and exit of the die, respectively. Substitution of these boundary conditions into Eq. (6.3) gives:

$$x = n \sin \varphi + n\left[\frac{a}{R_0} c\varphi - \sin \varphi\right]\left[3\left[\frac{z}{L}\right]^2 - 2\left[\frac{z}{L}\right]^3\right]$$

$$y = n \cos \varphi + n\left[\frac{a}{R_0} - \cos \varphi\right]\left[3\left[\frac{z}{L}\right]^2 - 2\left[\frac{z}{L}\right]^3\right] \tag{6.6}$$

$$z = z$$

Equation (6.6) can be rewritten in the following form:

$$x = n \sin \varphi + n\left[\frac{a}{R_0} c\varphi - \sin \varphi\right] f(z)$$

$$y = n \cos \varphi + n\left[\frac{a}{R_0} - \cos \varphi\right] f(z) \tag{6.7}$$

$$z = z$$

where $f(z) = 0$ at $z = 0$ (6.8)

$f(z) = 1$ at $z = L$

In the present analysis, function f is represented by the following cubic curve:

$$f = f(z) = 3\left[\frac{z}{L}\right]^2 - 2\left[\frac{z}{L}\right]^3 \tag{6.9}$$

Equation (6.7) describes not only the co-ordinates inside the plastically deforming region but also the relationship between the Cartesian and n, z

coordinate systems. Although the present analysis employs the cubic curve, represented by Eq. (6.9), for the description of the die profile and the assumed stream-lines of particles, it should be stated that, in Eq. (6.7) function f can be any general function of z provided the function satisfies the boundary conditions in Eq. (6.8).

Assuming that the plastically deforming zone is bounded by shear planes at the entry and the exit of die, and utilizing the determinant of the Jacobian of Eq. (6.7), the velocity components for incompressible material are determined as

$$V_x = \frac{n\left[\left(\frac{a}{R_0}\right)c\varphi - \sin\varphi\right]}{g(\varphi, z)} V_0$$

$$V_y = \frac{n\left[\left(\frac{a}{R_0}\right) - \cos\varphi\right]}{g(\varphi, z)} V_0 \tag{6.10}$$

$$V_z = \frac{1}{g(\varphi, z)} V_0$$

where

$$g(\varphi, z) = \left[(1-f)^2 + c\left(\frac{af}{R_0}\right)^2\right] + \left[c(1-f)\left(\frac{af}{R_0}\right)\right]\varphi \sin\varphi + \left[(1+c)(1-f)\left(\frac{af}{R_0}\right)\right]\cos\varphi \tag{6.11}$$

Strain-rate components, expressed by the following tensor form, can be easily found with the aid of coordinate transformation:

$$\xi_{i,j} = \frac{1}{2}\left[\frac{\partial V_i}{\partial x_j} + \frac{\partial V_j}{\partial x_i}\right] \tag{6.12}$$

It is apparent that the velocity field formed by Eq. (6.10) satisfies the velocity boundary condition, i.e., the volume continuity of material at the inlet and outlet of the die. This velocity (or strain-rate) field has also been proved to satisfy the incompressibility condition [25]. Hence the proposed velocity model fulfills the stringent requirement for the construction of a kinematically admissible condition.

6.5.4 Upper Bound Solution [26]

Since the streamlined die produces no velocity discontinuities at the velocity boundaries, the upper limit to the total power consumption (J^*), required to deform cylindrical billets to regular polygonal sections through this die, is represented as the sum of the power consumed due to the plastic deformation (W_i) and due to the die surface friction (W_s). Considering the symmetry of the die geometry, J^* is obtained as

$$J^* = 2N\,[W_i + W_s] \tag{6.13}$$

where

$$W_i = \frac{2}{\sqrt{3}}\sigma_0 \int_V \left[\frac{1}{2}\xi_{i,j}\xi_{i,j}\right]^{1/2} dV$$

$$= \frac{2}{\sqrt{3}}\sigma_0 \int_0^L \int_0^{\phi_m} \int_0^{R_0} \left[\frac{1}{2}\xi_{i,j}\xi_{i,j}\right]^{1/2} |\det J|\, dn\, d\varphi\, dz \tag{6.14}$$

$$W_s = m\frac{\sigma_0}{\sqrt{3}} \int_{S_3} |\Delta V_3|\, dS_3$$

$$= m\frac{\sigma_0}{\sqrt{3}} \int_0^L \int_0^{\phi_m} [V_x^2 + V_y^2 + V_z^2]_{n=R_0}^{n=1/2} \frac{1}{\cos\alpha}\frac{\partial(x,z)}{\partial(n,z)}\, d\varphi\, dz$$

Here ϕ_m is the maximum angle of inclination of the die surface element with respect to the projected surface of the element on to the x-z plane. ϕ_m is the maximum value of the angle determined by the symmetry of the die shape, and 'det j' is the determinant of the Jacobian of Eq. (6.7). Knowing the velocity and the strain rate components, and the coordinate transformation equation, the volume and surface integration are numerically carried out for given values of the yield stress of the material, and constant frictional factor (m) to obtain the total power consumption. The power computed can be converted to the average pressure (P_{ave}) and the relative stress (R_s), as follows:

$$P_{ave} = \frac{J^*}{\pi R_0^2 V_0} \tag{6.15}$$

$$R_S = \frac{P_{ave}}{\sigma_0} \tag{6.16}$$

The actual die length (L) is also reduced to the relative die length in the following sections, given by

$$R_L = \frac{L}{R_0} \tag{6.17}$$

6.5.5 Comparison with the Straightly Converging Die [26]

Using the above procedure, the power consumed due to plastic deformation (W_i) and die surface friction (W_s) are computed for the streamlined die as shown in Fig. 6.12 together with the total power consumption. The components of power consumed for the plastic deformation (W_i), velocity discontinuities at the entry (W_e) and the exit (W_f) boundaries, and die surface friction (W_s) are also demonstrated in Fig. 6.12. These power components were computed by the previous solution by Gunasekera and Hoshino [10], for the same extruding conditions as for the streamlined die. These figures explain the contribution as those for the streamlined die and the contribution of each power component to the total extrusion power.

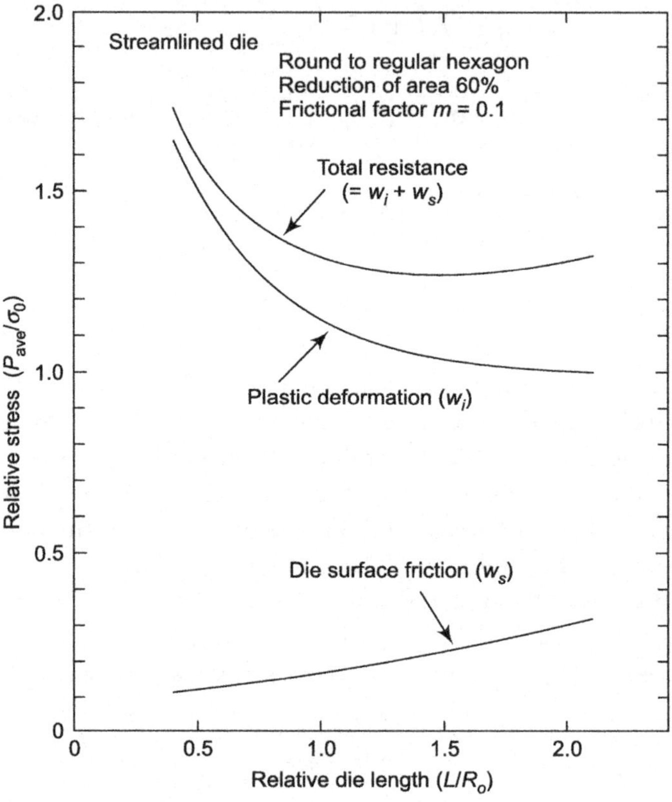

Fig. 6.12 Contribution of power components to the total extrusion power consumed inside the streamlined die [12, 26]

In the case of extrusion through the streamlined die, the comparison with the total power component due to plastic deformation is dominant, in particular for the shorter length of die. This power, due to plastic deformation, decreases drastically with increase of the die length. This change is attributed to the change of the flow direction inside the plastically deforming zone. In contrast, for the straightly converging die, the power component (W_i) remains almost stable at least in the region analyzed here. However, the effects of velocity discontinuities are quite appreciable. From Fig. 6.13, it is observable that the optimal die length, which requires the minimal total power, is primarily determined by the power components due to velocity discontinuities (W_e, W_f) and die surface friction (W_s).

Comparing the streamlined die with the straightly converging die, it is noted that the magnitude of the power component, W_i of the straightly converging die is smaller than that of the streamlined die for the same extruding conditions. However, with regard to the total power, the streamlined die always require less power as presented in Fig. 6.14. This figure shows the effect of the die profile upon the forming stresses in extrusion. It is clear, in the figure, that the streamlined die defined by cubic curves is superior to the die defined by straight streamlines in terms of the forming stress. The optimal die length of the streamlined die is shorter than that of the straightly converging die. These conclusions have also been verified by experiments performed on lead specimens at room temperature. Details of the experimental work on various die configurations will be reported in subsequent sections.

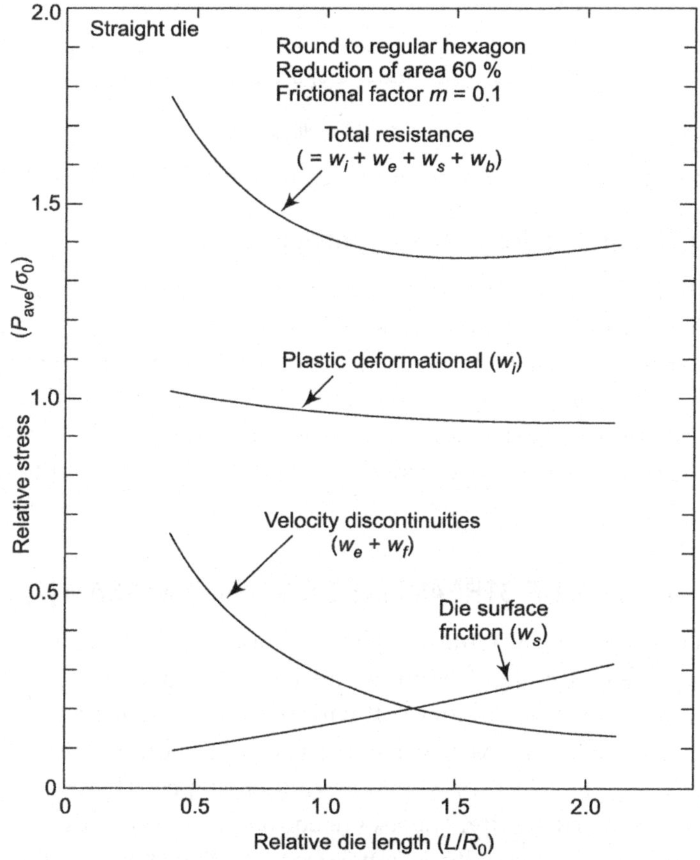

Fig. 6.13 Contribution of power components to the total extrusion power consumed inside the straightly converging die [12, 26]

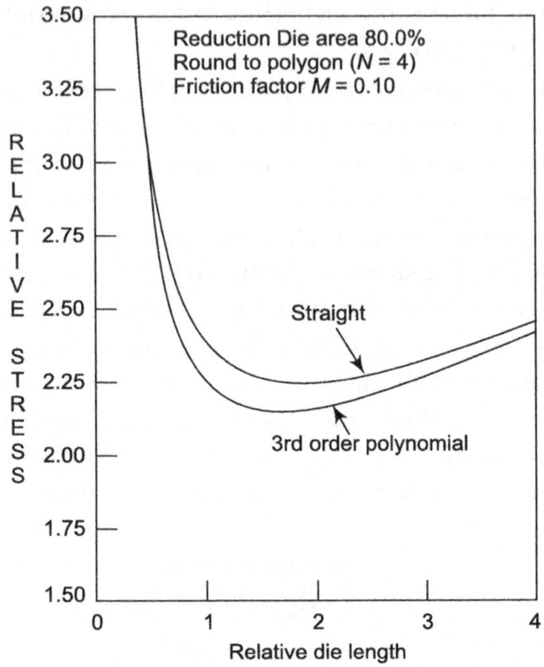

Fig. 6.14 Effect of die profile on the extrusion stresses for square sections [12, 26]

6.5.6 Application to Regular Polygonal Sections

The effect of die surface friction on the relative extrusion stresses required to extrude billets to a square product ($N = 4$, N is the number of sides to the polygon) are presented in Figs 6.15 and 6.16 with respect to the relative die length and the reduction of area, respectively. The optimal die length which produces the minimal extrusion stress decreases with the increase of friction, and the relative extrusion stress increases with the increase of the reduction of area.

6.6 DESIGN OF STREAMLINE DIES USING STREAM [1, 10]

STREAM is a software system, which is developed by Gunasekera [1] consists of three packages for the modeling and design of shaped extrusion dies. Although there have been other software packages available for the design of extrusion dies, STREAM is considered to be the only software package capable of designing dies for re-entry product shapes. The package is fully interactive and user-friendly. The user needs no prior knowledge of computer programming or advanced die design technology in order to design extrusion dies. The product geometry may be entered interactively through the keyboard, and can also be read from a data file.

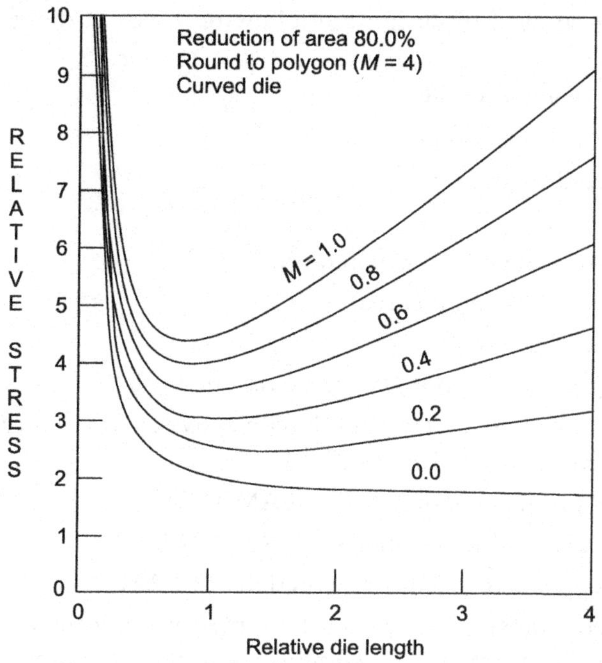

Fig. 6.15 Effect of die profile on the extrusion stresses for square section with the die length [12, 26]

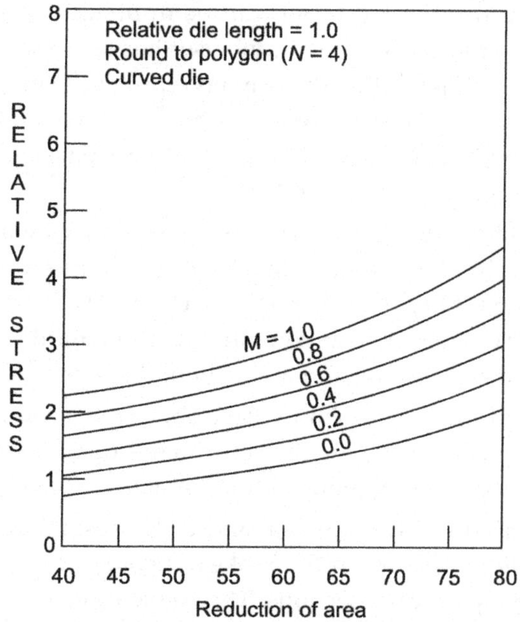

Fig. 6.16 Effect of friction on the extrusion stress for square section with the reduction of area [12, 26]

STREAM can be used to design the following types of die geometries (or shapes):

1. Straight-converging die
2. Convex-extrusion type
3. Concave-drawing type
4. Parabolic
5. Cubic streamlined (based on radius)
6. Streamlined (based on area)
7. Constant strain rate

The following product geometry can be handled:

1. Solid product geometry, round, with convex and re-entry geometries - STREAM 4.3 and STREAM STSQ
2. Hollow product geometry - STREAM 4.4
3. Multi-hole product geometry - STREAM 4.5

Three levels of geometric data are stored in STREAM:

(a) Product geometry, which can be typed through the keyboard using STREAM or using a digitizer attached to the computer. STREAM also offers two kinds of radii calculation of the product geometry, one is a 3-point approximation, and the other is a 5-point approximation. For larger radii, the latter one can be applied to obtain a smoother corner or leg radius. Moreover, the user has an opportunity to use the mapping coordinates directly to avoid repetitive keying-in. All product geometries and mapping geometries may be stored and retrieved using STREAM or CUTTER (CUTTER is a software package for the CAM of the dies designed using STREAM).

(b) Graphics compatible geometry can be used for viewing the geometry of the die. The data file is compatible with a number of popular 3D graphics and CAD packages such as AUTOCAD, ProEngineer, etc. The die geometry can be displayed with or without hidden lines, with color shading and smoothing, and at different angular or axial positions.

(c) A Programming Tool or Numerical Control (NC) compatible geometry is a data file containing the 3D coordinates of all the points of the die geometry stored either along splines or across splines.

STREAM is based on the most advanced die design technology. It uses a mapping technique (based on the Stokes Theorem) for the design of dies for complex re-entry product sections. The conventional die design concepts developed by various researchers, notably Nagpal and Altan [2], and Gunasekara and Hoshino [11], cannot be used to design dies for re-entry product sections.

Die design consists of mapping sections of the billet on to sections within the product on proportional area basis. Thus, points on the perimeter of billet can be mapped on to corresponding points on the perimeter of the product ensuring the same extrusion ratio is preserved within each sector. Thereafter splines (of any geometry) can be fitted from the billet to the product to define the surface of the die.

The estimation of the ram force, stress, strain, and strain rate distribution is based on the slab method. This provides approximate but adequate results for the selection of the press and other variables. The length of the die, which can be used to minimize the extrusion force, is still a user input variable. However, extensive physical and analytical modeling (using FEM) at the U.S. Air Force Material Laboratories has shown that for good results with Al alloy powders with SiC whiskers, the ratio of die length to billet diameter (L/D) of about one should be used for extrusion ratios of up to about 20. Larger extrusion ratios would require longer die lengths.

STREAM can be used for streamlined die design with a variety of I/M alloys (steel, Al, titanium, etc.) and Powder Metallurgy (P/M) alloys with or without whiskers. It is particularly useful for round to very complex geometry extrusions with either single or multi holes and for hollow products. STREAM is considered the only software package capable of designing dies for re-entry product shapes starting from any arbitrary shaped billet and/or multi-hole dies and/or extrusion of hollow products. Some examples of die surfaces generated using STREAM are shown in Fig. 6.17 [27].

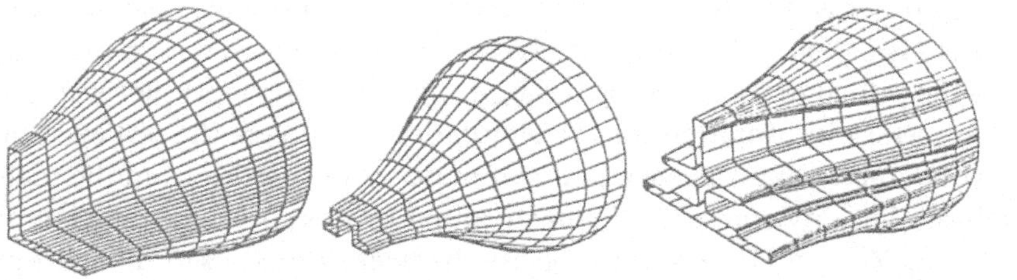

Fig. 6.17 Some examples of die surfaces generated using STREAM [10, 27]

6.7 ADVANCED NUMERICAL METHODS FOR SIMULATION METAL FORMING PROCESSES

The descriptions in this chapter are based on the commercial metal forming simulation software Simufact.forming, used by companies, universities and research institutes worldwide for simulating a wide variety of metal forming processes. The software contains two approaches to simulate metal flow, a finite element (FE), and a finite volume (FV) method. Typically, the FV solver is best suited for flash forgings, such as a crankshaft hot forging. The FE solver is best used in all other cases. Both approaches are discussed in this chapter.

Numerical modeling for simulation of metal forming processes, in general, consists of three major steps, namely, pre-processing, analysis and post-processing. In the pre-processing step, it will be asked to provide the required information of the forming process interested through the forming GUI (Graphical User Interface). Based on the information, analyses will be carried out by the solvers. There are two solvers, one can select based on the forming models. These two solvers are the finite element solver and the finite volume solver. The analysis by the solver is the most time consuming step. The solver will generate results which can then be reviewed in the post-processing step. In the post-processing step, one will use the forming GUI to view results and create reports that complete a modeling iteration. Very often some decision will be made based on the post-processing report and one will start another modeling iteration until satisfactory results are achieved. The analysis step and the post-processing step can happen at the same time. This means you can view results before the analysis step finishes. This is because the analysis is done incrementally and many intermediate results are generated before the solver completes the analysis. These intermediate results can be investigated while the analysis is being carried out by the solvers. The analysis can be terminated if so desired.

6.7.1 Pre-Processing Basics

Pre-processing is the first step in the modeling. You need to use your knowledge of the forming process to create a model that approximates the real forming process. You will import geometry files of tools, dies and workpieces into the GUI. You will define the characteristics of machines or presses that drive the punch or tool. The material properties of the workpieces and tools are important information for the simulation. You have the option to input your own material properties or insert material properties from the material database. The environment of your forming process is also important in the modeling. You will be asked to specify the heat properties of the workpiece, tools and dies. These include the initial temperature, heat transfer coefficient and radiation emissivity to the surrounding area. You will need to specify the friction properties and wear properties between the workpiece and tools. A volume mesh is required for the FE analysis, while a surface mesh is required for the FV analysis. In both cases, if the deformation is large, remeshing is needed in the analysis. All the meshing and remeshing are carried out automatically with some user control input in the pre-processing step.

The analysis step in simulation can consume a lot of computer time. The following are some tips that you can use to shorten the analysis time. You can use your knowledge of the forming process to simplify the model and to speed up the analysis step without compromising your analysis results. For example, you can use a two dimensional (2D) model instead of a three dimensional model (3D) in your simulation. If your forming process is axi-symmetrical such as cylinder upsetting, a 2D axi-symmetric model should be sufficient for your modeling need. You can also simplify your model further if you think the deformation of the tools and dies is insignificant compared to the deformation of workpiece. Thus, you can define your tools and dies as rigid bodies. In cold forming, the heat transfer in the rigid bodies may not be important. You can then define your rigid tools and dies as rigid dies without heat transfer. If the die and tool deformation is important in the simulation, there are methods for you to do a die stress analysis in two steps, in which the die stress analysis is carried out after the forming analysis is completed. This procedure also helps speeding up your simulation. Furthermore, you can make use of symmetry in your modeling so that instead of modeling the whole workpiece you only model a section of the workpiece. For example, in gear forming, you can simulate one tooth instead of the whole gear. You can also make use of the cyclic symmetry in your modeling to speed up analysis. For example, in gear forming, the tooth might not be straight but curved with certain angle. If your modeling involves many forming stages and in the early stages, the forming simulation can be simplified as 2D model, you can create your model in 2D for those early stages and 3D in the later stages. This 2D-to-3D procedure will significantly speed up your analysis. In multistage forming simulation, you can transfer results from previous analysis to the analysis of the next stage as the initial condition. You can also switch the modeling from FE solver to FV solver in a later stage if deformation is very large due to forming of a flash or fold. In those cases the FV method is very efficient and is a good solver choice. Using the parallel processing capability in the software is another way to speed up the analysis if your computers are equipped with multiple processors. The domain decomposition method (DDM) in the software allows you to make use of the multiple processors on your computer. During the pre-processing step you select how many processors you would like to use to simulate your forming process. The DDM technique divides your model automatically into different domains and assigns these domains to each processor so that they can be analyzed separately in parallel. Multi-threading is another way to speed up the analysis. It utilizes the multiple processors in your computer to solve equations in parallel.

6.7.2 Pre-Processing Terminologies

The following are explanations of some basic terminologies.

2D-Two dimensional model

A 2D model usually requires much less computational power to simulate. Most of the 2D simulations can be carried out in a few minutes. There are two types of 2D models: axisymmetric and plane strain.

Axi-symmetric model

If your model geometry and forming conditions are rotational around one axis, you can use this model and assume the simulation results are the same axi-symmetically. For example, a ring upsetting can be simulated as a 2D axi-symmetric model (Figs 6.18 and 6.19).

Fig. 6.18 Ring upsetting model

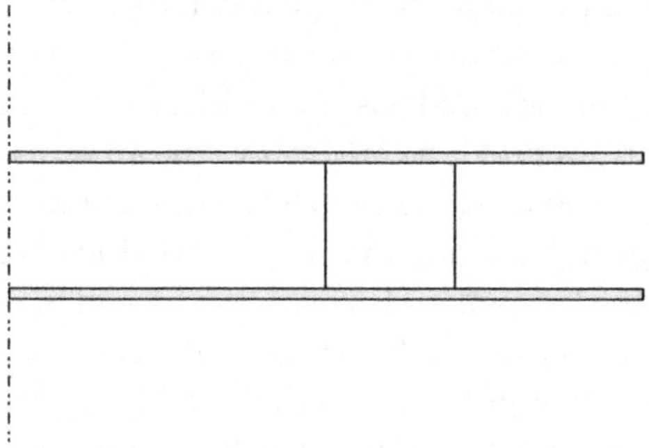

Fig. 6.19 2D axi-symmetric model

Plane strain model

If your model geometry and forming conditions are homogeneous in one of the three dimensions and the length of that dimension is much longer compared with the other two dimensions, you can use this model and assume the simulation results are the same along that dimension. For example, a flat rolling can be simulated as a 2D plane strain model (Figs 6.20 and 6.21).

Symmetry planes

If you can find symmetry in the model, you can simulate a section of the workpiece instead of simulating the whole part. You will need to use symmetry planes to mark its boundaries.

Cyclic symmetry model

If the model is symmetric periodically by rotating with a certain angle around a fixed axis, you can simulate your model with cyclic symmetry instead of

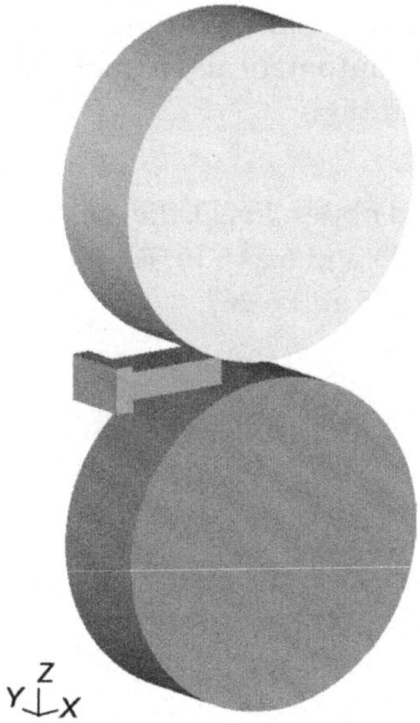

Fig. 6.20 Flat rolling

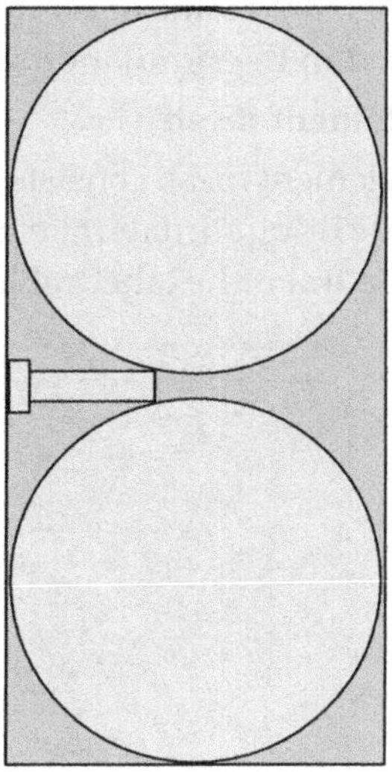

Fig. 6.21 Flat rolling with 2D plane strain model

simulating the complete part. In Fig. 6.22, the workpiece and die are cyclic symmetry in shape because the same geometry shape repeats itself by rotating 60° around the vertical axis.

Fig. 6.22 Cyclic symmetric part for gear forging

FE-The finite element method

The finite element method is one of the most powerful numerical methods in solving complex engineering problems. To use a FE solver, the internals of the geometry needs to be meshed (or discretized in small pieces called elements).

A good mesh is important to get good simulation results. During the forming simulation, the FE mesh is often "re-meshed" automatically to make sure the mesh used in the simulation is in a good condition.

Finite element mesh

A finite element mesh consists of elements and nodes. In 2D, the mesh consists of triangle or quadrilateral elements (Figs 6.23 and 6.24). In 3D, element can be tetrahedral or hexahedral in shape (Figs 6.25 and 6.26).

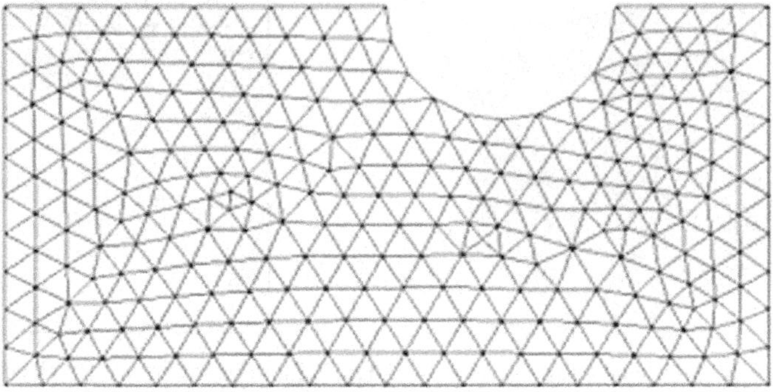

Fig. 6.23 2D triangle mesh

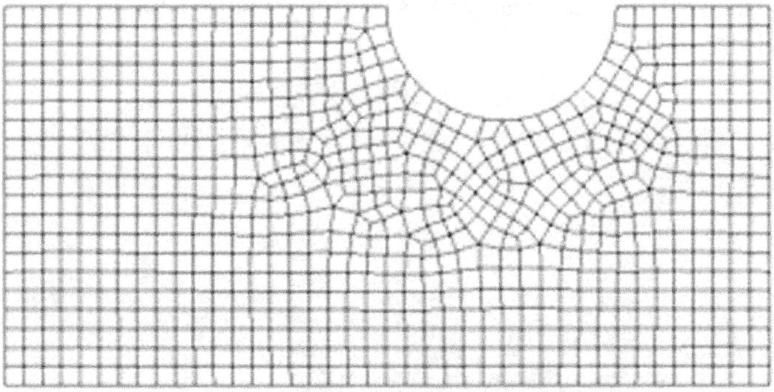

Fig. 6.24 2D quadrilateral mesh

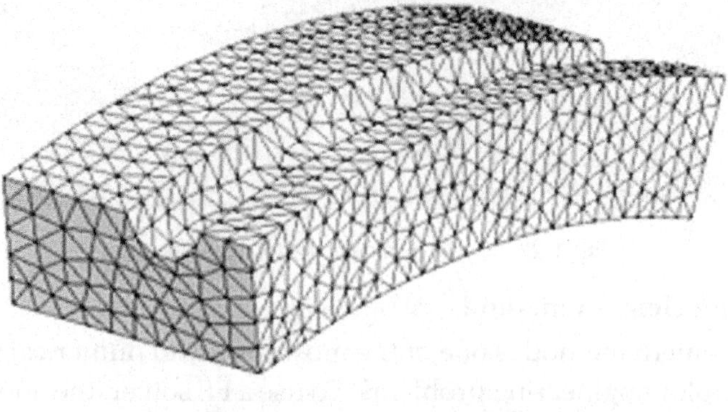

Fig. 6.25 3D tetrahedral mesh

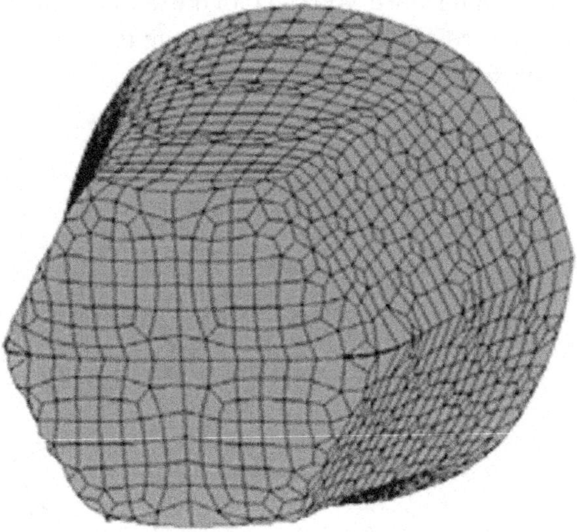

Fig. 6.26 3D hexahedral mesh

Re-meshing

In FE analysis, the mesh often becomes distorted due to large deformation in metal forming simulation. A distorted mesh generates poor results and often causes the analysis to stop prematurely. To avoid this problem, re-meshing is used from time to time to create a better mesh during the analysis.

FV-the finite volume method

In the finite volume method, the internals of the geometry are covered with a regular grid. The material flow simulation is done based on this fixed grid, called the Finite Volume Mesh. The outer surface of the part is meshed with triangular facets. Therefore, in the FV method there is only a need for "re-meshing" of the outer surface of the part, not the internals. This makes the FV solver uniquely suited for 3D hot forming with complex die geometry and parts. Because there is no internal mesh distortion problem, the FV analysis is fast and robust for this type of forming processes, even when laps or folds occur.

Flow lines

Flow lines are lines or curves used to visualize the material flow. By using flow lines, the material flow can be visualized in a better manner during the forming process.

Particles

Particles are material point markers that travel with the material as it deforms. As with the flow lines, you can use them to check material flow, but in addition, you are able to make history plots of the material variables at the location of the moving particle. For instance, you can plot the temperature history to study if the temperature stayed within the proper range for your forming process as shown in the Figs 6.27 and 6.28.

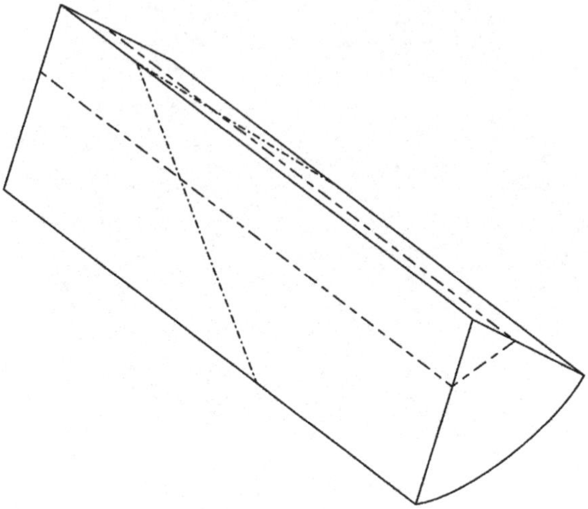

Fig. 6.27 Flow lines (– – – – – –) and particles (–·–·–·–·–·–) on a workpiece

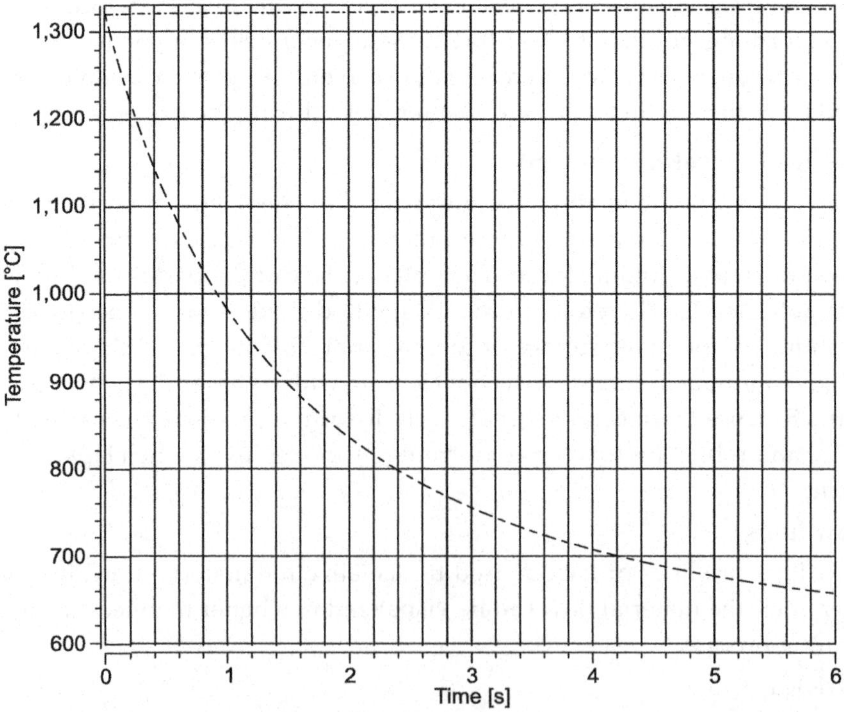

Fig. 6.28 Temperature plot of particles in the center (–·–·–·–·–·–) and surface (– – – – – –) of a deformed workpiece

6.7.3 Analysis Basics

This section gives a more detailed description of the underlying numerical techniques that are used during the simulation of metal forming processes. The descriptions are based on the solvers used inside the Simufact.forming solver.

FEM solver

The finite element solver used in Simufact.forming is an enhanced version of the MARC solver from MSC.Software Corporation. The finite element method is a numerical method that solves differential equations in a continuous field covered with a finite element mesh. A mesh consists of many elements of different shapes connected by nodes. MARC was developed on the basis of the displacement method. The stiffness methodology used in MARC addresses force–displacement relations through the stiffness of the system. The force–displacement relation for a linear static problem can be expressed as

$$Ku = f \tag{6.18}$$

where K is the system stiffness matrix, u is the nodal displacement, and f is the force vector. Assuming that the structure has prescribed boundary conditions both in displacements and forces, the governing Eq. (6.18) can be written as

$$\begin{bmatrix} K_{11} & K_{12} \\ K_{21} & K_{22} \end{bmatrix} \begin{Bmatrix} u_1 \\ u_2 \end{Bmatrix} = \begin{Bmatrix} f_1 \\ f_2 \end{Bmatrix} \tag{6.19}$$

u_1 is the unknown displacement vector, f_1 is the prescribed force vector, u_2 is the prescribed displacement vector, and f_2 is the reaction force. After solving for the displacement vector u, the strains in each element can be calculated from the strain displacement relation in terms of element nodal displacement as

$$\varepsilon_{el} = \beta\, u_{el} \tag{6.20}$$

The stresses in the element are obtained from the stress–strain relations as

$$\sigma_{el} = L\, \varepsilon_{el} \tag{6.21}$$

where σ_{el} and ε_{el} are stresses and strains in the elements, and u_{el} is the displacement vector associated with the element nodal points; β and L are strain–displacement and stress–strain relations, respectively.

The equations governing some other procedures are similar. For example, the governing equation of heat transfer analysis is

$$C\dot{T} + K_T T = Q \tag{6.22}$$

where C is the heat capacity matrix, K_T is the thermal conductivity matrix, Q is the thermal load vector (flux), T is the nodal temperature vector, and $\dot{T}$ is the time derivative of the temperature. Equation (6.22) reduces to

$$K_T T = Q \tag{6.23}$$

for the steady-state problem. Note that the equation governing steady-state heat transfer [Eq. (6.23)] and the equation of static stress analysis [Eq. (6.18)] take the same form.

The matrix equations for the thermal–mechanical problem are as follows:

$$K(T)\,u = f \tag{6.24}$$

$$C(T)\dot{T} + K_T(T)T = Q + Q^I \tag{6.25}$$

In Eqs. (6.24) and (6.25), the stiffness matrix $\boldsymbol{K}$, heat-capacity matrix $\boldsymbol{C}$, and thermal-conductivity matrix $\boldsymbol{K_T}$ are all dependent on temperature. $\boldsymbol{Q^I}$ is the internal heat generated due to inelastic deformation. The coupling between the heat transfer problem and the mechanical problem is due to the temperature-dependent mechanical properties and the internal heat generated. If an updated Lagrangian analysis is performed, K and k are dependent upon prior displacement.

The governing equations described above are either sets of algebraic equations or sets of ordinary differential equations. The time variable is a continuous variable for the ordinary differential equations. Selecting an integration operator reduces the set of differential equations to a set of algebraic equations. The final form of governing equations of all analysis procedures is, therefore, a set of algebraic equations.

The finite element system is an assembly of all its local element system. For example, the system stiffness matrix K is expressed in terms of the element stiffness matrix $\boldsymbol{K_i^{el}}$

$$K = \sum_{i=1}^{n} K_i^{el} \tag{6.26}$$

where n is the number of elements in the system. The element stiffness matrix can be expressed as,

$$K^{el} = \int_{v^{el}} \beta^T L \beta dv^{el} \tag{6.27}$$

where v^{el} is the volume of the element. The heat capacity matrix can be expressed as

$$C^{el} = \int_{v^{el}} N^T \rho c N dv^{el} \tag{6.28}$$

where N is the element shape function, ρ is material density and c is specific heat.

The nodal force vector f in Eq. (6.18) includes the contributions of various types of loading.

$$f = f_{\text{point}} + f_{\text{surface}} + f_{\text{body}} + f^* \tag{6.29}$$

where f_{point} is the point load vector, f_{surface} is the surface load vector, f_{body} is the body (volumetric) load vector, and f^* represents all other types of load vectors (for example, thermal strains).

The point load is associated with nodal degrees of freedom and can be added to the nodal force vector directly. The equivalent nodal force vector must be calculated from the distributed (surface/volumetric) load first and then added to the nodal force vector. In MARC, the computation of equivalent nodal forces is carried out through numerical integration of the distributed load over the surface area of volume to which the load is applied. This may be expressed as:

$$f_{\text{surface}} = \int \mathbf{N}^T \boldsymbol{p} \mathrm{d}\boldsymbol{A} \tag{6.30}$$

$$f_{\text{body}} = \int \mathbf{N}^T \boldsymbol{p} \mathrm{d}\boldsymbol{V} \tag{6.31}$$

where p is the pressure, A is the surface area and V is the volume.

In general, a forming process is governed by nonlinear equations. The nonlinear effects like material nonlinearities, geometric nonlinearities and nonlinear boundary conditions are taken into consideration in MARC. With material nonlinearities, material properties can be a function of temperature, plasticity, deformation rate, porosity, etc. The material can experience elastic or elastic–plastic deformation. The geometric nonlinearities are related to large deformation and rotation of parts. This happens in most of the metal forming applications. The nonlinear boundary conditions allow you to simulate die and workpiece contact, friction behavior and complex movement of forming equipments. To solve the nonlinear system, MARC does it incrementally. Many increments are needed in simulating a forming process. The updated-Lagrangian approach is used in MARC in a sense that the deformation of a workpiece can be represented by the distortion of the mesh. For the thermal–mechanical coupling problem, Eqs. (6.24) and (6.25) are solved in a staggered manner in that Eq. (6.25) is solved first and then Eq. (6.24). Both these are nonlinear equations. A Newton–Raphson method is used to solve these nonlinear equations. Many iterations are needed to solve the nonlinear equations until the solution is converged. Consider the following set of equations:

$$\boldsymbol{I}(u) = f \tag{6.32}$$

$$K(u)\,\mathbf{d}u = f - I(u) \tag{6.33}$$

where $\boldsymbol{I}(u)$ is the internal nodal-load vector and f is the external nodal-load vector.

$$\boldsymbol{I}(u) = \int \beta^T \sigma \mathrm{d}V \tag{6.34}$$

Suppose that the last obtained approximate solution is termed δu (i), where (i) indicates the iteration number. Equations (6.32) and (6.33) may then be written as

$$K(u^{(1)})\ \delta u = f - I(u^{(1)}) \tag{6.35}$$

This equation is solved for $\boldsymbol{\delta u}$ and the next appropriate solution is obtained by

$$u^{(i+1)} = u^{(1)} + \delta u \tag{6.36}$$

The solution of this equation completes one iteration. The same process can be repeated. The convergence is measured by either residual checking or displacement checking. The convergence criterion can be defined as follows.

Residual checking

$$\left(\frac{\| F_{\text{residual}} \|_{\infty}}{\| F_{\text{reaction}} \|_{\infty}} \right) < TOL_1 \tag{6.37}$$

$$\| F_{\text{residual}} \|_{\infty} < TOL_1 \tag{6.38}$$

where F is the force vector, and TOL_1 is the control tolerances. $\| F \|_{\infty}$ indicates the component of F with the highest absolute value. In some special problems, such as free thermal expansion, there are no reaction forces. MARC uses displacement checking.

Displacement checking

$$\frac{\| \delta u \|_{\infty}}{\| du \|_{\infty}} < TOL_1 \tag{6.39}$$

$$\| \delta u \|_{\infty} < TOL_1 \tag{6.40}$$

Where du is the displacement increment vector, δu is the displacement iteration vector. With this method, convergence is satisfied if the maximum displacement of the last iteration is small compared to the actual displacement change of the increment. A disadvantage of this approach is that it requires at least one iteration, regardless of the accuracy of the solution.

The algebraic linear equation [Eq. (6.35)] requires an equation solver. MARC provides a few solver options. They can be grouped into either direct or iterative solvers. In general, direct solvers need more memory with exact solution while iterative solvers require less memory but with approximate solutions.

The data flow diagram (Fig. 6.29) shows how analysis is carried out with MARC.

Element types for FE solver

Different element types are used when the FE solver is selected for the analysis. There are six types of commonly used elements for the FE forming analysis:

FE: 2D axi-symmetric

A four-node, isoparametric, arbitrary quadrilateral is used for 2D axisymmetric applications. It uses four integration points and a constant dilatation method to eliminate potential element locking when material becomes incompressible. Each node has two degrees of freedom (DOF) or unknown displacement variables (see Fig. 6.30).

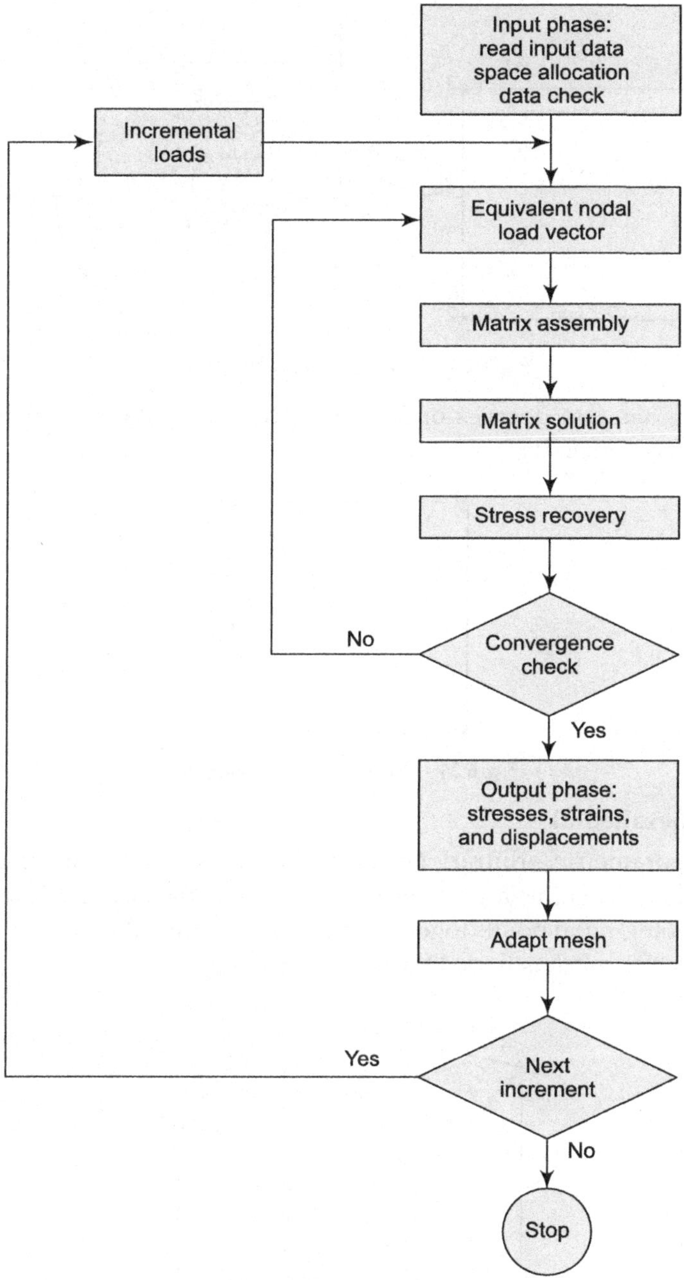

Fig. 6.29 Flow diagram of FEM analysis

FE: 2D plane strain

A four-node, isoparametric, arbitrary quadrilateral is used for 2D plane strain applications. It uses four integration points and a constant dilatation method to eliminate potential element locking when material becomes incompressible.

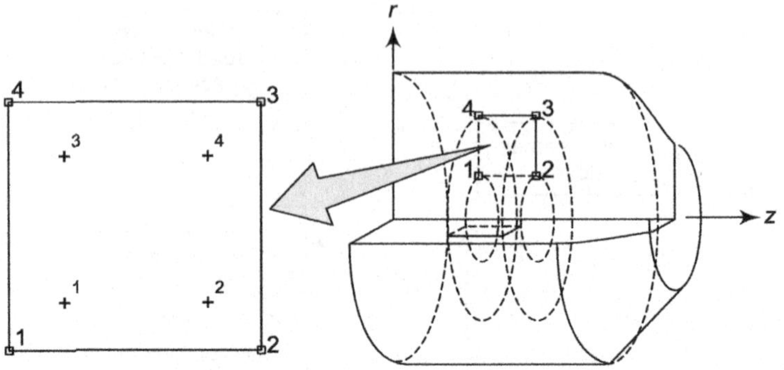

Fig. 6.30 2D axi-symmetric quadrilateral

Each node has two degrees of freedom (DOF) or unknown displacement variables (see Fig. 6.31).

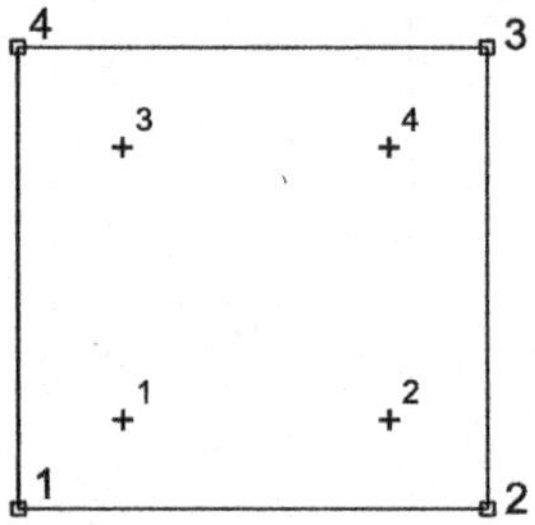

Fig. 6.31 2D plane strain quadrilateral

FE: 3D – hexahedral

A 3D, isoparametric, arbitrary hexahedral is the most accurate element type available for simulating large deformation and material flow in a 3D analysis. It uses four integration points together with constant dilatation to avoid element locking. Each node has three DOFs (see Fig. 6.32).

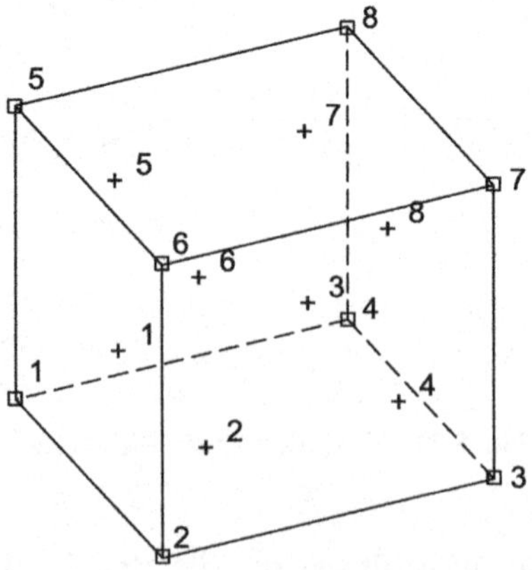

Fig. 6.32 A hexahedral

FE: 3D – large deformation tetrahedron

A 3D, isoparametric, 4 + 1-node, low-order, tetrahedron with an additional pressure DOF at each of the four corner nodes is an alternative element for simulating large deformation and material flow in a 3D analysis. It has four integration points and it is designed for solving incompressible material behavior such as metal forming with plastic deformation. In general, you need to use more tetrahedron elements in the simulation for the same accurate results as compared to using hexahedron elements (Fig. 6.33).

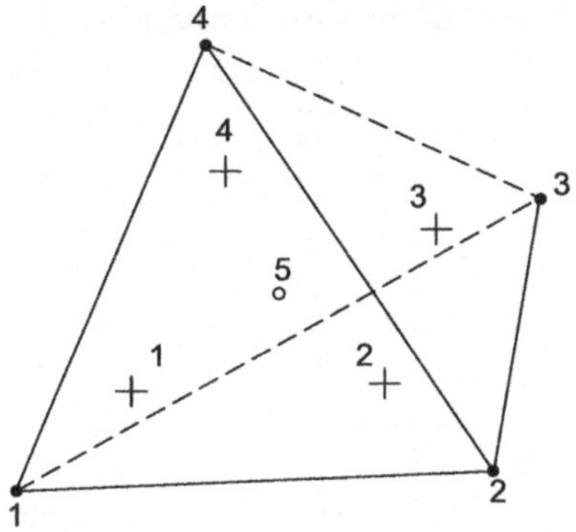

Fig. 6.33 A large deformation tetrahedron

FE: 3D – small deformation tetrahedron

A four-node linear isoparametric 3D tetrahedron is an alternative for analyzing linear elastic material behavior such as in the tool and die stress analysis with small deformation. It has only one integration point at the center. Each node has three DOFs. The computation cost for this element is small (Fig. 6.34).

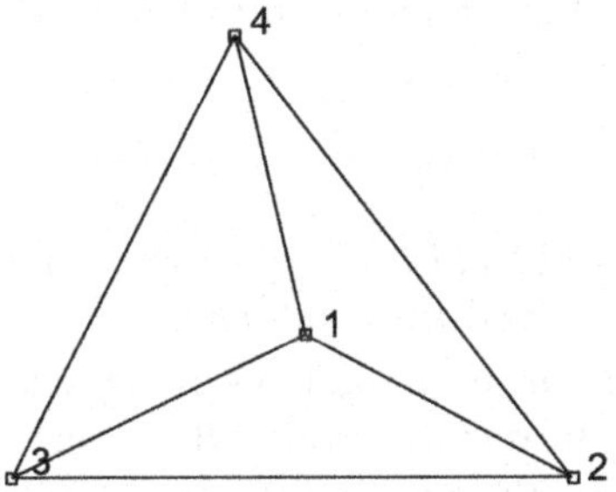

Fig. 6.34 A small deformation tetrahedron

FE: 3D – solid shell

A 3D solid-shell element with eight-node brick topology is the preferred element for simulating certain sheet metal forming processes (Fig. 6.35). The element uses enhanced assumed strain formulation for transverse normal component and assumed strain formulation for transverse shear components. The element uses one integration point in the element plane and a user defined number of the integration points through the element thickness. In this way the element can capture accurate material plasticity under bending load. An additional variationally consistent stiffness term is included to eliminate the hourglass modes that are normally associated with reduced integration. Compared to general shell elements, solid-shell element is better for double-sided contact as it has separate top and bottom element faces. It is also a better element for sheet forming analysis compared to the Hexahedral or Tetrahedron element, as it has more integration points in the thickness direction. With Hexahedral elements in the sheet forming simulation, more elements are needed in the thickness direction to capture the bending effect. This can be difficult if the thickness is relatively small in size.

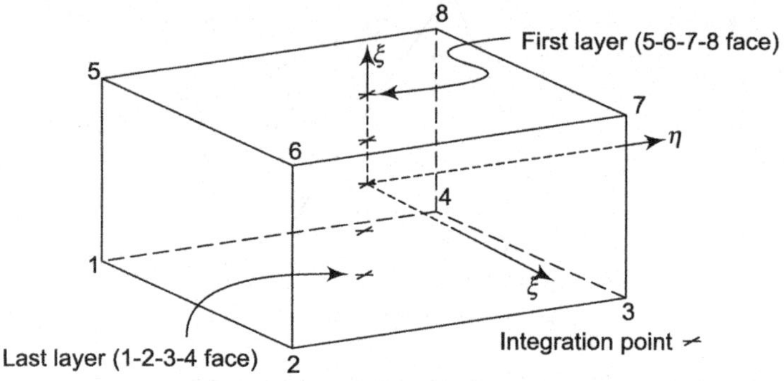

Fig. 6.35 Solid-shell element

FV solver

The finite volume solver used in Simufact.forming is an enhanced version of the DYTRAN solver from MSC. Software Corporation. In this section, we will explain the numerical methodology used.

The solver is using an Eulerian description, expressing the conservation of any physical quantity in the form:

$$\frac{\partial Q}{\partial t} + \nabla F = 0 \tag{6.41}$$

where Q is the variable and F $(= F_{\text{convection}} + F_{\text{diffusion}})$ is the corresponding flux, consisting of convection and diffusion terms.

The equation gives the conservation laws of mass, momentum and energy, in which the variables and the corresponding fluxes are listed in Table 6.1, where v_i = the velocity vector, h_i = the heat flux, e_i = the total energy per unit of mass.

Table 6.1 Variables and corresponding fluxes

Conservation law		Mass	Momentum	Energy
Q	Variable	ρ	ρv_i	ρe_t
F	Convection	ρv_i	$\rho v_i v_j$	$\rho e_t v_i$
F	Diffusion	0	$-\sigma_{i,j}$	$h_i - \sigma_{i,j} v_j$

By using the Gauss' divergence theorem, the above conservation law can be written in an integral form for a 'control volume' V, surrounded by a closed surface S, which is stationary within a total domain:

$$\frac{\partial}{\partial t}\int_V Q dV + \int_S F \cdot n dS = 0 \tag{6.42}$$

In Simufact.forming, the internals of the workpiece material is sub-divided in a finite volume mesh, where the finite volume elements are aligned with the coordinate system axes. This is illustrated in Fig. 6.36.

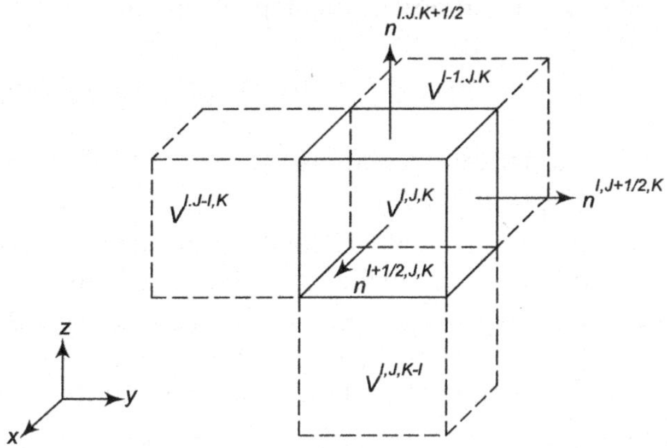

Fig. 6.36 Finite volume elements

A straightforward and simple discretization in Cartesian coordinates is then obtained by numerically integrating the integral form of conservation equations:

$$\Delta V^{I,j,K}\frac{\Delta Q^{I,J,K}}{\Delta t} = \left(F^{I+\frac{1}{2},J,K}\Delta S^{I+\frac{1}{2},J,K} - F^{I-\frac{1}{2},J,K}\Delta S^{I-\frac{1}{2},J,K}\right.$$
$$+ F^{I,J+\frac{1}{2},K}\Delta S^{I,J+\frac{1}{2},K} - F^{I,J-\frac{1}{2},K}\Delta S^{I,J-\frac{1}{2},K}$$
$$\left.+ F^{I,J,K+\frac{1}{2}}\Delta S^{I,J,K+\frac{1}{2}} - F^{I,J,K-\frac{1}{2}}\Delta S^{I,J,K-\frac{1}{2}}\right) \tag{6.43}$$

where $\Delta V^{I,J,K}$ = volume of the element; $Q^{I,J,K}$ = mean value of Q in the element, and located at the center of the finite volume element; $F^{I+\frac{1}{2},J,K}\Delta S^{I+\frac{1}{2},J,K}$ = flux normal to the surface $\Delta S^{I+\frac{1}{2},J,K}$ of the face of the finite volume element.

FV solver—standard vs. high order solver options

The accuracy of the solution is highly dependent on the accuracy with which the fluxes, like $F^{I+\frac{1}{2},J,K}$, across the faces of the elements are determined. The high order solver is using a second order approach, taking into account the gradients of the variables in the finite volume elements. This approach is the most accurate, and should be used by default. Another option is provided in the solver, called the standard solver, where the fluxes at the faces are approximated using a simple linear interpolation. The standard solver is faster, and very well suited for initial quick design studies.

FV solver—explicit time integration

The governing equations are integrated in time using an explicit dynamic procedure. All the variables in the governing equations are evaluated in a Runge–Kutta time integration scheme. The complete system of equations is solved by separating the diffusion flux and the convection flux respectively into an acoustic step and an advection step. There are three advantages of using this approach:

- Avoid the non-linearity due to convection terms when the impulse is evaluated
- Stabilize the numerical calculations
- Simplify the implementation

Furthermore, no iteration is employed for the interaction between the acoustic step and the advection step, due to the small time step in the explicit method.

In the acoustic step, the stress and impulse waves are assumed to propagate through the control volumes of the whole domain. The variation per unit time of the quantity in each volume is determined by the diffusion flux through its surrounding faces.

Using the assumption that the mass density is constant in each finite volume element during the acoustic step, the conservation law of mass is satisfied automatically.

The conservation law of momentum is written in volume integral form as:

$$\int_V \frac{\partial v_i}{\partial t}\,\mathrm{d}V = \frac{1}{\rho}\int_S \sigma_{ij} n_j\,\mathrm{d}S \tag{6.44}$$

where n_j is the unit normal vector on the boundary S if the finite volume element. The values on the faces of the finite volume element are obtained by solving the Riemann problem. With this, the new velocity in the finite volume element is updated.

The conservation law of energy is written in volume integral form as:

$$\rho\int_V \frac{\partial e}{\partial t}\,\mathrm{d}V - \int_V \sigma_{ij}\dot{\varepsilon}_{ij}\,\mathrm{d}V = -\int_S h_i n_i\,\mathrm{d}S \tag{6.45}$$

Solution of this integral provides a new temperature field, as affected by the heat generated from plastic deformation, the conduction of heat within the material and the heat transfer from the surface.

In the advection step, the new velocity as updated in the acoustic step is used to calculate the material flow from one finite volume element to another. The material transports with it the corresponding properties (stress/strain, energy, material characteristics, etc.). This convection flux is calculated without changing the value of the physical quantities that are transported with the material. This removes the divergence term, and the conservation laws for mass, momentum and energy simplify to:

$$\frac{\partial(\rho\varphi)}{\partial t} + \frac{\partial(\rho\varphi v_i)}{\partial x_i} = 0 \tag{6.46}$$

where φ is any physical quantity per unit mass. The integral form used in the advection step of the finite volume method is given as follows:

$$\frac{\partial}{\partial t}\int_V \rho\varphi \, dV = -\int_S \rho\varphi v_i n_i \, dS \tag{6.47}$$

Once the mass transport has been calculated, the new density in each finite volume element is updated. Furthermore, the strain rate tensor is calculated from the velocity increment, as:

$$\dot{\varepsilon}_{ij} = \frac{1}{2}\left(\frac{\partial v_i}{\partial x_j} + \frac{\partial v_j}{\partial x_i}\right) \tag{6.48}$$

With this new strain rate tensor, the strain is updated. Next, the stress values are determined from the elastic–plastic constitutive relations, which are dependent on temperature, strain and strain-rate. Also, when activated, the phase transformations and grain sizes are updated according to the new temperature and strain fields.

FV solver—automatic surface tracking

The shape of the deforming material is encapsulated by a tessellated surface comprised of triangular facets. These facets are purely geometric entities rather than finite elements. The faceted surface is constrained to move with the material in order to track the exact material surface and also precisely apply the boundary conditions to the material inside each finite volume element.

In order to accurately describe the shape of the workpiece, the faceted surface is automatically updated at certain times during the simulation. The surface re-meshing is extremely robust, and can capture very small features, like letters on a die, as well as laps and folds. In case of multi-stage processes, laps and folds are automatically carried over into the next stage, giving an accurate prediction of the final location of a lap. This allows you to determine if a lap ends up in the part or in the flash, and how deep the lap is. In case the lap ends up in the part and is too deep, the forging process must be re-designed.

6.7.4 Post-Processing Basics

At the post-processing step, the GUI will be used to view and examine analysis results. A report is then generated and makes decision on the forming process design issues. There are many standard result outputs such as, temperature, stress and strain, and die wear that can be viewed. The following are some basic plots obtained at the post-processing step.

Result plot

Result plot allows viewing certain result at a fixed time step or stroke. For example, you can examine the effective plastic strain (a total deformation measure) contour plot of a workpiece at a time instant (see Fig. 6.37).

Animation

Viewing animation allows you to see the complete forming process in a movie mode. You can play, forward and backward the movie or simply doing the step-by-step view of the results (Fig. 6.38).

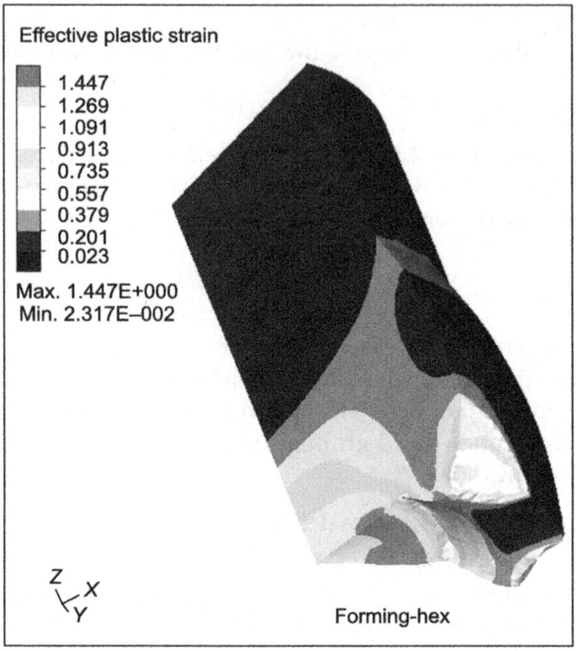

Fig. 6.37 Effective plastic strain result plot

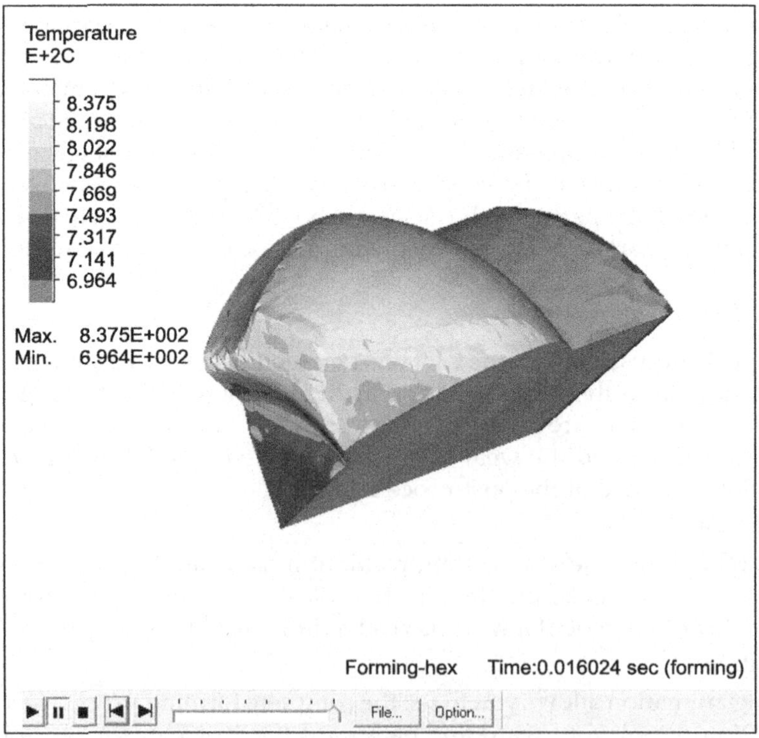

Fig. 6.38 Animation player of temperature contours

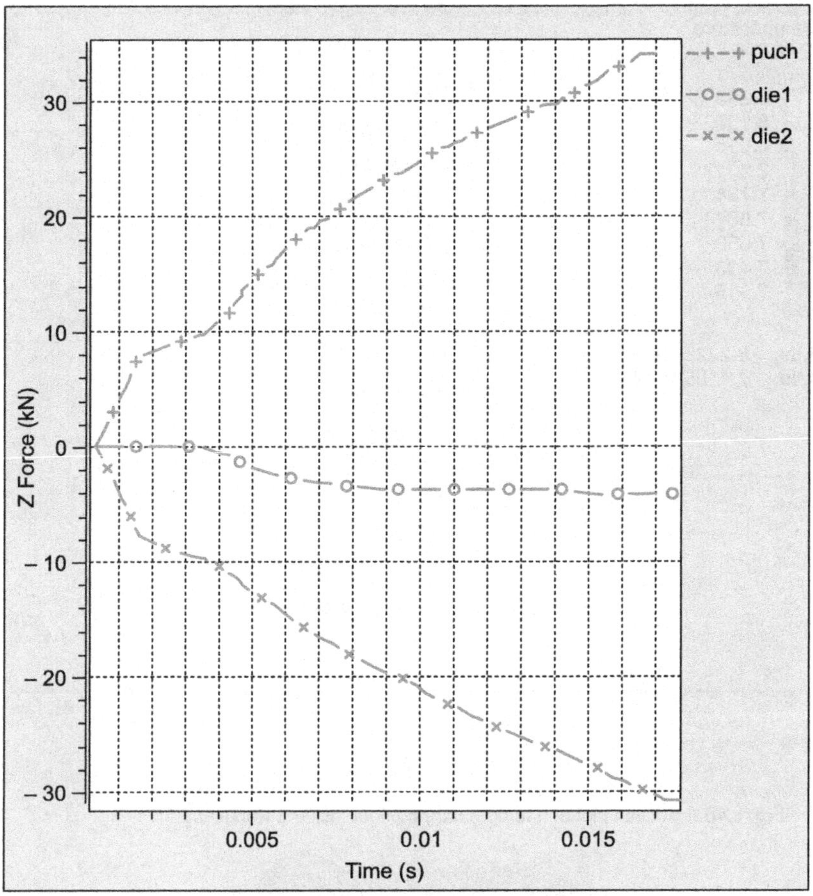

Fig. 6.39 History plot of forming forces

History plot

A history plot allows you to examine forming forces and movement of the tools. This gives you an overall picture of your forming process (Fig. 6.39).

Section plot

In both animation and result plot, you can examine the results inside a workpiece by using a section plot. In the section plot, you define a cutting plane and the results on the cutting plane will be shown (Fig. 6.40). This is a useful tool to examine any damage in the forming process that cannot be seen easily on the surfaces.

Flow lines and particles

If you insert flow lines and particles in a workpiece at pre-processing step, you can view them at the post-processing. Flow lines and particles allow you to see the material flow during the deformation (Fig. 6.41).

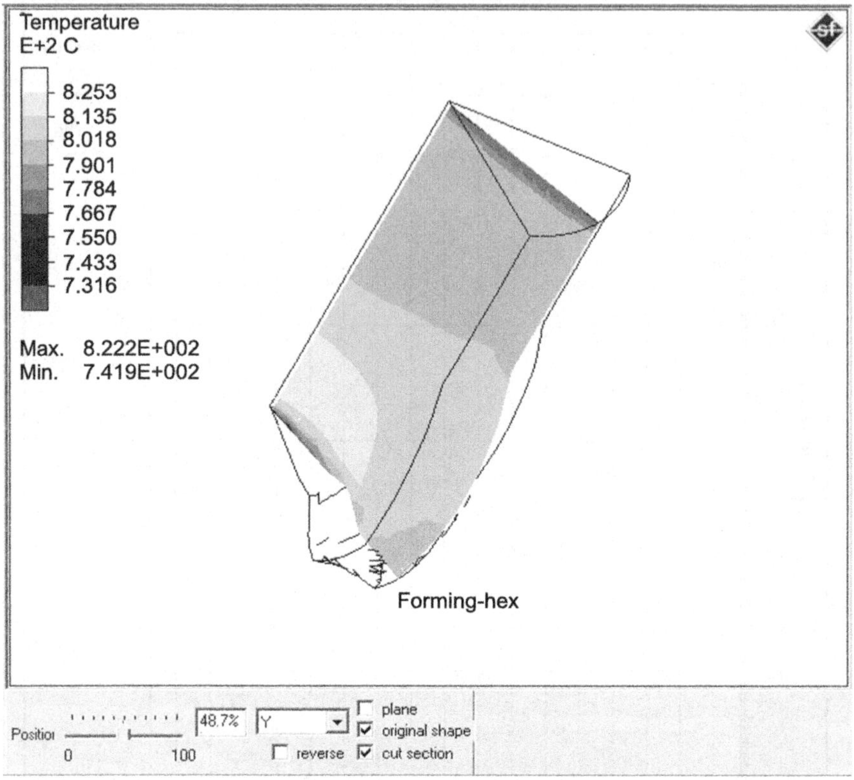

Fig. 6.40 Section plot of a temperature contour inside a workpiece

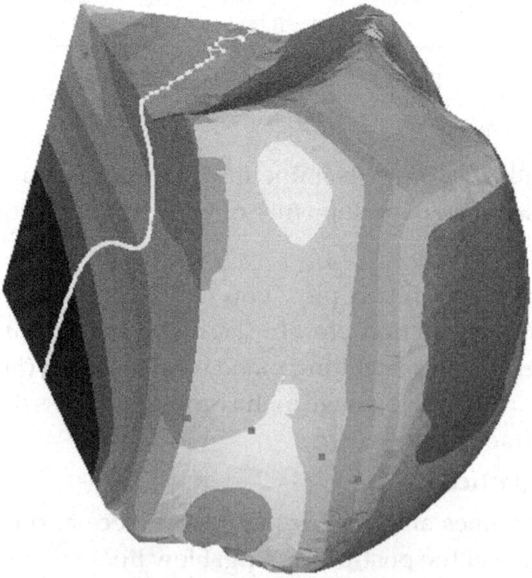

Fig. 6.41 Flow lines and particles in the result plot

6.8 APPLICATION TO FORGING DESIGN OF A SUPPORT ARM

In this section it will be described how the Finite Element and Finite Volume methods described in the previous section can be applied to the simulation of the manufacturing process of a support arm (Fig. 6.42).

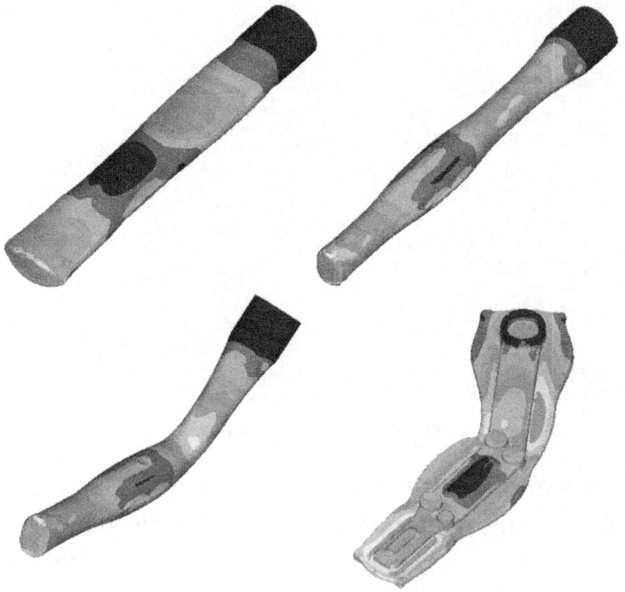

Fig. 6.42 Process steps to manufacture a support arm

6.8.1 Process Description, Model Description and Results

In this manufacturing process, a cylindrical billet will be forged to a support arm using a four-stage process:

1. Cross rolling for initial pre-distribution of the material and elongation of the billet
2. Cross rolling for further pre-distribution of the material
3. Bending of the billet
4. Finish forging to final shape

The aim of the simulation is to verify that the pre-form produced in process steps 1–3 result in complete die filling.

Initial cross rolling of the billet

One simplification of this model is the temperature distribution of the billet at the beginning of the process. It is assumed, that the temperature is constant (no cooling has taken place) prior to the first process stage. The thermal expansion of the workpiece due to its elevated temperature must be considered in the model. The dies can be simplified as rigid dies, but the heat transfer between billet and die must be taken into account. A spring loaded manipulator tool (here shown in yellow) can be used to guide the billet during the rolling process (Fig. 6.43). This process is best simulated using the FE solver.

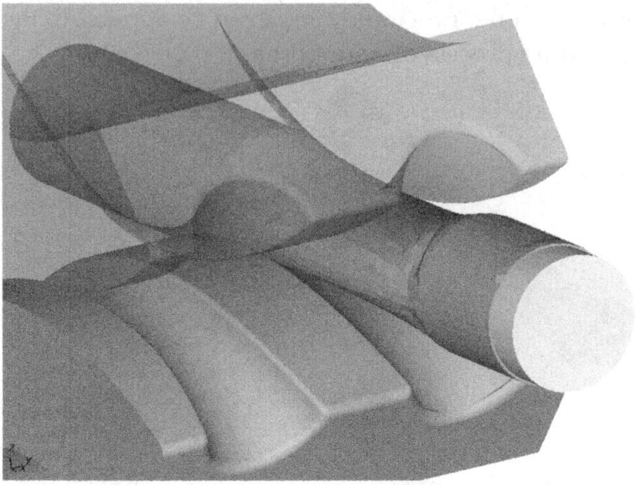

Fig. 6.43 Initial cross rolling

Additional cross rolling of the billet

A second cross rolling operation is used to further develop the material distribution as shown in Fig. 6.44. This process is best simulated using the FE solver.

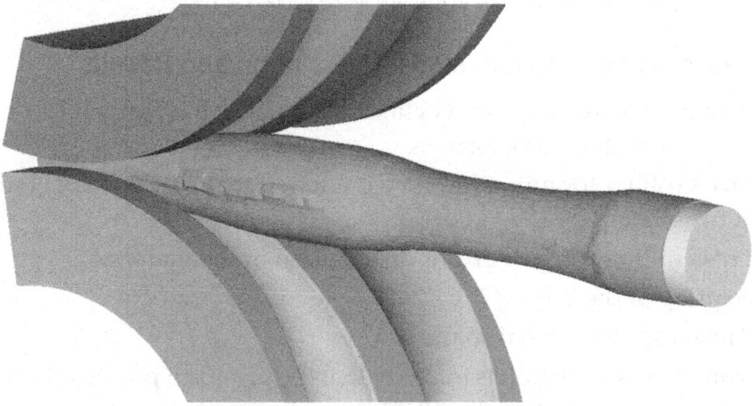

Fig. 6.44 Additional cross rolling

Bending to produce curved shape needed for forging process

The billet with adjusted material distribution in its longitudinal direction is then bent in an upsetting operation. This is done with two rigid dies. In the images, a third die is shown in the background for visually verification that the workpiece has been bent enough to fit into the die of the last process step. This process is best simulated using the FE solver as shown in Fig. 6.45.

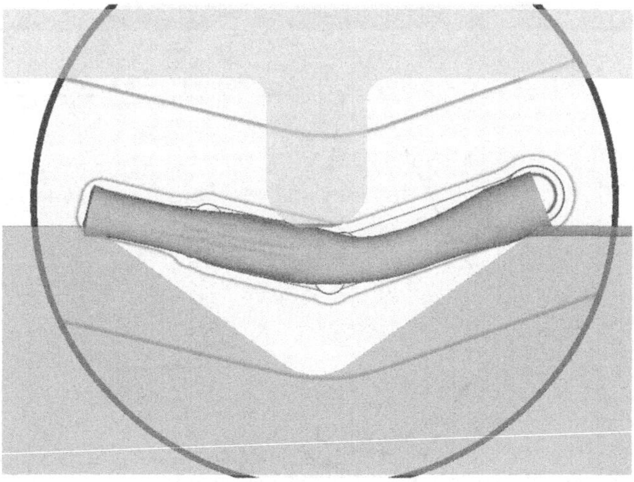

Fig. 6.45 Bending

Finish forging

Finally, the workpiece will be forged to its final shape in a crank press. Elasticity effects of the tools can be neglected for this process. As can be seen in the Figs. 6.46 and 6.47, the dies fill completely, and the required tonnage is predicted as depicted in Fig. 6.48. This process is best simulated using the FV solver.

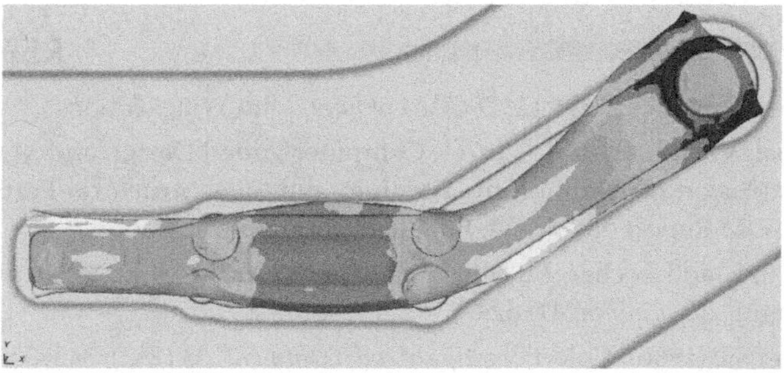

Fig. 6.46 Finish forging step—effective plastic strain during stroke

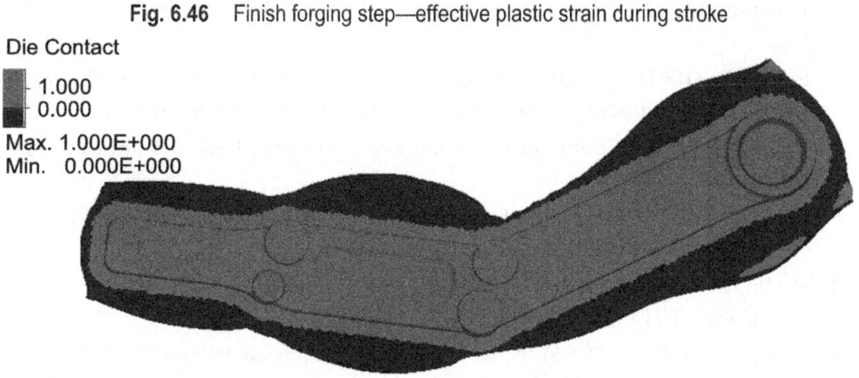

Fig. 6.47 Finish forging step—die contact at end of stroke

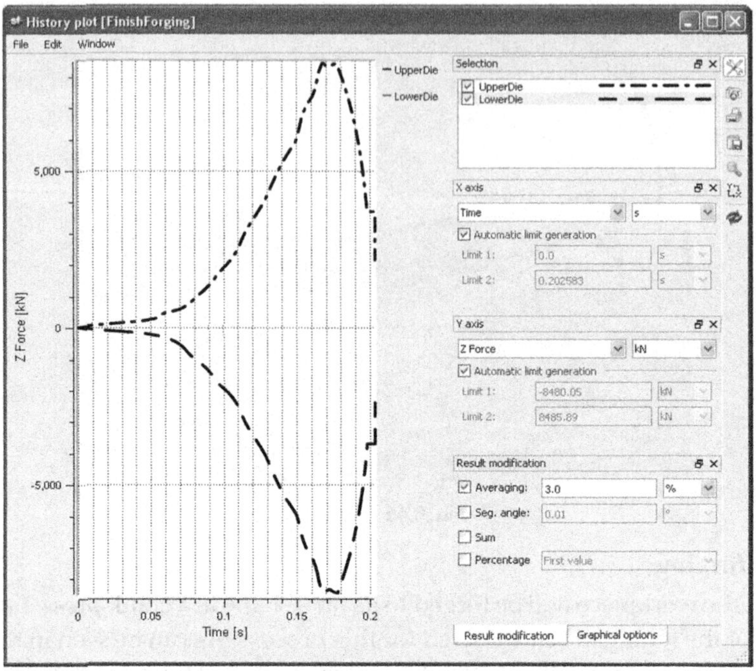

Fig. 6.48 Finish forging step—required tonnage

REFERENCES

1. Gunasekera, J.S. (1989). *CAD/CAM of Dies*, John Wiley & Sons.
2. Nagpal, V. and Altan, T. (1976). Computer-Aided Design and Manufacturing for Extrusion of Aluminium, Titanium and Steel Structural Parts, (Phase 1), AVSCOM Report No. 76–12, Battelle Columbus Labs.
3. Mockli, F. and Locher, M. (1965). State of the Art in Making Extrusion Dies (in German), *Aluminium*, **41**; 629.
4. Wilson, G. (1966). Cold Forging, *Metal Treatment*, **33** (252); 345–353.
5. Feldmann, H.D. (1961). *Cold Forging of Steel*, Hutchinson and Company, Ltd.
6. Sieber, K. (1964). Special Cold Forging Tools, Particularly for Solid Forming on Multi-Stage Transfer Presses, *Wire World International*, **6**(6); 165–178.
7. Bello, L. (1989). *Aluminum Extrusion Die Correction*, Fellom Publications, San Francisco.
8. Chadwick, R. (1959). The Hot Extrusion of Nonferrous Metals, *Metallurgical Reviews*, **4**(15); 189–255.
9. Perlin, I.L. (1964). Theory of Metal Extrusion. *Metallurgiya*, Moscow, (English Translation FTD-HT-23-616–67).
10. Metals Handbook (2005), ASM (American Society of Metals), Vol. 14A, Metal Working: Bulk Forming.

11. Gunasekera, J.S. and Hoshino, S. (1982). Analysis of Extrusion of Drawing of Polygonal Sections Through Strictly Converging Dies, ASME *Journal of Engineering for Industry.*, **104**; 38–45.
12. Gunasekera, J.S. and Hoshino, S. (1985). Analysis of Extrusion or Drawing of Polygonal Sections Through Straightly Converging Dies", ASME *Journal of Engineering. for Industry* **107**; 229–233.
13. Richmond, O., Devenpeck, M.L. (1962). A Die Profile for Maximum Efficiency in Strip Drawing, *Proceedings of 4th U.S. National Congress of Applied Mechanics,* **2**; 1053–1057.
14. Devenpeck, M.L. and Richmond, O. (1965). Strip Drawing Experiments with a Sigmoidal Die Profile, ASME *Journal of Engineering. for Industry.,* **87** (4); 425–428.
15. Hill, R. (1966). A Remark on Diagonal Streaming in Plane Plastic Strain", *Journal of Mechanics and Physics of Solids,* **14**; 245–248.
16. Richmond, O. and Morrison, H.L. (1967). Streamlined Wire Drawing Dies of Minimum Length, *Journal of Mechanics and Physics of Solids,,* **15**; 195–200.
17. Devenpeck, M.L. (1971), *Experimental Evaluation of Theoretically Ideal Drawing Dies,* Plenum Press, New York–London, pp. 215–234.
18. Chen, C.T. and Ling, F.F. (1968). Upper Bound Solutions to Axisymmetric Extrusion Problems, *International Journal of Mechanical Sciences.,* **10**; 863–879.
19. Avitzur, B. (1968), *Metal Forming: Processes and Analyses,* McGraw-Hill, New York.
20. Frisch, J. and Mata-Pietri, E. Experiments and Upper Bound Solution in Axisymmetric Extrusion, *Proceedings of the 18th International Machine Tools Design and Research Conference.,* pp. 55–60.
21. Mata-Pietri, E. and Frisch, J. (1977). "Metal Flow Through Various Mathematically Contoured Extrusion Dies", *Proceedings of the 5th North American Manufacturing Research Conference,* pp. 99–104.
22. Nagpal, V. and Altan, T. (1975). Analysis of the Three-Dimensional Metal Flow in Extrusion of Shapes with the Use of Dual Stream Functions", *Proceedings of the 3rd North American Manufacturing Research Conference,* Pittsburgh, Pa., pp. 26–40.
23. Nagpal, V. (1977). On the Solution of Three_Dimensional Metal_Forming Processes, ASME, *Journal of Engineering for Industry,* **99**(3); 624–629.
24. Yang, D.Y. and Lee, C.H. (1978a). Analysis of Three Dimensional Extrusion of Sections Through Curved Dies by Conformal Transformation", *International Journal of Mechanical Sciences,* **20**; 541–552.
25. Yang, D.Y., Kim and M.U., Lee, C.H. (1978b). An analysis for Extrusion of Helical Shapes from Round Billets *International Journal of Mechanical Sciences* **20**; 695–705.
26. Hoshino, S. (1981). Extrusion of Non-Axisymmetric Sections through Converging Dies. *Ph. D. Thesis,* Monash University, Australia.
27. Kraft, F.F. and Gunasekera, J.S. (2005), *ASM Handbook on Metal Forming,* Chapter on Hot Extrusion, Vol. **14A**; 421–439.

CHAPTER 7

Development of New Metal Forming Technologies Utilizing Computational and Experimental Approaches

Takashi Kuboki
Department of Mechanical Engineering & Intelligent Systems,
University of Electro-Communications,
Chofu_shi, Tokyo, Japan

7.1 THEORIES ON PLASTICITY

Research on plasticity of metal started in the mid 19th century, and theory for plasticity started to develop in the early 20th century and was integrated by Hill in the mid 20th century [1, 2]. Tresca (1864) stated that metal starts yielding when the maximum shear stress reaches to a critical value, based on experimental results on metal extrusion [3]. In 1913, von Mises proposed a hypothesis that metal starts to yield when invariant J_2 in stress tensor reaches to a certain value [4]. Hencky (1924) delivered to von Mises criteria a physical meaning by stating that metal starts to yield when elastic shear deformation energy reaches to a certain value [5].

Levy (1870) and von Mises (1913) proposed an equation on flow rule neglecting components for elastic deformation [4, 6]. Prandtl (1924) expanded

Levy–Mises equation considering elastic components [7] and Reuss (1930) completely generalized Prandtl's theory [8].

The theories on yield conditions and flow rules have been discussed and improved. Hill proposed yield locus for anisotropic material [2]. Further precise description of the locus has been tried [9] and experimental verifications for yield conditions have been done [10]. The history of deformation affects the stress–strain relationship. Bauschinger (1886) pointed out the Bauschinger effect as "after stress is reversed from tension to compression both the elastic range and a yield point for the reversed direction of straining have completely disappeared" [11]. In order to describe this effect, many equations have been proposed [12–14].

In order to practically utilize the theories on plasticity for metal forming, the theories should be implemented into analytical methods like slab method, energy method, upper-bound method, slip line field analysis, finite element method and boundary element method [1, 15].

In the slab method, the work piece would be divided into small elements, inside which stress and strain are assumed to be constant. The slab method gives a stable and concise analysis satisfying the equation of equilibrium and yield condition. This method was applied to analysis of plate rolling [16] and the theories derived from this method have contributed the development of the rolling technology.

The upper-bound method is based on a second variational principle which says that the actual working rate by external forces from tools is less than any working rate calculated from arbitrary admissible velocity fields. Therefore, the obtained results will show a larger force than actual one, resulting in a safer estimation in machine design or decision of working conditions. The analytical precision and applicability of the upper-bound method was considerably improved by the development of upper bound element technique (UBET) [1] which describes behavior of material deformation. This technique has been applied for the analysis of drawing, extrusion, forming, rolling, upsetting, and so on.

Nowadays the finite element method (FEM) is a strong and imperative technique for research and development of metal forming. The present-day concepts of FEM has been built up on many previous approximate methods [17], such as finite differences, variational finite differences, trial functions, variational methods, weighted residuals, piecewise continuous trial functions, structural analogue substitution, and direct continuum elements. The first paper on the finite element method is often said to be that by Turner, Clough *et al.* on a problem of stiffness and deflection in the aircraft industry [18]. The name of "finite element analysis" was coined by Clough [19]. Since the advent of the finite element methods, it has been improved by many researchers [20]. The 'geometric stiffness matrix' in conjunction with an up-dating of coordinates was adopted to consider geometric non-linearity [21–23]. A similar technique was applied to consider material non-linearity [24]. The iteration technique using Newton-Raphson method was used and recommended by many researchers to overcome the unquantifiable build-up of error in forward-Euler approach [24–28].

With the enhancement of computer ability, the finite element method has become useful and practical in metal forming. The finite element method is a general analysis that gives solution without constructing or revising the solver itself. Before the advent of finite element analysis, researchers and engineers had to construct an analysis code by themselves, whereby relationships between displacement and strain, strain and stress, stress and external forces, for example. In commercial codes of the finite element method, relationship from displacement to external forces are implemented in advance. Only a set of conditions of geometry, loading, constraints, contact, material, mesh, and other controlling data is needed. It gives users the distributions of stress and strain, which is difficult to obtain in other analytical methods. Engineers would be able to utilize the finite element method for designing machines based on the solutions, and optimize process in manufacturing line to deform material into target shapes, or to control the mechanical properties of the product. Researchers would be able to carry out a series of numerical experiments without fabricating actual experimental set-ups or tools. The numerical experiments could be those which cannot be conducted in the actual world, which would tell researchers genuine mechanism in metal forming.

While computational approaches are more powerful and efficient with the development of theories, experimental approaches would be more important for research and development in metal forming. Certainly with increase of the number of elements in the finite element method, the precision would be improved and it was theoretically proved that infinite number of elements would give the true solution for partial differential equations. However, the phenomena in the actual world are more complicated and it would inevitably be impossible to take all the relationship between cause and results into consideration in analysis. Theories, which should deliver partial differential equations, should be modified based on the facts in experiments. Only the experiment will convey the truth in actual world.

In this section, key sentences will be introduced based on the author's experience in metal forming, where computational and experimental approaches were used, and then some of examples will also be shown.

7.2 KEY SENTENCES FOR RESEARCHER AND DEVELOPMENT IN METAL FORMING

Here, the author would like to introduce to readers key sentences, which are useful and helpful in research and development. Some of them were taught by elders and experts, some were by colleagues. The authenticity of the key sentences might depend on situations.

Key sentence: Analysis is garnish of julienne radish served with raw fish.

Here, raw fish is laboratory experiments or phenomena in actual operation lines which is the main dish in projects, while julienne radish is analysis which just garnishes experimental results. This tells us not rely on analyses and not to conduct analyses without seeing experimental results.

Key sentence: Analyses are suspicious. Experiments tell the truth.

All the phenomena cannot be considered in analyses, and some errors must inevitably be generated in simulations. Experiment would tell us what really occur in the actual metal forming, and may suggest what are missing in simulations.

Key sentence: Experiments are suspicious. Analyses tell the truth (under ideal conditions).

Experimental results often change depending on situations. For example, some materials, whose chemical components are same, might bring different results, if the materials are supplied at different time. A slight change of temperature might affect machines or lubrication, and lead to different results. In many cases, the cause of difference cannot be identified. On the other hand, numerical experiments would be carried out under ideal conditions without disturbance.

Key sentence: Predictions of values are difficult for simulations. Simulations would be able to predict tendency or influence of parameters.

It would be difficult for simulations to predict the values accurately in actual phenomena, and it would not be practical. However, simulations tell us how we should change parameters for better metal forming.

Key sentence: Simulations will show you what you cannot see in actual phenomena.

It would not be easy to see stress or strain. Although, strain gauges will tell strain and some new techniques would even tell the distribution, it would not be easy to see stress or strain distributions inside bulk. Simulations will tell us easily the distributions, which often tell us the mechanism in metal forming.

Key sentence: Although simulations might tell you what you cannot see, you will fail to see it if you have no intention.

It is the responsibility and ability of researchers and engineers to know what simulations are telling them. If researchers and engineers do not have any intention or objective, they do not know what simulation results mean.

Key sentence: When you build up simulation models, then you completed half the projects.

It is important to consider deliberately what are essential in phenomena and should be considered when you build up projects. If you build up it properly, then you will find the solution efficiently.

Key sentence: Analysis enables you to conduct numerical experiment using dozens of different tools over a night.

You can try out different geometries of tool in simulations without fabricating actual tools. It will save time and cost.

Key sentence: It is important to see and feel the actual phenomena.

Using you eyes and hands will activate you brain to bring fruitful results.

Key sentence: Conduct concise experiments using papers, clay and gelatine.

Even some concise experiments using some materials, which is not similar to the target material, will suggest some truth. If you want to observe the mechanism of flattening or buckling in tube bending for example, you would know some mechanism by bending a paper tube. You would see exaggerated flattening and buckling which is caused by the distance from the neutral axis.

Key sentence: Try thought experiments.

Imagination of metal forming in ideal conditions would help you to brush up your sense of engineering and to observe actual phenomena.

Key sentence: Use your colleagues, and do not do everything by yourself.

Collaboration is important.

Key sentence: Do not say clever things, saying something is too difficult to do.

First you should try, before thinking what would be problems.

Key sentence: In metal forming, guide materials in a way where materials themselves want to deform.

Unreasonable forming will lead to failure. The best solution might exist among simple and concise deformation.

Key sentence: Draw auxiliary lines to solve problems.

As in geometry, drawing auxiliary lines or putting something new into some phenomena will bring you new vision and lead you to solution. A leap of imagination will lead to a new invention.

The rest of the chapter will introduce some examples of problems solved by computational and/or experimental approaches.

7.3 DRAWING FORCE IN BAR DRAWING

The calculation of drawing forces during bar drawing may be one of famous problems, and a precise analysis method was proposed by Geleji using the slab method [29]. Here, the finite element method is used instead of the Geleji's method. Figure 7.1 shows the schematic illustration of the drawing process. The head of bar, which is pushed through into the die hole, is clamped and pulled so that the bar cross section might be almost identical to the shape of die hole. In many cases, the die hole is the axisymmetric circular shape as in this case. In this numerical experiment, a stress–strain curve, which was measured in tension test of medium carbon steel AISI1045, was adopted.

The stress–strain diagram is shown in Fig. 7.2. Table 7.1 shows the drawing condition and die geometry. Only the die half angle α is changed during the numerical experiment. It would be important to guess what would occur in metal forming processes for improvement of sense of engineering. Here is a question for readers, if they do not know the answer.

Question 1: How will the drawing force change upon increase of die half angle α?

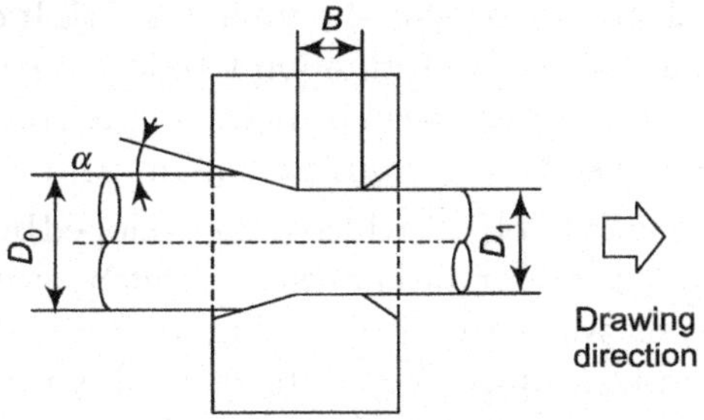

Fig. 7.1 Schematic illustration of bar drawing

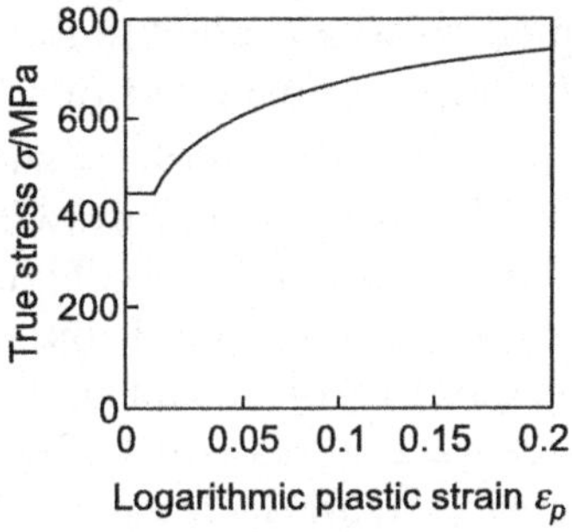

Fig. 7.2 Stress–strain diagram of AISI 1045

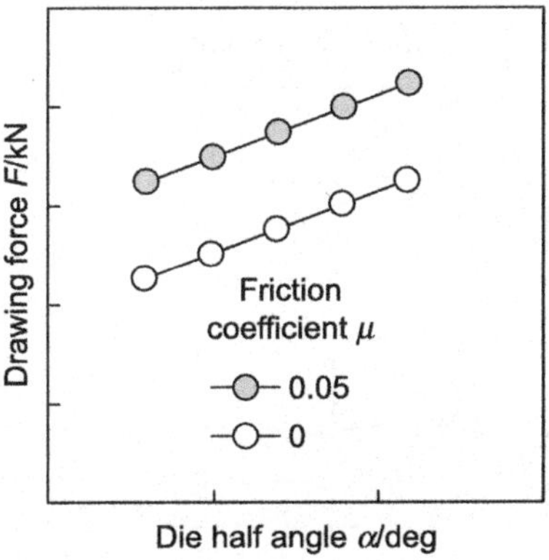

Fig. 7.3 Imagined drawing force behavior guessed by many people who are not familiar with drawing

Table 7.1 Drawing condition and die geometry

Parent bar	Stress-strain Relationship	Measured value of annealed AISI1045
	Diameter D_0/mm	33.0
	Friction coefficient	0.05, 0.0
Die geometry	Die half angle α/°	3–11
	Inside diameter D_1/mm	30.0
	Reduction R_d (%)	17.4
	Bearing length B/mm	9.0

Many of readers, who are not familiar with the drawing process, might guess the behavior of the drawing force as shown in Fig. 7.3. It might be natural to guess that the total amount of deformation would increase with increase of die half angle α, leading to increase of drawing force. The following part will show some analytical results of the finite element method.

The commercial code ELFEN [30], which was developed by Rockfield Software Limited, Swansea, was used for the analysis of stretch forming. Axisymmetric elastic–plastic analysis was carried out using an implicit method. The von Mises' yield criterion was adopted, and the normality principle was applied to the flow rule. Constraints were dealt with by the penalty function method. A quadrilateral element was used because of the simplicity of the material deformation. The F-bar method was applied to the element for overcoming volumetric locking with simple four-noded quadrilateral elements [31].

Figure 7.4 shows the finite element mesh. A sufficiently fine mesh was adopted for bar for the stable calculation of drawing force. A coarse mesh would cause a large variation of drawing force. Figure 7.5 shows the contact pressure during drawing, and this result would give another chance for readers to change their decision on the effect of die half angle, α, on the drawing force, *F*. One of the important characteristics is that the contact length, [N]–[X], between die and bar increases with decrease of die half length, α, which may have been noticed without analysis in advance. Another important characteristic is that contact pressure increases at the entrance [N] and exit [X] of the taper part of die, called "approach".

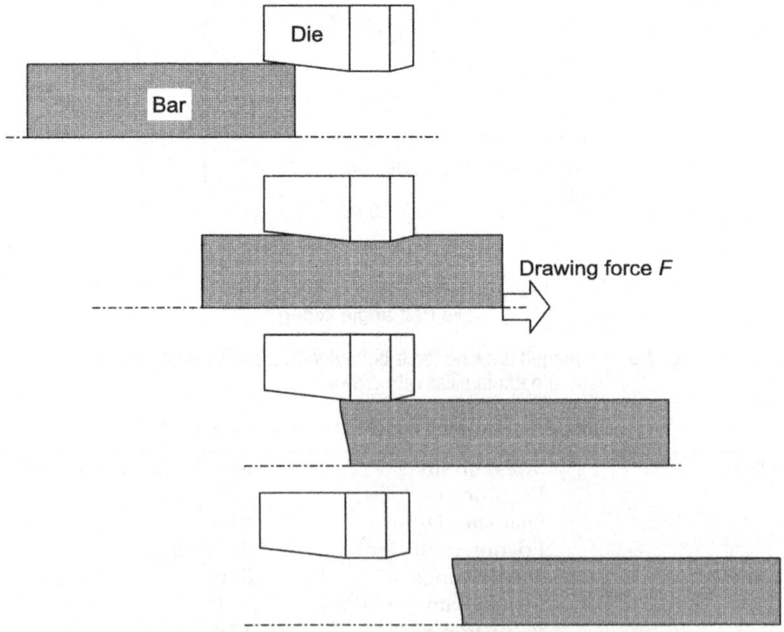

Fig. 7.4 Mesh in the finite element method for drawing

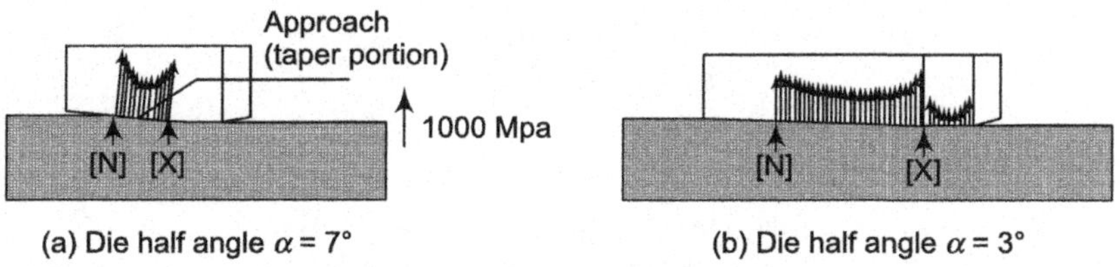

Fig. 7.5 Contact pressure during drawing.

It would be meaningful to calculate an approximate value of forming force in metal forming concisely without using the finite element analysis or some analytical methods such as the slab method.

The assumptions are as follows:

1. The contact pressure is equal to the true stress corresponding to the plastic strain to which the material deforms. This is actually not true due to some reasons including the existence of hydraulic pressure.
2. The equivalent plastic strain is calculated based on the change in the cross-sectional area rate. This is actually not true due to some reasons including nonuniform deformation.
3. Friction is neglected. This is also not true in actual phenomena.

The normal pressure at approach p_n will be calculated by

$$p_n = \sigma = f(\varepsilon_p) \tag{7.1}$$

where $f(\varepsilon_p)$ is the stress obtained from the equivalent plastic strain referring Fig. 7.2. The axial component of normal pressure P_n is

$$p_z = p_n \sin(\alpha) \tag{7.2}$$

The integration of the axial pressure P_z will give the drawing force F as

$$F = 2\pi \int r p_n \sin(\alpha)\, ds \tag{7.3}$$

where s = the position along the taper slope. At the inclined surface infinitesimal length ds is expressed in terms of infinitesimal change in radius dr as

$$ds = \frac{1}{\sin(\alpha)} dr \tag{7.4}$$

From the above equations, the drawing force will be expressed as

$$F = 2\pi \int_{r_1}^{r_0} r f(\varepsilon_p)\, dr \tag{7.5}$$

Even though ε_p is not constant, $f(\varepsilon_p)$ does not change drastically with change in ε_p as shown in Fig. 7.2. Therefore, the drawing force will approximately and concisely be calculated by

$$F \approx A \cdot f(\varepsilon_R) \tag{7.6}$$

where A = projection of the contact area to a plane, which is normal to the direction of the force. In the case of drawing, it would be

$$A = \pi(r_0^2 - r_1^2) \tag{7.7}$$

There are some ways to decide the representative equivalent plastic strain ε_R, like

$$\varepsilon_R = -\ln\left(\frac{r_m^2}{r_0^2}\right) \tag{7.8}$$

$$r_m = \frac{r_0 + r_1}{2} \tag{7.9}$$

Equation (7.6) gives us a very concise and useful method to calculate the forces in metal working, and also a general solution, which would be applicable to other metal working processes, such as rolling. As it is derived by neglecting hydrostatic stress, nonuniform deformation, and friction, Eq. (7.6) suggests a lower forming force than the actual one. However, it would be helpful to know the approximate forming force needed for the machine, and to check the forming force calculated by more complicated and precise methods, such as the finite element analysis. Considering the drawing conditions defined by Table 7.1 and the mechanical properties shown in Fig. 7.2, Eqs (7.6)–(7.9) would give the drawing force F of 98 kN.

Figure 7.6, which was obtained by the finite element analysis, gives the answer to Question 1. The drawing forces range from 93–177 kN, which change depending on the die half angle α and friction coefficient μ. The drawing force F of 98 kN from Eqs (7.6)–(7.9) would support the magnitude of the drawing forces calculated by the finite element method. The answer to Question 1 is now answered just as Geleji explained [29].

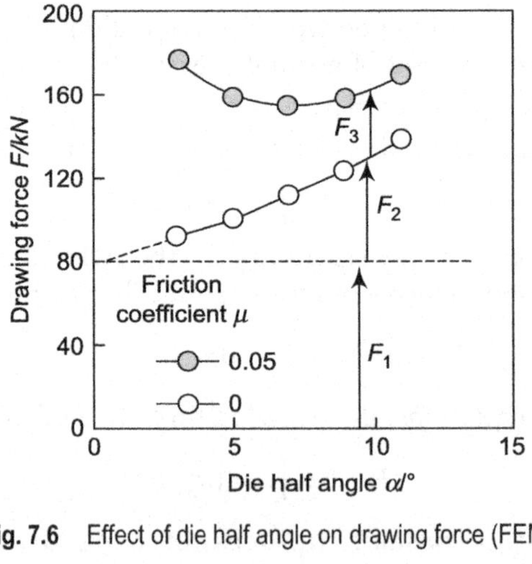

Fig. 7.6 Effect of die half angle on drawing force (FEM)

Answer 1: when friction coefficient μ is zero, the drawing force F monotonously increases with increase of the die half angle α. When friction coefficient μ is 0.05, the drawing force has the minimum value around $\alpha = 7°$. If $\alpha < 7°$, F increases with decrease of α. If $\alpha > 7°$, F increases with increase of α.

Drawing force F would be divided into three components; the minimum bare force F_1 to elongate the bar length and reduce the diameter, the force F_2 for shear deformation which increases with increase of the die half angle α, the force F_3 for pulling the bar against friction force.

The minimum bare force F_1 is almost equal to the force F calculated by Eqs (7.6)–(7.9). This minimum bare force F_1 is constant regardless of die half angle α. Some readers might have thought that F_1 increases because of $\sin\alpha$ term in Eq. (7.2). This is not the case because the effect of $\sin\alpha$ is cancelled out by the infinitesimal length expression of Eq. (7.4).

The reason why F increases with α is that the shear deformation increases with increase of the die half angle α. Figure 7.7 shows some stress and strain distributions. The local deformation of bar would be schematically described as in Fig. 7.7(a). While a small element would be just elongated along the bar axis z, an element near surface would be sheared in negative direction at the approach entrance [N] and sheared in the opposite direction at the approach exit [X]. This behavior is also observed as the shear stress distribution in Fig. 7.7(b). Negative shear stress appears at an area starting at point [N] and positive shear stress appears at an area starting at point [X]. The shear strain has a rather gradual change than the shear stress as in Fig. 7.7(c), because strain would be an integrated value of stress, as suggested by Reuss [8] as follows:

$$\frac{d\varepsilon_x}{s_x} = \frac{d\varepsilon_y}{s_y} = \frac{d\varepsilon_z}{s_z} = \frac{d\gamma_{xy}}{\tau_{xy}} = \frac{d\gamma_{yz}}{\tau_{yz}} = \frac{d\gamma_{zx}}{\tau_{zx}} = d\lambda \tag{7.10}$$

where s_x, s_y, s_z are deviatoric stresses and $d\lambda$ is a scalar factor of proportionality which changes depending on stress and stress state and history. Due to the shear deformation in the vicinity of the bar surface, the equivalent strain near surface is larger than that in the center part as shown in Fig. 7.7(d). As the material is undergoing shear deformation at the entrance [N] and the exit [X] of the approach, the contact pressure is large at [N] and [X] as shown in Fig. 7.5.

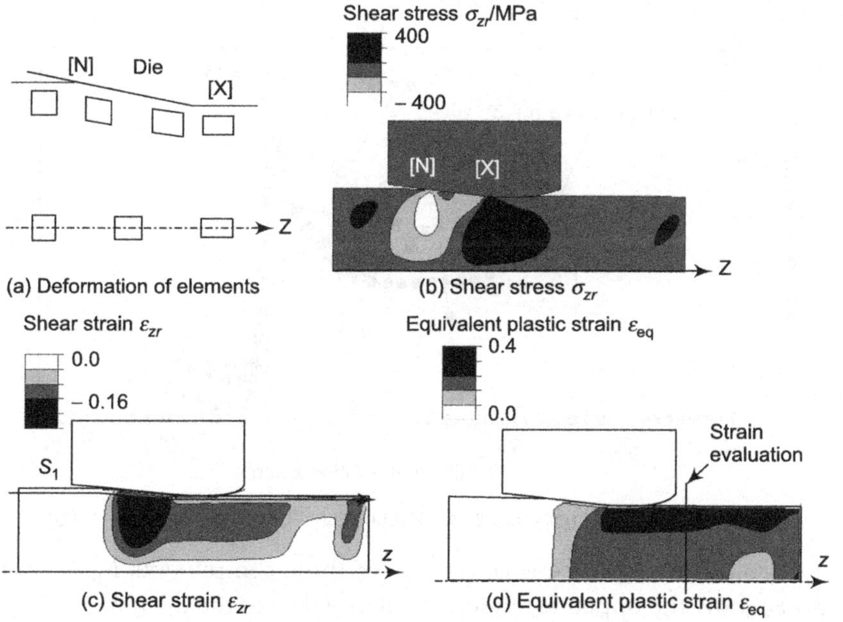

Fig. 7.7 Stress and strain distribution (die half angle α = 7°)

Figure 7.8 shows the effect of the die half angle α on the equivalent plastic strain ε_{eq}. As the shear deformation at [N] and [X] increases with increase of α, ε_{eq} increases with α. This increase of ε_{eq} leads to the increase of the contact pressure p at [N] and [X]. This is the mechanism for the phenomenon that the drawing force component F_2 increases with increase of die half angle α.

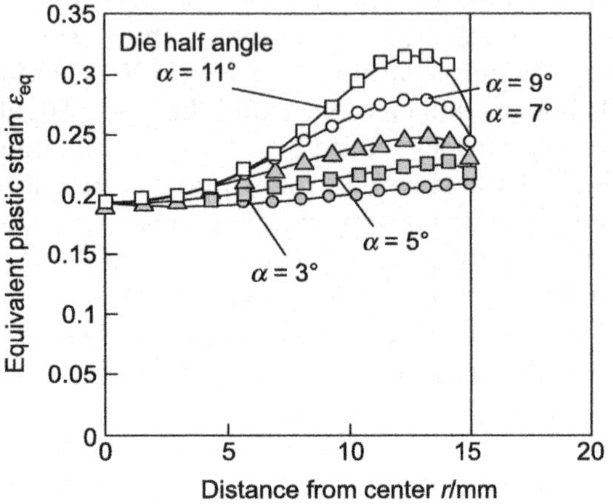

Fig. 7.8 Effect of die half angle on equivalent plastic strain (friction coefficient $\mu = 0.05$)

The third component of drawing force F_3 increases with decrease of the die half angle. Some readers might have noticed the reason after seeing Figs 7.5 and 7.9. The decrease in α causes increase of contact area, which leads to the increase of F_3 (see Fig. 7.6).

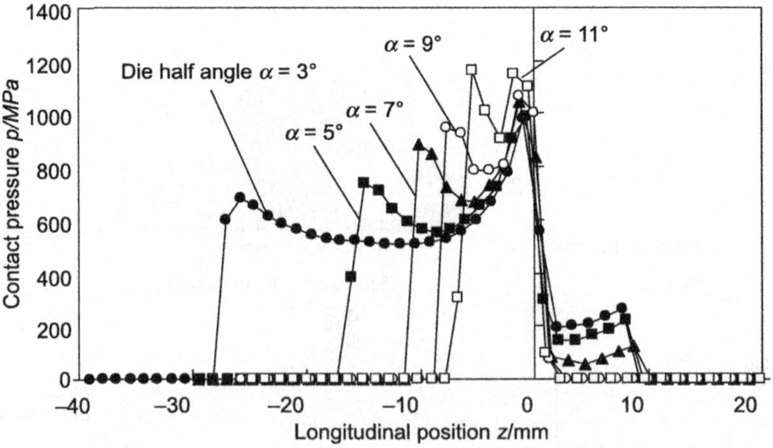

Fig. 7.9 Effect of die half angle on contact pressure (Friction coefficient $\mu = 0.05$)

The total drawing force is composed of three components F_1, F_2 and F_3. F_2 and F_3 show the opposite tendency with respect to the die half angle α. As a result, the drawing force F has the minimum value around $\alpha = 7°$.

The detailed analysis of the metal forming processes is possible by finite element method (FEM). Some key sentences would be supported here in the context of FEM analysis:

Key sentence: Analyses tell the truth (under ideal conditions).

The finite element method enables some numerical experiments by neglecting the effect of some factors, which real experiments would not be able to conduct. For example, the friction coefficient will not be zero in actual experiments.

Key sentence: Simulations will show you what you cannot see in actual phenomena.

Although stresses cannot be measured in actual experiments, they can easily be seen in the finite element analysis.

Key sentence: Although simulations might tell you what you cannot see, you will fail to see it if you have no intention.

For example, the shear stress and strain was easily focused upon as shown in Fig. 7.7(b) based on the observation and examination by previous researchers in this case. If it is a new metal forming process, we have to brush up our own sense of engineering and to determine what kind of phenomenon should be examined by the analysis.

Besides drawing forces, there are other phenomena which should be focused upon for the tool design. The less the drawing force, the less the machine ability will be required. Based on the result of drawing force in Fig. 7.6, the die half angle, α, is usually determined in industry in many cases. However, α should be determined considering other problems in the process line. One of the problems during drawing is the fractures that occur in the center, called "central burst" or "Chevron Cracks", which is schematically shown in Fig. 7.10(a). If a die design is carried out for suppression of central burst, smaller die half angle is preferable [32]. One of the most influential factors on the central burst would be stress; especially the axial stress must be focused upon in the case of drawing, which is shown in Fig. 7.10(b). The axial stress is found to be maximum at [C] in the figure.

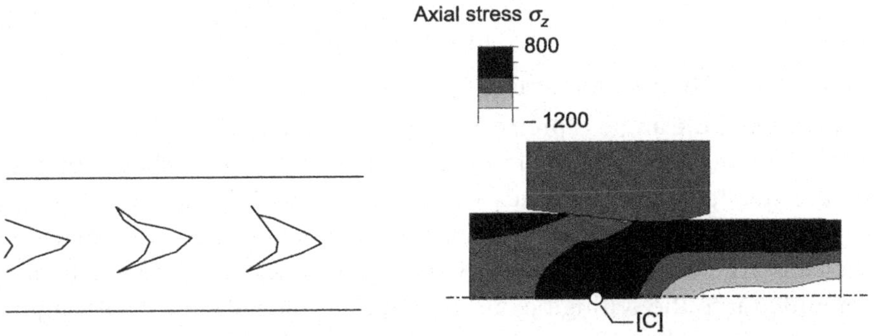

(a) Schematic illustration of central burst

(b) Axial stress distribution calculated by the finite element analysis.

Fig. 7.10 Central burst and axial stress distribution

Figure 7.11 shows analytical results on the effect of drawing conditions on the axial stress and void index [32]. Here, the reduction in area R_d is defined by

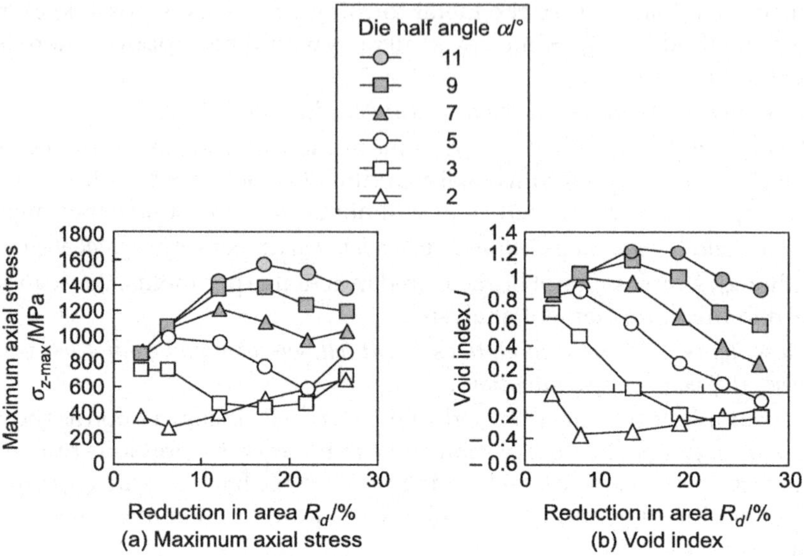

Fig. 7.11 Effect of drawing conditions on stress and void index calculated by the finite element method

$$R_d = 1 - \left(\frac{D_1}{D_0}\right)^2 \tag{7.11}$$

Void index J was defined by

$$J = \frac{1}{\ln((D_1/D_0)^2)} \int_0^{\varepsilon_p} \frac{\sigma_m}{\sigma_{eq}} \, d\varepsilon_p \tag{7.12}$$

which derives from Oyane's Eq. [33] by dividing per pass reduction $\ln((D_1/D_0)^2)$

$$J = \frac{1}{\ln((D_1/D_0)^2)} \int_0^{\varepsilon_p} \frac{\sigma_m}{\sigma_{eq}} \, d\varepsilon_p \tag{7.13}$$

where σ_m is the hydrostatic stress.

The smaller die angle is preferable for suppression of central burst than $\alpha = 7°$ which would give the minimum drawing force. Here, it should be noted that the smaller die angle needs the longer die, which may be difficult to fabricate or to mount into conventional setting boxes in drawing machines. When the target diameter of drawn wire is changed, a die of different diameter is mounted into the setting box in drawing machine. As the axial length of the conventional drawing dies are small, the length of the setting box is also limited, leading to the difficulty to mount a die with small die angle and long axial length.

Another problem in drawing is residual stresses. If residual stresses are well-balanced inside a drawn bar, there would not be any geometrical distortion.

However, if the residual stresses are unbalanced due to some machining or other forming, it would cause unfavorable deformation. Even if residual stresses are well-balanced, there is a possibility that fatigue endurance may deteriorate.

Figure 7.12 shows an analytical result of axial residual stress distribution under a typical drawing condition (Table 7.1) with die half angle $\alpha = 7°$. The Sachs method was used for the experimental measurement [34]. The main cause of the residual stress is nonuniform deformation during drawing [35]. As the material near surface is constrained by the die while the center part is drawn from forward side, the center undergoes more plastic deformation (elongated longer) than the surface under the approach. However, the total longitudinal strain, composed of plastic and elastic strains, must become constant from the center to the surface after drawing to ensure compatibility. As a result, the surface must be elongated at the bearing, and the residual axial stress becomes tensile at the vicinity of the surface and compressive at the center after drawing.

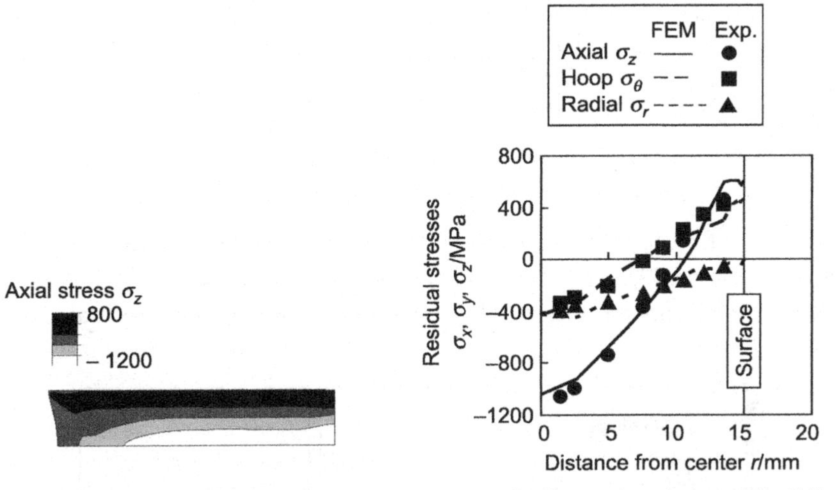

(a) Distribution of axial residual stress calculated by the finite element method

(b) Comparison between the finite element analysis and experiment

Fig. 7.12 Residual stress after drawing in a typical condition (die half angle $\alpha = 7°$, friction coefficient $\mu = 0.05$)

For the suppression of residual stress, one of the effective solutions is to give uniform deformation at the end of the deformation. With decrease of die half angle, the deformation becomes uniform across the cross section. However, the total length of die becomes unrealistically long for to level the residual stress sufficiently. Double-tapered die in Fig. 7.13 would be an effective method for suppression of residual stresses within a realistic die length [36, 37].

Table 7.2 Drawing condition for double-tapered die

Parent bar	Stress-strain relationship	Measured value of annealed AISI1045
	Diameter D_0 /mm	33.0
	Friction coefficient μ	0.05
	Die half angle, 1st α_a/°	0.1–5.0
	2nd α_b/ °	11.0
Die geometry	Inside diameter, 1st D_a/mm	30.0
	2nd D_b/mm	29.98–29.0
	Bearing length B/mm	9.0

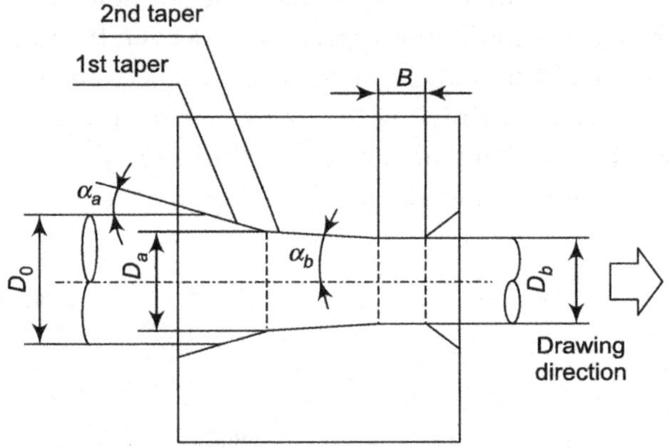

Fig. 7.13 Schematic illustration of double-tapered die

While the first taper reduces the bar diameter, the second taper levels the residual stresses. Figure 7.14 shows an optimization result of die geometry by the finite element method. The ordinate is intensity of residual stress I_x calculated by

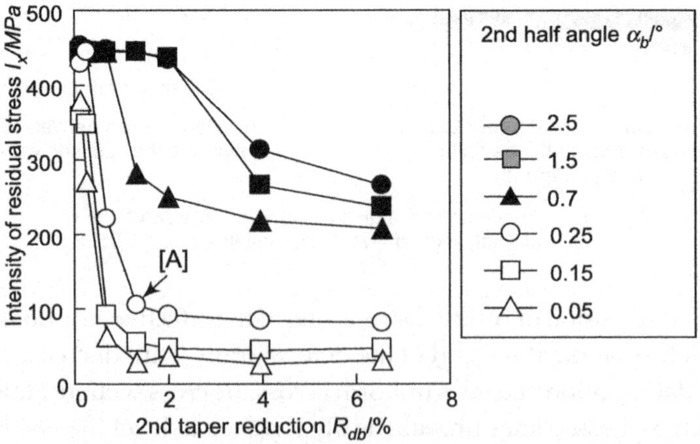

Fig. 7.14 Effect of die angle and reduction at the second taper on intensity of residual stress

$$I_x = \frac{4}{\pi D_b^2} \int_0^{\frac{D_b}{2}} 2\pi r \left| \sigma_z \right| dr \qquad (7.14)$$

With increase of second taper reduction R_{db} and decrease of second taper angle α_b, the residual stress can be suppressed. However, the die is required to be very long at the same time. Therefore, the optimum condition would be the point denoted by [A], where the total die length would be suppressed around twice the size of the bar diameter. Figure 7.14 is an example of a key sentence:

Key sentence: Analysis enables you to conduct numerical experiment using dozens of different tools over a night.

It would have taken a couple of months and be costly to find an optimum die design by experiment, because dozens of die must have been designed and fabricated. However, the analysis has shortened the time for finding the optimum design.

The optimum designs of tools and process would be different depending on the objective; minimizing the drawing force [29], suppression of void [32], levelling residual stresses [35, 36, 38], straightness of drawn bars [38], and so on. Unfortunately no die geometry or drawing conditions were found to satisfy all the requirements. Therefore, it would be essential to determine the problems in the actual operation lines and to select the solution method.

7.4 DIE-LESS TUBE BENDING

Tube bending is one of the most popular metal-forming processes. Figure 7.15 shows a typical bending method, called draw bending. The head of a straight tube is clamped by the clamp. The tube is pushed into the groove of the bending die by the pressure die. While the bending die rotates, the pressure die normally move in the tangential direction of the bending die. When the target bending angle is 180° for example, the bending angle must exceed 180° to over-bend the tube due to springback. Springback occurs due to the stress distribution caused by bending. During bending, tensile and compressive stresses exist at the extrados and intrados of the bending arc, respectively. The integration of the stresses must make the same force and moment imposed by tools. After the tools release the tube and the force and moment from the tools become zero, springback occurs to make the internal force and moment zero.

Draw bending is a precise and stable method of tube bending, because the tube is constrained by bending and pressure dies. However, there are some disadvantages in the method. It would be important and difficult to sort out the problems in conventional metal forming method, and to solve the problems and make improvements.

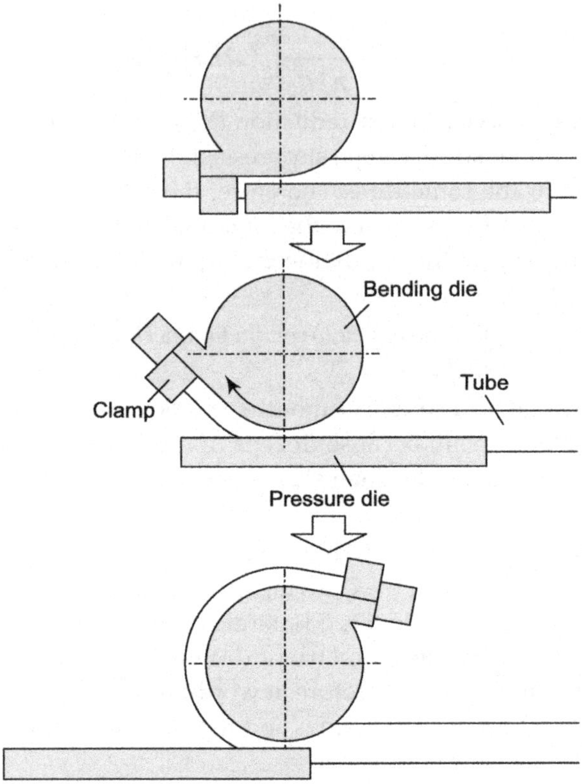

Fig. 7.15 Schematic illustration of drawing bending

Question 2: what are disadvantages in draw bending when it is applied for manufacturing U-bent tubes for electric power plant?

U-bent tubes are used as steam generators in electric power plants. In order to install many tubes to obtain sufficient heat transfer efficiency, the bending radius of the U-tubes approximately ranges from 50 to 2000 mm as shown in Fig. 7.16. The precise bending radius is required for the U-tubes because high durability is mandatory for the devices on the electric power plant. The tolerance of the bending radius is 1.5 mm. If the precision is not sufficient, the vibration and repetitive deformation during operation would cause fracture at the connection part. Draw bending process has been used as a suitable manufacturing method due to its high bending precision. However, there are still problems in draw bending for manufacturing U-bent tube.

Answer 2: (1) A set of many bending dies with different radii is needed, and it leads to increase of maintenance and facility cost.

(2) Changing bending dies for each bending radii needs labour forces and time, which leads to the cost increase.

In order to solve the problems, a new bending method, called die-less U-bend method has been developed [39]. Those who have learned the subject of "Strength of materials" might think that four-point bending is a potential

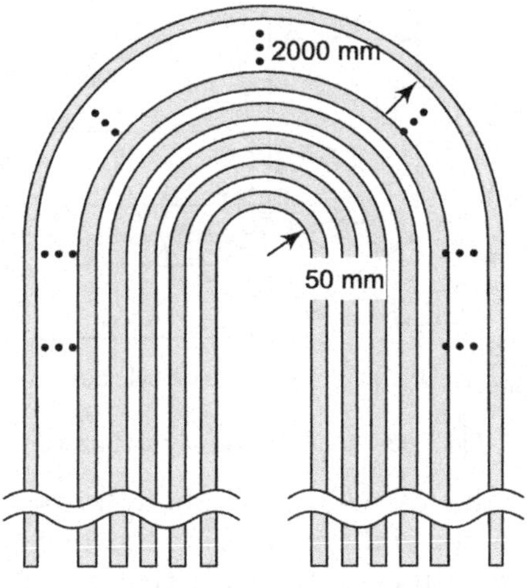

Fig. 7.16 Application image of U-bent tube

method without using dies, because four-point bending yields uniform moment during a certain length ideally as shown in Fig. 7.17. However, it is actually very difficult to realize uniform moment for tube bending due to local crash and flattening which appears at the contact area between the tube and the tools. The moment becomes easily nonuniform with slight change of the tube condition. An innovative 3-dimensional bending method, which is called MOS (Murata Ohashi Suzuki) bending and does not use bending dies, was invented by Murata [40]. While a tube is pushed through a moving die from a guide cylinder, the die position u and angle ϕ is controlled as shown in Fig. 7.18. While the direction of the die position u decides the bending direction, the combination of u and ϕ decide the bending radius.

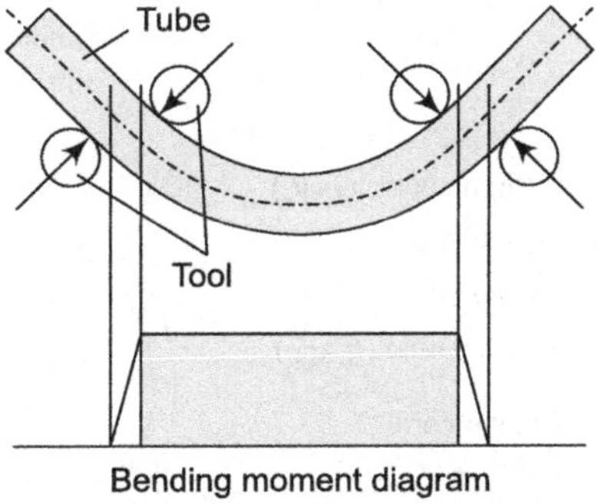

Fig. 7.17 Four-point bending

Although MOS bending is a splendid method for 3-dimensional bending, it would not be applicable for precise U-bent tube in electric power plants. Because the long straight parts at the both ends of the U-bent tube might affect the moment at the bending point P due to its inertia and gravity, and MOS bending has inevitably a radius transition region v in Fig. 7.18.

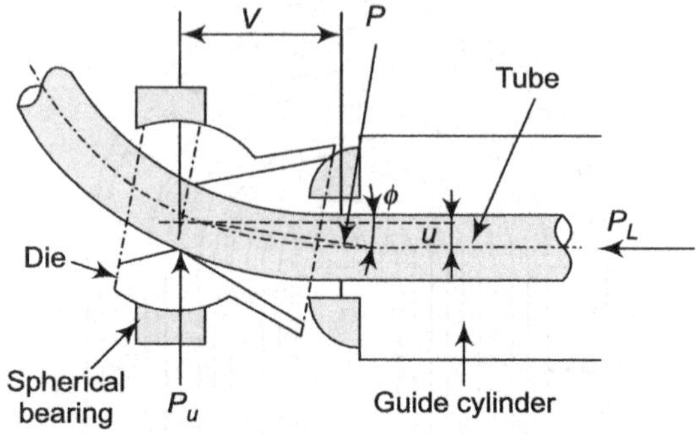

Fig. 7.18 MOS (Murata Ohashi Suzuki)

Figure 7.19 shows die-less U-bend method. In die-less U-bend method, the influence of the inertia and gravity from the long straight parts is small because the ends of the bend arc are firmly clamped. A straight tube is held by a clamp and feeding rolls at the beginning of bending. While rotating arm rotates on an arc of an arm radius R_0 at travelling velocity of V_ϕ, the feeding rolls synchronously feeds the tube at a feeding velocity V. The tube is ideally bent at the bending position [P]. The bending radius is easily changed by changing the arm radius R_0, i.e., the position of the clamp on the rotating arm and the distance between the center of the rotating arm and the unit of the bending roller and the feeding rolls.

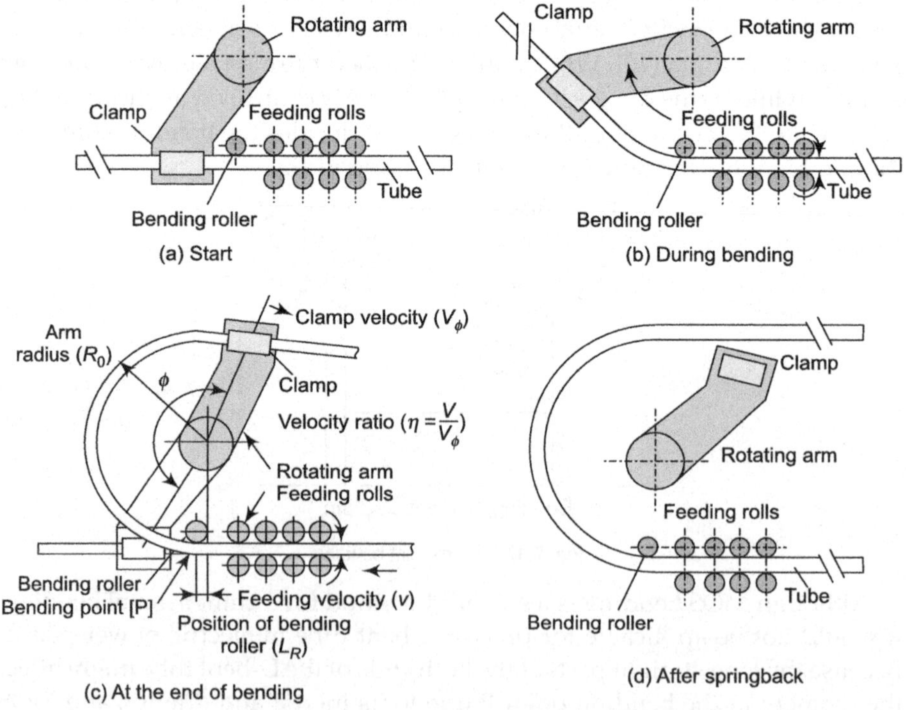

Fig. 7.19 Schematic illustration of die-less U-bend method

The arm radius R_0 determines the approximate bending radius. It seems that the bending radius becomes constant when the velocity ratio η, which is defined by

$$\eta = \frac{V}{V_\phi} \tag{7.15}$$

is 1.0, it is actually not uniform when $\eta = 1.0$. In order to adjust the bending radius, the velocity ratio η and the position of bending roller L_R might be used.

The stress–stain state in the tube is complicated during U-bend tube. Elastic and plastic deformations coexist. Some parts are subjected to increasing plastic deformation, while others are under elastic recovery. Furthermore, some might be subjected to plastic deformation after elastic recovery. Therefore, it might be difficult to determine the appropriate condition only by experimental trial and error. Some analytical method was needed. In the case of this project, a fundamental analytical method based on a deformation theory of curved beam was used instead of the finite element method. The fundamental analytical methods based on curved beam theories would be efficient under the following conditions:

1. The deformation of material is not severe. The shape of cross section is assumed to be constant without flattening, so the distance from the neutral axis will be constant before and after bending.
2. Stress in axial direction is dominant in the deformation. Radial and circumferential stresses are not assumed to affect deformation. Only axial stress is considered for moment calculation in numerical analysis.

For a sheet metal bending, Imai *et al.* successfully applied a fundamental analytical method for plate V-bending to accomplish a precise process method in actual process lines [41].

In the analysis in die-less U-bend method, stress–strain diagram is assumed to be described by

$$\sigma = \begin{cases} E\varepsilon & \text{when } \varepsilon \le \varepsilon_e \\ E\varepsilon + H(\varepsilon_e) & \text{when } \varepsilon > \varepsilon_e \end{cases} \tag{7.16}$$

where E = Young's modulus, H = work hardening modulus and ε_e = limit of elastic strain.

This assumption leads to a simple equation to calculate the moment M from curvature $1/\rho$ as

$$M = f\left(\frac{1}{\rho}\right) \tag{7.17}$$

The details of the function f are described in the reference [39]. The bending shape after deformation is calculated from the curvature distribution before and after deformation. According to deformation theory of curved beam [42], explained in Fig. 7.20, displacement of a certain point R on the curved beam or tube is calculated from

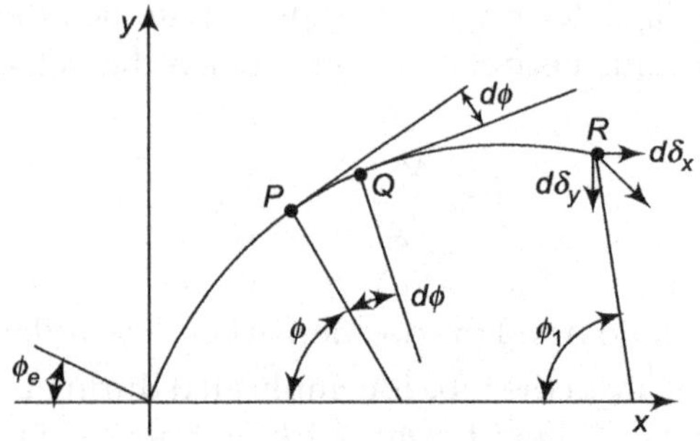

Fig. 7.20 Deformation of curved beam

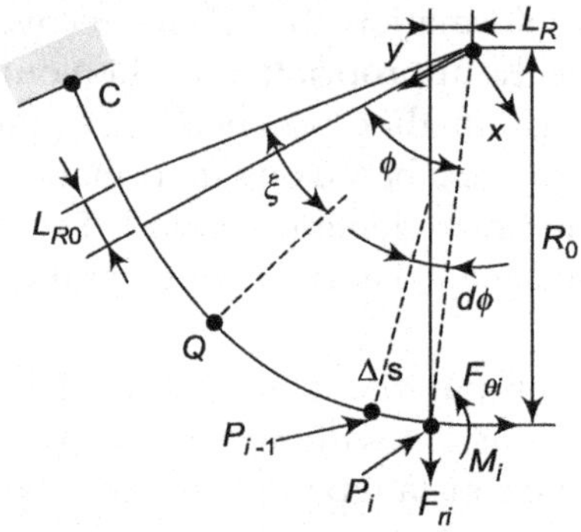

Fig. 7.21 Model simulating die-less bending processes at *i*th step

$$\delta_x = \int_{\phi_0}^{\phi_1} \left(\frac{1}{\rho} - \frac{1}{\rho_0} \right) \rho_0 \, (y_0 - y) \, d\phi \tag{7.18}$$

$$\delta_y = \int_{\phi_b}^{\phi_1} \left(\frac{1}{\rho} - \frac{1}{\rho_0} \right) \rho_0 \, (x_0 - x) \, d\phi \tag{7.19}$$

$$\delta_\phi = \int_{\phi_b}^{\phi_1} \left(\frac{1}{\rho} - \frac{1}{\rho_0} \right) \rho_0 \, d\phi \tag{7.20}$$

As these equations are general and useful, they might be applicable for many bending problems of longitudinal parts. Based on these equations, the die-less bending process is simulated by a model shown in Fig. 7.21. In the real bending process, the tube is bent in a continuous manner, with the tube feeding, the arm rotating and the bending point P moving on the tube gradually. In the model, the tube is discretely bent step by step, with the tube feeding by distance ΔS, the arm rotating by angle $\Delta\phi$ and the bending point P moving by distance ΔS. When the bending process progresses at ith step, a moment M_i, a force on the tube axis $F_{\theta i}$ and a force on the rotating arm radial direction F_{ri} are loaded at the bending point P.

The bending conditions of the die-less bending process are as follows:

1. Cross-sectional size of tube: 17 mm in diameter, 1 mm in thickness
2. Radius of rotating arm R_0: 725 mm

3. Position of the bending roller L_R: 0–10 mm
4. Velocity ratio of the tube-feeding to the clamp on the rotating arm η: 0.970–1.0
5. Material of tube: Nickel alloy.

Figure 7.22 shows the results of theoretical analysis and experiment under conditions of $L_R = 0$ and $\eta = 1.0$. Analytical results are in agreement with the experimental values. Although these conditions seem to be optimum due to the complete synchronism of tube-feeding and clamp velocity, bending radius becomes large both on the clamp side and on the bending roller side, as shown in Fig. 7.22(a). This phenomenon is caused by moment distribution while bending. Figure 7.22(b) shows moment distribution on the curved tube when the rotation angle ϕ =70°, 140°, and 210°. For each rotation angle, the maximum position occurs in the middle of the curved tube. As a result, bending phenomenon becomes unstable and bending radius becomes large at both ends of the curved tube.

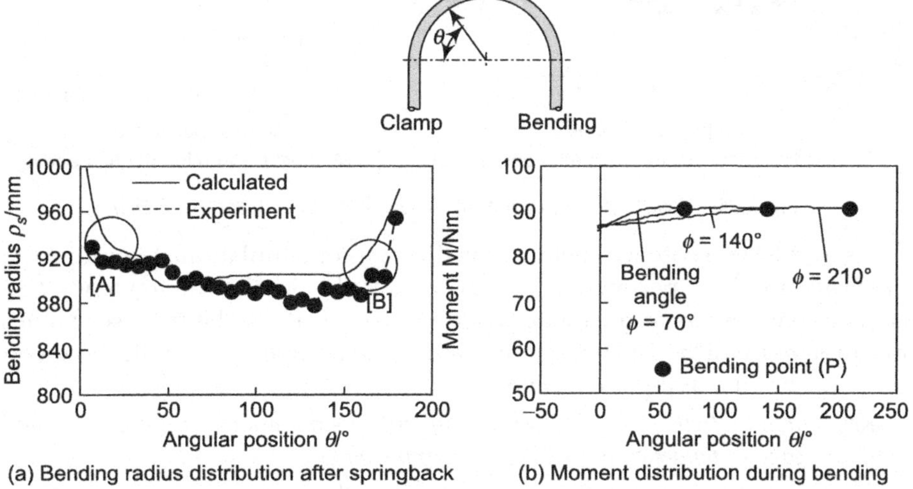

Fig. 7.22 Results of analysis and experiment (when $L_R = 0$ and $\eta = 1.0$)

Figure 7.23 shows results under conditions of $L_R = 0$ and $\eta = 0.970$. Compared with Fig. 7.22, bending radius on the bending roller side becomes smaller, i.e., the influence of η is large on the bending roller side. For each rotation angle ϕ, moment M is maximum at the bending point P, shown in Figure 7.23(b). As a result, bending phenomenon becomes stable and bending radius is constant at angular position $\phi > 50°$.

Figure 7.24 shows results under conditions of L_R = 10 mm and η = 0.985. Compared with Fig. 7.22, bending radius on the clamp side and on the bending roller side becomes smaller, i.e., the influence of L_R is large on the clamp side. At each rotation angle ϕ, moment M is maximum at the bending point P, as shown in Fig. 7.24(b).

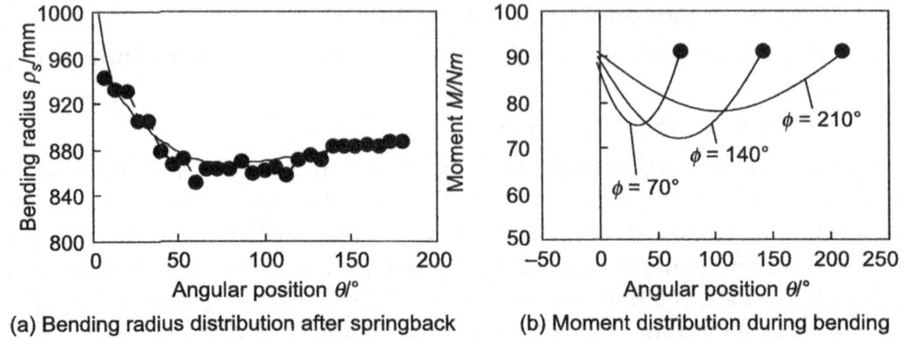

Fig. 7.23 Results of analysis and experiment (when $L_R = 0$ and $\eta = 0.970$)

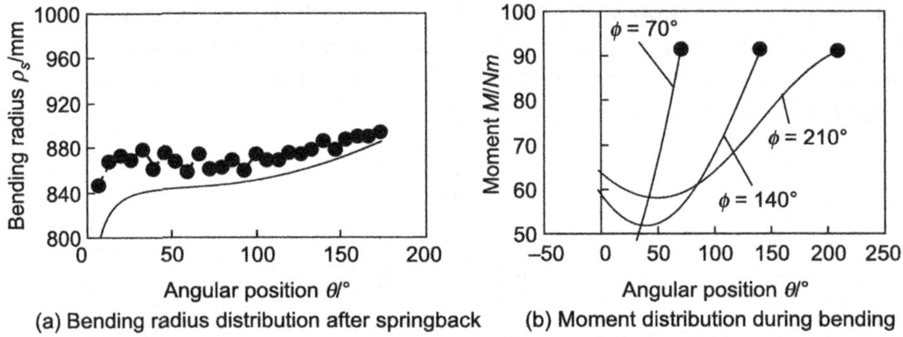

Fig. 7.24 Results of analysis and experiment (when $L_R = 10$ and $\eta = 0.985$)

As in the key sentences below, it is difficult for simulations to predict the precise value in experiments. However, it could predict the tendency and effect of parameters in the metal working process. It also helps to see values like moment which is difficult to measure in experiment, and will show you the concept for solving the problems.

Key sentence: Predictions of values are difficult for simulations. Simulations would be able to predict tendency or influence of parameters.

Key sentence: Simulations will show you what you cannot see in actual phenomena.

The concept obtained here for the precise bending was that the position of the bending roller L_R should be controlled in the former half of bending, while the velocity ratio η should be controlled in the latter half of bending considering the influence magnitude on the bending radii. A further simulation was composed to calculate the optimum bending conditions to obtain the target radii by iteration convergence.

Figure 7.25 shows experimental results of the optimum conditions and bending radius distribution after spring back. The optimum conditions are characterized as follows:

1. L_R should increase at the beginning of the bending.
2. L_R should be constant in the middle and at the end of the bending.

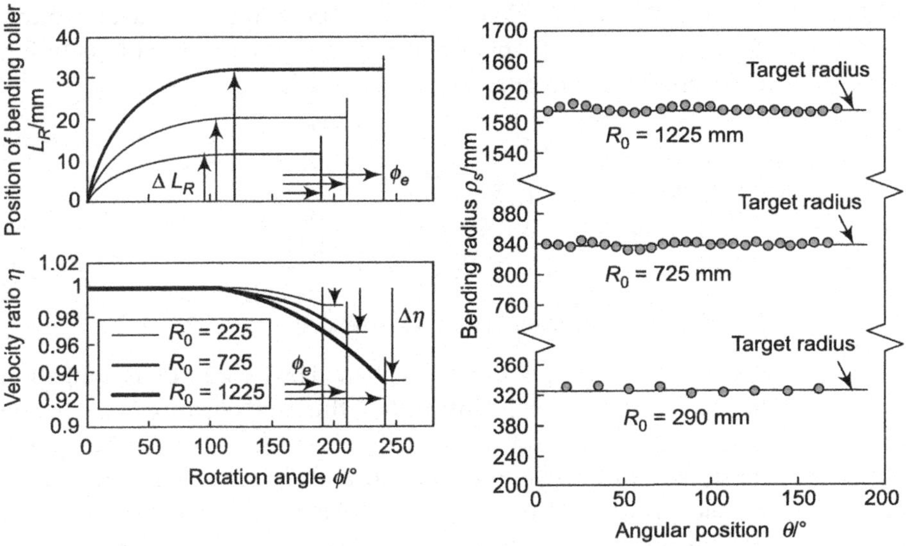

(a) Optimum bending conditions calculated by simulation

(b) Bending radii distribution under optimum conditions in experiment

Fig. 7.25 Optimization of bending conditions and experimental verification

3. η should be constant in the first half of the bending.
4. η should decrease in the latter half of the bending.

The bending radii were constant and the bending shape satisfied the specifications for U-bent tubes used in the electric power plant.

As we can see in this example, fundamental analyses are useful to consider the mechanism and optimize the working conditions and finite element analysis. It would be difficult for computational methods, including both fundamental analyses and the finite element analysis, to predict the value with 100% accuracy. There would still be various phenomena which cannot be described numerically. However, computational methods would tell us the concept of improvement, if we have the ability to see what the computational methods are saying.

7.5 TUBE BENDING METHOD WITH AN EXTREMELY SMALL BENDING RADIUS

Enhancement and prediction of forming limit would be another concern academically and industrially. Forming limit in tube bending includes reduction of bending radius as well as thinning of tube wall. Small bending radii would decrease the space for automobile engine parts, air conditioning appliances, water-supply equipment, and so on.

When a thin-wall tube is bent in a small radius, a few defects like buckling, wrinkle and rupture would occur as shown in Fig. 7.26. The minimum radius without defects which is realized in conventional bending is said to be around 1.5 times of tube diameter [43]. These defects would be resumed in bending of clay or paper tube by our hands.

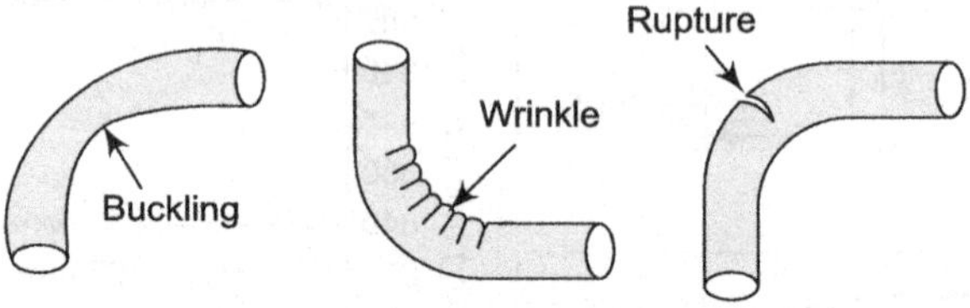

Fig. 7.26 Defects in tube bending

The occurrence of the defects would be explained by longitudinal elongation of the extrados and shrinkage of the intrados of the bending arc caused by the distance from neutral planes, and volumetric constancy. When a tube is bent, the extrados of the bending arc should elongate and the intrados should shrink depending on the distance from the neutral plane, where the longitudinal material length is constant. If the cross-sectional stiffness or strength is small, the cross section becomes oval or flattened so that the distance from the neutral plane might be small. The flatness or ovality increases with decrease of bending radius. Once the flatness becomes large at some specific part due to variation of the bending moment, the flatness acutely increases at the same point because of decrease of second moment of area about the neutral axis of the section. This results in buckling.

The application of external constraints by tools such as mandrels and die grooves would be an effective method for suppression of buckling. However, further small bending radius would cause wrinkle and rupture. If the tube circularity is maintained by external constraints, the tube should longitudinally shrink at the intrados of the bending arc. Due to the volumetric constancy, the tube wall should be thick. Otherwise, wrinkle occurs with maintaining the longitudinal length and thickness. If the work for wrinkle is smaller than that of thickening, wrinkle occurs. Wrinkle tends to occur with decrease of tube wall thickness. Similarly, if the tube circularity is maintained by external constraints, the tube should longitudinally elongate at the extrados of the bending arc. If the elongation exceeds the limit of the material, rupture would occur. Therefore, if the following condition is satisfied, tube bending with a very small bending radius is feasible without wrinkles and ruptures.

Condition-1 for tube bending with an extremely small bending radius:

The elongation of the extrados and shrinkage of the intrados of the bending arc should be suppressed.

A shear bending method using hydraulic pressure had been proposed [44], which satisfies the above Condition-1. The principle is shown in Fig. 7.27. A tube is inserted in the holes of dies and the hydraulic pressure P_i is applied on tube-inside surface. The moving die slides at shearing velocity V_s by shearing force F_s. At the same time, tube is pushed into the die at pushing velocity V_p by pushing pressure P_p. The tube will undergo a shearing deformation continuously at shearing zone which is located 45° in diagonal direction. The tube is formed into a crank shape product. The die corner radius r_c determines the inside bending radius of tube, while the outside bending radius r_o would be determined by other parameters like inner pressure P_i. It is noteworthy that the length along Route *T*, *B* and *M* are ideally constant during deformation satisfying the Condition-1.

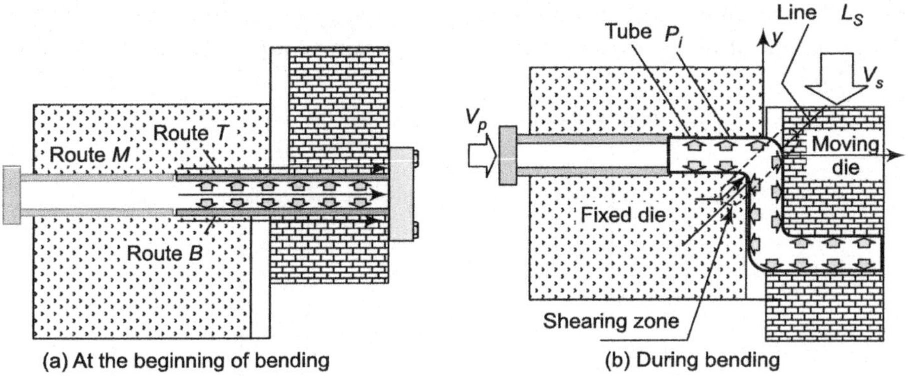

Fig. 7.27 Schematic illustration of shear bending with hydraulic pressure

These lines are kept constant due to shear deformation along line L_S. This line L_S is exactly and literally the auxiliary line in the following key sentence.

Key sentence: Draw auxiliary lines to solve problems

Other engineers drew an auxiliary line, which means a leap of imagination, to invent a more concise shear bending method as shown in Fig. 7.28. Two mandrels are used instead of the inner hydraulic pressure [43, 45, 46]. One of the two mandrels is attached to the moving die and it moves during bending. The velocity ratio V_p/V_s, which is the ratio of the pushing velocity at tube end V_p to the moving velocity of moving die V_s, becomes the only dominant and controllable parameter.

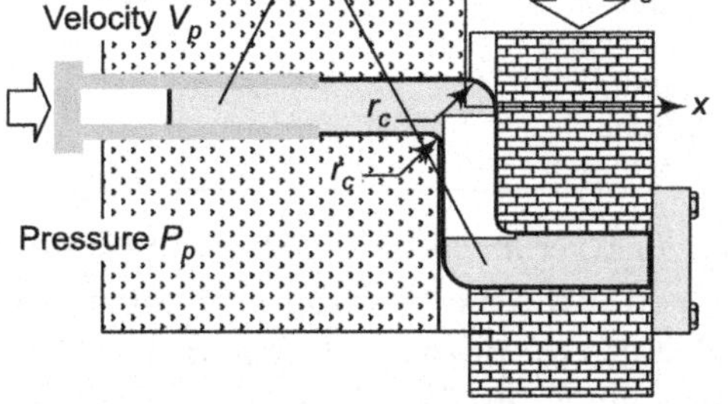

Fig. 7.28 Schematic illustration of shear bending with mandrels

The mechanism in shear bending would be examined with sense of engineering and a slight help from the finite element method. The commercial code ELFEN used in the modelling of processes described in this chapter, was used for the analysis of shear bending with mandrels in the conditions shown in Table 7.3. The elastic–plastic analysis was carried out using a hexahedron solid element. The explicit scheme was used for shortening the calculation time. The Coulomb friction law was assumed and the friction coefficient μ was determined as 0.05. The stress–strain diagram of the parent tube is shown in Fig. 7.29. The diagram was obtained by tension test of aluminum A1050 and extrapolated after rupture occurred as the strain would reach a value larger than 1.0.

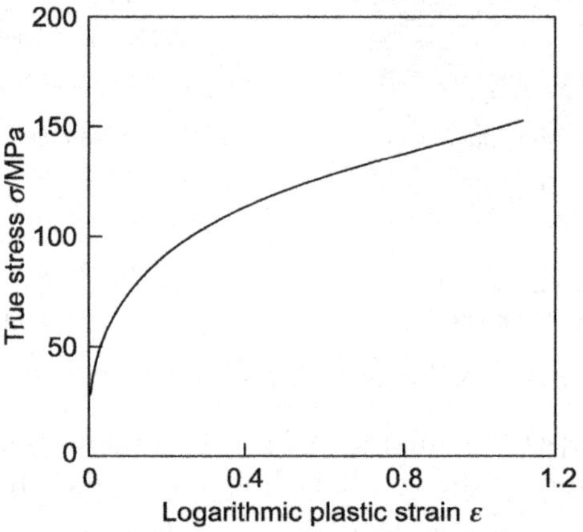

Fig. 7.29 Stress strain diagram for aluminum 1050

Table 7.3 Bending conditions

Bending	
Die corner radius r_C /mm	3.0–5.0
Radial clearance c /mm	0.1
Friction coefficient μ (FEM)	0.05
Velocity ratio V_p/V_s	0.6–1.0
Inner pressure P_i /MPa	0–14
Tube	
Material	A1050
Young's modulus E /GPa	70
Yield stress Y /MPa	30
Poisson's ratio ν	0.34
Diameter D_0 /mm	30.0
Thickness t_0 /mm	1.5–2.0

Here let's *try thought experiments* of ideal bending states in order to observe the difference between ordinary bending and shear bending. Figures 7.30 and 7.31 show ideal states for pure bending and shear bending, respectively. In pure push bending, when material comes from left-up side and go out to right-down side, the following characteristics would be picked up.

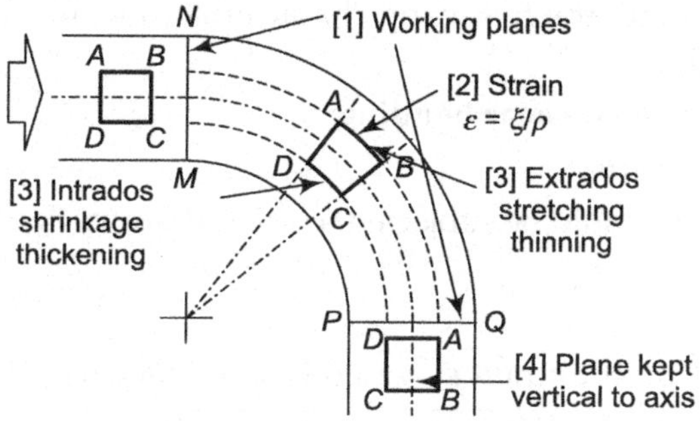

Fig. 7.30 Ideal pure bending

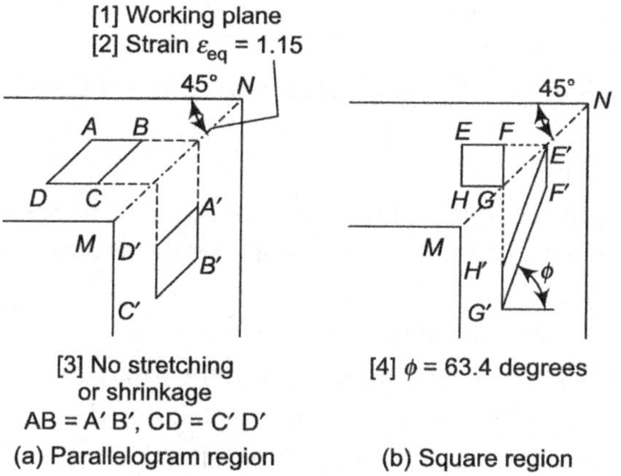

Fig. 7.31 Ideal shear bending

Characteristics in ideal pure bending

1. Deformation area

   The material would be subjected to bending at the plane MN and unbending at the plane PQ.

2. Strain

   Axial strain ε would be expressed as

$$\varepsilon = \frac{\xi}{\rho} \tag{7.21}$$

   where ξ = the distance form the neutral plane and ρ = bending radius.

3. Deformation

   At the extrados of the bending arc, as the tube wall stretches in the axial direction and the tube wall should be thin due to the volumetric constancy. On the other hand, at the intrados of the bending arc, as the tube shrinks in the axial direction and the tube wall should be thick.

4. Plane vertical to tube axis

   The plane BC, which is initially normal to the tube axis, continues to be normal to the tube axis.

On the other hand, the following characteristics would be picked up for ideal shear bending.

Characteristics in ideal shear bending

1. Deformation area

   The material would be subjected to shear deformation at the plane MN once.

2. Strain

   Equivalent plastic strain ε_p would be expressed as

$$\varepsilon_p = \sqrt{\frac{2}{3}(\varepsilon_x^2 + \varepsilon_y^2 + \varepsilon_z^2) + \frac{1}{3}(\gamma_{xy}^2 + \gamma_{yz}^2 + \gamma_{zx}^2)} \tag{7.22}$$

$$= 1.15$$

   applying Hencky's theory [5] considering $\gamma_{xy} = 2.0$ and other strains are zero.

3. Deformation

   The tube will be subjected to neither elongation nor shrinkage. The line segment AB or CD will keep its initial length. As a result, thickness must be constant.

4. Plane vertical to tube axis

   The plane FG, which is initially vertical the tube axis, will tilt by an angle ϕ of 63.4° [45], which is geometrically calculated.

These thought experiments would be checked by the finite element methods and experiments. Figure 7.32 shows an example of the finite element method on equivalent plastic strain distribution, deformation and thickness strain distribution. The characteristics of actual shear bending would be as follows.

Characteristics in actual ideal shear bending

1. Deformation area

   The material would be subjected to shear deformation gradually at a broad bending area rather than a plane.

2. Strain

   Equivalent plastic strain ε_p would be near 1.0 at the centre route *M*, while it is much less value around the route *T* or *B*.

3. Deformation

   The change of the tube wall thickness is much smaller for the extremely small bending radius. However, some thickness deviations are observed at the bending corner of [H], [I], [J] and [K].

4. Plane vertical to tube axis

   The plane, which is initially vertical the tube axis, will tilt by an angle ϕ of 63.4° at the centre route M. The tilt angle *f* becomes much smaller around the routes *T* and *B*.

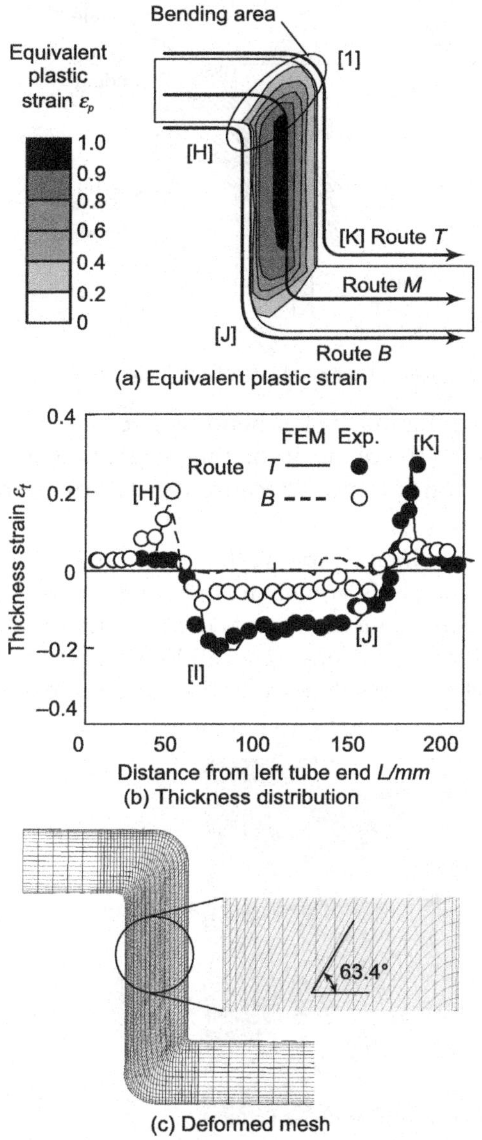

Fig. 7.32 Characteristics in shear bending (die corner radius r_c = 5 mm, tube thickness t_0 = 1.5 mm, velocity ratio V_p/V_s = 0.8, hexahedron solid element for FEM)

The deformation behavior in the actual shear bending is slightly different from that in the ideal one. In particular, the difference is remarkable at the top route *T* and bottom route *B*, and the deformation is similar to that of pure bending. Therefore, the deformation zone is schematically divided into three zones as shown in Fig. 7.33 [47].

The material is subjected to deformation similar to shear bending at the zone 2 of the centre part, while it is subjected to deformation similar to pure

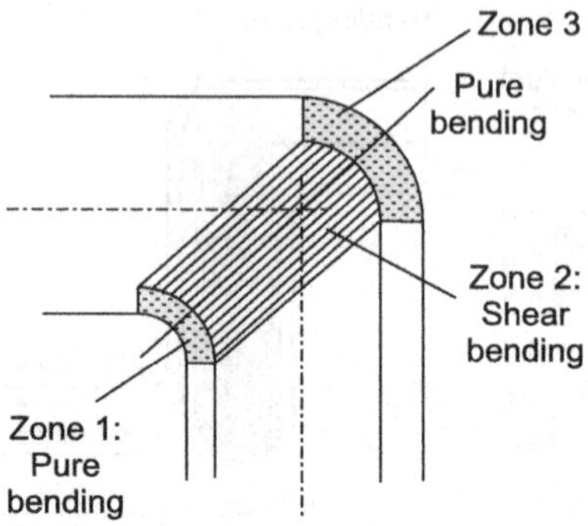

Fig. 7.33 Schematic illustration of type of deformations in actual shear bending

bending at the zones 1 and 3. Shear bending becomes dominant because the deformation occurs in the plane of the tube wall at the zone 2. Pure bending is dominant at the zones 1 and 3 because the deformation occurs out of the plane of the tube wall.

The effect of the velocity ratio V_p/V_s on the bending behavior in shear bending using mandrels is shown in Fig. 7.34. A triangle Morley's shell was used for the finite element analysis instead of a solid element. The shell element tends to show exaggerated wrinkle than actual phenomena, while the solid element shows less than the actual. The finite element analysis and the experiment showed a good agreement.

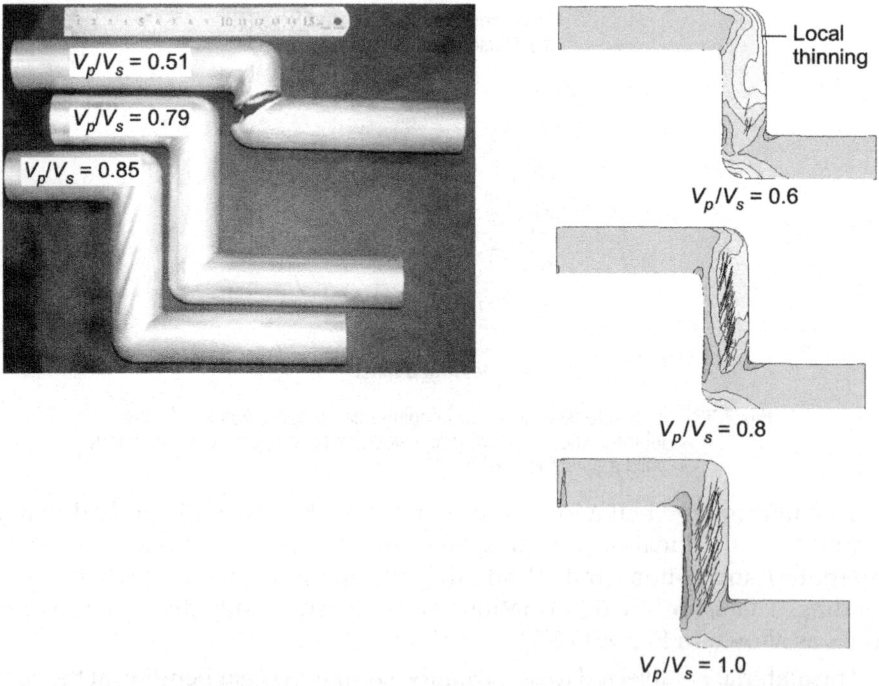

Fig. 7.34 Effect of velocity ratio on bending behavior (triangle Morley's shell for FEM)

When the pushing velocity at tube end V_p is equal to the moving velocity of moving die V_s, i.e., $V_p/V_s = 1.0$, wrinkle appears due to the deformation similar to pure bending at zone 1. With decrease of V_p/V_s, the number and the height of wrinkles decrease because of decrease in the supply of material to bending zone. However, excessive decrease of velocity ratio V_p/V_s would lead to local thinning causing rupture, in particular, at zone 3, where the pure-bending-like deformation occurs. At an optimum condition, the tube was successfully bent with an extremely small bending radius.

Shear bending with mandrels are applied in actual production line for practical use for manufacturing automotive parts by the inventors [43]. Both of the shear bending methods with hydraulic pressure and mandrels are splendid inventions, which would probably have been done through a number of trial and errors, which must have led the leap of imagination, i.e., *drawing auxiliary lines.* Computational approaches would have been nothing at this stage of invention, even if the inventors have had some methods.

In the complex deformations like shear bending, it would be meaningful to conduct thought experiments to examine the essence in the deformation. Computational approaches would help you examine the results of thought experiments and would lead to discovery of some unexpected principle. It would be difficult for the analyses to predict the phenomena with 100% accuracy. It would be important to put computational methods to practical use rather than trying to enhance the accuracy unnecessarily.

7.6 PREDICTION OF MECHANICAL PROPERTIES IN METAL FORMING

Even though the current level of ability of the finite element method is satisfactory for practical use in many cases in industry, there would be other cases where more precise ability is required for the numerical method. For example, spring-back in tube and plate bending, residual stresses after metal forming and strength of products could be affected by a slight change of mechanical properties including anisotropy and the Bauschinger effect.

The methods for treating the mechanical properties would be categorized into two points of view. One is a phenomenological method whereby the constitution equations are used to numerically describe the complex behavior of the stress–strain relationship obtained empirically. The other is to predict the stress–strain behavior from rather fundamental hypotheses based on information of microstructure. This section focuses upon the latter methods.

Various trials have been made for clarification of relationship between phenomena in micro and macro scales [48, 49], including methods which utilize the finite element analysis [50, 51]. Advantages in application of the finite element methods are that these methods can consider actual microstructures in metals. However, many of these studies are unilateral methods to predict macroscopic behaviors from microscopic characteristics. Multi-scale modelling by the homogenization method is known as a method to completely couple the problems on mechanics in micro scale and material behaviors in macro scale [52, 53]. It was a turning point for the homogenization method when Terada and Kikuchi proposed a method to treat the boundary conditions of the whole structure and an algorithm formulation of convergence [54, 55]. Since then, the homogenization method was expanded for the finite element method and stated to be applied for metal forming [56–61].

The following is an example for application of homogenization method for prediction of stress–strain diagram during tension–compression test of carbon steel, which was verified by experiments [62]. The chemical compositions of the steel are shown in Table 7.4. Two patterns of heat-treatment were adopted to change the shape of the carbide in the experiments. The microstructures of the materials are shown in Fig. 7.35. The heat-treatment has controlled the microstructure. One is the ferrite–pearlite structure; the pearlite includes alternately arrayed thin plates of ferrite and carbide. The ferrite and carbide are respectively dark gray and light gray parts in the SEM image. The other is the structure with spheroidized carbide. The existence of carbide increases with increase of carbon. Pearlitic or spheroidized carbides exist uniformly when the carbon content is equal to 0.080, which is the condition of eutectic carbon steels.

Table 7.4 Chemical composition (mass %)

	C	Si	Mn	P	S	Cr
Low carbon	0.06	0.03	0.32	0.017	0.014	0.04
Medium carbon	0.45	0.19	0.73	0.014	0.010	0.12
High carbon	0.81	0.18	0.50	0.070	0.004	0.04

	Pearlite	Spheroidized
Low-carbon steel $C = 0.06$ %		
Medium-carbon steel $C = 0.45$ %		
High-carbon steel $C = 0.80$ %		

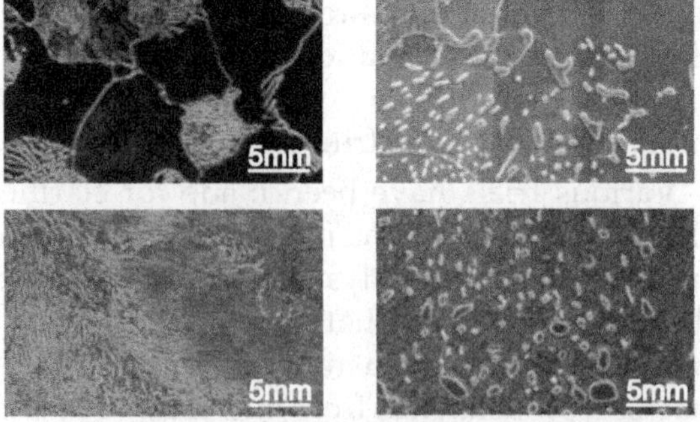

Fig. 7.35 Microstructures in SEM

Figure 7.36 shows the finite element meshes for the homogenization method. Some readers, who are interested in detailed theories in the homogenization, should refer to some references [54, 55]. In the homogenization methods, following assumptions are conceptually made.

1. While the microstructures exist at random in the actual material, representative volume elements (RVEs) exist periodically. Therefore, periodic boundary condition was applied.
2. While the microstructure has a finite length ε in the actual material, the side length of the RVE is infinitely small ($\varepsilon \to 0$).
3. The mechanical properties of the ferrite and carbide are the same to those of bulk materials. While the mechanical properties of ferrite could be measured by tension test, that of carbide, the bulk of which is difficult to fabricate, was found in a literature [63].

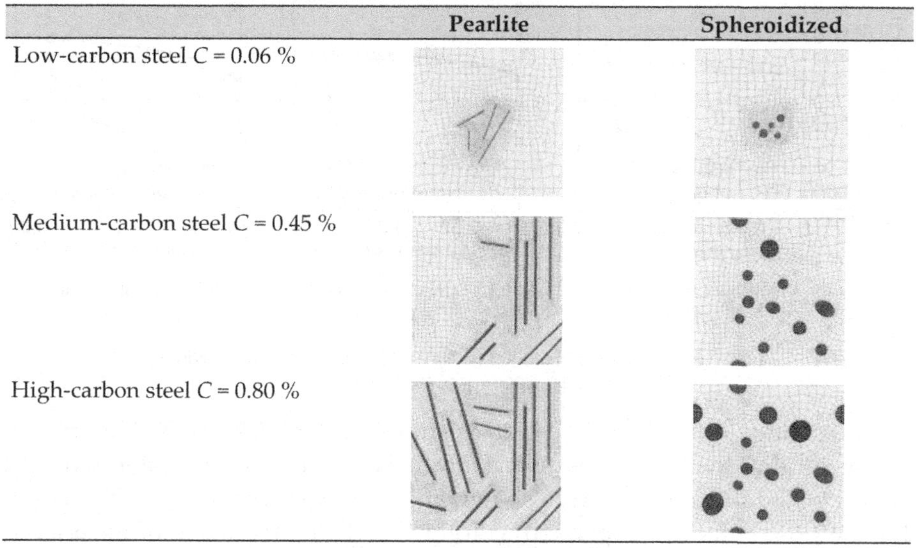

Fig. 7.36 Microstructures in FEM

The experimental and numerical results are respectively shown in Figs. 7.37 and 7.38. The two results are qualitatively in good agreement on the behavior of the Bauschinger effect, whereby both the elastic range and a yield point for the reversed direction of straining have decreased after stress is reversed from tension to compression. With increase of carbon content, the Bauschinger effect increases. The structures with pearlite have stronger Bauschinger effect than those with spheroidized carbide.

The mechanism of the Bauschinger effect in the dual-phase structure would be explained based on the stress–strain distributions in the analysis. The changes of equivalent strain distributions in the RVE are shown in Fig. 7.39, taking high-carbon steel with spine-shaped carbide as an example. The change during tension is described in Fig. 7.39(a). It is important to note that strain concentration is observed in the vicinity of the carbide tip as X_1.

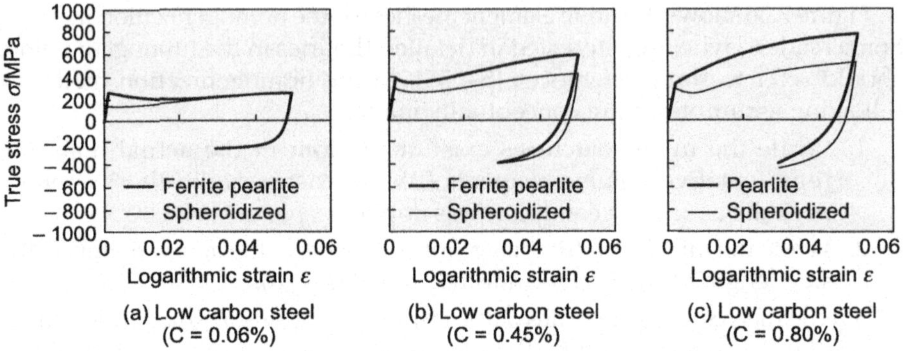

Fig. 7.37 Influence of carbon content and morphology of carbide on stress–strain diagram, experimental results

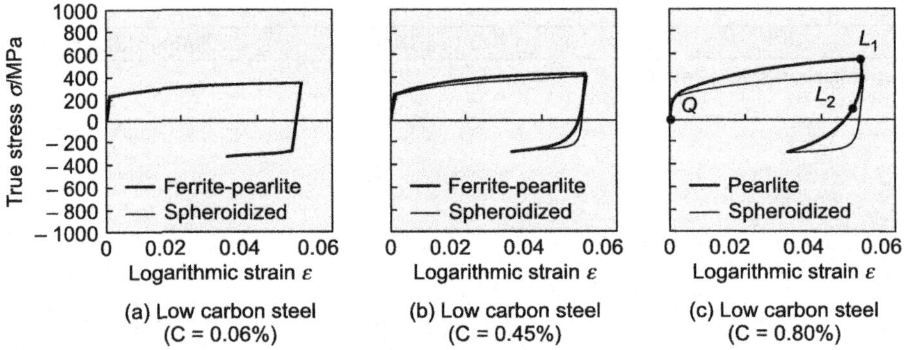

Fig. 7.38 Influence of carbon content and morphology of carbide on stress–strain diagram, numerical results

The change during compression is shown in Fig. 7.39(b). Noteworthy is the result that the strain does not concentrate during compression where it concentrated during tension as shown at X_1 and vice versa at X_2. The reason for the phenomena is explained as follows. At the first stage of tension, as deformation accelerated around the region at the tip of cementite, tensile stress appeared around the tip. At the second stage of compression, elastic deformation lasted longer at the tip of cementite because of tensile stress, which occurred at the first stage. The numerical analysis may be the only method which suggests this kind of deformation in microstructure.

It would be interesting to note that the analysis with simple assumption realized Bauschinger effect numerically, whereby the mechanical properties of bulk were used as the properties in microstructure. The analysis here does not consider the crystal plasticity or dislocations. The simple assumption would be satisfactory in the analysis of dual-phase metals. Consideration of the crystal plasticity would not enhance the precision in the analysis in this case, where stress and strain distributions exist in a wide area along the cementite, and this "long range stress" causes the Bauschinger effect. If the material is a single phase metal, where the "long range stress" and strain do not exist,

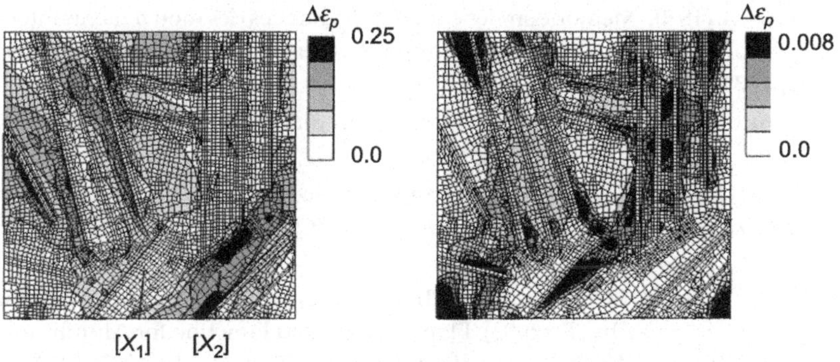

(a) During tension (O to L_1) (b) During compression (L_1 to L_2)

Fig. 7.39 Calculated change of equivalent plastic strain

consideration of the crystal plasticity would be useful [61]. Therefore, it would be important to spot the essence of phenomena in metal forming process, and to decide the appropriate and required specifications of the analysis for effective application of the numerical methods.

SUMMARY

In this chapter, key sentences were introduced based on the author's experience. Hopefully it would become of some help for readers in the future. For the verification of the concept, some research works and examples were introduced, which include bar drawing, die-less bending, bending with an extremely small radius and prediction of mechanical properties in metal forming.

With the recent development of computers, numerical analyses have become a convenient tool for the application to metal forming. Analyses would tell us what we cannot detect in easy ways in actual processes. They would conveniently propose the optimum working conditions and best tool geometries. However, they would not be able to predict with 100% accuracy. Experimental verifications are absolutely imperative. We need to know the actual phenomena and problems in the actual process lines before construction of numerical models. We need to keep brushing up the sense of engineering in order to utilize the powerful computational tools.

REFERENCES

1. Kiuchi, M. (2007), *Plastic Forming*, in Japanese, Maruzen.
2. Hill, R. (1950), *The Mathematical Theory of Plasticity*, Oxford University Press.
3. Tresca, H. (1864). sur l'écoulement des corps solides, soumis à de fortes pressions, *Comptes Rendus Acad. Sci. Paris*, **5**; 754.
4. Mises, V. (1913). Guttinger Nachrichten, Mathematisch-physische Klasse, 582.
5. Hencky, H. (1924). Zur Theories plastischer deforma nen und der hierdurch im material hervorgerufenen Nachspannuger, Zeitschrift für angewandte Mathematik und Mechanik, **4**; 323

6. Levy, M. (1870). Memorie sur les equations generales des mouvements intervieurs des corps solides ductiles, Comptes rendus de l'Académie des sciences, **70**; 1323
7. Prandtl, L. (1924). Spannungsverteilung in plastischen Körpern, *Proceedings of 1st International Congress on Applied Mechanics*, 43.
8. Reuss, A. (1930). Berücksichtigung der elastischen Formänderung in der Plastizitätstheorie, Zeitschrift für angewandte Mathematik und Mechanik, **10**; 266.
9. Barlat, F., Brem, J.C., Yoon, J.W., Chung, K., Dick, R.E., Lege, D.J., Pourboghrat, F., Choi, S., and Chu, E. (2003). Plane Stress Yield Function for Aluminum Alloy Sheets - Part 1: Theory, *International Journal of Plasticity*, **19**; 1297–1319.
10. Kuwabara, T., Ishiki, M., Kuroda, M., and Takahashi, S. (2003). Yield Locus and Work-hardening Behavior of a Thin-walled Steel Tube Subjected to Combined Tension-internal Pressure *Journal de Physique IV*, **105**; 347–354.
11. Bauschinger, J. (1886). Über die Veränderung der Elastizitätsgrenze und die Festigkeit des Eisens und Stahls durch Strecken und Quetschen, durch Erwärmen und Abkühlen und durch oftmals wiederholte Beanspruchungen, Mitteilungen aus dem Mechanisch-technischen Laboratorium, **13**; 1-116.
12. Armstrong, P.J. and Frederick., C.O. (1966). A Mathematical Representation of the Multiaxial Bauschinger Effect, *CEGB Report* RD/B/N731.
13. Chaboche, J.L. and Rousselier, G. (1983). On the Plastic and Viscoplastic Constitutive Equations, Part I and II, *Journal of Pressure Vessel Technology*, Transactions of the ASME, **105**; 153–164.
14. Yoshida F. and Uemori, T. (2002). A Model of Large-strain Cyclic Plasticity Describing the Bauschinger Effect and Workhardening Stagnation, *International Journal of Plasticity*, **18**; 661–686.
15. Oyane, M. (1983), *Theory of Plasticity*, Mar., in Japanese, Yokendo.
16. Karman, T.V. (1925). Zeitschrift für angewandte Mathematik und Mechanik, **5**; 139.
17. Zienkiewicz, O.C. and Taylor, R.L. (2006), *The Finite Element Method Set*, Sixth Edition, Elsevier.
18. Turner, M.J., Clough, R.W., Martin, H.C., and Topp, L.J. (1956). Stiffness and Deflection Analysis of Complex Structures, *Journal of Aeronautical Sciences*, **23**; 805.
19. Clough, R.W. (2004). Early History of the Finite Element Method from the View Point of a Pioneer, *International Journal for Numerical Methods for Engineering*, **60**; 283–287.
20. Crisfield, N.A. (1997), Non-linear Finite Element Analysis of Solids and Structures, Volume 1: *Essentials*.
21. Turner, M.J., Martin, H.C. and Melosh, R.J. (1960). Large Deflection of Structures Subject to Heating and External Load, *Journal of Aeronautical Sciences*, **27**; 97–106.
22. Argyris, J.H. (1964), *Recent Advances in Matrix Methods of Structural Analysis*, Pergamon Press.

23. Argyris, J.H. (1965). Continua and Discontinua, *Proceedings of Conference on Matrix Methods in Structural Engineering,* Air Force Inst. of Tech., **69**; 633–636

24. Zienkiewicz, O.C. (1971). *The Finite Element in Engineering Science,* McGraw-Hill, London.

25. Mallet, R.H. and Marcal, P.V. (1967). Finite Element Analysis of Non-linear Structures, *Proc. ASCE, Journal of the Structural Division,* **93**; ST3, 221–234.

26. Oden, J.T. (1967). Numerical Formulation of Non-linear Elasticity Problems in *Proc. ASCE, Journal of the Structural Division,* **93**; ST3, paper 5290.

27. Oden, J.T. (1969). Finite Element Applications in Non-linear Structural Analysis in *Proceedings of the Conference on the Finite Element Methods,* Vanderbilt University Tennessee.

28. Haisler, W.E., Stricklin, J.E. and Stebbins, F.J. (1971). Development of Evaluation of Solution Procedures for Geometrically Non-linear Structural Analysis by the Discrete Stiffness Method in *AIAA/ASME 12th Structure, Structural Dynamics & Materials Conf.,* Anaheim, California.

29. Geleji, A. (1955), Die Berechnung der Kräfte und des Arbeitsbedarfs Bei der Formgebung im Bildsamen Zustande der Metalle, *Akademi-Verlag, S.,* Budapest, pp. 262–263.

30. Mullet, P. and Rance, J. (1996). Applied Discrete Element Technology: The Assessment and Strengthening of Masonry Arches, *Bench Mark: Report from CINTEC Worldwide,* no. 1,17–24.

31. de Souza Neto, E.A., Peric, D., Dutko, M., and Owen, D.R.J. (1996). Design of Simple Low Order Finite Elements for Large Strain Analysis of Nearly Incompressible Solids, *International Journal of Solids and Structures,* **33** (20–22); 3277–3296.

32. Kuboki, T., Abe, M. Neishi, Y. and Akiyama, M. (2005). Design Method of Die Geometry and Pass Schedule by Void Index in Multi-pass Drawing, Transactions of ASME, *Journal of Manufacturing Science and Engineering,* **127** (1); 173–181.

33. Oyane, M., Sato, T., Kunio, O., and Shima, S. (1980). Criteria for Ductile Fracture and Their Applications, *Journal of Mechanical Working Technology,* **4**; 65–81.

34. Sachs, V.G. (1927). Der Nachweis Innerer Spannungen in Stangen und Rohren," *Zeitschrift für Metallkunde,* **19**; 352–357.

35. Kuboki, T., Akiyama, M., Neishi, Y., and Kuroda, K. (2000). Effect of Final Drawing with Light Reduction on the Levelling of Residual Stress Distribution in Cold Bar Drawing, Proceedings of the Institution of Mechanical Engineers, Part C: *Journal of Mechanical Engineering Science,* **214** (12); 1389–1400.

36. Kelly, T.A., Landis, M.N., Kyle, F.A., and Nachtman, E.S. US Patent-3157274.

37. Kuboki, T., Akiyama, M., Neishi, Y., and Kuroda, K. (2001). Effect of Die Geometry on Residual Stress Level Present after Bar Drawing, *Ironmaking Steelmaking,* **28** (1); 65–71.

38. Kuboki, T., Furuta, H. and Akiyama, M. (2004). Suppression of Bending after Steel Bar Drawing by Heavy-reduction Pass, Proceedings of the Institution of Mechanical Engineers, Part B: *Journal of Engineering Manufacture,* **216** (12); 1461–1470.

39. Kuboki, T., Furugen, M., Osaka, S., and Ono, T. (1998). Development of Die-less Bending Process for Precision U-bent Tube, *The 7th International Conference on Steel Rolling*, Chiba, pp. 981–987.

40. Murata, M., Yamamoto, O. and Suzuki, H. (1994). Development of CNC Machine for Tube Bending, *Journal of the Japan Society for Technology of Plasticity*, **35** (398); 262–267, in Japanese.

41. Imai, K., Koyama, J., Jin, Y., Murata, M., and Kuboki, T. (2008). Computational Method for Calculation of Flange Dimension in V-bending, *Steel Research International*, **79**; 209–216.

42. Timoshenko, S. (1940), *Strength of Materials*, D. Van Nostrand Company, Inc.

43. Powel, G. and Avitzur, B. (1971). Forming of Tubes by Hydraulic Pressure, *Report of Institute for Metal Forming*, Department of Metallurgy and Material Sciences, Lehigh University.

44. Tanaka, M., Michino, M., Sano, K., and Narita, M. (1994). Pipe Bending Technology With Zero Bending Radius, *Journal of the Japan Society for Technology of Plasticity*, **35** (398); 232, in Japanese.

45. Goodarzi, M., Kuboki, T. and Murata, M. (2005). Deformation Analysis for the Shear Bending Process of Circular Tubes, *Journal of Materials Processing Technology*, **1** (162–163), 492–467.

46. Kuboki, T., Goodarzi, M. and Murata, M. (2007). Forming Behavior of Shear Bending With Mandrel as Substitution Method of Hydroforming, *Tubehydro 2007*, Harbin, pp. 115–120.

47. Goodarzi, M., Kuboki, T. and Murata, M. (2007). Effect of Die Corner Radius on the Formability and Dimensional Accuracy of Tube Shear Bending, *International Journal of Advanced Manufacturing Technology*, **10**; 66–74.

48. Watanabe, I. (2007). *Doctoral thesis*, Tohoku University, in Japanese.

49. Taylor, G.I. (1938). Plastic Strain in Metals, *Journal of the Institute of Metals*, **62**; 307–324.

50. Gotoh, M. (2005) A Finite Element Formulation for Large Elastic–plastic Deformation Analysis of Polycrystals and Some Numerical Considerations on Polycrystalline Plasticity, *International Journal for Numerical Methods in Engineering*, **12**; 101–114.

51. Harren, S.V. and Asaro, R.J. (1989). Nonuniform Deformation in Polycrystals and Aspects of the Validity of the Taylor Model, *Journal of the Mechanics and Physics of Solids*, **37**; 192–232.

52. Benssousan, A., Lions, J.L. and Papanicoulau, G. (1978). *Asymptotic Analysis for Periodic Structures*, Amsterdam–New York–Oxford; North-Holland.

53. Cioranescu, D. and Donato, P. (2000). *An Introduction to Homogenization*, Oxford Lecture Series in Mathematics and Its Applications 17, Oxford Univ. Press.

54. Terada, K. and Kikuchi, N. (1995). Nonlinear Homogenization Method for Practical Applications, *Computational Methods in Micromechanics*, **212**; 1–16.

55. Terada, K. and Kikuchi, N. (2001). A Class of General Algorithms for Multi-scale Analyses of Heterogeneous Media, *Computer Methods in Applied Mechanics and Engineering*, **190**; 5427–5464.

56. Akiyama, M., Kuboki, T., Oikawa, K., Matsui, K., and Terada, K. (2002). Influence of Carbon Content and Carbide Morphology of Carbon Steels on Stress–strain Curve in Vicinity of Yield Point," *Materials Science and Technology,* **18**; 1272–1278.

57. Terada, K., Matsui, K., Akiyama, M., and Kuboki, T. (2004). Numerical Re-examination of the Micro-scale Mechanism of the Bauschinger Effect in Carbon Steels, *Computational Materials Science,* **31**; 67–83.

58. Miehe, C., Schroder, J. and Schotte, J. (1999). Computational Homogenization Analysis in Finite Plasticity Simulation of Texture Development in Polycrystalline Material, *Computer Methods in Applied Mechanics and Engineering,* **171**; 387–418.

59. Matsui, K., Terada, K., Akiyama, M., Kuboki, T., and Oikawa, K. (2002). Mechanism of the Bauschinger Effect by the Multi-scale Modeling, *Transactions of the Japan Society of Mechanical Engineers,* Part A , **68** (675); 71–78, in Japanese.

60. Nakamura, Y., Tam, N.N., Mori, T., Ohata, T., and Nakamachi, E. (2008). Virtual Dual-Phase Steel Modeling and Its Experimental Verification by Combination Method and Crystal Plasticity Finite Element Method, *Journal of the Japan Society for Technology of Plasticity,* **49** (569); 86–90, in Japanese.

61. Watanabe, I., Terada, K., Terada, K., Akiyama, M., and Neishi, Y. (2003). Multi-scale Analysis for Polycrystalline Metals, *Journal of Applied Mechanics: Japan Society of Civil Engineers* **6**; 239–246, in Japanese.

62. Kuboki, T., Akiyama, M., Matsui, K., and Terada, K. (2002). Influence of Carbon Content and Carbide Morphology on the Stress–strain Curve and Bauschinger Effect, *7th International Conference on Technology for Plasticity,* Yokohama, pp. 1201–1206.

63. Umemoto, M. (2001). Current Advances in Materials and Processes, *Journal of Metastable and Nanocrystalline Materials* **14** (3); 554, in Japanese.

CHAPTER 8 Sheet Metal Forming: Issues and Recent Developments

P.P. Date
Department of Mechanical Engineering, IIT Bombay,
Powai, Mumbai 400076, India

ABSTRACT

Sheet metal is the form of raw material that is extensively used in millions of tones across the world starting with large macro sized parts to the micro-sized ones. The products encompass the consumer product manufacturing sectors and much more. The cost of manufacture of sheet metal products involves a significant contribution from the raw material costs. The domain of sheet metal forming is very different from that of machining which necessitates a different class of technologies, processes, tools and machine tools to be developed. To those who hold machining as synonymous with manufacturing (that being often the case, given the number of machine shops around), the domain of sheet metal forming is a completely different world.

This chapter aims at bringing out the distinction between the processing of sheet metal and machining, and goes on to discuss some of the major issues in sheet metal forming and reviews some of the latest developments that attempt to address these.

Developments in new materials, processes, tools and machine tools is an ongoing process and this chapter looks back into the recent past for the trends in research in sheet metal manufacturing and manufacture of sheet metal products. It is seen that new developments have taken place at all levels of manufacturing, namely, product design, prototyping, materials, processing, tools and machine tools, by application of intelligent manufacturing paradigms and achieving robustness in manufacturing for better product quality.

8.1 INTRODUCTION

In the modern times emphasis on productivity and consistent product quality places great value on consistency in the quality of raw material and process control. Producers of raw material seek to reduce the variability in the properties of the material. Improvements in machine tools enable use of novel technologies and developments in automation enables better process control. Quality variables are continuously monitored and corrective actions are often taken through a feedback control loop. Given a high rate of production, a 100% inspection of the product is often achieved by 'in process inspection' cutting down on the cost of inspection, and making inspection feasible after every step in manufacturing. This concept of total quality effectively arrests variability in the product quality thereby reducing wastage and enhancing the volume of quality production.

When large production volumes are involved reworking of defective parts is practicable only as long as rework volume is small enough. This means 'chance defectives' (which might occur very rarely) despite efforts to minimize variability can be economically reworked manually wherever feasible. This underscores the need for a control over the quality and mechanical behavior of raw material and that over the processing conditions.

The above has a strong effect on assembly processes. With smaller product variability selective assembly would no longer be required. Variations in the product quality at the assembly level would be smaller, enabling sub assembles to fit together with ease and without rework into a complete assembly giving consistent performance in service. Assembly of sheet metal parts in many manufacturing sectors is however still manual to take care of the (legitimate) variability in the quality of the components being assembled.

There are basic differences in the entire process of manufacturing when a metallic sheet is used as the raw material, vis-à-vis that in machining, where the form of raw material is in the form of a block of material having comparable size in the three dimensions. These differences begin with the very nature of the raw material, namely, an undeformed 'flat' sheet, its properties and the purpose of testing (attributes one looks for). The differences extend through the process of designing and dimensioning of a sheet metal product, design and selection of tools, dies and machine tools to inspection, quality control and assembly.

In particular, attention is drawn to the large number of mutually interacting variables influencing the quality of sheet metal product (Table 8.1). In contrast, the main control variables in machining, for a given workpiece geometry and tool design, are as few as speed, feed and depth of cut. The quality parameters in machining are also few, namely, the surface roughness, form error, and the dimensional error. In contrast, quality issues in the different sectors of industry

Table 8.1 Applications, technologies and quality issues of sheet metal in different sectors of manufacturing industry

Sector	Sheet metal parts	Sheet metal technologies	Sheet metal grades	Quality issues
Automotive and transport	Skin panels, Structural members Brackets, and braking system parts Bicycle parts — mud guards, wheel rims, sprocket wheels, and chain links Railways — coaches, wagons, containers Bus, and truck bodies Aerospace parts — fuselage, aircraft interiors	Shearing and fine Blanking, Bending, Deep drawing Stretch forming, Stretch bending Roll forming, hot stamping Superplastic forming and diffusion bonding Welding and joining	Steels: IF, EDD, DD, DP, CP, HSLA, BH; Aluminium: Alloys like Al 2024, 5083, 7075, 6016 Stainless steel	Uniformity in thickness, thinning, surface blemishes, springback, burr, flatness of the flange
Electrical and electronic	Connectors, embedded inserts Switchboard parts, switchgear parts, covers of electrical products, switch bodies, etc. Transformer laminations Stator and rotor laminations	Shearing Blanking Bending Curling Slide forming Deep drawing	Copper Steels: Cold rolled grain non-oriented (CRGNO), EDD	Springback Burr, wrinkles
White goods	Furniture — tables, chairs, filing cabinets, security equipment, cupboards, refrigerators, kitchen sinks, fans, washing machines, dryers and dish washers	Shearing, fine blanking Bending Roll forming Deep drawing Welding/Joining	Steels: DD, EDD, IF, precoated and prepainted sheet Stainless steel	Burr, heat affected zone (corrosion resistance), springback
Brown goods	Ovens, electric irons, electric kettles, toasters, geysers	Shearing, deep drawing Bending Welding/Joining	Stainless steel Copper, brass	Burr, heat affected zone, springback
Miscellaneous items	Locks, utensils, construction and farm implements, cooking gas cylinders, oxygen cylinders, garment clips	Deep drawing Roll forming Ironing Bending	Steel – Hot rolled, spring steel Stainless steel, brass	Burr, heat affected zone, minimum thickness, Thickness variation, springback

wherein the sheet metal is formed using a number of technologies are listed in Table 8.2. Hence it would be of interest to understand the distinction between the manufacture of machined products and that of sheet metal products.

The objective of this chapter therefore is to highlight these differences, bring out the manufacturing related issues in achieving sheet metal product quality vis-à-vis that of the machined counterparts. Developments that have materialized in the grades of sheet metal, processes, machine tools and sheet metal product manufacturing scenario as a whole to overcome these difficulties, have also been reviewed in the subsequent sections.

8.1.1 How Sheet Metal Processing Differs from Machining

A sheet has a large surface area and a small thickness, so that one must pay attention to restrict further thinning. Non-uniform thinning is worse, and manufacturability of a sheet metal part strongly depends on the interplay between the process and the material properties, making process selection a crucial step. A machined product on the other hand, like a gear for example, is manufacturable using a number of machining processes.

The very process of dimensioning in sheet metal products differs from that of machined parts. Unlike machined parts the feature (usually a hole) that is most unlikely to get deformed/displaced from its original position or unlikely to change its form or dimension during processing is taken as the reference. All dimensions are specified in relation to this feature, and also inspected using the same as the reference. Use of pilot holes as references is known to be crucial for part manufacturing and in-process inspection.

The large diversity of shapes to which a sheet is deformed is largely asymmetric, while a large volume of machined products is axisymmetric. Process design (design of the processing route) is therefore crucial in sheet forming. One pays attention to the distribution of strain so as to monitor thinning of the sheet.

Unlike machining, sheet forming is a non-steady state process. The zone of deformation continuously evolves over time. The points of maximum and minimum strain shift with deformation. Similarly, the cross section encompassing these two zones can change as deformation progresses.

The purpose of testing machinable raw material for its properties is largely for the in-service requirements, as the machinability of a material is not as sensitive to raw material properties as the sheet metal formability is. Hence, wherever the properties to satisfy in-service requirements are in conflict with those needed for formability, special processing techniques and long processing routes are required. Issues like pressure welding of work material to tool (BUE or material pickup) during process are damaging to both machine and sheet metal forming.

Table 8.2 Desired quality attributes of sheet formed products emerging out of different manufacturing processes

Desired quality attributes of sheet metal product	Basic process	Materials used	Machine tool	Name of the forming process	Implications in downstream mfg/assy
Precise dimensions of sheared feature	Shearing	Ferrous and non-ferrous	Triple action press	Fine blanking	Deburring of the net shape part essential
Shearing/separating without burr	Shearing Laser cutting	Ferrous and non-ferrous formable sheet metal, coated and uncoated All formable sheet metal	Triple action press Or Turret punch press CNC laser cutting	Fine blanking + deburring Laser cutting	Burr may cause improper seating in mfg and assembly Square cut edges, high precision in dimensions
Bending to long continuous sections	Bending	All formable sheet metal	Roll forming line	Roll forming Press brake	High rate of productivity, low springback possible Low/Medium volumes, large variety
Reduction in material input	Superplastic forming	Ultrafine grained non-ferrous material	Superplastic forming press	High strain rate Superplastic forming	Complex shaped parts with microformed features in a single stage
Diffusion Bonding	Welding	Fine and ultrafine grained non-ferrous material	Superplastic forming press	Diffusion bonding	Joining process eliminated
Obtaining accurate dimensions (overcoming) springback, accuracy of dimensions	Hot forming Hot forming Fluid pressure forming (cold/warm)	Boronized steel Non-ferrous materials – Al, Mg, Ti alloys of certain compositions and grain size < 10 microns Sheet and tubes, ferrous/ non-ferrous	Hot forming press Superplastic forming press (controlled strain rate forming) Hydroforming Press or injection moulding machines	Press hardening Superplastic forming Hydroforming or viscous pressure forming (with calibration)	Thermal induced dimensional variations to be controlled
Products of high complexity	Drawing	Stainless steel	Press with multiple point blank holding Segmented blank holder	Deep drawing with multiple point blank holding	Control over strain paths

There is no finishing operation for sheet metal and the surface finish can be improved to a very limited extent. Hence handling and storage of sheet metal coils as well as formed parts is important. Sheet metal parts with errors cannot be salvaged. They have to be discarded.

Unlike machining wherein the machining occurs only on the surface being machined and bulk of the raw material remains unaffected, the entire section of a sheet metal blank participates in deformation.

Selection of the machine tool (stiffness, the drive, etc.) for a sheet metal operation depends on the material and the process being performed. For instance, the variation / variability of velocity over the stroke is required for some processes (long stroke processes like deep drawing to ensure fast approach and return and slow working stroke) and materials (the rate sensitive ones in particular). Hence what can be successfully formed on a hydraulic press might fail on a mechanical press. Stiffness of the machines used for separating processes (like blanking and punching) is required to be much higher than those for long stroke operations. Comparatively, the machining process is not so sensitive to machine tool characteristics as the sheet forming process is. The machinability does change with machining speed, but is not as sensitive to the machine tool characteristics as the formability is.

Dies designed for steel as work material seldom work for aluminum as work material, other parameters (product design, machine tool, etc.) remaining the same. Hence a change in raw material might call for a complete change in the tool design or at least substantial rework of the existing dies. A change in work material to be machined may call for a change in tool geometry, but once standardized, all it requires is a change of tool.

Elastic after-effects in the form of springback and residual stresses are prominent in sheet formed parts leading to inconsistent dimensions. Tolerances on the rolled sheet thickness and those on the yield strength introduce unpredictable variations in the springback, depending on the magnitudes of these two variables.

Change in constraints due to in-process trimming for example is usually accompanied by a change in shape of the formed part.

It is therefore not feasible to achieve in sheet metal parts the tolerances that can be achieved in machining. The material properties interact with every element of the forming system. Defects introduced at the initial stages cannot be eliminated in the subsequent stages. The inconsistency in dimensions and thickness distribution introduced in one step often get aggravated in the subsequent steps in processing leading to rejections. Material properties that continuously change along the processing route can lead to severely inconsistent dimensions on the part.

Table 8.3 Distinction between sheet formed parts and the machined parts

S. no	Point of comparison	Machined components	Sheet metal parts
1.	Rigidity of parts	Non-compliant parts	Compliant parts
2.	Designing and dimensioning	Orthogonal machined datum planes for dimensional reference	Features that will not shift/ get deformed required as reference. Holes usually serve as reference — edges / blank boundaries which no longer remain straight go into trim scrap.
3.	Generation of the desired shape	Tool movement generates the geometry, one tool to generate different machined geometries	Tool movement linear, tool design (die design) specific to the shape of the product
4.	Zone (volume) of metal affected by the process at a given time	Only the surface or a thickness of metal close to the surface participates in the machining process, the bulk of the material is relatively unaffected	The entire sheet metal blank gets deformed (e.g., flows into the die).
5	Shape of the zone of deformation	Remains relatively unchanged — a small shear zone of a stable shape is formed (steady state process)	The shape and size of the zone of deformation evolves continuously
6.	Dimensional gradients / deviations	Unintended dimensional gradients can be avoided	Unintended dimensional gradients e.g., thinning of sheet metal and those from springback are a part and parcel of the process
7	Tolerances achievable	Close tolerances (in microns or even nanometers) can be achieved	Elastic after-effects, part compliance and variables not under complete control prevent close tolerances from being achieved.
8	Surface finish and dimensional integrity	There are several stages of finishing operation	There is no finishing operation. The part must be used (after trimming, etc.) as it is.
9	Rework and ease of rework	Out of tolerance parts can be salvaged	Salvaging an out-of tolerance part is not possible.
10.	Diversity of shapes	Any shape can be made with appropriate tool path	Large diversity of tool designs and control variables

(Cont

S. no	Point of comparison	Machined components	Sheet metal parts
11.	Purpose of testing of raw material	Machinability, in service requirements.	Formability and in-service requirements.
12.	Tool materials	Machinability is strongly associated with tool life, which in turn strongly depends on tool material.	Tool material change does not ensure formability. Tool geometry and sheet metal properties together determine formability.
13.	Standardization of tool geometry and control parameters	Easy, standards on tool angles, standard tool geometries of single and multi-point cutting tools are widely available.	Only guidelines are feasible. Standardization across components is impractical. However, classification of components based on similar features
14.	Cost of tooling	Much lower. Single tool works to generate many shapes irrespective of complexity of the tool. Smaller lead time in procuring tools.	Tools are expensive as they are not mass produced. Change of tool is expensive and difficult. Lead time in making a new tool is large.
15.	Machine tools	Machine tool drives are mechanical drives which generate the requisite tool paths that can be very complex. Accuracy of tool path is crucial. Cutter compensation, collision checks are important. Forces involved are much lower than in forming.	Machine tool drives are primarily built for delivering power (heavy duty drives). The forces involved are large and machine tool frame stiffness is a major consideration. Availability of drive force, speed and energy must be ensured after tool contact.
16.	Springback and residual stresses	Extremely small. Machining relieves surface residual stresses	Un-released elastic stresses (springback) on account of product geometry appear as residual stresses. It is shape dependent and processing history dependent.
17.	In-process stability of shape	Stable shape throughout. Better form accuracy of holes as well as external surfaces.	In-process stability of shape is determined by an interplay between the residual stresses and separating operations like trimming and punching. Holes punched with a circular punch might not necessarily be round.

(Contd.)

S. no	Point of comparison	Machined components	Sheet metal parts
18.	Inspection	Form and dimensional errors of individual parts can be inspected. Assembly occurs by selective assembly.	Inspection is performed at well defined critical points only. Non-ideal, untoleranced parts are required to get the assembly to within dimensional tolerances.
19.	Fixture design	1-2-3 principle used for location in fixtures	n-2-1 principle (where $n > 3$) is required for sheet metal parts. Clamping sequence is more critical due to compliance of sheet metal parts.

Inspection of sheet metal parts is unlike those of machined parts. Sheet metal parts are constrained in a fixture (applying constraints similar to those the part would experience in an assembled condition) and inspected for dimensional errors only at certain critical points (also referred to as Key Product Characteristics). In fact, principles of inspection and assembly used for machined parts fail on account of compliance of sheet formed parts. Difficulty in controlling dimensions of compliant assemblies calls for a functional build approach wherein one attempts to get assemblies to within limits of tolerance.

Machining fixtures are designed based on the 3-2-1 principle. It is usually not feasible to locate and support a sheet metal part on just three supports. The principle of directing the clamping force acting towards the locator (to avoid sagging/bending of the product under clamping force) is not practicable with sheet metal parts. The quality of a sheet metal part is very sensitive to the sequence of clamping and the sequence in which spots welds are made at various points to join two sheet metal parts.

The foregoing makes standardization (something which has revolutionized mass production in machining) of processing variables difficult in sheet metal working. It is possible to get, for a given material standard machining conditions, whereas forming conditions tend to be material, product design and tool design specific. Hence manufacturing concerns evolve their local standards which are used captively by the respective manufacturer.

Table 8.3 summarizes the differences between a machined part from that formed from sheet metal.

8.2 IMPLICATIONS ON RAW MATERIAL MANUFACTURE

Sheet metal properties interact with almost all the elements of sheet forming as discussed above. Hence material is selected first and product is designed for that material. Sheet metal manufacturing processes that include numerous stages of rolling and heat treatments, together with the chemical composition determine the properties and their consistency.

Properties of steel sheets (in particular, drawability) for instance depends on the chemical composition, the hot rolling variables (finishing and coiling temperatures), cold rolling variables (about 70% reduction), and the annealing conditions. Typical conditions for batch annealing and continuous annealing of steel are given in Table 8.4.

Raw material (sheet metal) manufactured for conventional forming technologies must have a grain size of about ASTM 7-9 or finer in case of thin sheets to avoid 'orange peel effect', control in processing being required for the following control variables influencing the final product quality.

Table 8.4 Annealing conditions in the manufacture of steel sheet for deep drawn products

Variable	Box annealing	Continuous annealing
Heating rate	20–200 Kh^{-1}	~500 $Kmin^{-1}$
Annealing temperature	650–700°C	650–900°C
Time at annealing temperature	2–20 h	1–5 min

8.2.1 Chemical Composition

Minimizing variability in the chemical composition enables exercising control over numbers and distribution of inclusions arising from the presence of elements like P and S in steel. An Al/N ratio close to 10 in steel (meant for achieving high drawability through batch annealing after cold rolling) would contribute to consistency in drawability. Given the fact that some variability in chemical composition is inevitable on account of the very nature of the steel making practice, processes downstream must be robust enough to accommodate these variations so as to produce sheet metal of.consistent mechanical behavior.

8.2.2 Grain Size [1-3]

Control over grain size is the outcome of chemical composition and thermo-mechanical processing. In addition to the above therefore, control is necessary over temperature and the speed of working of the mechanical working devices like rolling mills, forging presses, and extrusion presses, etc. for minimizing the variability in the metallurgical characteristics and the mechanical behavior of the sheet. A reduced variation in grain size contributes to consistency in yield strength. Similarly, through control over processing variables, a consistent work hardening rate (which influences the formability particularly in stretch forming) and better press shop performance can be obtained.

Of late, ultrafine grained sheet metal, particularly of non-ferrous material, is becoming popular for sheet formed parts. This means freedom from Luders bands and possibility of using the superplastic forming process for large volume production as in automobiles to decrease the material input. Such materials enable reduction in the weight of the product being made.

Material is subjected to severe plastic deformation like high pressure torsion, Equi Channel Angular Pressing (ECAP), friction stir processing and accumulated roll bonding (ARB). Of these, ARB and sheets rolled from ECAPped billets is suited to grain refinement in sheet metal. A setup for ECAP is shown in Fig. 8.1. Tables 8.5(a)-(c) [3] indicate the enhancement in tensile properties as a result of ECAP in aluminum. It is seen from Table 8.5(a) that the ECAPped material showed a two stage work hardening behaviour (indicated by two work hardening rates n_1 and n_2 and the corresponding strength coefficients K_1 and K_2). A negative strain hardening rate after initial mildly positive strain hardening would only be expected in view of the severe plastic deformation undergone by the material prior to testing. The coarse grained material which was not ECAPped (CG material) showed a single stage hardening behavior [Table 8.5(b)] with work hardening rates much greater than those of the ECAPped aluminium alloy Al 1070.

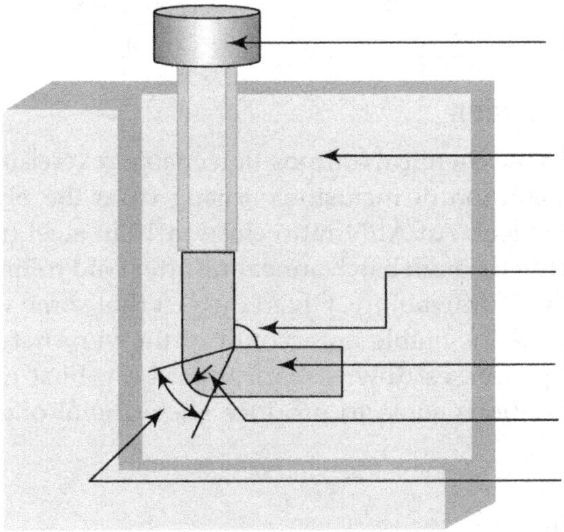

Fig. 8.1 Schematic of the ECAP setup

The overall enhancement in the mechanical properties by virtue of ECAP is summarized in Table 8.5(c), wherein comparison is based only on the positive work hardening range of the ECAPped material with that of the CG material. Hence while the work hardening stage (stage 1) in the ECAPped material ends well before failure, that in the CG material ends with instability after a single stage of work hardening. The UFG material, on an average, is seen to score over the CG material in respect of yield strength and the uniformity in strain. It shows similar UTS for both the materials and a lower work hardening rate in UFG than the CG material as would be expected. The planar anisotropy of the property (defined as $\Delta X = [X_0 + X_{90} - 2X_{45}]/2$), which should ideally be zero, seems to increase in case of the UTS, the YS and true uniform strain.

Table 8.5 (a) Tensile properties of the UFG Al 1070 [3]

Direction	YS	n_1	K_1	End stage 1	n_2	K_2	True uniform strain	UTS
0	161.67	0.287	524	0.051	–0.05	167.9	0.059	199.3
45	104	0.287	435	0.048	–0.107	129.5	0.053	171
90	151.67	0.139	286.9	0.046	–0.026	166.2	0.061	186
Average	130.33	0.250	420.22	0.048	–0.072	148.3	0.0567	181.8
Planar anisotropy of the property	52.67	–0.074	–29.55	0.0003	0.0687	37.53	0.0075	21.67

Table 8.5 (b) Tensile properties of the Coarse Grained (CG) Al 1070 [3]

Direction	YS	n_1	K_1	End stage 1(true strain percent)	UTS
0	108.33	0.462	1028	0.038	189.6667
45	111.53	0.420	892	0.024	182.3333
90	140	0.204	378	0.028	182
Average	117.85	0.377	797.94	0.028	184.08
Planar anisotropy of the property	12.63	–0.087	–189.45	0.010	3.5

Table 8.5 (c) UFG properties normalized by the corresponding Coarse Grained (CG) material property [3]

Direction	YS	n_1	K_1	End stage 1 (True strain percent)	True uniform strain	UTS
	Ratio of the properties of the UFG material to the corresponding property of the CG material					
0	1.49	0.62	0.51	1.342	1.26	1.051
45	0.93	0.68	0.49	2.028	1.51	0.94
90	1.08	0.68	0.76	1.602	1.78	1.022
Average	1.11	0.66	0.53	1.692	1.495	0.99
Planar anisotropy of the property	4.17	0.85	0.156	0.035	1.27	6.19

Thickness

Thickness, or more generally shape control of the sheet influences downstream processing considerably. Thickness variations introduced in the initial stages of rolling cannot be corrected downstream. The thickness of the sheet emerging after several stages of hot and cold rolling is the total outcome of the measures of thickness control (passive roll gap setting, roll camber, front and back tensions) taken at each stage. A uniform thickness, and consistent yield strength ensure consistency in springback when the sheet is bent around a given radius. Once this is achieved, tools compensated for springback can be made and sheet components manufactured to closer tolerances than before.

Springback is sensitive to shape/contour of the line of bend [Figs 8.2(a) and (b)]. Figure 8.2(a) shows an 'S-Rail' that is made by drawing a blank and springback varies from section to section along the channel by virtue of the geometry (curved line of bend). Figure 8.2(b) shows a 'C tray', wherein two bends are made at two different radii of curvature of the line of bend. The central portion provides blank holding surface in addition to that around the flange. The corresponding variation in springback along the bend line [Figs 8.3(a) and (b)] calls for spatially varying processing variables and hence a more sophisticated process control. Details of this study are reported elsewhere [4].

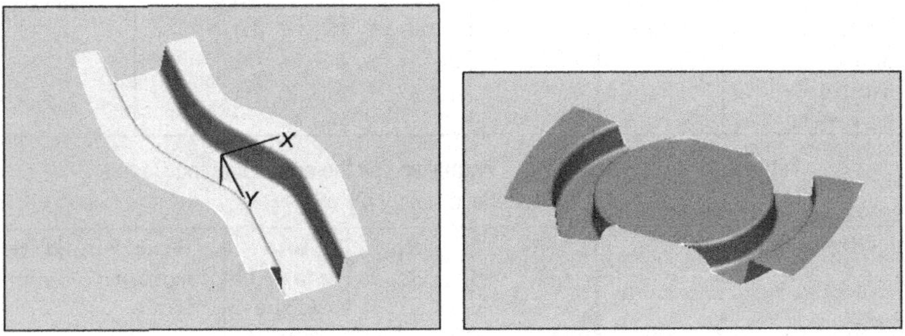

(a) (b)

Fig. 8.2 Products having curved lines of bend [4]

The assembly process is influenced by the tolerances to which the sheet metal products can be manufactured.

The assembly philosophy of 'Functional Build' brings in the necessary robustness to assemble sheet products despite inconsistent product dimensions, arising inevitably out of inconsistent springback, which in turn has roots in factors mentioned earlier. This is because unlike in machining inconsistent springback and its magnitude both make it difficult to control the dimensions away from the line of bend, e.g., at the opening of deep drawn components. Hence rather than attempt producing individual parts to tolerance (expensive) and then assemble them, one makes assemblies of sheet metal parts by introducing into the assembly non-ideal, non-critical intolerance parts to absorb the dimensional variations so that the dimensions of the assembly is within tolerance specifications.

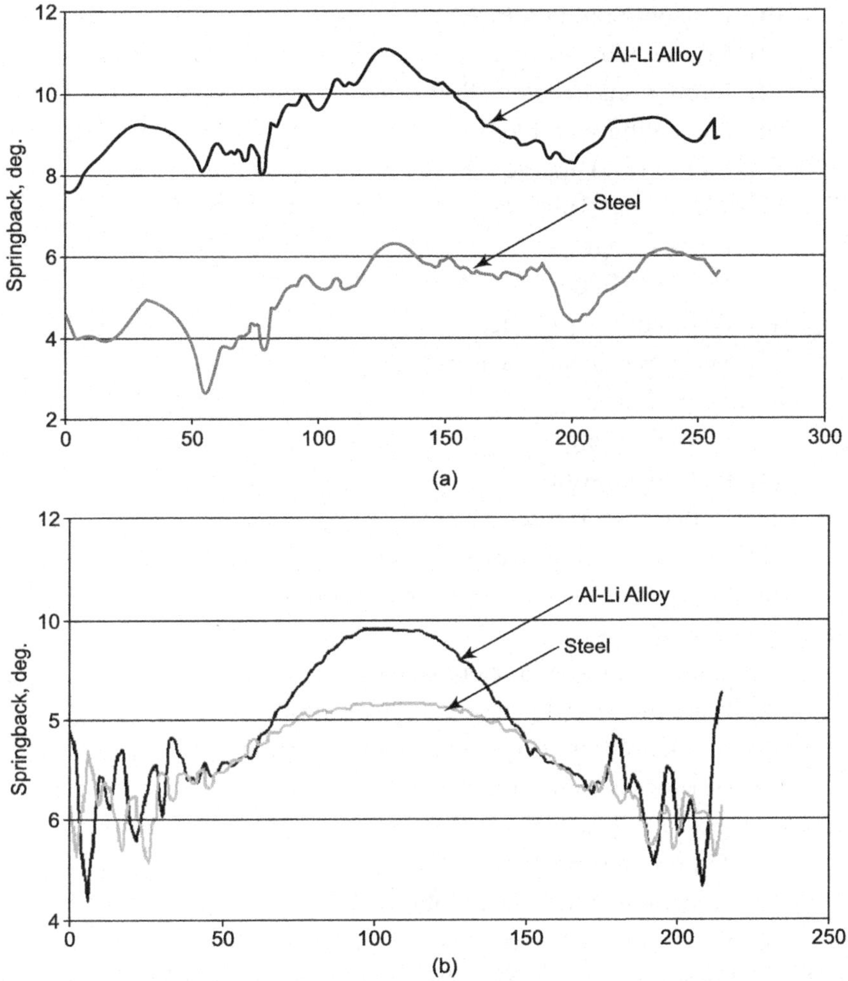

Fig. 8.3 Variation of springback along the curved line of bend (a) S-rail and (b) C tray with constant blank holding force [4]

However, now, if it becomes possible on account of precise control over the various processing steps, and with consistent material quality, one might not have to resort to functional build assembly! Hence thickness control in conjunction with material and processing variables becomes important. A list of major variables influencing the final product quality at every step in manufacturing is given below:

Variables affecting the quality of a sheet metal formed product

1. *Sheet metal properties*
   (a) Chemical composition
   (b) Heating temperature of ingot
   (c) Hot rolling temperature

   (d) Finishing temperature
   (e) Coiling temperature
   (f) Reductions in cold rolling
   (g) Annealing variables
   (h) Storage and ageing
2. *Product shape and size*
   (a) Blank design
   (b) Product shape
   (c) Depth of the product
   (d) Sharp radii/sharp features, etc.
3. *Tool design variables*
   (a) Blank holding area ('addendum', draw beads if any)
   (b) Die entry radius
   (c) Punch nose radius
   (d) Punches for holes as per product design (cam punching, if any)
   (e) Sharp features consistent with product design
4. *Processing variables*
   (a) Lubrication—coefficient of friction
   (b) Blank holding force—spatial variation
   (c) Blank holding force—variation with time
   (d) In-process trimming
5. *Process and Machine tool selection*
   (a) Manufacturing rate
   (b) Accuracy of dimension obtainable
   (c) Controllability of processing variables

If the thickness tolerances are small and unidirectional then it might become practicable to completely and consistently compensate for springback, and thereby control the product dimensions better.

8.3 SHEET METAL PRODUCT DESIGN

Product design influences the tool design, the initial shape of the raw material (blank design), the processing variables like friction and lubrication, the blank holding, etc., and hence the strain path. This important parameter (strain path) is the outcome of all complex interactions among these variables (see Table 8.1) and is defined as the locus of the states of strain a point in the deforming material experiences as it gets deformed from its initial flat undeformed shape to the final deformed one. Hence different points in a product experience different strain paths. Similarly, if a product is made in several stages (that usually is the case), the strain path over the entire processing route influences the formability of the sheet metal. Formability is therefore strongly dependent on history, i.e., prior processing a sheet might have undergone. Hence sheet metal must be characterized so as to identify the critical strain paths (those deleterious to formability) and the forming limits established for such strain paths. Points or regions in a formed product undergoing such critical strain paths must be identified as critical regions and mechanisms of failure like thinning, necking, wrinkling, etc. monitored continuously.

Distortion of features during subsequent processing can be avoided by taking into account elastic after effects like springback and using appropriate product design guidelines.

8.4 MODELING OF SPRINGBACK [5, 6]

A number of models are available to predict springback in sheet metal about straight lines of bend [5–12]. These are of use in most of the straight line bends made in press brakes, and are often applied to bending in roll forming machines as well [12]. Many phenomenological models account for kinematic hardening in addition to isotropic hardening during bending. Material models, i.e., the constituitive equations and yield criteria (both isotropic and anisotropic) have been used to model the effect of processing variables on the process of bending and springback [8–11]. One of the simpler models accounting for strain hardening is described in the next section.

8.4.1 Model to Predict Springback [5, 6]

The model is based on Johnson's approach for estimating the springback.

Following assumptions are made:

(a) The material obeys the isotropic work hardening rule and its true stress–true strain relationship is described by the Hollomon equation in the plastic range, i.e.,

$$\begin{aligned} \sigma &= E\varepsilon \quad \sigma < \sigma_p \\ \sigma &= K\varepsilon^n \quad \sigma > \sigma_p \end{aligned} \tag{8.1}$$

where K is the strength coefficient and n the strain hardening exponent.

(b) The material has uniform mechanical properties, i.e., the material is homogeneous.

(c) The material has a stable microstructure throughout the deformation process.

(d) Deformation occurs under isothermal conditions

(e) Plane strain conditions prevail.

(f) The neutral axis lies in the mid-plane of the sheet

(g) Bauschinger effect is absent; isotropic hardening is assumed

(h) Effect of normal anisotropy (R value) is neglected

(i) Shift in the neutral axis is neglected in the absence of a superimposed tensile stress.

(j) Friction at the die entry radius as well as the punch nose radius is neglected. Given the geometry of the setup, the sheet does not contact the tool anywhere except the entry radius (line contact) and the punch nose radius (small surface contact).

(k) The surface roughness increases with true strain, and is proportional to the initial grain size, i.e.,

$$r = r_0 + k\varepsilon d \tag{8.2}$$

where k is the degree of roughening, d the grain size and ε the true strain.

This assumption is made to check for the development of unacceptable levels of surface roughness. This would be the case particularly when the d/t ratio is large and the roughness measured as depth of the trough with reference to the crest of the surface profile on the rough (due to orange peeling) surface would be unacceptable. The local decrease in sheet thickness due to surface roughness profile however, is not taken into account while determining the springback.

8.4.2 Analysis of the Process

A schematic showing a sheet bent using a die and a punch is given in Fig. 8.4. The neutral plane is midway through the sheet thickness and is assumed to remain at this position (Fig. 8.5).

From the geometry (Fig. 8.4), the stroke, h, may be related with the angle of bend θ by

$$h = (B \sin \theta + A \cos \theta - A)/\cos \theta \tag{8.3}$$

where $A = r_p + r_d + t$ and $B = 0.5 \times w + r_d$

Here the sheet is taken to roll over the die entry radius. The point of contact with the die radius changes continuously and this in turn changes the effective die width (die gap), w.

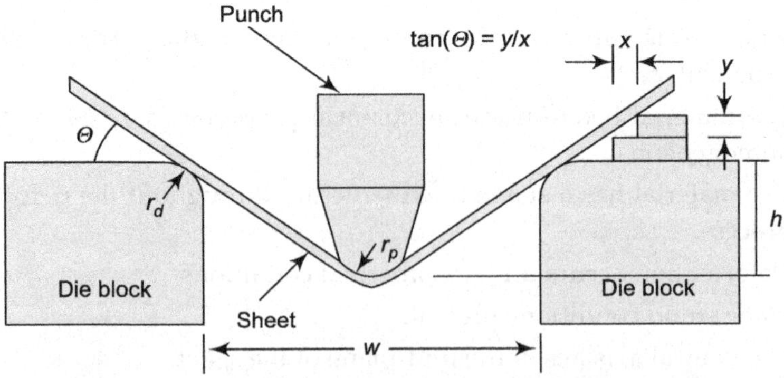

Fig. 8.4 (a) Setup for studying springback: θ = angle of bend, w = die gap r_d = die radius, h = punch travel [5, 6]

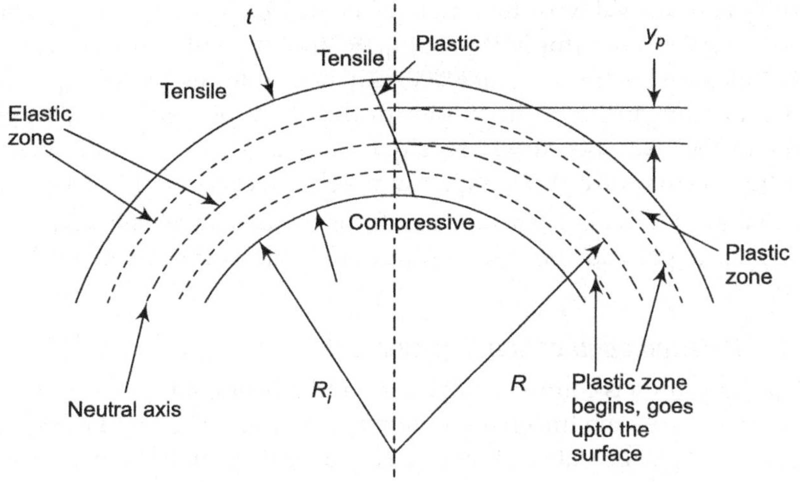

Fig. 8.5 Stress distribution over the cross section in bending [5, 6]

The rate of change of the angle of bend (and hence angle of contact, i.e., wrap with the punch and die radii) with stroke is given by

$$d\theta/dh = \cos\theta/(B\sec\theta - A\tan\theta) \tag{8.4}$$

which is obtained by differentiating Eq. (8.3).

The instantaneous radius of curvature of the bend for a given punch travel is determined as given in the following section.

8.4.2.1 Determining the radius of curvature

The radius of curvature of the sheet varies from infinity (when flat) to $(r_p + t/2)$ when completely wrapped around the punch through 180 degree. This would mean a value of $\theta = 90$ degree, which, according to Eq. (8.3) would call for infinite travel. This difficulty is overcome by taking the value of $\theta_{max} = \pi/2 - 10^{-7}$ radians so that the sheet might be visualized to wrap completely around the punch nose radius at a large but finite travel. This defines the farthest the punch would need to possibly travel for complete wrapping around the radius during the stroke.

The radius of curvature of the neutral axis at a given punch travel is calculated backwards from this point using the following relations:

$$dR = -(R/\theta)\, d\theta/dh\, (\delta h) \tag{8.5}$$

Taking any intermediate position during the stroke as the '*m*th. position', i.e., *m*th. increment of δh, the radius of curvature at the m-1 st. position may be written as

$$R_{m-1} = R_m + (dR/dh)_m * \delta h, \tag{8.6a}$$

since

$$-dR/dh = R/\theta * d\theta/dh \tag{8.6b}$$

The maximum value of the angle of bend (θ_{max}) is taken to be slightly lower than $\pi/2$ radians [for finite h in Eq. (8.3)] as $\pi/2 - 10^{-7}$ radians. The value of $d\theta$ is taken to be $(\pi/2 - 10^{-7})/20000$ radians and for every $\theta = \theta_{max} - m * d\theta$ [m corresponding to the "m"th position during the stroke, Eq. (8.6a)] the punch travel and the radius of curvature is determined and this is continued until the radius of curvature at the required punch travel is determined.

The bending moment per unit length of the bend and subsequently the springback angle can then be determined. This is described in the next section.

8.4.2.2 Determination of bending moment

The bending strain is comprised of elastic and plastic components. The stress distribution across the thickness is shown in Fig. 8.5. The point of transition from elastic to plastic state is shown at a distance y_p from the neutral axis. This point is determined by equating the elastic bending stress with that at yield point. The elastic stress distribution may then be given by

$$\sigma = \sigma_p\, y/y_p, \tag{8.7a}$$

and in the presence of a tensile stress, σ_T (as would be the case if the sheet were clamped down by a blank holder),

$$\sigma = \sigma_p\, y/y_p - \sigma_T \tag{8.7b}$$

The value of y_p may be determined by the condition at incipient yielding that

$$\sigma_p/y_p = E/R$$

so that

$$y_p = \sigma_p R/E \tag{8.7c}$$

so that

$$dy_p = \sigma_p\, dR/E \tag{8.7d}$$

This increment of y_p is added to the previous value for every incremental travel until the requisite bend angle under load is achieved.

The elastic component of the bending moment is given by

$$B.M._{\text{elastic}} = (2\sigma_p - 3\sigma_T)\, y_p^2/6 \tag{8.8}$$

The presence of a tensile stress reduces the bending stress for yielding and thus the moment applied for yielding. The expression for bending moment the plastic regime may be written as

$$B.M._{\text{plastic}} = 2[{}_{y_p}\!\int^{t/2} (\sigma_p - \sigma_T)\, y dy + {}_{y_p}\!\int^{t/2} K\,[\ln(1 + y/R)]^n\, y dy] \tag{8.9}$$

so that

$$\begin{aligned} B.M._{\text{plastic}} = 2\{&(\sigma_p - \sigma_T)(t^2/8 - y_p^2/2) + KR^2\,\{(\ln(1 + t/2R))^{n+2}/(n+2) \\ &+ 1.5\,(\ln(1 + t/2R))^{n+3}/(n+3) + 7/6\,(\ln(1 + t/2R))^{n+4}/(n+4) \\ &+ 15/24\,(\ln(1 + t/2R))^{n+5}/(n+5)) - (\ln(1 + y_p/R))^{n+2}/(n+2) \\ &- 1.5\,(\ln(1 + y_p/R))^{n+3}/(n+3) - 7/6(\ln(1 + y_p/R))^{n+4}/(n+4) \\ &-15/24\,(\ln(1 + y_p/R))^{n+5}/(n+5))\} \end{aligned}$$

where n is the strain hardening exponent. In this expression, $\varepsilon \sim \varepsilon_1$, in view of plane strain.

Total bending moment = $B.\ M_{\text{total}} = B.\ M_{\text{elastic}} + B.M._{\text{plastic}}$

8.4.2.3 Evaluation of springback

Following the Johnson's method, the springback is visualized as an elastic moment applied in the opposite sense. Thus,

$$\begin{aligned} Et^3/R_E = [&3(\sigma_p - \sigma_T)t^2 - 4\sigma_p\, y_p{}^2 + 24KR^2\,\{(\ln(1 + t/2R))^{n+2}/(n + 2) \\ &+ 1.5\,(\ln(1 + t/2R))^{n+3}/(n + 3) + 7/6\,(\ln(1 + t/2R))^{n+4}/(n + 4) \\ &+ 15/24\,(\ln(1 + t/2R))^{n+5}/(n + 5)) - (\ln(1 + y_p/R))^{n+2}/(n + 2) \\ &- 1.5\,(\ln(1 + y_p/R))^{n+3}/(n + 3) - 7/6\,(\ln(1 + y_p/R))^{n+4}/(n + 4) \\ &- 15/24\,(\ln(1 + y_p/R))^{n+5}/(n + 5))\}] \end{aligned} \tag{8.10}$$

Substituting y_p as $\sigma_p\, R/E$ from Eq. (8.7c) and simplifying the expression for the ratio R/R_F (where R_F is the radius of curvature after springback and R the one before springback, i.e., under load) the final expression for springback may be obtained from the difference in curvature of the sheet before and after unloading. Hence, given that

$$1/R - 1/R_F = 1/R_E, \text{ so that } 1 - R/R_F = R/R_E$$

$$\begin{aligned} R/R_F = 1 - R/R_E = 1 - &3(\sigma_p - \sigma_T)R/Et + 4\,((\sigma_p - \sigma_T)R/Et)^3 \\ &- 24KR^3/Et^3\{(\ln(1 + t/2R))^{n+2}/(n+2) + 1.5\,(\ln(1 + t/2R))^{n+3}/(n+3) \\ &+ 7/6\,(\ln(1 + t/2R))^{n+4}/(n + 4) + 15/24\,(\ln(1 + t/2R))^{n+5}/(n + 5)) \\ &- (\ln(1 + y_p/R))^{n+2}/(n + 2) - 1.5\,(\ln(1 + y_p/R))^{n+3}/(n + 3) \\ &- 7/6\,(\ln(1 + y_p/R))^{n+4}/(n + 4) - 15/24\,(\ln(1 + y_p/R))^{n+5}/(n + 5))\} \end{aligned} \tag{8.11}$$

From the above expression the ratio R/R_F (R being the radius of curvature under load) or alternately, θ_f/θ (Since $\dfrac{R_f}{R} = \dfrac{\theta_f}{\theta}$, θ_f being the bend angle after springback and θ the angle before springback) may be determined. The springback then may be determined as

$$\Delta\theta/\theta = \theta_f/\theta - 1$$

8.4.2.4 Correlation of the predictions with experimental data

Experimental data from literature was used to compare the results of this model. Fig. 8.6 shows a comparison between the predictions by Altan *et al.* [8], the present model and the experimental data. It is observed that the predictions from the present model correlate well with the experimental data. Except for relatively low punch travel, the correlation appears to be excellent. A similar observation can be made from Table 8.6, which gives a comparison of the experimental data with the predictions from the present model for various tool geometries.

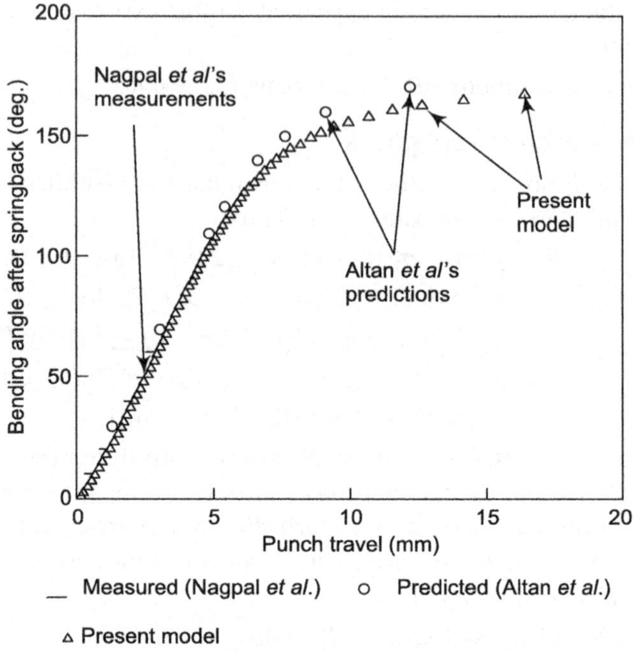

Fig. 8.6 Correlations of the model with experimental data of Nagpal *et al.* [9], and predictions of Altan *et al.* [8]

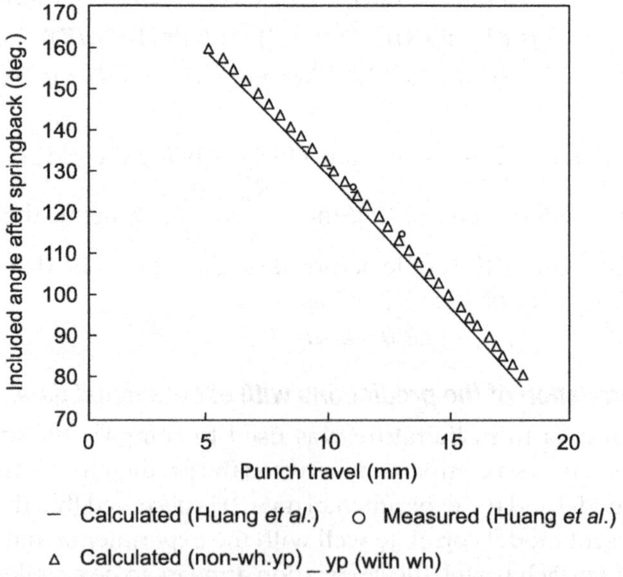

Fig. 8.7 Predictions from the present model with and without work hardening vis-à-vis the calculations and experiments of Huang *et al.* [11]

Table 8.6 Comparison of the predictions of the present model with that in the literature [11]

R_p (mm)	R_d (mm)	w (mm)	Measured angle under load [11] (degree)	Measured angle after springback (degree) [11]	Predicted angle after springback (degree) (present model)	Predicted punch travel (mm) (present model)	Measured punch travel [11] (mm)
2.286	1.83	9.53	92.54	90	89.84	4.488	4.394
2.286	1.52	9.5	100.3	97.42	97.56	4.466	4.39
2.540	1.52	9.53	92.12	90	89.09	4.17	4.14
2.54	1.52	15.9	94	90	91	7.67	7.10

It may be inferred from Table 8.6 that the predictions of the present model with regard to both the angle after springback as well as punch travel agree well with the experimental values from literature [11].

A comparison of the angle of springback plotted against punch travel is given in Fig. 8.6. The present model shows excellent agreement with the experimental measurements of Nagpal *et al.* [9]. Similar comparison with the experimental values and the FEM predictions of Huang *et al.* [11] again shows close agreement of the present model with the experimental result (Fig. 8.7).

8.4.2.5 Sensitivity of springback to the n, K and the superimposed tensile stress

Equation (8.11) was further used to determine the dependence of springback on the work hardening parameters and the superimposed tensile stress. Effect of the material properties on springback was also studied.

8.4.2.5.1 Effect of the work hardening parameters

The effect of n, K and σ_T on the bending moment is shown in Fig. 8.8. Here also, the effect of K in increasing the bending moment is observed. Also, a low n value leads to a higher bending moment since for the same value of K, a lower n value means a higher flow stress at a given value of true strain and consequently a greater bending moment. A low n value and a high K value leads to the highest bending moment (and correspondingly bending load for the same die geometry).

The sheet thickness was taken as 1 mm, $\nu = 0.3$, $n = 0.242$. It is observed that a higher strain hardening exponent *reduces* springback. The difference between the springback values is highly sensitive to the value of K. At high K values the springback is very sensitive to the n value (Fig. 8.8). In view of these observations the sensitivity of springback to n value was studied as described in the next section.

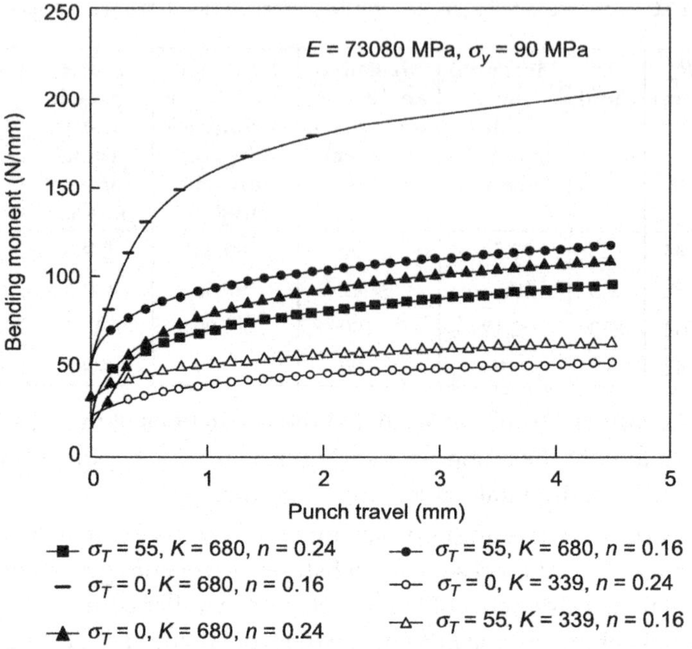

Fig. 8.8 Effect of strength coefficient (K), strain hardening exponent (n) and superimposed tensile stress on the bending moment (and hence the springback)

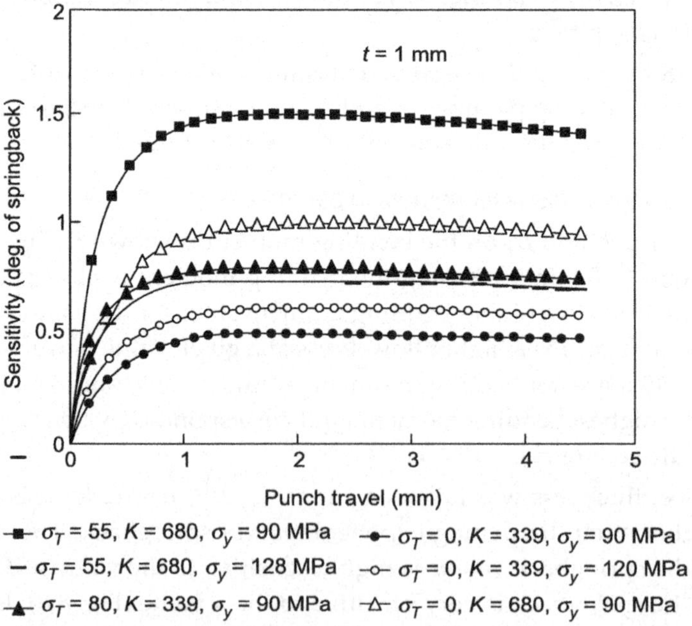

Fig. 8.9 Sensitivity of springback (change in springback) to yield strength, strength coefficient and superimposed tensile stress

8.4.2.5.2 Sensitivity of springback to the n value

Figure 8.9 shows the effect of strain hardening exponent (n) and the strength coefficient (K) on the springback in the presence and absence of a superimposed tensile stress. The figure shows the high sensitivity of K and superimposed tensile stress. The values of the input parameters to the model are listed in Table 8.7. Combinations of control variables is labeled as *A, B, C, D, E* and *F*.

For every punch travel (and hence every angle of bend), included angle after springback for an attempted included angle of 90 degree was determined for $n = 0.24$ and $n = 0.16$. The difference between the two was taken as the sensitivity of springback to the n value. This was plotted for different values of superimposed tension σ_T, strength coefficient, K and yield strength, σ_p (Fig. 8.9).

The sensitivity is found to increase initially with punch travel. This is due to the cross section progressively becoming plastic with increasing punch travel. Once the entire section has become plastic, the sensitivity remains unchanged.

It may be seen that the sensitivity is large for a high value of K and σ_T, while it is minimum when $\sigma_T = 0$ and K is low. The significant influence of K is brought out by comparing the curves when $\sigma_T = 0$ and $\sigma_p = 90$ MPa. An increase in the yield strength is found to increase the sensitivity of springback to the n value though this effect is not as prominent as that of K. The effect of σ_T is seen to be significant from curves pertaining to $\sigma_p = 90$ MPa and $K = 339$ MPa.

Since K depends upon the microstructure and the processing conditions, the same material will show enhanced springback if K were to be large, despite identical difference between the n values. This could be a reason why sheets otherwise conforming to standard specifications show inconsistent springback.

Table 8.7 Different combinations of values of the material properties influencing springback

	A	*B*	*C*	*D*	*E*	*F*
σ_T (MPa)	55	0	80	55	0	0
K (MPa)	680	339	339	339	339	680
σ_p (MPa)	90	90	90	120	120	120
E (MPa)	73080	73080	73080	73080	73080	73080

The effect of a superimposed tensile stress (like the one when the sheet is clamped with a blank holder) also depends upon K. For the same difference in the superimposed tensile stress, the sensitivity of springback to work hardening turns out to be higher for a higher value of K.

8.4.2.6 Effect of other parameters on springback

Results pertaining to the effect of other parameters like sheet thickness and Young's modulus were in line with the expectations. The punch nose radius together with the sheet thickness was found to have a pronounced effect on springback.

Surface texture is spoilt due to orange peeling which is often a problem in bent parts. A coarse grain size relative to the sheet thickness (a high *d*/*t* ratio) is known to cause this defect. The initial grain size, sheet thickness, initial surface roughness, rate of roughening, and true strain determine the surface roughness of a bend. The effect of grain size and initial surface roughness for a sheet thickness of 1.27 mm is seen to be significant . An increase in the grain size from 50 μm to 200 μm shows substantial roughening, while a wide range of initial surface roughness has little effect.

8.5 SHEET METAL SPECIFICATIONS AND SHEET METAL CHARACTERIZATION

Sheet metal specifications must therefore take into account these complex interactions and must be achievable consistently. An Analysis of Variance (ANOVA—Which is a statistical method whereby one can establish the effects of different control variables on the variance of the response variable) will enable identifying the most important parameters in a sheet metal specification. For instance, such an analysis of drawn components shows plastic strain ratio (*R* value) to be the most influential variable, while the yield strength does not seem to have much of an effect. Hence the permissible variations in yield strength could be large while a minimum *R* value will have to be specified.

Formability criteria notwithstanding, a consistent outcome is required from sheet whose properties lie within the limits of tolerance on the nominal specifications. The robustness in the process can be brought in using intelligent tools which will be able to predict successful forming of the material of a given specification for a given drawn component manufactured under given forming conditions of blank holding and lubrication. Such a tool must therefore be able to (a) predict if given material, tool and processing variables would ensure successful forming and (b) if not, suggest probable changes in values of the control variables that would lead to a successfully drawn product.

Such an ANN based robust tool was developed recently for drawn products and verified analytically [13–15]. A similar approach is being used to develop an Artificial Neural Network (ANN—a 'black box' that is used for pattern classification function approximation etc. particularly when an exact functional relation between the response and control variables is either unavailable or impractical) based tool to enable sheet metal manufacturers to minimize variation in product quality despite inevitable variations in the processing conditions (during hot rolling, cold rolling, and annealing) [15].

Initially, about 18 different variables were identified as control variables to act as an input to the ANN. These were:

1. Rolling temperature 2. finishing temperature 3. coiling temperature 4. Reduction in cold rolling 5. heating rate in annealing, 6. annealing temperature 7. annealing atmosphere 8. method of annealing (batch or continuous), and 9. chemistry of the steel (*C, Mn, Si, Al/N ratio,* O_2*, B, P, S, N Al*).

The entire effort consisting of (a) incorporating robustness into the sheet manufacturing process aiming at minimizing property variation (b) developing a tool to enable product specific material specifications and ensure robustness in the forming process for consistent product quality (c) developing strain distribution based failure criteria for a more realistic and consistent formability criterion and hence more consistent assessment of sheet metal quality, are all aimed at minimizing the variability in the product quality.

Product specific sheet metal specifications are suited to critical components as they ensure that the sheet will withstand the critical strain paths developing during the manufacture of that product. For other non-critical products, conventionally available grades of steel sheet (IF, EDDQ, DDQ, DQ, etc.) may be specified. Overspecification of sheet properties enhances the cost of manufacture of sheet part, leading to serious loss of cost competitiveness. Hence an intelligent material specification will be useful to

(a) advise a vendor as to whether a given coil will be suitable to make a given product, (so that wastage due to rejections may be avoided), and if not suitable, what specification would do the job.

(b) advise a designer as to what specification would be required consistent with the product/tool design and processing variables to form the part

An intelligent ANN based system was developed and verified to arrive at material specifications for a family of parts. A sheet metal product shape was parametrized and simulations performed for a variety of material specifications and upscaling of dimensions. Hence an ANN catering to rectangular components, another catering to cylindrical ones, etc., was trained so as to encompass a wide variation in dimensions in parts belonging to a given 'shape family'. Two levels of each processing variable were considered. The number of simulations was minimized using design of experiments so that infinite combinations of material, tool and processing variables could be encompassed using few and finite number of simulations.

This enabled the use of an ANN to predict the peak strains if the material, geometry, the processing conditions or their combination were altered. In order to obtain product specific material specification, an ANN trained for that product shape was used to predict the peak strains. If the outcome was unacceptable, the algorithm would alter the processing variables first, that being the cheapest way to correct the situation. Thereafter a change in the sheet thickness (if permissible) would be used to reduce the peak strains. The material properties and the tool design would be addressed by the algorithm the last so as to minimize the cost of bringing the peak strains to acceptable levels.

It is therefore important to be able to measure, record, and control the processing variables so as to obtain the result predicted by the ANN. Sheet metal characterization together with these measurements can then be used to train and deploy the ANN for effecting intelligent process control. The ANN based control would be instrumental in making the process more robust and less sensitive to sheet metal characteristics by making the forming parameters adapt to sheet metal property variation. Such an approach will be important in manufacturing sheet metal parts of high level of geometric complexity as well as the critical parts on account of processing complexity (those prone to developing defects due to a small operating window of processing variables, like drawn parts made from very thin sheets).

For the above mentioned procedure to succeed, reliable characterization of sheet metal behaviour in pressworking is necessary for acceptable design and control of the process.

Characterization of sheet metal also needs to be consistent. It is often found that the same test performed in different laboratories (or even different testing machines) yields variations of the order of 15% in the quantity being measured. Uncertainties in characterization of sheet material contribute to variations in the product quality. Genuine differences in sheet quality interact with the uncertainties associated with the standard procedures of characterization and hence one often finds it difficult to correlate the press shop performance with the measured characteristics of a number of coils being processed on the shopfloor.

Most of the tool designs and process designs are based on numerical simulations. These enable determining the right process parameters and help anticipate problems on the shopfloor and enables finding potential solutions. While thumb rules and design guidelines might be followed, simulations help a better visualization and hence optimization of the process. For instance, Blank holding force for axisymmetric components should be about 30% of the punch load. Similarly, nomograms suggest that for deep drawn cylindrical cups, the maximum reduction in diameter in the first draw is restricted to about 48%. These are guidelines which are relatively independent of material variables, and therefore require fine tuning through simulation, to suit given material properties and processing conditions. For rectangular products, the blank–holder pressure is given by Sommer *et al.* [16] as:

$$\text{BHP} = \text{km}((A_b/A_p) - 1)^*\text{UTS} \tag{8.12}$$

where BHP-is the blank holder pressure at the start of drawing.

A_b—plan area of the sheet blank

A_p—plan area of punch

k—material factor

m—shape factor

The material factor k is given as:

$$k = 1+((r_{\max} - r_{\min}) \times 0.49 \times 10^{-3})/r_m n_m \tag{8.13}$$

where: $r_{\text{max-}}$ is the maximum r-value in the plane of the sheet

r_{min}—minimum r-value in the plane of the sheet

r_m—average normal anisotropy = $(r_0 + 2r_{45} + r_{90})/4$

n_m—average strain hardening exponent = $(n_0 + 2n_{45} + n_{90})/4$

The factor m is given by

$$m = 1 + ((d_{fo}/t_o) - 175) \times 0.17 \times 10^{-2} \qquad (8.14)$$

where d_{fo}- is the equivalent punch diameter (calculated for rectangular punch)

t_o - original sheet thickness

The above equation considers a number of factors affecting the blank holder force. This is the force applied uniformly over the blank holder area by conventional blank holders. For multiple point blank holding [17], however, simulation would be required to obtain the spatial distribution and time dependent variation in the blank holding force at different points over the blank. Hence as complexity of the components increases, their criticality increases and the importance of simulations for such components in particular cannot be underestimated.

The reliability and usability of these designs strongly depend upon the input data. Springback compensation on the tools for instance would be inadequate if the test data is not consistent. Deviations from the expected shape from the desired are therefore inevitable despite availability of CAE tools. Minimizing these deviations then becomes the objective of the design and manufacturing activities.

Sheet metal characteristics are often used to establish formability criteria. Several such criteria are in use. Uniaxial tensile criteria include the *nxR* product, while biaxial criteria like the FLD (Forming Limit Diagram) have been in use for decades. Formability characterization in the form of the Erichsen cup height (since the olden times) and Forming Limit Diagrams (FLDs, Fig. 8.10) have become a standard practice in sheet metal forming. Material specifications are often formulated on the basis of these.

Analytical prediction of the Forming Limit Diagrams (FLDs) led to development of analytical models to explain the deformation and failure of sheet metal due to biaxial stretching. Initial efforts of Hill [18] to propose localized necking (for negative minor strains, e.g., drawing) and Swift [19] who proposed diffused necking (for positive minor strains, e.g., in stretching) as mechanisms of failure were developed upon further by Chan *et al*, [20] and Marciniak *et al.* [21].

Marciniak *et al.* explained the failure by localized necking even in stretching regime (where it is theoretically not possible), on account of growth of an initial imperfection in the sheet leading to a change in strain path towards plane strain and consequent failure. A number of analytical models based on this hypothesis of Marciniak *et al.* (the M-K hypothesis) were developed taking in to account additional phenomena like void growth and coalescence, shear band formation, surface roughness effects, those of different yield criteria, etc. in an attempt to predict the FLD for a number of materials.

An experimental verification of the M–K hypothesis [22] confirmed the deviation of the strain path towards plane strain, although the high imperfection sensitivity of the M–K model remained unaddressed. All models considered only one mechanism of failure to operate. However, Date *et al.* [23] proposed a M–K based model taking into account the non-quadratic yield criterion, void growth, surface roughness, and two concurrently operating mechanisms of failure, namely, geometric softening and shear band formation each contributing half of its limit strain potential. This model correlated very well with the experimentally determined FLDs, and the excessive imperfection sensitivity of the M–K model was considerably lowered. The incorporation of two failure mechanisms was motivated by the difference in the mechanism of failure in steel from that in aluminum, wherein shear bands start from the surface of an aluminum sheet and move inwards nucleating a void where they intersect each other.

Research on formability criteria notwithstanding, traditionally, Erichsen cup test and of late, the FLD have been viewed as standard tests to characterize the formability of sheet metal. However, use of either the Erichsen cup height or the FLD in the form of a Keeler–Goodwin band as an acceptance criterion permits plenty of variation in the quality (mechanical behavior during processing) of sheet metal and therefore provide an unreliable, uncertain estimate of formability particularly for critical drawn industrial components. For instance Erichsen cup test involves more stretching than drawing in, the extent of which is uncertain. It therefore, gives an unreliable estimate of drawability.

These established criteria have serious limitations. For example, the uniaxial criteria are indicative and do not guarantee that two materials with the same/similar *nxR* product will show similar behavior in the laboratory or the shopfloor. The FLD is highly sensitive to strain path and it is impractical to have a bunch of FLDs to cater to the diverse strain paths experienced by different regions in the drawn product. One therefore uses an envelope, namely the Keeler–Goodwin band as an approximation. Moreover, the strain at a point cannot be independent of strains in the rest of the deforming sheet. Hence two strain distributions having similar peak strains might refer to two completely different situations, although the FLD, which considers only the magnitude of strains at the peak would indicate similar severity of deformation. Therefore, the strain distribution based criteria would be more realistic but are more qualitative at this point.

A Forming Limit Stress Diagram (FLSD) was defined as a robust formability tool, devoid of the sensitivity of the FLD to strain path. The FLSD, however, is too insensitive to strain path. Hence one has two extremes to contend with – a high sensitivity of the FLD and nearly complete insensitivity of the FLSD to the strain path.

Hence there is a need for a quantitative, strain distribution based criterion for formability, which overcomes the limitations of the criteria available to date.

It is therefore important to evolve a sufficiently sensitive formability criterion that will be able to distinguish between two materials differing widely in quality from those likely to exhibit similar press shop performance. Development of intelligent 'product specific' material specifications has already been described in the previous section and a similar empirical approach has been adopted by the shopfloors. This is aimed at consistent success in forming critical drawn sheet metal components.

A similar result can be achieved using strain distribution based formability criteria. These would characterize the quality of strain distribution quantitatively and a critical value based imminent failure could be used as the failure criterion [24–29]. By this method the formability characterization is product specific, accounting for the complex interplay among the measurable as well as immeasurable variables (like the coefficient of friction, interface stress distribution, etc.), leading to better consistency in product quality. A failure criterion based on Fourier analysis of strain distribution was recently proposed. It was verified on the basis of a uniaxial tensile test [24, 25], and experimental investigations under biaxial conditions [32] were in reasonable agreement with the analytical predictions. The strain distribution in drawn cups undergoing uniform deformation could be fitted to a lower order Fourier series, while that for a failed cup required higher order Fourier fit. This is because fitting sharp strain peaks (small radii of curvature at such peaks) required greater contribution from higher order terms, notably the sine coefficients. It may be mathematically shown that the SNI (strain non-uniformity index) is proportional to the rate of change of strain gradient, and hence the curvature at the strain peak. Higher the SNI, higher would be the curvature at the peak, which is borne out by relatively high values of the sine coefficients. The experimental study showed that lower order sine coefficients were much smaller in magnitude than the higher order sine coefficients in failed cups. In successful cups, the cosine coefficients were significantly greater than the sine coefficients [32]. However, more work remains to be done to be able to generalize the conclusions to components formed on the shopfloor.

8.6 ANALYSIS OF SPATIAL STRAIN DISTRIBUTION [26–32]

The spatial thickness strain distribution in a drawn sheet metal holds within it information about the deformation. Decoding the strain distribution profiles as the process progresses, one can probably obtain enormous information on the formability, imminent failure and enable the user to make "informed" decisions on a quantitative basis.

Some efforts have been made by the author [27–29] and others, for instance, [26] to extract meaningful information from the strain distribution (based on half peak width of the strain distribution profile), which is the outcome of a complex interaction among material, processing, tool design, and product design related variables.

In this section, strain distribution is characterized using two quantities, namely the constraint factor (CF) and the strain non-uniformity index (SNI). The advantage of these parameters vis-a vis the strain path and the FLD will readily become apparent. Prediction of the FLD using these new formability criteria will be compared with those using conventional criteria, in particular, for complex strain paths.

Hence the present section

(a) brings out the physical significance of the formability based criteria, namely the Constraint Factor (CF) and the Strain Non-Uniformity Index (SNI)

(b) correlates the above criteria to the conventional concepts of formability like the Forming Limit Diagram (FLD) and the strain path.

(c) illustrates the use of the criteria in predicting the FLD for complex strain paths

8.6.1 Characterization of Strain Distribution [27–29]

The strain distribution (Fig. 8.10) is characterized by two variables, namely, the "Constraint factor", CF and the "strain non-uniformity index", SNI. The details of evaluation of these variables are given elsewhere [27–29].

The Constraint Factor is defined as

Constraint factor = (Peak minor strain – Pole minor strain)/ (Peak major strain – Pole major strain) (8.15)

where pole refers to the region of minimal deformation and the peak the region of maximum deformation.

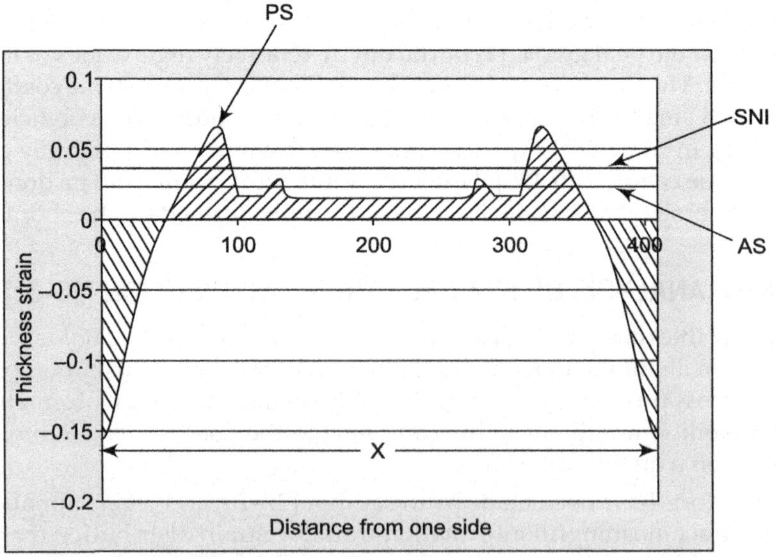

Fig. 8.10 A schematic thickness strain distribution indicating the peak strain (PS), average strain (AS) and the S–N–I.

With increasing punch travel the spatial strain distribution becomes increasingly uneven as indicated by an increasing strain non-uniformity index as in Fig. 8.11. If the SNI rises steeply it amounts to rapid strain localization ending in failure. On the other hand, a slow rise of SNI indicates a relatively mild variation in strain over the critical cross section leading to a successfully formed product.

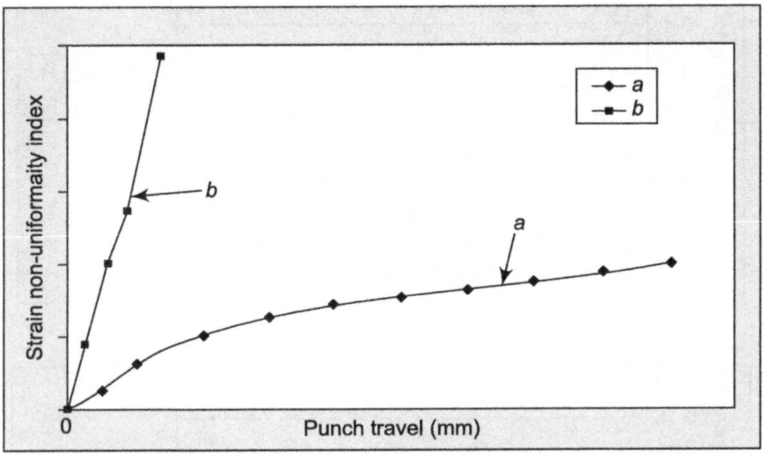

Fig. 8.11 Strain Non-uniformity Index (SNI) for (a) successful forming and (b) high SNI leading to failure

The degree of strain non-uniformity is a combined outcome of the material, tool, product geometry, and forming conditions.

A highly negative value of CF refers to drawing, while a positive CF to stretching. A CF near zero would mean plane strain. Hence a highly negative value of CF are desirable for large formability while a small (near zero) value of CF would be undesirable as plane strain permits minimum formability.

8.6.2 Effect of Material and Processing Variables on the CF and SNI

As seen from Fig. 8.12(a), a high R value (high plastic strain ratio) material shows a more negative CF for the same shape drawn under identical conditions, indicating high drawability. The corresponding SNI Fig. 8.12(b) is seen to be lower for a high R value material, which indicates a more uniform thickness strain distribution for such a material. This is in line with expectations and normal practice.

Figures 8.13(a)-(b) show the effect of blank holding on the CF and SNI. It is seen that the blank holding force has practically no influence on the variation in the CF for a wide range of BHF Fig. 8.13(a). So is the case with the SNI Fig. 8.13(b). However, an excessively large blank holding force causes the CF to rise towards plane strain and the corresponding SNI is seen to rise rapidly and stands out in Fig. 8.13(b). This shows that the deep drawing process is relatively insensitive to blank holding force, as long as it is not excessive. A sensitivity analysis also confirms this observation [29].

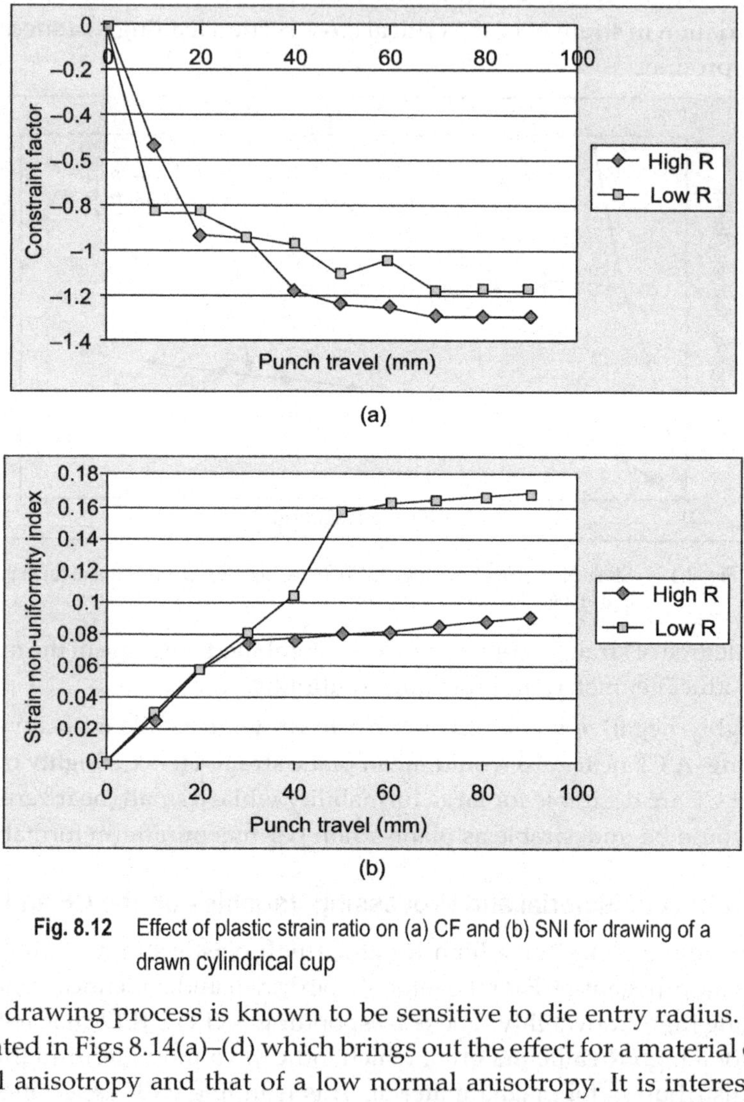

Fig. 8.12 Effect of plastic strain ratio on (a) CF and (b) SNI for drawing of a drawn cylindrical cup

The drawing process is known to be sensitive to die entry radius. This is illustrated in Figs 8.14(a)–(d) which brings out the effect for a material of high normal anisotropy and that of a low normal anisotropy. It is interesting to notice that the die entry radius has a significant influence on the formability of low *R* value material, while for high *R* value the range of variation studied showed no influence. For low *R* value material a larger die entry radius is seen to promote greater formability (negative CF and lower SNI). This explains why aluminum alloys need greater die entry radius compared to that in steel.

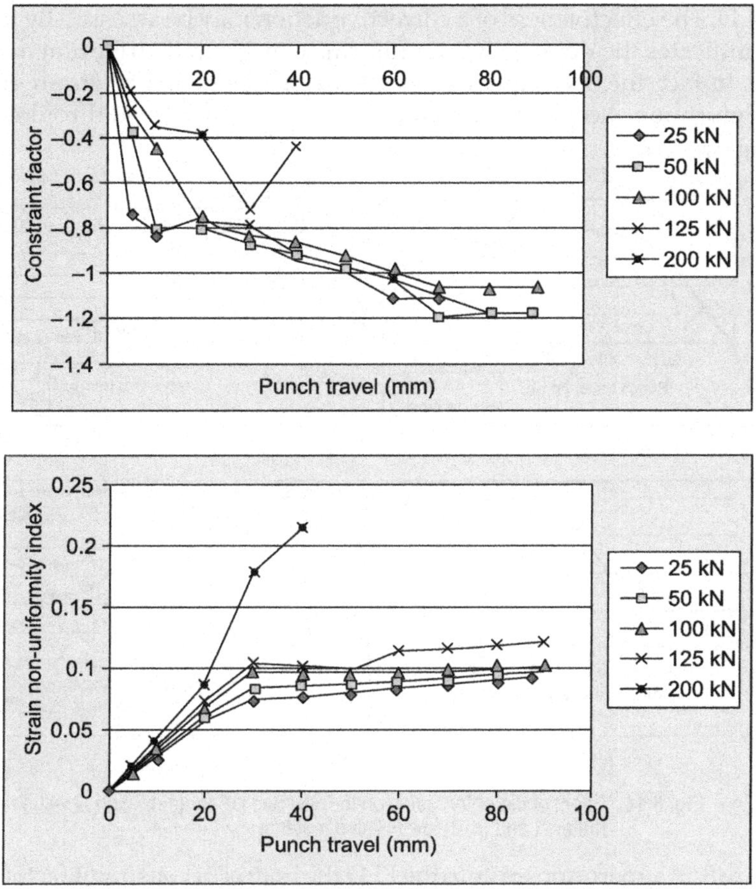

Fig. 8.13 Effect of blank holding force on (a) CF and (b) SNI [29]

8.6.3 Constraint Factor and SNI vs. FLD and Strain Path

The CF and SNI are both determined based on the strain distribution in a particular cross section, whereas the strain path refers to a point. It is much easier to monitor the variation of CF during a punch travel than track strain paths of individual points, particularly when the critical region shifts during the forming process.

CF reduces to the strain ratio when the strain at the pole (minimum deformation zone) is zero. Thus, strain ratio (also often called as strain path) is a special case of CF. The strain states for the peak and the pole (required to calculate the CF) are indicated on the FLD and the corresponding CF values determined for drawing and near plane strain deformation [Figs 8.15 (a) and (b)].

Just as the superimposition of the FLD with the strain path prompts the process designer to troubleshoot a process by changing appropriate constraints, the same is possible using the constraint factor WITHOUT the need of an FLD. The effectiveness of a corrective action may be assessed by the SNI, which indicates the degree of non uniformity in strain distribution in a cross section. In fact, the SNI signifies the rate of change of spatial strain gradient in the deforming sheet metal, and like CF it can be monitored throughout the forming process.

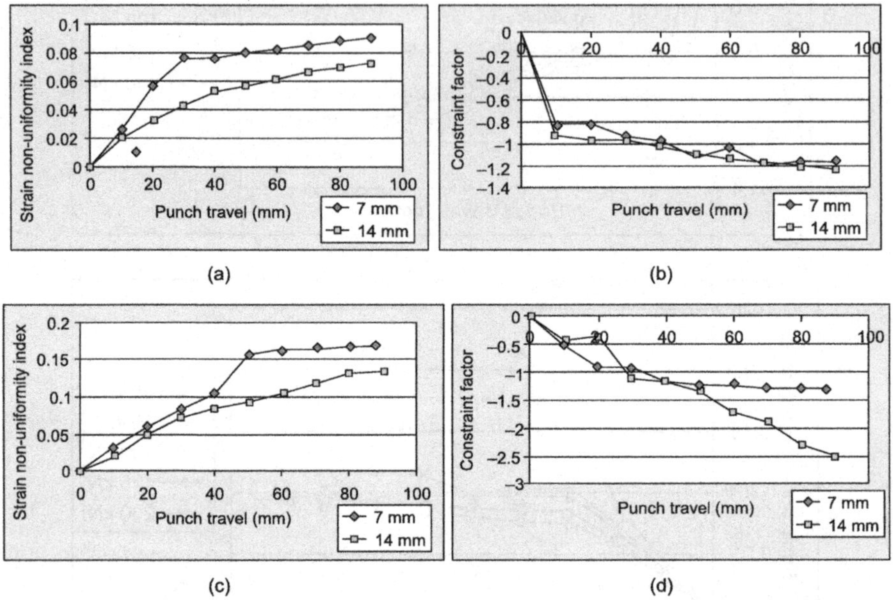

Fig. 8.14 Effect of die entry radius on the SNI and CF for (a, b) high *R* value material and (c, d) low *R* value material

Figure 8.16 superimposes onto the FLD the matrix of constraint factor values corresponding to different strain states at the peak and a given strain state at the pole. It is seen that the Forming limit line passes through different CF values, which would change depending on the magnitude of polar strains. While there is no change in the FLD due to a change in polar strains, the CF at failure changes accounting for a change in the strain path at the peak due to a change in lateral constraints. The strain paths have also been superimposed on the FLD and the constraint factor matrix. It is easy to note that a change in the strain path can be effected by changing the constraint factor.

The forming limit refers to a demarcation between successful forming and failure. While there cannot be a single line of separation a zone with a probability of failure varying across it characterizes the forming limit. Such a limit is very sensitive to strain path, and will vary from point to point as different points on the deforming sheet metal experience a different strain path. Moreover, if one knew apriori that failure will occur at a particular point then one will know exactly which point one must track for determining the critical strain path. This, however, is not practicable. Hence one resorts to using an FLD based on linear strain paths, which provides an approximate tool for assessment of limit strain in the deforming product.

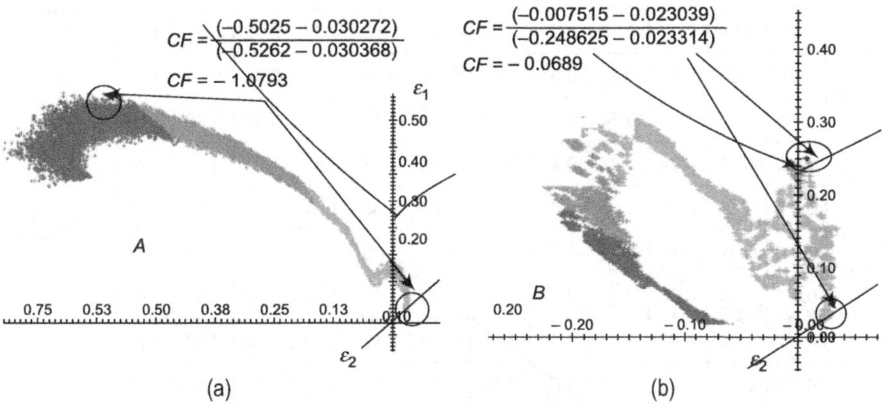

Fig. 8.15 Peak and pole strain states in the critical cross section superimposed on the FLD. The corresponding CF values are illustrated for (a) drawing regime (b) near plane strain deformation

The CF on the other hand shows a certain discernible deviation towards plane strain and indicates failure. One needs to observe only a single cross section encompassing the point of minimum and maximum strains. It is easier to work with and monitor cross sections than monitor points changing location. At what punch travel CF will indicate failure for a given set of material properties and processing conditions depends on the product geometry. This accounts for the differences in strain path. Similar would be the case for other factors causing strain path variation.

8.6.4 FLD and Constraint Factor

Figure 8.16 shows CF values together with an FLD in the $\varepsilon_1 - \varepsilon_2$ space. Fig. 8.16(a) shows the positive constraint factor matrix together with the FLD. In this region the CF > 0. Equibiaxial tension (stress ratio = 1) passes through cells representing a CF = 1. Figure 8.16(b) shows the corresponding CFs on the LHS of the FLD. A sample calculation to obtain the CF is also shown.

In other words the new formability criteria defined perform the functions of the FLD and strain path, but do not have the drawbacks of these two conventional criteria of formability.

8.6.5 SNI vs. nxR Product

Figure 8.17 shows a correlation between the SNI and the *nxR* product as an established index of formability. The strain distribution profiles have been taken from the work of Melander *et al.* [26] and analyzed as described earlier. An increase in the *nxR* product shows a decrease in the SNI, i.e., better formability, which is qualitatively in line with expectations. Three materials,

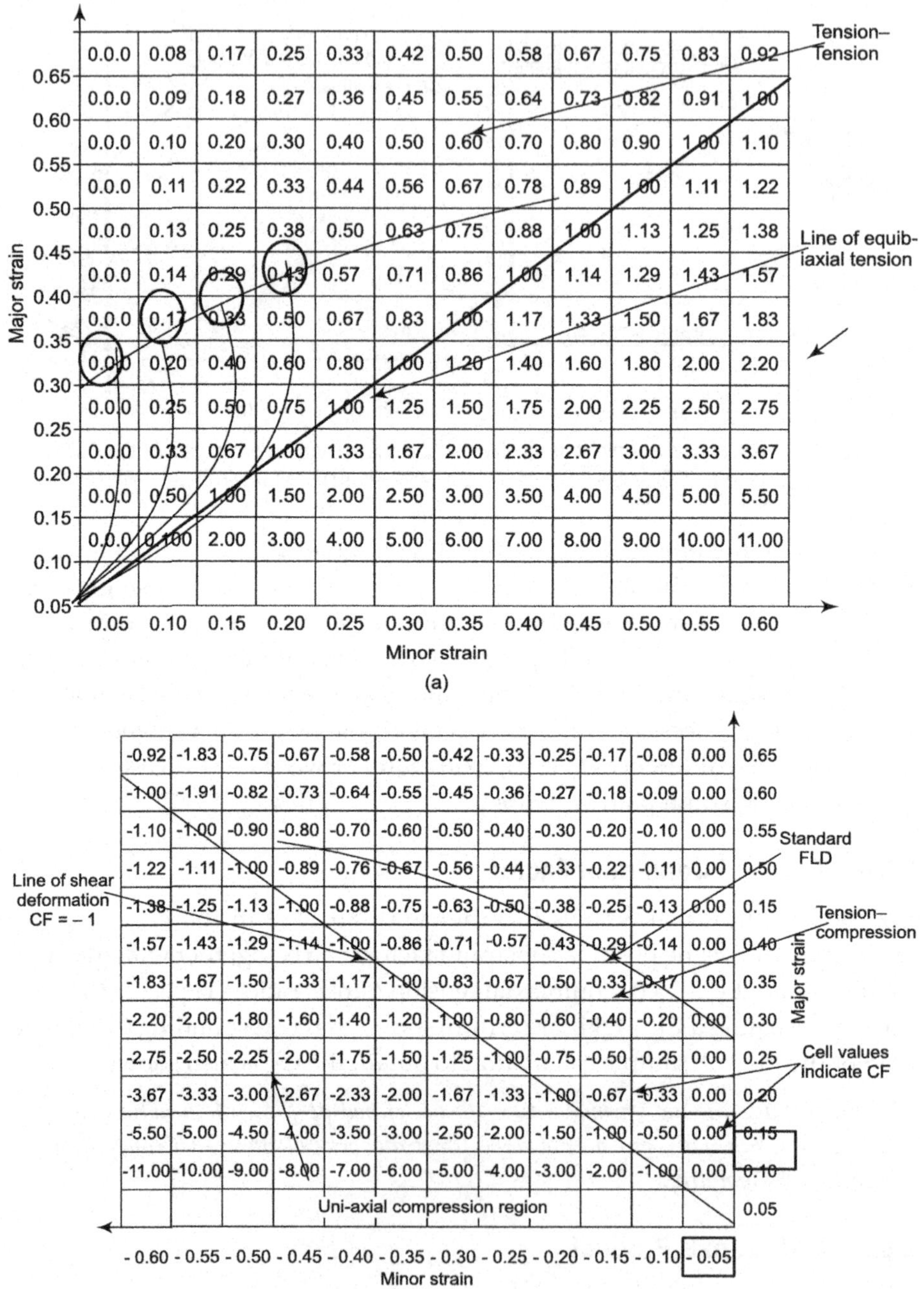

Sample calculation: $CF = \dfrac{(\varepsilon_{2\,peak} - \varepsilon_{2\,pole})}{(\varepsilon_{1\,peak} - \varepsilon_{1\,pole})} = \dfrac{((-0.05) - (-0.05))}{(0.15 - 0.05)} = 0$

(b)

Fig. 8.16 CF values superimposed over the FLD in the $\varepsilon_1 - \varepsilon_2$ space (a) Stretching regime (b) drawing regime [20]

DP 450, HSLA and SS1147 show similar SNI despite widely different *nxR* products. Here it is likely that a similar magnitude of SNI developed at different punch travels in the three materials. As for SS 1147 and P370, the *nxR* product seems to be similar, but the SNI value brings out the significant difference in the formability of the two materials.

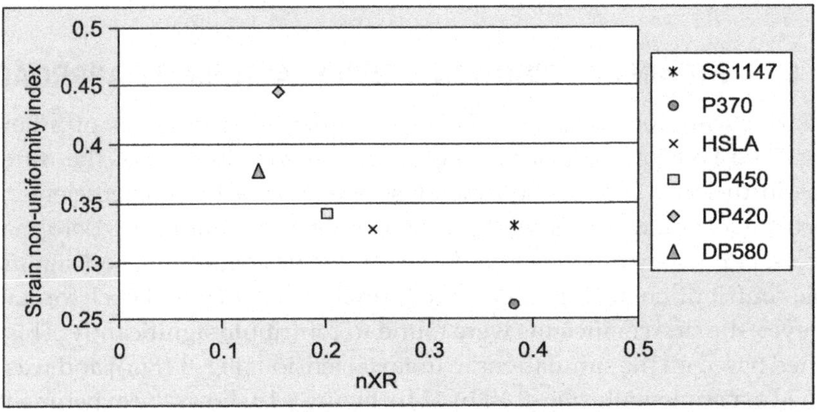

Fig. 8.17 Correlation between *nxR* product and SNI. Strain distribution data was taken from published literature [26, 30–32]

8.6.6 Some Important Observations

From the foregoing it may be observed that

(a) Both CF and SNI have a physical significance–CF represents the strain ratio when polar strains are zero, while SNI represents the rate of change of spatial strain gradient. A higher SNI would mean a sharper peak (smaller radius of curvature at the peak).

(b) Inferences based on the Strain distribution based formability criteria defined in this work confirm those from the existing criteria. In particular,

$CF < 0$ for drawing, $CF > 0$ for stretching, $CF \sim 0$ for plane strain; $CF = -1$ for pure shear

(c) CF is sensitive to material properties, tool geometry and the strain path

(d) High n and R values leads to a lower SNI; high *nxR* product also leads to a low SNI

(e) The same *nxR* product value may lead to different SNI values and vice versa, in line with common experience

(f) CF and SNI are useful as tools for comparison of material behavior of two materials given a product geometry material response to two different geometries material response to different processing conditions

(g) CF and SNI are useful as formability tools for process control
(h) The newly defined formability criteria, namely the CF and SNI overcome the shortcomings of the FLD and FLSD.
(i) CF and SNI can be used independent of the FLD which is tedious to determine

8.7 STRAIN DISTRIBUTION STUDIES: ALTERNATE APPROACH

Spatial distribution of strain may also be quantified in terms of a Fourier series. The Fourier coefficients determine the contribution of the different terms in the sine and cosine functions. Behavior of these coefficients with punch travel could throw light on the events occurring in the deformation process. For instance, it was found that the cosine coefficients had significant stable contribution during uniform deformation. At incipient neck formation, however, the sine coefficients were found to contribute significantly. This was verified based on the simulations in uniaxial tension [Fig. 8.18(a)] and was also verified experimentally (Fig. 8.18(b), [24]). Figure 8.19 shows the experimentally determined thickness strain distribution. Such a strain distribution is useful in calculating the strain non-uniformity index.

Similarly, predictability of the strain distribution of a large product using that of a geometrically similar scaled down component could enable

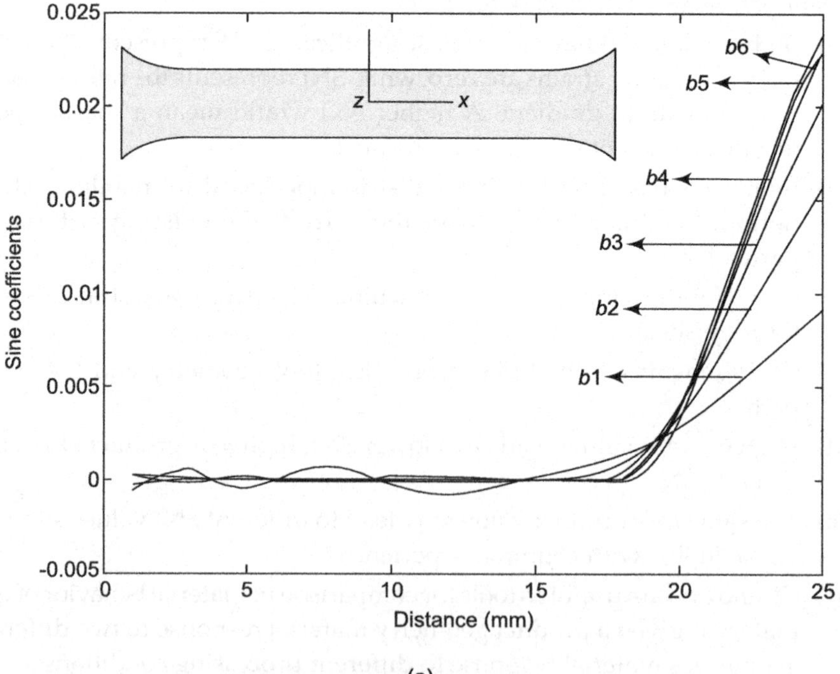

(a)

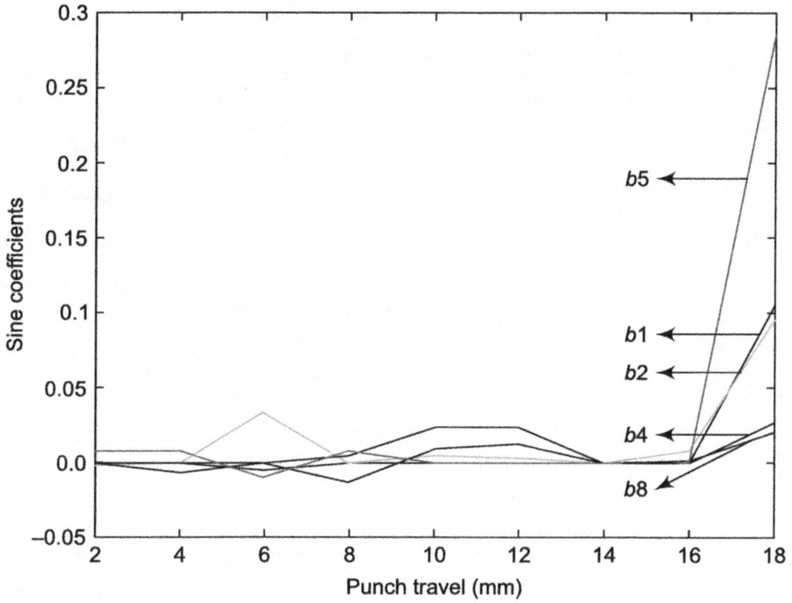

(b)

Fig. 8.18 Variation in the sine coefficients representing uniaxial tensile strain distribution: (a) analytical (based on PAMSTAMP simulation) and (b) experimental

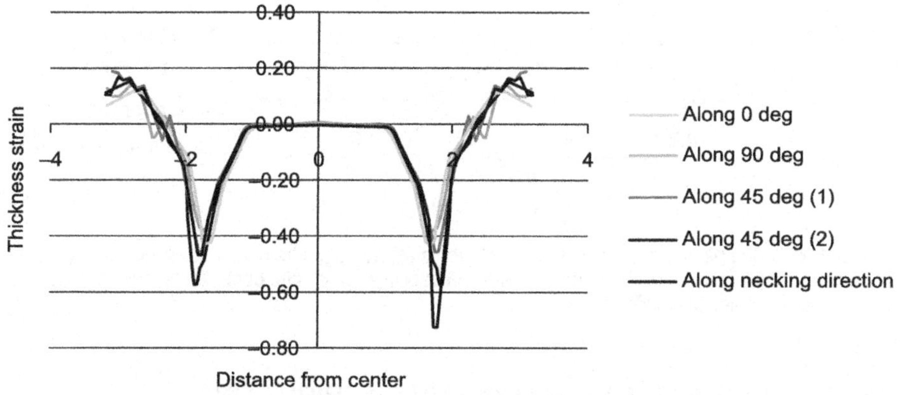

Fig. 8.19 Thickness strain behavior along different cross-sections

prototype development using small sized workpieces saving on time and cost of development. Here again, one recognizes that the friction is not scaleable and that the r/t ratios have to be maintained for a meaningful interpretation. In view of this, Fourier coefficients of the strain distribution in two geometrically similar drawn sheet metal components were examined for their variation during punch travel. It was found that the corresponding lower order cosine coefficients from thickness strain distribution in the two drawn products showed parallelism in their variation with punch travel indicating scaleability, as seen in Fig. 8.20.

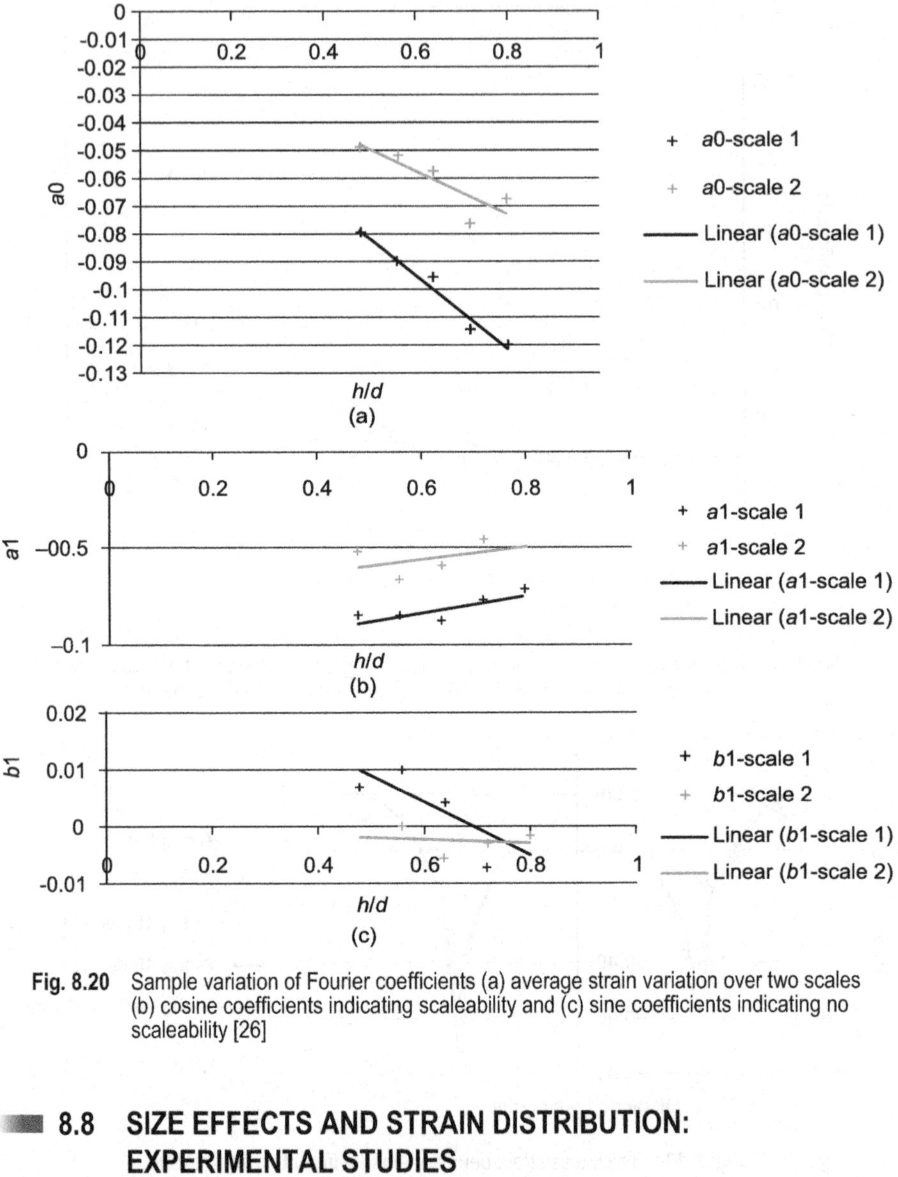

Fig. 8.20 Sample variation of Fourier coefficients (a) average strain variation over two scales (b) cosine coefficients indicating scaleability and (c) sine coefficients indicating no scaleability [26]

8.8 SIZE EFFECTS AND STRAIN DISTRIBUTION: EXPERIMENTAL STUDIES

Effect of product size has a significant effect on the strain distribution and hence the overall quality of the product. Smaller sizes (relative to sheet thickness) are relatively simpler, while the larger ones tend to fail. For instance, drawn cups of four different diameters were attempted. The single draw ratio below the 'limiting draw ratio' was used, and it remained unchanged in each of the cases. Similarly, the r/t ratio (radii at the die entry and punch nose for instance) was maintained constant, equal to 6 at the die entry and equal to 4 at the punch nose. This means that all four cups should have been successfully drawn. This however, was not the case and the largest size ($h/d = 0.8$) would fail consistently. There was considerable difficulty in drawing the second-largest size ($h/d = 0.72$) and inconsistent success was obtained. The smaller sizes could be successfully drawn consistently.

Scaleability of the strain distribution, if possible, would enable prediction of strain distribution of a larger sized product based on that obtained in the geometrically similar scaled down model. This was investigated by fitting spatial strain distribution profiles from the drawn cups into a Fourier series and examining the variation of Fourier coefficients with strain, and comparing this variation obtained from different sizes of formed cups. The scaleability investigated earlier using numerical simulations showed just about 60% scaleability [25], but the experimental work recently completed [26] shows scaleability to a scaling factor of 2 (100% upscaling).

The inferences from the strain distribution reported here have been validated at the laboratory level. The proposed procedures for addressing the issues which originate in the shop floor have to date, not been verified on an industrial scale. The procedures described are logical tools to improve the strain distribution in favor of the customer expectations. A better strain distribution means better quality of the product. The next section reviews some of the emerging technologies and new materials in improving quality of sheet metal parts.

8.9 PRODUCT QUALITY THROUGH NEW TECHNOLOGIES AND NEW MATERIALS

New technologies are being developed to address the following issues in sheet metal formed products:

(a) Precision (like laser cutting, forming and welding)
(b) Speed (like high speed presses for punching and coining, slide forming transfer presses, automation, etc.)
(c) Minimizing springback (like press hardening, electromagnetic forming, warm forming, superplastic forming, etc.)
(d) Decreasing material input (like tailor welded blanks, tailor welded tubes, high strength materials, use of design features for better stiffness, etc.)
(e) Enhancing performance in service (DFM/design for safety related paradigms, design methodology, better manufacturing practices)
(f) Better tool life through new tool materials and cryogenic heat treatment of existing materials

8.9.1 Precision

CNC technology has been combined with most of the sheet metal forming machine tools to increase the precision in positioning of the workpiece, minimizing errors in location of the features of the workpiece and thereby enhancing the fittability of the sheet metal component into an assembly. Laser based techniques together with CNC are being used for localizing the energy input in cutting and welding. Hence, even the conventional spot welding process is being replaced by laser spot welding process on account of better controllability of the process. Needless to say, a better technology demands better quality of fixturing to harness full benefits.

8.9.2 Speed

An increase in cost is the outcome of using technologies leading to a high level of repeatability and hence consistent performance of the product. Hence production volumes need to be high to justify the cost and defects must be kept down to ppm levels as reworking of the product is impractical at high production volumes. Similarly, automation of part handling and inspection would be a necessity for consistency in each step.

Product and process design therefore go together. Data defined in the design of the sheet metal blank must be used in orientation and placement of sheet/partially formed product onto the dies, in-process inspection (to ensure rapid, consistent 100% inspection as opposed to inspection by random sampling) and assembly.

8.9.3 Minimizing Springback

Technologies for minimizing springback have become important on account of the recent trend to use high strength materials so as to decrease the material input and hence the weight of the product using a thinner sheet. The need to control springback is more severe because thinner gauges of high strength sheet together promote higher springback.

Technologies like 'press hardening' have been introduced wherein sheets of high in-service strength are deformed at relatively high temperature (lower forming forces, lower/negligible springback,) and are cooled in the die itself, thereby obtaining more consistent dimensions but inducing thermal stresses and strains. Such steels have boron added to them. Alternately, materials which behave soft in forming and acquire strength in further processing, e.g., after painting (during baking of the paint), as in bake hardening steels which achieve a similar effect, but to a milder degree. The chemical compositions of these two materials are very different and they highlight two different non-conventional approaches to address the issue of springback.

Springback may be reduced using high energy rate forming processes like the electromagnetic forming, explosive forming, electrohydraulic forming, etc., or processes like hydroforming which permit 'calibration' (equivalent of bottoming in conventional springback control) so that fine details (and sharp bend radii) may be achieved to the desired form and dimensions. In hydroforming the pressure of the working fluid that is transmitted equally in all directions and absence of friction together ensure uniform distribution of strain. This enhances the forming limits as well as the pressures that can be used for calibration in hydroforming.

Roll forming is another process wherein productivity in the manufacture of long bent parts is facilitated, as opposed to the restricted length of bend that is possible in a press brake, with minimal springback by virtue of the flexibility in configuring the roll forming line. Here again, efficient in-process inspection and timely process correction may be performed to ensure product quality.

Another approach that might help overcoming springback would be high strain rate superplastic forming. Usual process of superplastic forming is rather slow, although it leads to large strains of the order of few thousand percent. The latter part is attractive to automotive products in several ways, namely,

(a) non-ferrous alloys, and in particular, the light alloys show structural superplasticity. Use of such alloys would reduce the weight of the autobody by virtue of their low density.

(b) large strains (and hence a uniform strain distribution) would mean a lower material input to the product being made, thereby saving on the weight of the product.

(c) since the part would be made in a single stage by superplastic forming, multistage processing, part handling, and in-process part storage would be avoided. In fact the existing machines could be deployed to make several pieces of a single part simultaneously, overcoming the deficiency in the rate of manufacture.

The slow speed of the process usually renders it commercially unviable. However, if the strain rate can be increased, the productivity could rise to acceptable levels. Use of ultra-fine grained (UFG) materials would be able to enhance the strain rate at which a part could be manufactured. Several techniques of grain refinement based on severe plastic deformation (SPD) have been deployed, namely, ECAP, HPT, ARB, etc., to name a few. A schematic of the ECAP process and the typical results obtained have already been discussed earlier.

In all these processes which attempt to minimize springback as described above, the difficulty in getting the right dimension on account of springback is transformed into the similar difficulty due to temperature and temperature gradients causing locked in thermal stresses.

8.9.4 Decreasing Material Input

The objective of decreasing material input is to make the product lighter and as far as possible, cheaper without sacrificing on the functionality. The use of thinner gauges of high strength sheet metal, use of light metals like magnesium (and its alloys) for warm forming, use of tailor welded blanks and laser joining are all attempts at decreasing the material input. The present section will highlight the use of tailor welded blanks. The contour along which the weld may be placed is an important issue discussed in the subsequent section.

8.9.4.1 Tailor welded blanks [33–42]

A tailored blank consists of blanks of different sheet thicknesses or different alloys, which are welded together along a predetermined contour prior to forming. The TWB may have two or more different blanks welded together using a butt joint into a single blank. A lap joint is not very popular on account of its geometry (requiring edge preparation and post weld planishing), an inferior corrosion resistance and inherent difficulties in the use of such TWBs in forming.

The constituent blanks of a TWB may differ in grade, gauge thickness, strength, and presence/absence of coating. With tailor welded blanks (TWBs) one can use thinner gauge of a stronger sheet, or thinner gauges in less demanding areas of the product, as might be necessary to achieve correct distribution of weight (material) and overall weight reduction.

Hence, a thicker (or stronger) sheet would be used in load bearing regions while a thinner one would suffice elsewhere. For instance, a thicker gauge near the hinge and a thinner one away from it for a door panel would shift the centre of gravity of the door closer to the hinge thereby decreasing its weight as well as sagging of the door over a period of time. As of now, the TWBs are in use for the inner panels which are aesthetically less critical (not being visible to the customer), but time is not far when they would be used for skin panels as well.

An advantage of these blanks is that the number of sheet metal parts which need to be fixtured and joined after forming, may be reduced, needing fewer number of dies and better dimensional consistency. In addition, a reduction in expenditure on assembly fixtures, manpower and time in assembling a number of parts adds up to cost savings.

However, in view of the different thicknesses in the TWBs one requires stepped tools or significant, expensive modifications in the existing tools. The manufacturing process must be adapted to the use of TWBs to account for thickness variation and the presence of the weld. In that sense, hydroforming is a very suitable process for forming a TWB or a Tailor welded tube (TWT), wherein the step resulting from different thicknesses being welded can be easily accommodated on the fluid side, so that the die side of the product looks continuous with no evidence of the presence of a joint.

It is important that the weld is not severely deformed as the formability of the weld would be inferior to the materials used in the TWBs. In particular, the weld must not experience high tensile stresses perpendicular to the weld line. The implications of using a TWB are documented and discussed in the literature [33–42]. A number of technologies like that of multiple point blank holding, segmented dies, and laser cutting and welding have been developed to enable manufacture and processing of theTWBs [43–45]. In this context, hydroforming helps as the strains are more uniformly distributed in this process.

During forming the thinner (or weaker) sheet deforms more leading to movement of the weld line towards the thicker sheet. This could move the weld into a severely deforming zone. One may either prevent weld line movement by mechanically arresting the weld line [33–35] or accommodate the movement into the design of the part. The important issue therefore is to control the weld line movement so that the final product is safe and weld line is in non critical part of product.

Investigations on the effect of drawbead dimension on the weld line movement in tailor welded blank [40] showed that weld line movement decreases as the size and height of the drawbead increase. Thus weld line movement can be controlled by appropriate drawbead design. Use of differential blank holding pressure for reducing weld line movement was proposed by [39]. The thinner/weaker material is subjected to higher blank holding pressure, which results into a more uniform flow of material in the thicker as well as the thinner regions. The weld line movement is thus curtailed.

In the present study, tailor welded blanks of CP aluminum with different thicknesses were used in deep drawing experiments as well as simulations. Unlike the existing methods three novel methods for reducing weld line movement by design were used, namely, the 'back projection method', 'minimum strain contour method' and 'pretensioning the thinner (weaker) portion'. Of these, the back projection method permits weld line movement into the desired location, while the latter minimizes the movement by suitably designing the weld contour.

Finally, the thinner portion of a TWB made from two different sheet thicknesses of aluminum was prestrained until it work hardened to the yield strength of the thicker one. Following this, the prestrained TWB was deformed to the desired shape and the weld line movement with reference to its initial position measured. The methods are evaluated in terms of the deviation of the location of the weld line from the desired one after forming.

8.9.4.1.1 Finite element analysis for forming a TWB cup

In order to examine the effectiveness of the two methods, namely, the back projection method and the minimum strain contour methods (for reducing the weld line movement), elasto-plastic FEM simulations were conducted using

PAMSTAMP 2G software. The tools namely, punch, blank holder and the die are assumed to be perfectly rigid with surface to surface contact between the blank and the tooling. The commercially pure aluminum tailored blank of 1 mm and 1.2 mm thicknesses was considered for weld line movement analysis using the two methods. A constant friction condition ($\mu = 0.12$) corresponding to an unlubricated state was assumed at all tool interfaces. The deformation of the tailored blank was thus simulated for cylindrical cup and rectangular box shapes (Figs 8.21–8.22).

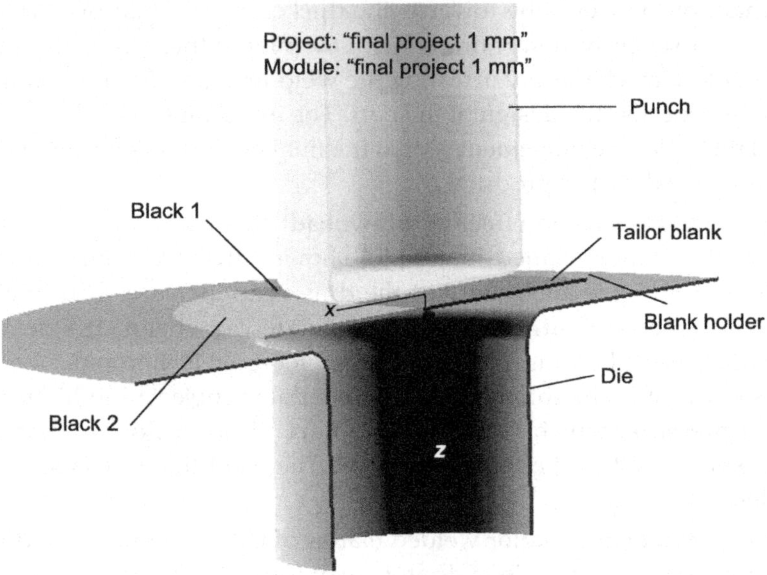

Fig. 8.21 Finite element sectional tool model for tailor welded cup simulated for back projection method (Part 1) [42]

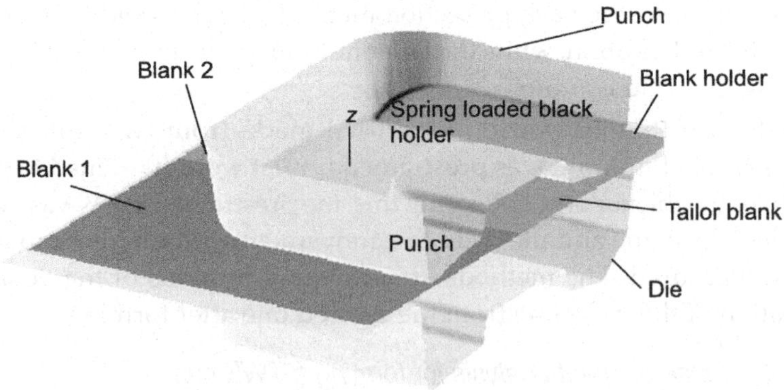

Fig. 8.22 Finite element sectional tool model for tailor welded box shaped part (Part 2) [42]

Figure 8.21 shows the sectional tool set up model for tailor welded cup and Fig. 8.22 shows the sectional tool set up for the box shaped part. Identical material and thickness combinations were used for both the cases. The mechanical properties of the base material are shown in Table 8.8(a). The processing conditions for the circular cup and the rectangular part are shown in Table 8.8(b) and in Table 8.8(c) respectively. The back projection method is illustrated in the context of a deep drawn cylindrical cup (Figs 8.23–8.24) while the minimum thickness strain contour method is illustrated for a rectangular box shaped part (Figs 8.25–8.26).

Table 8.8 (a) Material properties

Blank Material	Aluminum
Yields strength	70 MPa
Strength Coefficient	151 MPa
Strain hardening exponent	0.264

Table 8.8 (b) Process conditions for example part simulated for back projection method (Part 1)

Punch size	320×232 mm^2
Blank size	480×360 mm^2
Blank holding force	25 kN
Sheet thicknesses welded	1 mm and 2 mm

Table 8.8 (c) Process conditions for example part simulated for tracing the line on minimum strain method (Part 2)

Punch size	Diameter = 48 mm
Blank size	Diameter = 85 mm
Blank holding force	10 kN
Sheet thicknesses welded	1 mm and 1.2 mm

8.9.4.1.2 Back projection method

This method is based on the concept of simulating the deformation of the tailored blank and identifying the position where exactly the weld line would be required in final formed part This desired weld line contour is projected back onto a flat blank from the formed part. This gives the position of weld line in the flat blank for minimum weld line movement compared to the desired position. Simulation of deep drawing of a circular cup of punch diameter 48 mm was performed to verify the back projection method. Figure 8.23 shows the formed cup and required position of weld line (center of blank) in formed cup which is back propagated to flat blank.

Fig. 8.23 Nodes where weld line intended to be positioned after deformation in formed Part 1 [42]

Nodes as shown in Fig. 8.23 are projected back to a flat sheet so that the contour of the weld line for minimum weld line movement is obtained in the flat blank. Figure 8.24 shows the flat tailor welded blank with the weld line after back projection. If material properties, process parameters and intended position of weld line in a formed part are all known this method can give optimum weld position and blank shape design for minimum error in weld line position after forming.

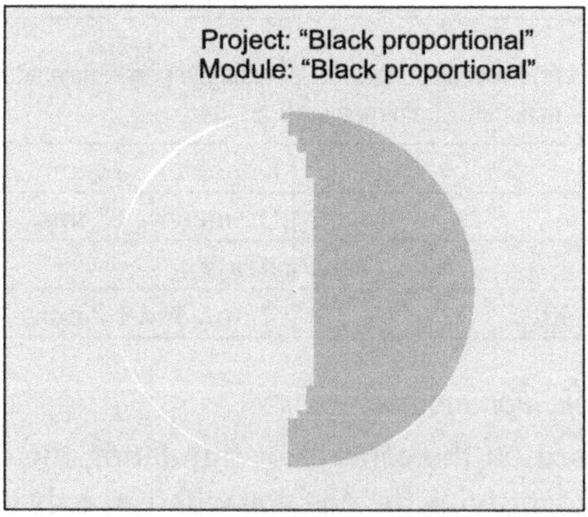

Fig. 8.24 Weld line position after back projection in flat blank of Part 1 [42]

8.9.4.1.3 Minimum thickness strain contour method

This method is based on concept that optimum position of weld line will correspond to the contour of minimum deformation, given by the minimum thickness strain. It is realized that the required position of weld line in the formed part and its position corresponding to the contour of minimum thickness strain may differ considerably.

In this method, the deformation of the part is simulated and the points of minimum thickness strain on the part are identified. These are projected back onto a flat blank to get the contour of weld line at minimum strain. The box shaped part in Fig. 8.25 shows one half of the formed part (to save on the computational time) with the nodes corresponding to the minimum thickness strain points marked on it.

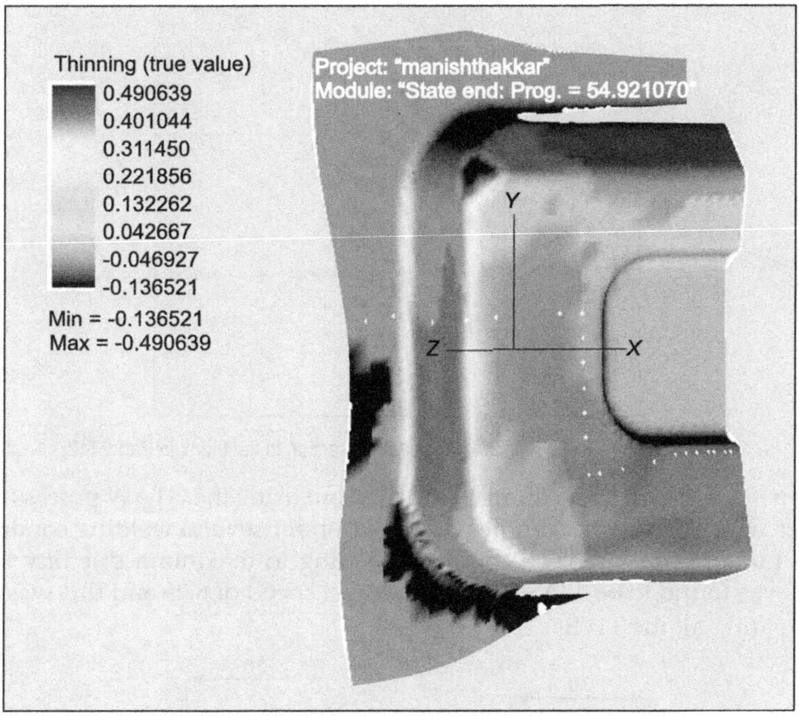

Fig. 8.25 Line of minimum thickness strain in a formed Part 2 [42]

The nodes are back projected onto a flat blank in order to find the position of proposed weldline on the flat blank. Figure 8.26 shows the position of the weldline in the flat blank.

This method is useful where one needs to reduce the chances of cracking in the weld that has comparatively poor ductility. In contrast, the back projection method is useful where aesthetics are important and the weld line needs to fall in the region designated by the designer. Since there is no consideration to the magnitude of deformation experienced by the weld line, the back projection method is suited to shallow drawn parts, those having very generous radii and where strain levels are generally very low, as in skin panels.

8.9.4.1.4 Prestraining of the thinner sheet

Initially the optimal welding parameters were worked out based on several experiments to weld longitudinally two thicknesses into a tensile sample, as

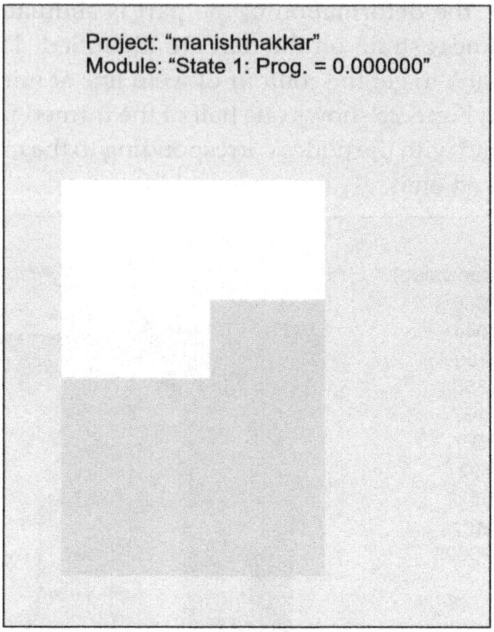

Fig. 8.26 Weld line position after back projection in flat blank in Part 2 [42]

shown in Fig. 8.27. The welding was carried out using the GTAW process using a filler wire. The tensile samples prepared under several welding conditions were pulled to failure. The condition leading to maximum ductility in the weld was found to be 47A current at a travel speed of 6.25 and this was used for welding all the TWBs.

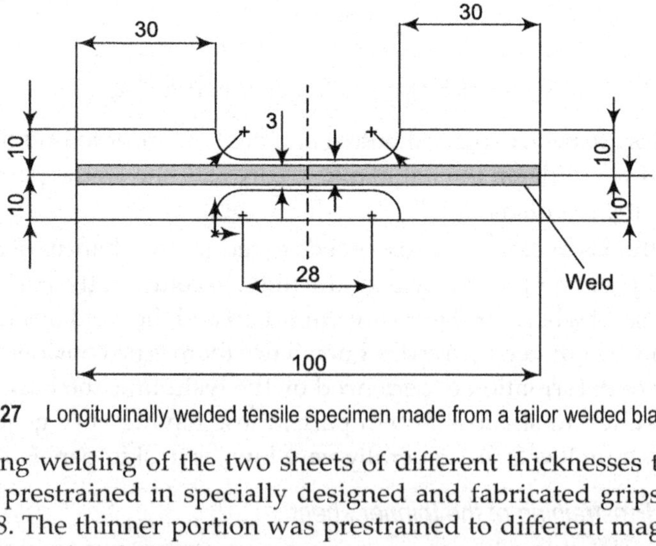

Fig. 8.27 Longitudinally welded tensile specimen made from a tailor welded blank [42]

Following welding of the two sheets of different thicknesses the thinner sheet was prestrained in specially designed and fabricated grips as shown in Fig. 8.28. The thinner portion was prestrained to different magnitudes of tensile strain and such TWBs were drawn using the tool shown in Fig. 8.29 into cylindrical cups (Fig. 8.30). Thereafter the weld line movement was measured in each case and a correlation obtained between the degree of prestrain and the weld line movement.

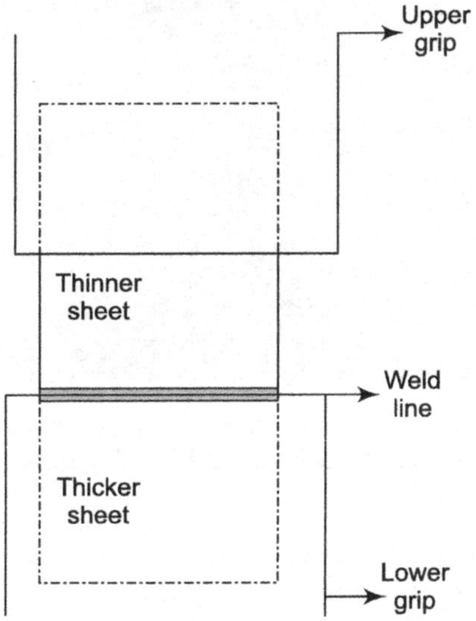

Fig. 8.28 Schematic showing the TWB sample gripped to prestrain the thinner portion of the sheet. Note that the thicker sheet is completely inside the grips (to avoid unintended tensile strain in the thicker sheet [42]

Fig. 8.29 Tooling used for tailor welded blanks. Shims of appropriate thickness were used to account for the difference in thickness [42]

The results are presented in the next section.

8.9.4.2 Results for back projection method

The back projected position of weld line as shown in Fig. 8.24, with a tailor welded blank of 1 and 1.2 mm thickness is simulated to investigate weld line movement

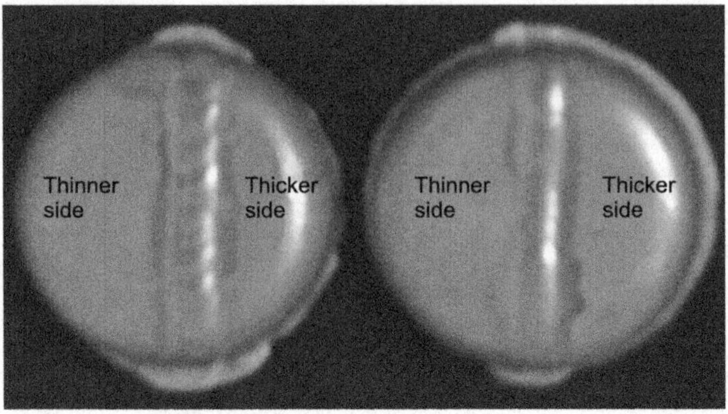

Fig. 8.30 Reduction in weld line movement in formed cups on welding along the contour determined by back projection method. The desired location of the weld line is at the center [42]

The formed cup with weld line obtained from back projection method is compared with formed cup with weld line at the center. The flat blanks are shown in Fig. 8.31 and formed cups are shown in Fig. 8.32. It can been seen from Fig. 8.32 that the deviation of the weld line from its desired position in the formed part is greatly reduced by applying back projection method.

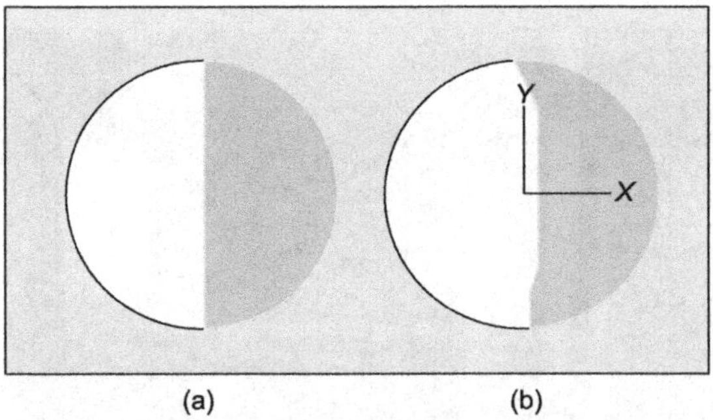

Fig. 8.31 Weld line position in flat blank.(Part 1) (a) weld line at center (b) Weld line position by back projection method [42]

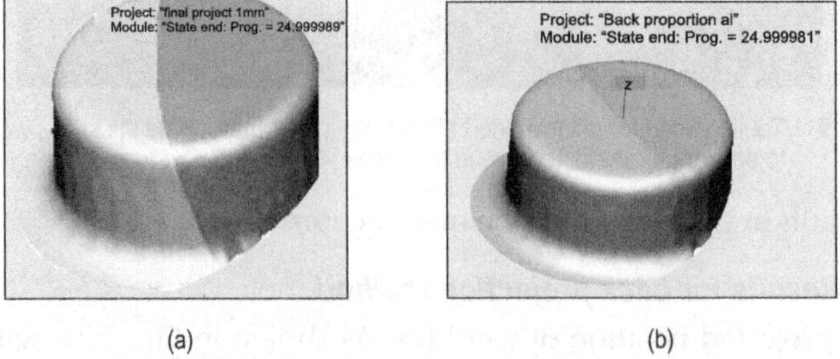

Fig. 8.32 Weld line movement in formed Part (a) Center line weld (b) Back projection method [42]

Figure 8.33 shows deviation of the weld line position and it can be seen that the maximum weld line movement in weld line at the center is around 4 mm while that in a cup formed from the TWB designed using the back projection method is 1 mm.

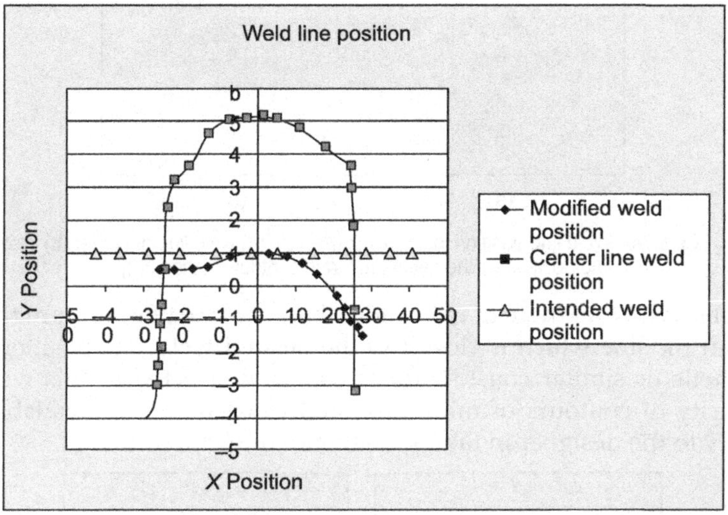

Fig. 8.33 Comparison of weld line movement [42]

As shown in Fig. 8.31 (b) in back projection method, weld line is at the offset towards the thinner blank side, this shift will accommodate extra drawing of thinner portion of the tailor welded blank. Therefore, this method gives designer a freedom to reduce weld line movement by simulating and changing the blank shape.

The weld was modeled as a line, i.e., the mechanical properties of weld bead are not considered assuming that it does not lead to any significant loss of accuracy in the result. This would be more appropriate when laser welding is used as the method of joining the two sheets. In contrast, friction stir welding or a method involving deposition of metal to form a weld bead will have to be modeled as a zone, accounting for the mechanical properties of the welded bead.

8.9.4.3 Results for the minimum strain contour method

In order to examine the effectiveness of this method the weld line position after forming a TWB designed using this method is compared with a TWB whose weld line is at the center. It is emphasized here that this method determines the position of weld line for minimum strain in contrast to the back projection method where the intended weld line position is determined by the designer. The weld position in a flat blank for two different cases is shown in Fig. 8.34. Figure 8.35 shows the weld line movement in the formed part for the two cases mentioned above. It can be seen that this method successfully determines a weld line position in a flat tailored blank so as to orient the weld to ensure minimum strain on it.

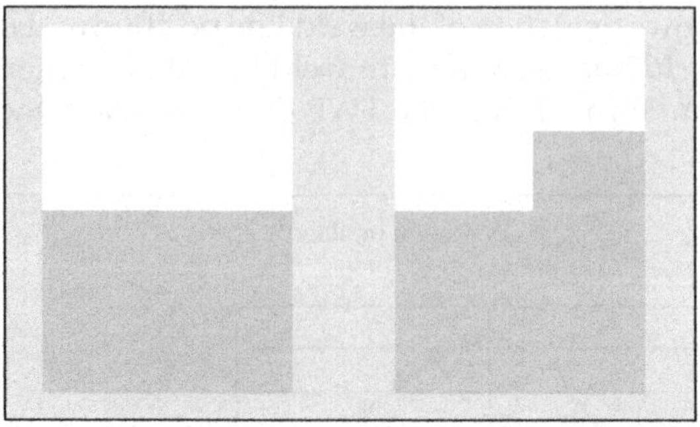

Fig. 8.34 Weld line position in flat blank (Part 2). (a) weld line at center (b) Weld line position by the minimum strain contour method [42]

It is observed that there may be more than one line of minimum thickness strain but the line which is closest to the desired weld line position (based on aesthetic or similar consideration) has been selected in this work. The multiplicity of contours of minimum strain that might be available offers flexibility to the designer in taking aesthetics into account.

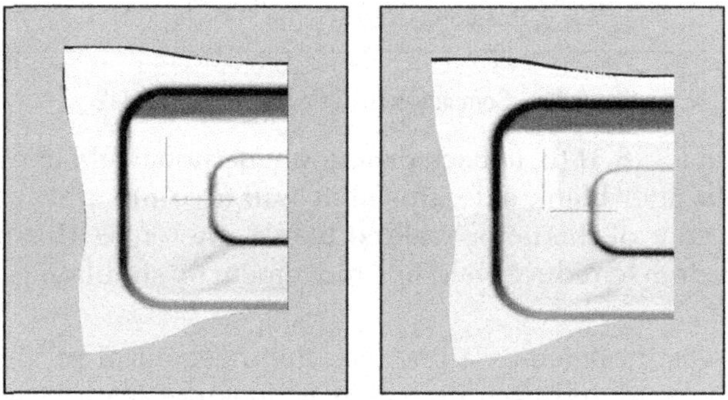

Fig. 8.35 Weld line position in formed blank (Part 2). (a) weld line at center (b) Weld line position by tracing the line of minimum strain method [42]

Figure 8.36 compares the weld line position after deformation with the contour of minimum strain in the part. It is seen that the deviation in the final weld line position is greatly reduced on using this method. Figure 8.36 shows that the maximum weld line movement is reduced from 11 mm to 7 mm.

8.9.4.4 Results of prestraining the thinner sheet

A parameter '*f*' ($=K_B t_B/K_A t_A$) was defined, with subscript '*A*' referring to the thicker/stronger portion of the sheet, and '*B*' the thinner/weaker one. The corresponding ratio of strains (strain in the thicker sheet to strain in the thinner one) is defined as 'α' ($= \varepsilon_A/\varepsilon_B$).

Calculations were performed for seven '*f*' values namely, $f = 1$, 1/1.01, 1/1.05, 1/1.15, 1/1.2, 1/1.25 and 1/1.3 each for three values of n_A, namely, $n_A = 0.15$, 0.20 and 0,25.

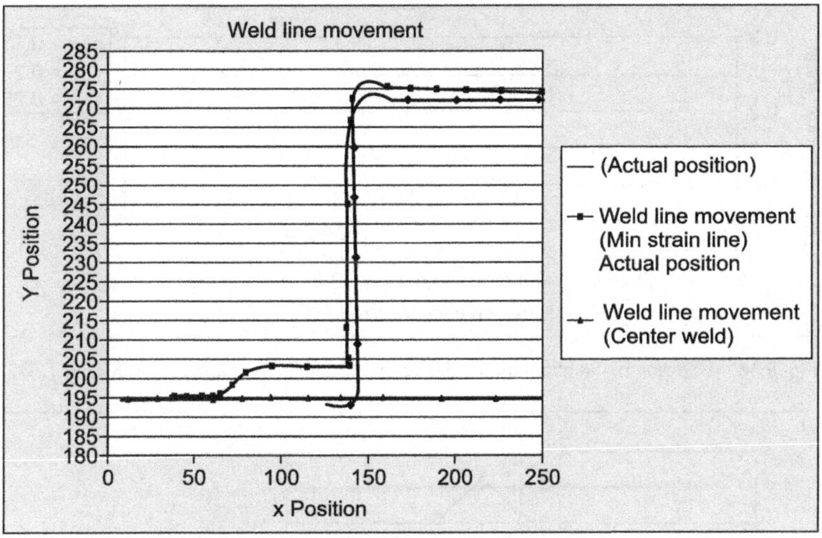

Fig. 8.36 Comparison of weld line position in a rectangular box using minimum strain contour method [42]

Based on the calculations above, the relation between the strain ratio α and f for different values of n_A is given in Fig. 8.37. The strain in the thicker sheet for a TWB wherein the strength differential across the weld is minimized by prestraining the thinner sheet to a maximum of $\varepsilon_B = n_B$ is shown in Fig. 8.38. As expected, a higher value of n_B leads to a higher value of α.

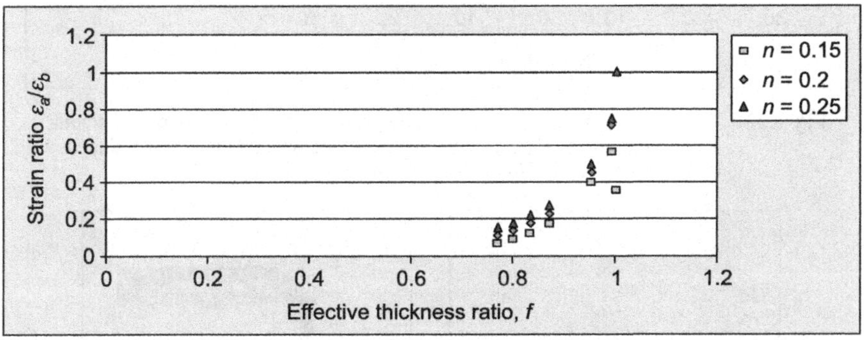

Fig. 8.37 Variation of the strain ratio with increasing effective thickness ratio K_At_A/K_Bt_B [42]

8.9.4.5 A comparison of proposed methods

The three simulation based methods proposed are compared to identify the scope and application of these methods in actual stamping industry. For comparing these methods back projection method is also applied to Part 2 (box shaped part) which is symmetric but not axi-symmetric. It is assumed that intended position of weld is at the centre of part. The intended position of the weld line and that obtained from the contour of minimum strain in a flat blank is shown in Fig. 8.34. Modified weld position obtained from back projection method is shown in Fig. 8.40(a).

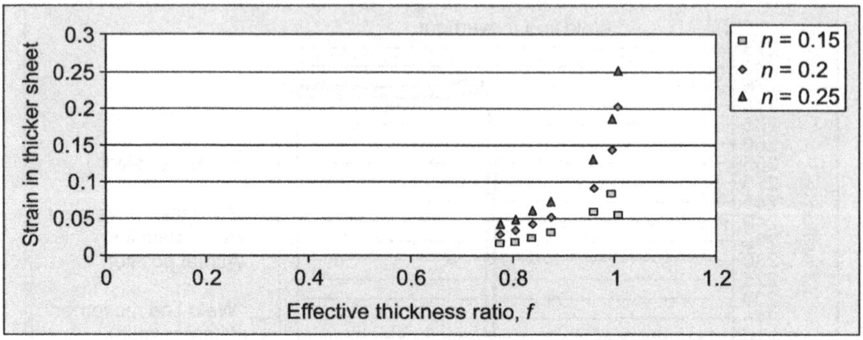

Fig. 8.38 Variation of the strain ratio with increasing effective thickness ratio K_Bt_B/K_At_A [42]

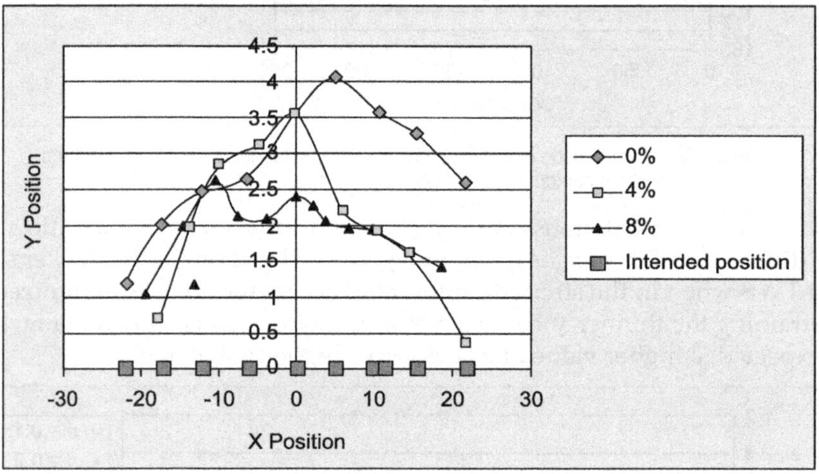

Fig. 8.39 Weld line movement after different degrees of prestrain in the thinner sheet. The maximum weld line displacement from the desired position is reduced from 4 mm to 2.5 mm for an 8% prestrain [42]

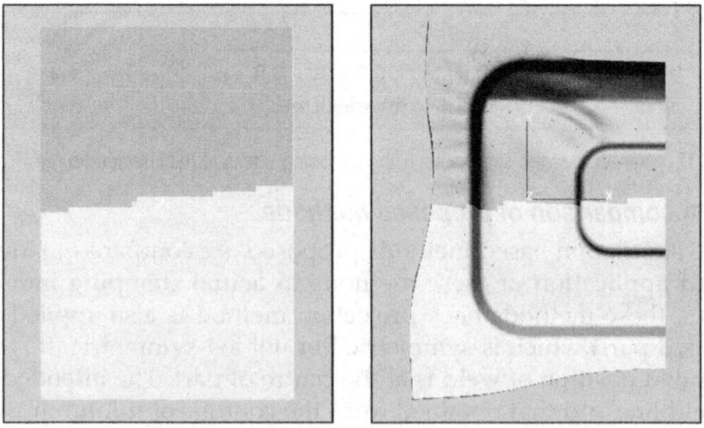

(a) (b)

Fig. 8.40 Weld position (a) flat blank (b) formed Part 2 by back projection method for Part 2 [42]

The final weld line position in formed part 2 is shown in Fig. 8.40(b). Figure 8.41 summarizes the deviation of the weld line from the intended center line position using the back projection method and its displacement from the intended location when welding is performed along the minimum strain contour.

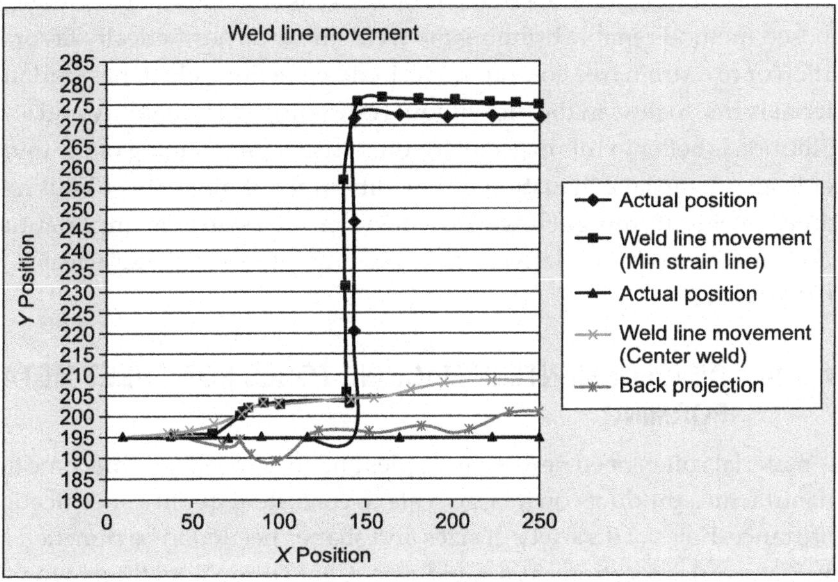

Fig. 8.41 Comparison of weld line movement/ displacement from the desired location [42]

A technological comparison of the three methods is given in Table 8.9. The domain of application of each of these methods has been identified in light of the technological advantages and limitations of each.

Table 8.9 Comparison of the proposed methods

Method	Prestrain in the thinner sheet	Back projection	Contour of minimum strain
Position of weld line	Depends on the magnitude of prestrain and the strain hardening rate of the thinner sheet	Follow designer's intent (aesthetics governs the position)	Technological advantage determines position of the weld line
Reduction in weld line movement weld failure	Small but depends on the relative work hardening rates Weld may fail if not stronger than the two sheets	More for same no. of iterations Weld might get severely stressed. Weld failure a distinct possibility.	Less as compared to the Back projection method Less possibility of weld failure because weld line follows minimum strain line
Application	Lightly curved panels where dent resistance is an issue.	Low forming severity and aesthetics overwhelmingly important	High forming severity in the formed product

It is seen that by using simulation based methods the weld line movement / displacement from its desired location can be greatly reduced. Iterations using these methods will further reduce the deviation of the weld line from its intended position. The existing methods to reduce weld line movement are based on restraining the flow of the thinner/weaker material. In contrast, the proposed methods enable bringing the weld line to an aesthetically favorable position or to a strain free contour as desired, using finite element simulation. Material is free to flow in the die without restraining so formability and strain distribution is better in this method. In comparison, prestraining of the thinner sheet leads to partial exhaustion of ductility in the thinner sheet limiting its ductility during forming. However, when dent resistance is important and the forming severity is relatively low, prestraining the thinner sheet is the best alternative.

8.10 DEVELOPMENTS IN MACHINE TOOLS FOR SHEET METAL FORMING

New materials often need new technologies and these need new machine tools to manufacture products on a large scale to consistent quality specifications. For instance, holes of a variety of sizes and shapes needed to be punched into a sheet at precise locations at a rapid rate. Conventional methods would at least require substantial tool changing time. Development of turret punching machines for example enabled substantial reduction in the punching times. Wherever a specific array of holes of specified dimensions was required, a process involving an array of punches and dies fitted onto punch and die plates, respectively, made the required number of holes simultaneously leading to savings in time. A similar objective could be achieved by placing these tool elements around the periphery of two rolls.

Machine tools dedicated to hydroforming have been developed to enable different configurations of the process (hydraulic assisted deep drawing, hydro-mechanical forming, sheet and tube bulging, etc.) to be performed on a single machine tool.

Several modifications have been made to conventional machine tools to improve upon their precision, accuracy and adapt them to the numerous technologies being developed.

(a) Incorporation of numerical control (CNC) has enabled measurement and control of forming variables, automation, lower dimensional variability and hence better product quality.

(b) Segmented blank holders/segmented dies have enabled the use of multiple point blank holding and the use of tailor welded blanks, which, by virtue of different thicknesses/properties often require different blank holding forces in different regions [37, 38].

Machine tools have been designed to perform hot forming operations leading to either superplastically formed or press hardened products to be repeatedly manufactured. Press brakes fitted with feedback control of the bend angle so as to control springback and hence deviation of the angle of bend to within close tolerances enable great enhancement in the quality of the formed parts. This is particularly so for the brackets formed from sheets. Errors in (angles of) bending cause tolerance cascading and lead to difficulties in assembly of parts.

Indeed, if the benefits of UFG materials were to be realized, presses incorporating furnaces to heat the sheet metal rapidly to the desired temperature would be required to cut down on the cycle time. Such materials will have to be formed superplastically at a higher strain rate compared to that in conventional superplasticity

In order to achieve the benefits of precision at large volumes, process control and hence measurement of all control variables must be incorporated into the equipment. For instance, measurement of punch force and the blank holding force variation with time, deflection of the press, sensors for monitoring die wear, and flow of material into the dies requires extensive instrumentation to enable controlling these variables to within acceptable ranges so as to obtain product properties within tolerance limits. Troubleshooting and optimization of the process is enabled when the data on the different events in a process are available.

In view of increasing 'stroking rate' (no. of strokes per minute to enhance productivity), two deleterious consequences emerge:

(a) increase in heat generated due to friction at the sliding surfaces
(b) increase in punch velocity at the time of contact

Temperature rise is inevitable and the issue is addressed incorporating better cooling of the sliding surfaces and by providing thermal compensation on the guideways so that loss of parallelism through thermal expansion, wear and tear can be minimized or overcome. This is crucial for maintaining precision in components made on presses operating at high stroking rates of the order of 700–1000 strokes a minute.

Short stroke length finds favor with the high stroking rates mentioned. The increase in speed of the punch (as a consequence) makes the punch 'hit' the sheet at the time of contact. This causes a sudden acceleration of the material leading to a rapid increase in the tensile stress. Many a failure in components drawn at high speeds have been rectified by reducing the speed. However, one may continue at high speed using drives that enable approach of the punch over larger crank angles in a mechanical press, and use a mechanism like the 'drag link' that reduces the speed at which the punch would contact the sheet. The speed increases in the idle part of the stroke.

Machine tools for continuous, uninterrupted production runs help eliminate the down time needed for changing the coil of sheet. This is achieved by in-process joining of the ends of the two coils, i.e., the leading edge of the subsequent coil to the trailing edge of the pervious coil. A coil based manufacturing process would therefore be expected to be more efficient and can be made more consistent than the cut-to length form of raw material, which requires elaborate equipment to feed the sheet and safety systems to prevent double feeding (feeding of two sheets at a time).

Servomachine tools (like servo presses and press brakes) involving feedback control enable real time measurement of control variables for an 'on-line' (i.e., in-process) corrective action. These are amenable to 'intelligent' machine tool control since the control parameters can be measured and changed in real time. For instance, a servo press brake could sense the force required to bend it (which would be influenced by the hardness, i.e., yield strength and the sheet thickness), determine the stroke length, bottoming force, etc., and take corrective action to minimize the error in the bend angle.

In-process handling of partially formed sheet metal components, if performed manually, is slow, inconsistent and often deteriorates quality due to in-process storage, and non-repeatable positioning, non-repeatable deformation in handling, etc. This is addressed by incorporating a transfer mechanism into the press or the die so as to enable consistent handling and consistent part quality. This requires the various rates of the different activities being performed, like feeding of the sheet, forming, and lifting the component from the previous stage and positioning it onto the tool in the next stage and probably, in-process inspection, to be synchronized.

Machine tools which permit precise synchronization of the various elements in the press drives which are often programmable lend themselves to high volume production as well as flexibility. Such machine tools combined with paradigms like SMED (single minute exchange of dies) could enable in future a substantial reduction in the press working cost since the setup time for a transfer press/die would be reduced substantially.

An ever increasing complexity and geometric diversity in drawn sheet metal parts could be expanded with the help of machine tools enabling variable blank holding, machine tools which permit variation in the holding force in space as well as in time.

From the foregoing it is amply clear that the machine tools have a significant role in bringing in high precision at a high rate of production, where any kind of re-working would be impractical.

8.11 LUBRICATION, TOOL WEAR, AND TOOL LIFE

Friction and lubrication in sheet metal forming remains an important issue, particularly while drawing coated or pre-painted sheets. Tool marks (usually oriented circumferentially, i.e., perpendicular to the direction of flow of material) or in general, high frictional resistance causes the coating to get 'picked up' onto the tool. This induces large shear strains as the particles from the surface of the coating material gets transferred to the tool. These hardened particles further scrape out material from the next blank being deformed and propagate the damage. Such a process spoils the aesthetic appearance of the pre-painted sheet in addition to the technological damage that it causes.

In order to avoid this, it is important to wipe the tool surface after drawing each component to clean out the debris from the previous blank.

In case of an uncoated sheet, cold pressure welding occurs due to friction between the sheet and the tool surface. Material from the surface of the sheet gets picked up onto the tool surface similar to that of a built up edge in machining. Another source of wear on the tools is the dust particles (chemically SiO_2) that get into the system with the sheet, being mixed up with the lubricant and similar means. Such particles get pressed between the sheet and the tool surfaces leading to scoring marks (scratches) during forming. The sheet must therefore be washed free from dust so as to prolong die life.

Use of lubricants is effective in reducing the friction and enhancing the die life. These prevent cold pressure welding and prevent material from the sheet being picked up onto the tool. The coefficient of friction between the tool and the sheet may be reduced by a wear resistant coating like that of TiN on the tool surface. New tool materials for drawing dies like polyurethane cast on the tool steel surface are instrumental in reducing the friction between the sheet and the tool. The backing of the steel surface imparts the necessary stiffness. These tool angles can be more easily shaped as they are more easily machinable and do not require hardening.

Lubrication with extreme pressure additives is a must for a successful operation. In addition to preventing tool wear and protecting the sheet surface from getting scratched during the forming operation, lubricants reduce loads required for forming, reduces tool and machine tool deflection thereby enhancing the overall quality of the product.

8.12 NON-CONVENTIONAL SHEET FORMING PROCESSES

8.12.1 Electromagnetic Forming/Joining [46–48]

The present section focuses on the numerical simulation of the electromagnetic welding process. A weld is formed in a fraction of a second at low temperatures, i.e., without melting. The time and temperatures involved in electromagnetic joining relative to resistance welding and fusion welding processes is shown in Fig. 8.42. Explosive forming/joining is another similar process requiring explosives to drive the deformation, generating more heat than electromagnetic forming process. Tables 8.10 and 8.11 bring out the similarities and differences between the two processes.

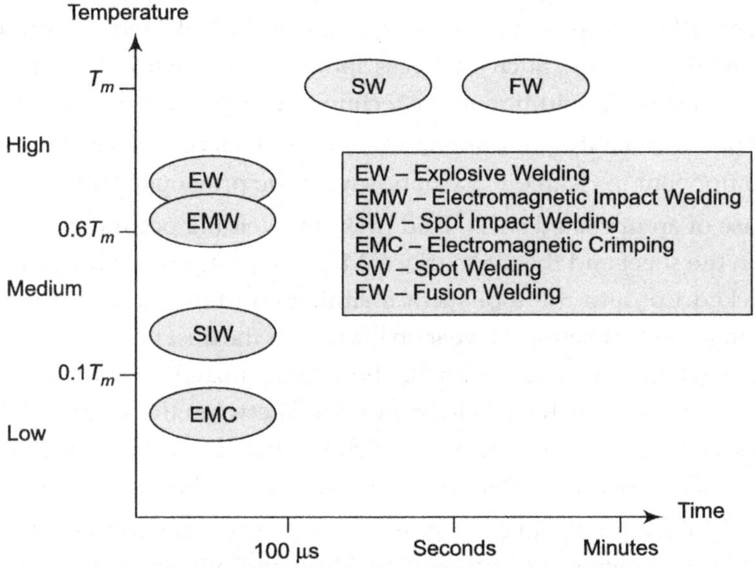

Fig. 8.42 Temperature and time required to perform solid state welds relative to those requiring fusion at the interface [47]

Table 8.10 Similarities between explosive welding and electromagnetic welding processes [47]

1.	Both of them are in principle impact welding processes
2.	Load are applied for small period of time (~ 50 μs)
3.	Due to the shock-like loading, the welding conditions depends on the velocity of impact and the angle of impact
4.	Same material models may be used to simulate both the processes

In the electromagnetic forming process, the current in the coil changes rapidly inducing a current in the workpiece (sheet or a tube) in a direction so as to oppose the electric field created by the current in the coil. The two opposing fields create a Lorentz force that repels the sheet (tube) away from the coil and acts as the driving force to cause deformation. In the present

Table 8.11 Differences between electromagnetic and explosive welding techniques [47]

	Explosive welding	Electromagnetic joining
1.	The load is progressive in nature, i.e., it is travelling along the sheet	The load has a nearly sinusoidal distribution over the sheet
2.	The impact angle depends entirely on the stand-off and is constant for all contacting points on the sheet	The impact angle depends both on the stand-off and the distance from the point of first contacts, and hence varies in nature.
3.	Energy spread over a large area	Energy is localized, just below the coil

context two coils were used to drive two sheets towards each other so as to collide and form the joint. The use of two coils meant a high relative velocity at the time of collision.

To achieve this, capacitors are initially charged to a high voltage (of the order of 20,000 volts to an order of magnitude higher potential difference) using a standard power supply. The capacitor stores the charge and hence the electrical energy (= 0.5* C^*V^2, where C is the capacitance and V the voltage). This energy is discharged in a short duration pulse of high current, so that the induced current will depend upon the rate of rise in current. The shape of the current pulse and the time duration of its existence are shown in Fig. 8.43(a) while the relationship between the energy of the capacitor bank and the magnitude of the discharge current is shown in Fig. 8.43(b). The Lorentz force being proportional to the induced current, which in turn is proportional to the discharge energy of the capacitor, the velocity of forming achieved increases with the discharge energy and the discharge current [Figs 8.44(a) and (b), respectively]. The initial (stand-off) distance between the two sheets determines the relative velocity at the time of collision.

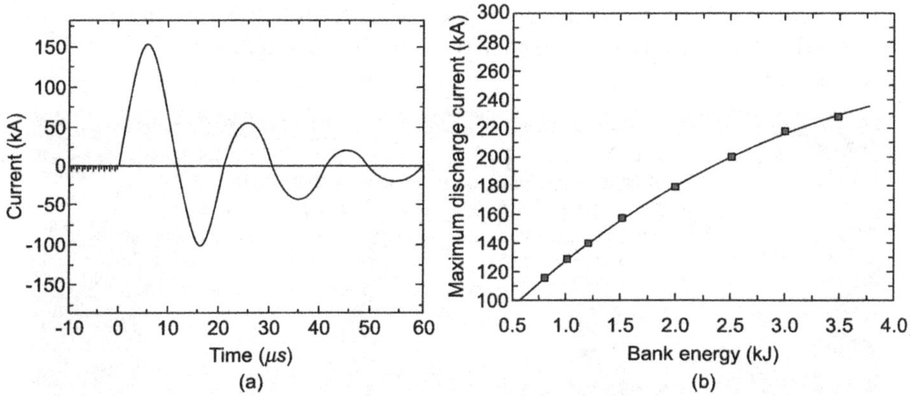

Fig. 8.43 (a) The shape of the current pulse and (b) the non-linear dependence of bank energy on the peak discharge current [46–48]

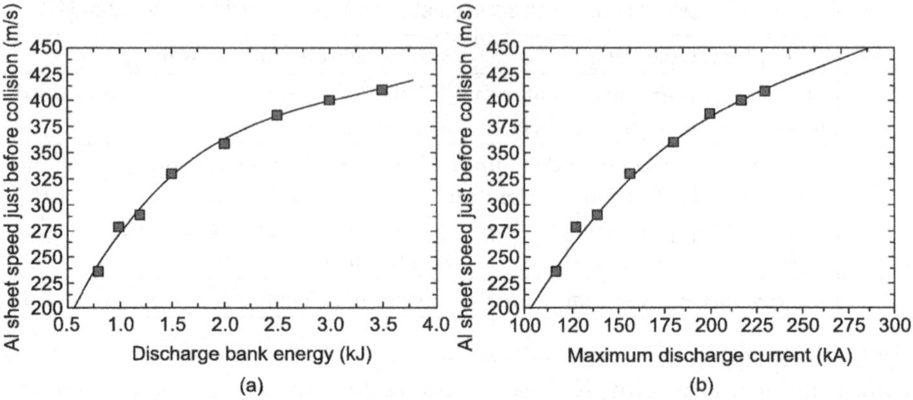

Fig. 8.44 Speed of the Al sheet just before collision: (a) Effect of bank energy discharged and (b) effect of maximum discharge current [46–48]

The sheets of aluminum welded under different conditions of stand-off and peak current are seen in Fig. 8.45. It is important to minimize the inductance of the overall circuit which is a sum total of the machine inductance which is generally very small and the inductance of the load, i.e., the workpiece–coil assembly, including the connecting cables. A simulation of two plates driven towards each other by the Lorentz force can simulate their collision, the resulting impact pressures and the stresses and strains developed as a consequence. Whether these effects are enough to form a solid state weld or not must be inferred by the user through a criterion established to correlate these results of collision with the actual outcome in the form of formation of the weld or otherwise. The issue here is that of establishing a numerical criterion of welding. A few criteria were identified, and by correlating the simulated weld interface with the experimental outcome (Fig. 8.45), an attempt was made to come up with a single criterion to predict the formation of the weld. Each quantity, as a criterion of weld formation was evaluated vis-à-vis the experimental results for some of the welds to detect errors in prediction, i.e., prediction of weld formation where there was actually none and vice versa [46, 47].

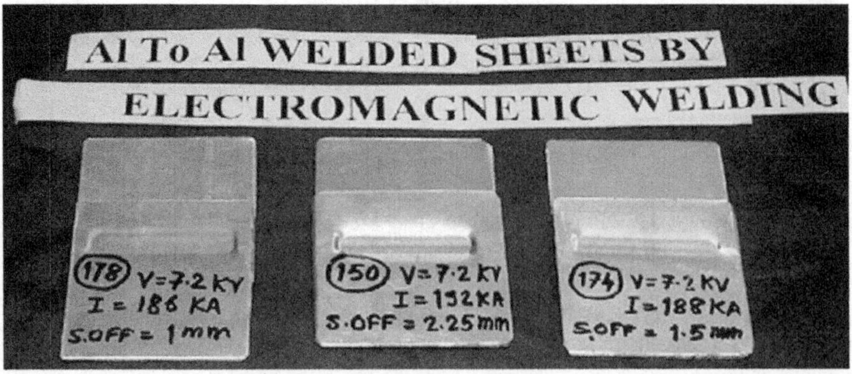

Fig. 8.45 Welded Al–Al sheets welded under different conditions [46–48]

To this end, simulations of the collision between the plates were performed using ABAQUS to obtain the quantities characterizing consequences of the collision. The configuration of the two plates together with the discretisation is shown in Fig. 8.46. The necessary distribution of the Lorentz force was obtained from Ansys-EMAG using the experimental values of the control variables, namely the coil design, experimental configuration, voltage, current, capacitance, and inductance of the circuit. The material behavior is described by the Johnson–Cook equation as given below:

$$\sigma = \left(A + B\left(\bar{\varepsilon}\right)^{n}\right)\left(1 + C\ln\dot{\varepsilon}_{p}\right)\left(1 + T^{*m}\right) \qquad (8.16)$$

where A, B, C, n and m are constants, while T^{*} represents $(T - T_{\text{room}})/(T_{\text{melt}} - T_{\text{room}})$

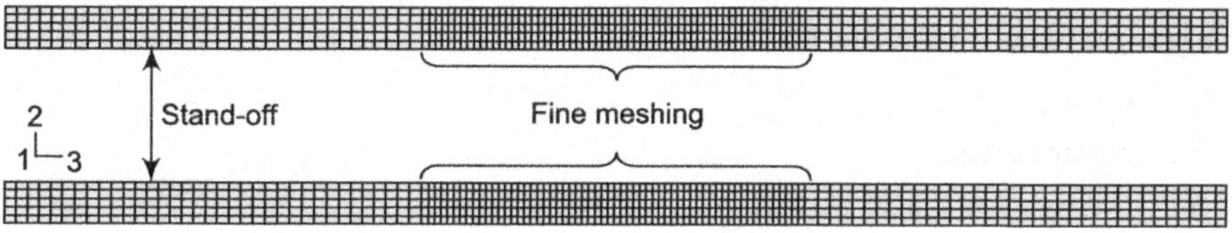

Fig. 8.46 Initial configuration of the sheets [46–48]

The following section discusses the outcome of the simulations. The details of the simulations may be found elsewhere [47].

The outcome of the simulation of the electromagnetically formed weld is shown in Fig. 8.47(a) and the corresponding joint experimentally obtained in Fig. 8.47(b). The different regions at the interface of the joint are shown in Fig. 8.48. Another view of the welded zone is compared with the experimental joint in Figs 8.49 (a) and (b). Both the figures emphasize the favorable relation between the prediction and the experimentally observed weld contour.

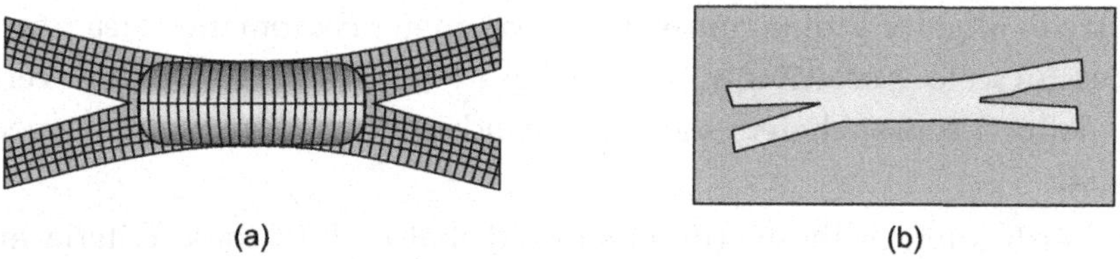

Fig. 8.47 Simulation (a) and the actual weld (b) between two aluminum sheets [46–48]

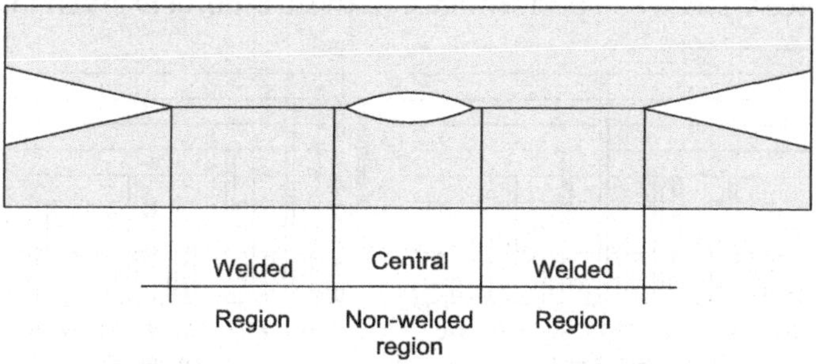

Fig. 8.48 A schematic close-up of the interface. The welded and non-welded regions are clearly demarcated. The weld is stronger than the parent material(s) [47]

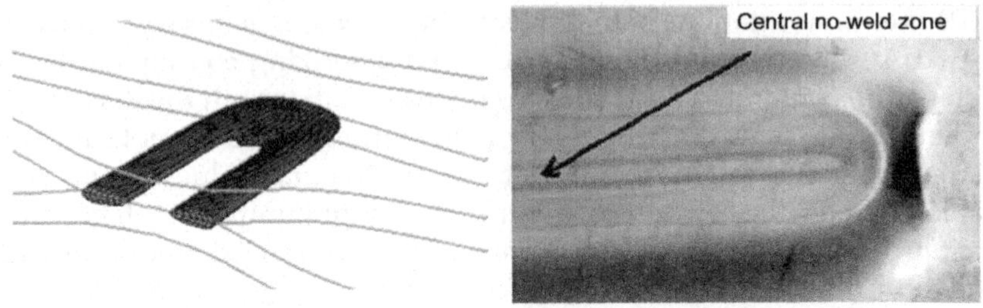

Fig. 8.49 Another view of simulation (a) and the actual weld (b) between two aluminum sheets [47]

Some of the criteria so as to interpret the consequences of the collision of the plates were chosen to be

(a) impact velocity
(b) impact angle
(c) plastic strain and strain distribution
(d) incremental plastic strain
(e) ratio of incremental plastic strain to total plastic strain
(f) specific pressure
(g) normalized normal stress

A frequency distribution of the different quantities listed above for welded and unwelded elements (determined by comparing the experimental result with that from the simulation) are shown in Figs 8.50 and 8.51. It was clear that no single quantity among those listed above could be used as a criterion for weld formation. Hence an attempt was made to correlate a pair of quantities and try to identify a zone consisting of parameter combinations leading to a successful weld. Accordingly, plots in Figs 8.52, 8.53 and 8.54 do show certain well defined zones wherein successful welds could be consistently expected [46-48].

A verification of these criteria showed that each of these criteria made incorrect predictions to a varying degree (predicting a weld where there was none and vice versa). The performance of these criteria is given in Table 8.12.

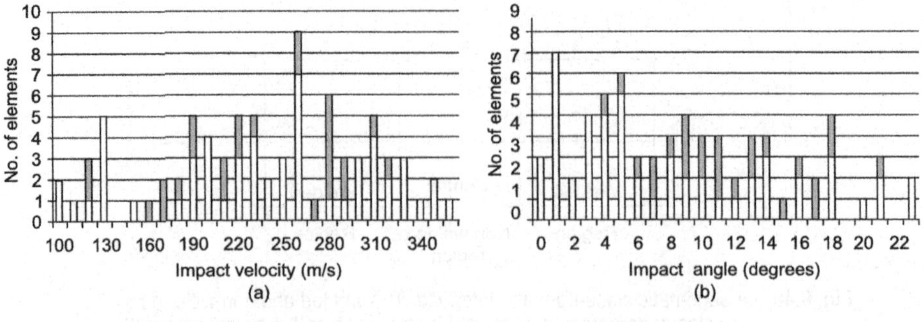

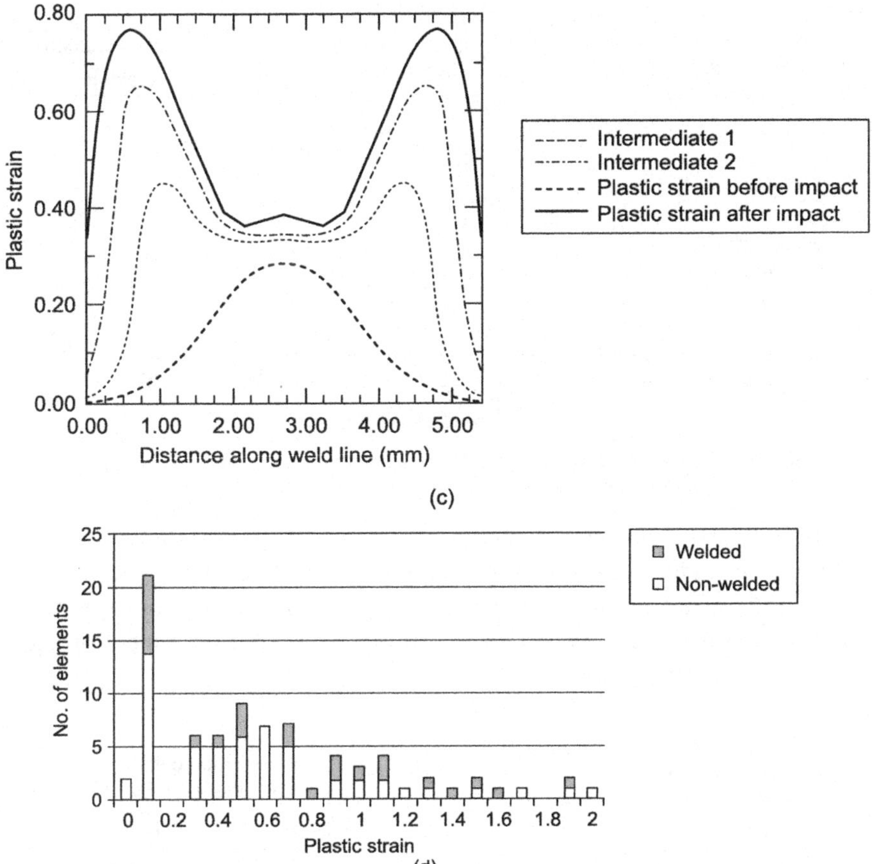

Fig. 8.50 Effects of (a) impact velocity, (b) angle (c) plastic strain distribution and (d) plastic strain on the frequency distribution of welded elements

It may be inferred that the first and the fourth criteria (based on impact velocity-impact angle relation and that based on the ratio of incremental strain to plastic strain, respectively) are reasonable. These are seen to be conservative in predicting slightly incorrect weld location or incorrect weld length, but a correct welded area.

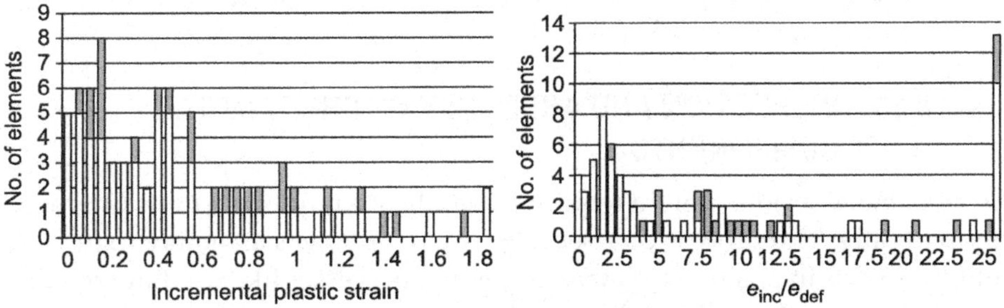

Fig. 8.51 Effects of (a) incremental plastic strain and (b) ratio of incremental strain to the total plastic strain on the frequency distribution of welded elements [47]

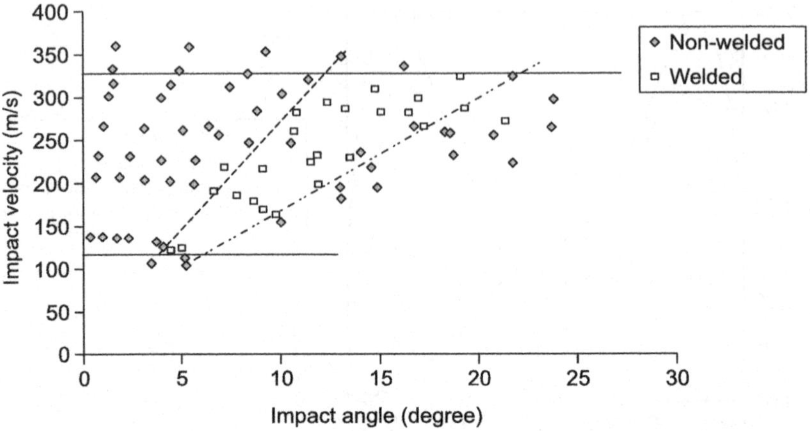

Fig. 8.52 A map of impact velocity with impact angle showing the operating zone promoting welding [47]

Moreover, the numbers of instances of a weld being predicted when there was none, and vice versa are almost equal in these two cases compared to the second and the third criteria. Hence these two relationships, from among those studied, are found to show promise as numerical criteria for electromagnetic welding of metallic sheets.

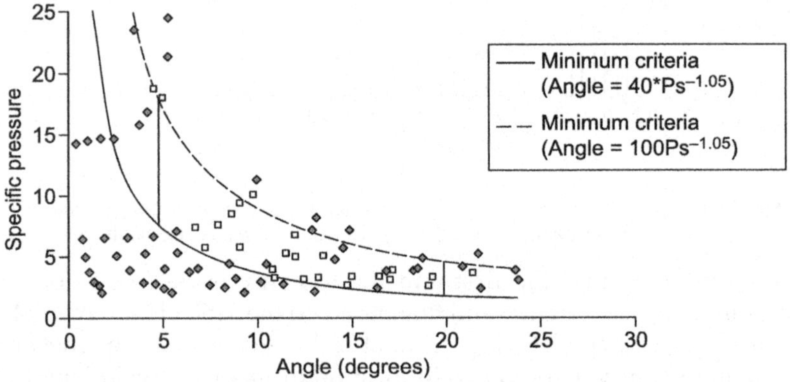

Fig. 8.53 A map of specific pressure with impact angle showing the operating zone promoting welding [47]

8.13 INSPECTION AND ASSEMBLY OF SHEET METAL COMPONENTS

A sheet metal product might require assembly of a number of parts produced individually. The process of assembly usually involves assembly fixtures to hold the parts in position followed by joining processes like welding, riveting or bolting together. The difference between the inspection and assembly processes for compliant sheet metal parts vis-à-vis those for the relatively rigid

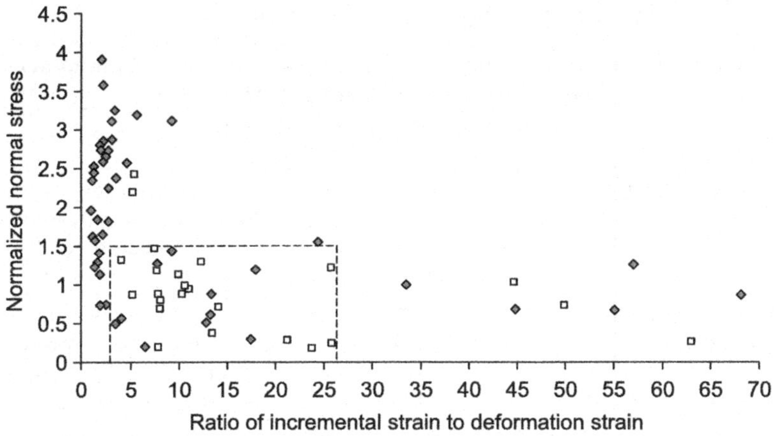

Fig. 8.54 A map of normalized normal stress with ratio of incremental strain to deformation strain showing the operating zone promoting welding [47]

machined parts have been brought out in Section 1.1 and Table 8.3. The present section brings out the considerations in sheet metal inspection and assembly. Tolerances are maintained at sub-assembly levels rather than individual component levels. Key product characteristics (dimensions) achievable by controlling the key control characteristics (dimensions) is the strategy used to ensure conformance of the assembly to design specifications.

Key product characteristics (KPCs) and Key Control Characteristics (KCCs) are identified at the product design stage. KPCs are the dimensions on the product which deserve to be held within tolerance. The key control characteristics are the dimensions, which when controlled to within a certain tolerance ensure that the key product characteristics are within designated limits of quality. Unlike a machined component wherein the dimension of a feature may be measured directly by go-nogo gauging, the sheet metal part must be placed in its 'natural position', i.e., the position in which it would get assembled and the constraints it would face due to joining processes (fastening, clamping, welding, etc.). This is necessary due to inherent compliance of the sheet metal part. Measurements (dimensional deviations from the desired values) are taken at pre-defined points to confirm conformance of the formed surface with the design. The number of points that form the quality characteristic determine the maximum number of 'control characteristics', i.e., locating and clamping points, punch curvature, measuring points, i.e., points at which measurements on deviations in position of tool elements to achieve the desired product characteristics may be performed. Actually, the number of measurement points (KPCs) inspected must at least be equal to the KCCs so as to get an idea about which of the KCCs is contributing to a given variation in a set of KPCs, so that corrective action can be taken at the appropriate point.

Table 8.12 Performance of different weld criteria [47]

Criterion	No of welded points shown as non-welded	No of non-welded points shown as welded	Comments
Impact velocity and Impact angle	6	8	The only criteria which has under-predicted the weld length with correct location of weld
Specific pressure and impact angle	2	13	Location of weld predicted is correct, but the weld length highly over-estimated
Ratio (e_{inc}/e_{def}) and input velocity	3	8	Predicts correct weld locations, but over predicts weld length
Ratio (e_{ine}/e_{def}) and normalized normal stress	6	9	Predicts displaced weld locations, but fairly correct weld lengths

8.14 CONCLUSIONS

From the various issues discussed, some of the important points raised in this chapter may be noted as listed below.

1. Sheet metal product goes through a very long processing route that is sensitive to quality of the raw material. There is a strong interaction between the processing parameters and the material properties.
2. Sheet metal forming differs greatly from machining. Hence issues in the sheet metal shops are very much unlike those arising in machining. The differences in the manufacturing practices are the outcome of these differences.
3. Grain refinement to an ultra-fine grain size enhances the properties of sheet metal. Fine grains prevent orange peel effect.
4. Springback and elastic after-effects are an inevitable part of sheet metal forming processes. Inconsistent springback makes it difficult to achieve designed dimensions to accurate levels. Unreleased springback causes residual stresses which are released during processing with every change in the mechanical constraint on the product. Springback, therefore influences the stability of shape of the product.
5. A model for predicting springback in plane strain bending of metals taking into account the work hardening behavior, the superimposed tensile stress and surface roughness is presented.The model shows good agreement with experimental results from the literature.

6. A new method of calculating the radius of curvature of the sheet consistent with the geometry of deformation at a given punch travel has been used in the analysis.
7. The model predicts that a high value of n, a low value of K and high superimposed tensile stress serve to decrease springback for a given bend geometry.
8. The effect of K in enhancing the effects of n and the tensile stress has been predicted to be significant. The difference in the springback of two materials having similar material properties was predicted to be significant if the value of K differed significantly.
9. The erratic springback performance of sheet metal in the shopfloor could be attributed to inconsistent values of K, the yield strength and the sheet thickness (despite other properties being similar) causing inconsistent springback in the material of the same grade.
10. New technologies to minimize the effect of springback, have been discussed.
11. Sheet metal characterization is a crucial activity. Accuracy of prediction of the numerical models strongly depends upon these.
12. Formability of sheet metal is strongly history dependent (strain path dependent)
13. Strain distribution based metrics of formability have been discussed in light of the established metrics like the FLDs.
14. The quantities characterizing the spatial strain distribution in a drawn component in the critical section have been defined. The variation of these during a drawing operation is well explained and conforms to established facts.
15. The effect of the major material and process variables, namely, R and BHF as predicted by the strain distribution based parameters like the CF and SNI are in line with expectations.
16. The CF and SNI serve as useful quantitative formability criteria for material comparison, process control and more.
17. They satisfy the requirements of formability criteria. They may be used in place of an FLD for all practical purposes.
18. New technologies have been developed to reduce the material input, enhance precision, accuracy, and enhance speed of production.
19. Weld lines move towards the thicker (stronger) regions in a tailor welded blank.
20. Three methods have been proposed to determine the weld contour in tailor welded blanks, which aim at decreasing the material input. The methods aim at either minimizing weld line movement by design or

pre-designing the position of the weld line after its movement, so that after movement, it occupies the position intended by the designer.

21. While the back projection method permits weld line movement to occur (into the desired location), the minimum strain contour permits little weld line displacement with respect to this contour.
22. Each of the three methods has a specific domain of application.
    (i) Contour of minimum strain method leads to the position of weld line in a stress free region in the blank.
    (ii) Back projection method helps the designer to determine the weld line contour on the flat blank when the final position of weld line is fixed, as in aesthetically important parts.
    (iii) Prestraining of the thinner sheet reduces by work hardening, the strength differential between the thicker and the thinner sheet across the weld line. A greater value of n for the thinner sheet will lead to increased strain in the thicker sheet and hence promote greater ductility of the TWB.
24. High energy rate electromagnetic forming/joining methods show a lower springback generate less heat and are suited to joining of heat sensitive materials like the UFG materials.
25. A few criteria for weld formation in electromagnetic forming have been studied.

    A couple of relations based on well defined, identifiable weld zones showed promise as criteria for weld formation. No single quantity amongst those studied could qualify as a criterion for the occurrence of a weld. Numerical predictions of the weld zone, both size and geometry compared favorably with observations from the experiments.

REFERENCES

1. Milind, T.R. and Date, P.P. (submitted). Analytical and Finite Element modeling of Strain generated in Equal Channel Angular Extrusion, *Int. J. Mech. Sci.*
2. Padhye, R.S. Milind, T.R. and Date, P.P. (2010). Texture and grain refinement in equal channel angular pressing using round corner dies, *International Conference on Advances in Materials and Processing Technologies (AMPT2010)*, Paris, France, October 24–27.
3. Date, P.P. and Rosochowski, A. (Unpublished work).
4. Vij, S. and Date, P.P. (2004), Studies of Variation of Springback with curved line of bend and variable blank holder forces, in Ghosh, S., Castro, J.M., and Lee, J.K. (Eds.), *NUMIFORM 2004,* Ohio State University, Columbus, Ohio, June 13–17, 2004.
5. Inamdar, M.V., Date, P.P., Desai, U.B. (2000). Studies on the prediction of springback in air vee bending of metallic sheets using an artificial neural network, *J. Mater. Processing Technol.* **108**; 45–54.

6. Inamdar, M.V. (1999). Springback Prediction in Air Vee Bending Process using ANN, *M. Tech. Dissertation*, Department of Mechanical Engineering, IIT Bombay, January 1999.

7. Queener, A. and De Angelis, R.J. (1968). Elastic Springback and residual stresses in sheet metal parts formed by bending, *Trans. ASME*, **61;** 757–768.

8. Wang, C., Kinzel, G. and Altan, T. (1993). Mathematical modelling of plane strain bending of sheet and plate, *J. Mat. Processing Technol.*, **39**; 279–304.

9. Nagpal, V., Subramanian, T.L. and Altan, T. (1979). Mathematical modeling of sheet metal formability indices and sheet forming processes, *Final Report*, AFMI-TR-79-4168, Airforce Materials Laboratory, Ohio.

10. Leu, D.K. (1997). A simplified approach for evaluating bendability and spring-back in plastic bending of anisotropic sheet metals, *J. Mater. Processing Technol.*, **66**; 9.

11. Huang, Y-M and D-K Leu (1998). Effects of process variables on vee die bending process of steel sheets, *Int. J. Mech. Sci.*, **27**; 631–649.

12. Stelson, A.K. and Gossard, D.C. (1992). An adaptive press brake control using an elastic–plastic material model, *Trans. ASME, J. Engg. for Industry*, **104**; 389–393.

13. Varma, S. (2004). "Limits strains prediction in deep drawn sheet metal components using ANN, *M. Tech. Dissertation*, Department of Mechancial Engineering, IIT Bombay.

14. Date, P.P. Intelligent materials selection for quality improvement in deep drawn products, *Final Report*, Department of Science and Technology, India.

15. Kumar, S., Sontakke, P.V. and P.P. Date, Unpublished work.

16. Doege, E. and Sommer, N. (1987). Blank holder pressure and blank holder layout in deep drawing of thin sheet metal; Advanced Technology of Plascticity, in *Proceedings of the Second International Conference on Technology of Plasticity*, pp. 1305–1314.

17. Chary, R.L. and P.P. Date, Development of a variable pressure blank holder, in *All India Manufacturing Technologies Design and Research Conference Proceedings.*

18. Hill, R. (1952). On discontinuous plastic states with special reference to localized necking in thin sheets, *J. Mechanics and Physics of Solids*, **1**; 19–30.

19. Swift, H.W. (1952). "Plastic instability under plane stress", *J. Mechanics and Physics of Solids*, **1**; 1–18.

20. Chan, K.S., Koss, D.A. and Ghosh, A.K. (1984). Localised necking of sheets at negative minor strains, *Met. Trans. A*, **15A**; 323–329.

21. Marciniak, Z., Kuczynski, K. and Pokora, T. (1973). Influence of plastic properties of a material on the forming limit diagram for sheet metal in tension, *Int. J. Mech. Sci*, **15**; 789–800.

22. Date, P.P., Swaminathan, K. and Padmanabhan, K.A. (1988). Room temperature forming limit diagram and tensile behaviour upto 200°C. of an Al–Ca–Zn superplastic alloy, *J. Materials Science*, **23**; 1351–1359.

23. Date P.P. and Padmanabhan, K.A. (1992). On the Prediction of the Forming Limit Diagams (FLDs) of sheet metals", *Int. J. Mechanical Sciences*, **34**; 363–374.

24. Ukhande, M. and Date, P.P. (2007). A strain distribution based failure criterion in drawing of sheet metal, *Proc. IDDRG 2007 Conference*, Audi plant, Gyor, Hungary.
25. Ukhande, M. (2006). Effect of geometric scaling on strain distribution in deep drawn sheet metal parts, *M. Tech. Dissertation*, IIT Bombay, Mumbai.
26. Schedin, E. and Melander, A. (1987). On the strain distribution during the stretch-forming of low-and high-strength sheet steels, *Journal of Mechanical Working Technology*, **15**; 181–202.
27. Date, P.P. (1996). Constraint Factor: Quantitative measure of strain distribution in sheet metal formed component, *Advances in Mechanical Engineering*, 651–656.
28. Desai, S.G. and Date, P.P. (2006). On the Quantification of Strain Distribution in Drawn Sheet Metal Products, *J. Mater. Processing Technol.*, **177**; 439–443.
29. Desai, S.G. (2008). An alternative method of quantitative characterization of non-uniformity in the strain distribution of drawn sheet metal parts, *Ph.D. Thesis*, IIT Bombay.
30. Ravi Kumar, D. (2002). Formability analysis of extra-deep drawing steel, *Journal of Material Processing Technology*, **130–131**; 31–41.
31. Narayanasamy, R. and Sathiya Narayanan, C. (2006). Forming limit diagram for Indian interstitial free steels, *Materials and Design*, **27**; 882–899.
32. Varma, N. (2010). Experimental studies on size effects in deep drawing, *M. Tech. Dissertation*, IIT Bombay.
33. Kinsey, B.L. and Cao, J. (1996). Forming of Tailor Welded Aluminium Blanks, *Journal of Minerals, Metals & Materials*, **59**; 69–83.
34. Ahmetoglu, M.A., Brouwers, D., Shulkin, L., Taupin, L., Kinzel, G.L. and Altan, T. (1995). Deep Drawing Of Round Cups From Tailor Welded Blanks, *Journal of Material Processing Technology*, **53**; 684–694.
35. Choi, Y., Heo, Y., Kim H.Y., and Seo, D. (2000). Investigation of Weld Line Movements for the Deep Drawing Process of Tailor Welded Blanks, *Journal of Material Processing Technology*, **108**; 1–7.
36. Chan, S., Chan, L. and Lee, T. (2003).Tailor Welded Blanks of Different Thickness Ratios Effects on Forming Limit Diagrams, *Journal of Material Processing Technology*, **132**; 95–101.
37. Eisenmenger, M., Bhatt, K. and Shi, M. (1995). Influence of laser welding parameters on formability and robustness of blank manufacturing: An application to a body side frame *Transactions of the SAE*, SAE paper no.950922, 877–888.
38. Shakeri, H., et al., (2002). Study of damage initiation and fracture in aluminum tailor welded blanks made via different welding techniques, *Journal of Light Metals*, **2**; 95–110.
39. Kinsey, B., Liu, Z. and Cao, J. (1999). New Apparatus and Method for Forming Tailor Welded Blanks, *Journal of Materials and Manufacturing*, **108**; 653–664.
40. Heo, Y., Wang, S. Kim, H., and Seo, D. (2001).The effect of the drawbead dimensions on the weld line movements in the deep drawing of tailor welded blanks, *Journal of Materials and Manufacturing*, **113**; 686–691.

41. Abdullah, K., Wild, P., Jesweist, J. and Ghasempoot, A. (2001). Tensile testing for weld deformation properties in similar gage tailor welded blanks using the rule of mixtures, **112**; 91–97.

42. Thakkar, M. (2005). Development of methods to reduce weld line movement in tailor welded blanks, *M. Tech. Dissertation*, Department of Mechanical Engineering, IIT Bombay, Powai, Mumbai.

43. Siegert, K., Hohnhaus J. and Wagener, S. (1998). Combination of hydraulic Multipoint cushion system and segment-elastic blank holders, *Transactions of the SAE*, SAE paper 980077, 1322.

44. Siegert, K., and Ziegler, M. (1997). Pulsating Blank Holder Force, *Transactions of the SAE*, SAE paper 970987.

45. Siegert, K. (1997). World Status Report on Metal Forming, Processes and Product Design, *International Metal Forming Congress*, 6–8 November 1997, Mumbai.

46. Kore, S.D., Dhanesh, P.C., Kulkarni, S.V., and Date, P.P. (2010). Numerical modeling of electromagnetic welding, *Int. J. Applied Electromagnetics and Mechanics*, **32**; 1–19.

47. Dhanesh, P.C. (2007). Analysis of electromagnetic joining", *M. Tech thesis*, Department of Mechanical Engineering., IIT Bombay.

48. Kore, S.D., Date, P.P. and Kulkarni, S.V. (2008). Electromagnetic impact welding of aluminium to stainless steel sheets, *J. Materials Processing Technol.*, **208**; 486–493.

CHAPTER 9 Flange Wrinkling in Deep Drawing

Ravindra K. Saxena*, Prakash M. Dixit#
*Department of Mechanical Engineering
Sant Longowal Institute of Engineering and Technology, Longowal – 148 106
#Department of Mechanical Engineering
Indian Institute of Technology, Kanpur – 208 016

9.1 INTRODUCTION

Sheet metal parts are used extensively in various engineering applications like automobile manufacturing, aerospace technology, electrical equipment *etc*. Deep drawing is a common manufacturing process to produce these sheet metal parts. In deep drawing process, an initially flat sheet (called a blank), usually controlled by a blank holder, is forced through a die by means of a punch to form a hollow product. The schematic diagram of a deep drawing process is shown in Fig. 9.1.

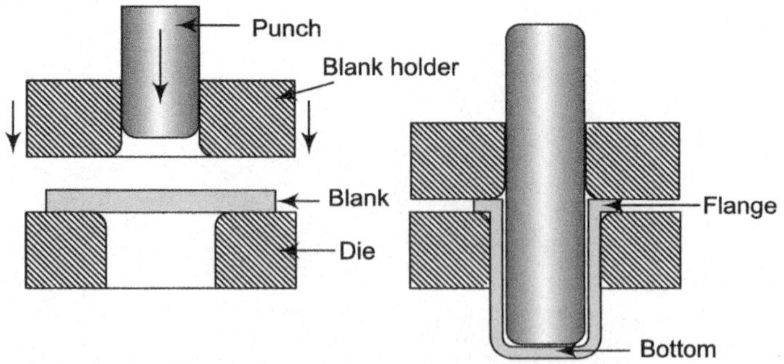

Fig. 9.1 Schematic of deep drawing process [26]

The circumferential stress in flange region of a deep drawn product is found to be compressive. As a result, at a certain state of drawing, the flange region buckles developing waviness or wrinkles (Fig. 9.2). This phenomenon is called wrinkling [1]. This usually happens if the sheet is thin and the blank-holder is either not used or the blank-holder force is not sufficient. In thick sheets, usually wrinkling does not occur even if the blank-holder force is not applied.

Fig. 9.2 Flange wrinkling in deep drawn cup (With permission from Wiley, copyright 2011) [20]

Wrinkling is one of the major defects in sheet metal products and also plays a significant role in the wear of the tooling. Study of wrinkling has become important with the increasing use of thinner gauge sheets in the automobile industry. The initiation of wrinkling is influenced by many process parameters like material properties of the blank, geometry of the sheet and tooling, other process parameters like the blank-holder force, friction at the die-blank interface, working temperature *etc*. From the view-point of designing the deep drawing equipment, it is necessary to know the amount of blank-holder force required to draw a cup of certain height without flange wrinkling when the other process parameters are specified. For this purpose, from the analysis point of view, it is important to determine the maximum cup height that can be drawn without flange wrinkling for the given set of process parameters. Since, the effects of the process parameters on the initiation of wrinkling are very complex, a general wrinkling initiation criterion applicable to all situations has not been proposed so far. As a result, the study of wrinkling is normally carried out case by case.

Equations governing the deformation in deep drawn products are non-linear possessing more than one solution. Upto a certain level of deformation, there is only one solution in which the flange of the deep drawn product remains flat. After that, depending on a combination of the process parameters, these equations possess more than one solution. One of these solutions, of course, corresponds to the flat configuration of the flange. However, this solution becomes unstable due to disturbances like geometric imperfection of the sheet, non-alignment of the blank centre with the punch/die axes *etc*. As a result, it is not observed in practice. On the other hand, the solution in which the flange develops waviness is a stable solution and is observed in practice. Thus, wrinkling is an *instability* in the flange region of the drawn product.

The phenomenon in which, at a certain level of primary variable, a second solution emerges from the first solution is called *bifurcation*. The first solution is called the *primary* or *fundamental* solution and the second solution is called the *secondary* or *bifurcated* solution. Usually, the first solution becomes unstable at this stage and the second solution, being stable, is observed in practice. In deep drawing process, initially, the bifurcated solution possesses a simple wavy pattern. However, as the deformation continues, even this solution becomes unstable giving rise to a more complicated wavy pattern. Thus, wrinkling is a phenomenon of *repeated* bifurcation.

Two broad approaches have emerged to analyze the flange wrinkling: (i) *energy method*, and (ii) *bifurcation method*. These methods predict the critical values of certain process parameters at the initiation of wrinkling. In the energy method, a certain wave-form of the flange is assumed and the critical values are determined from the condition that this wave-form must be in equilibrium. In the bifurcation method, the critical values are obtained from the condition at which the solution of the (non-linear) governing equations becomes non-unique. The bifurcation method also proposes the path for the secondary or bifurcated solution. There is a third approach in which the *contact algorithm technique* is used, in conjunction with the FE analysis, to *simulate wrinkling*. However, this approach does not predict the critical values of any process parameter at the onset of wrinkling.

9.2 ENERGY METHOD

In the *energy method*, the flange wrinkling is analyzed by extending the energy method of elastic instability to the case of plastic deformation. The early attempts to analyze flange wrinkling were based on this method [2, 3]. Senior [2] studied the flange wrinkling in circular cup by considering the flange as a narrow, one-dimensional, infinite strip. On the other hand, Yu and Johnson [3] analyzed it as a two-dimensional problem by treating the flange as an annular plate. In 2-D case, the flange wrinkling is analyzed by considering the flange as a (circular/square) plate with a (circular/square) hole, free at the outer edge, supported at the inner edge and subjected to lateral distributed force exerted by the blank-holder. Then, a certain wave-form is assumed in which the flange is considered to be in equilibrium. Some amount of work is required to create this deformation. Further, additional work is needed to overcome the blank-holder force. If the sum of these two works can be provided by the membrane part of the in-plane stresses, then the equilibrium of the flange in the assumed wave-form is possible. The balance of these three works gives the critical value of the desired process parameter at the initiation of wrinkling.

For the flange wrinkling in *circular cup*, it is possible to develop approximate analytical expressions for these three works using cylindrical polar coordinates

(r, θ). For a circular cup, lateral displacement $w\,(r, \theta)$ of the mid-surface of the flange is assumed to be the product of a linear function of r-coordinate and some trigonometric function of θ in terms of the assumed number of wrinkles n. The lateral displacement is assumed to be zero at $r = r_d$ where r_d is the inner radius of the flange. (Note that r_d is the sum of the die opening radius and the die profile radius.) When the flange is in *elastic* state, the work required to create this assumed waveform $w\,(r, \theta)$ is given by [4],

$$\Delta U = \int_0^{2\pi}\int_{r_d}^{r_0}\left\{\frac{D}{2}\left(\frac{\partial^2 w}{\partial r^2}+\frac{1}{r}\frac{\partial w}{\partial r}+\frac{1}{r^2}\frac{\partial^2 w}{\partial \theta^2}\right) - D(1-\nu)\frac{\partial^2 w}{\partial r^2}\left(\frac{1}{r}\frac{\partial w}{\partial r}+\frac{1}{r^2}\frac{\partial^2 w}{\partial \theta^2}\right) + D(1-\nu)\left(\frac{1}{r}\frac{\partial^2 w}{\partial r\partial\theta}-\frac{1}{r^2}\frac{\partial w}{\partial \theta}\right)^2\right\} r\,dr\,d\theta \tag{9.1}$$

where r_0 is the (current) outer radius of the flange, ν is the Poisson's ratio and D is the bending stiffness:

$$D = \frac{Et^3}{12(1-\nu^2)} \tag{9.2}$$

Here, E is the Young's modulus and t is the flange thickness (assumed constant). The prefix Δ is used to represent this work as it represents the *incremental* work done on the flange after it loses the straight configuration. The symbol U is used for this work as it represents the bending strain energy. When the flange is in *plastic* state, the above expression is assumed to be valid if D and ν are replaced by D_0 and ν_0 [2]:

$$D_0 = \frac{E_0 t^3}{12(1-\nu_0^2)},\quad \nu_0 = 0.5,\quad E_0 = \frac{4EE_t}{(\sqrt{E}+\sqrt{E_t})^2},\quad E_t = \frac{d\sigma_{eq}}{d\varepsilon_{eq}^{pL}} \tag{9.3}$$

Here, E_t is the tangent modulus, σ_{eq} is the equivalent stress and ε_{eq}^{pL} is the equivalent plastic strain. Thus, when the flange is in *plastic* state, the work required to create the assumed waveform $w\,(r, \theta)$ is given by [3]

$$\Delta U = \int_0^{2\pi}\int_{r_d}^{r_0}\left\{\frac{D_0}{2}\left(\frac{\partial^2 w}{\partial r^2}+\frac{1}{r}\frac{\partial w}{\partial r}+\frac{1}{r^2}\frac{\partial^2 w}{\partial \theta^2}\right) - D_0(1-\nu_0)\frac{\partial^2 w}{\partial r^2}\left(\frac{1}{r}\frac{\partial w}{\partial r}+\frac{1}{r^2}\frac{\partial^2 w}{\partial \theta^2}\right) + D_0(1-\nu_0)\left(\frac{1}{r}\frac{\partial^2 w}{\partial r\partial\theta}-\frac{1}{r^2}\frac{\partial w}{\partial \theta}\right)^2\right\} r\,d\,r d\theta \tag{9.4}$$

If the blank-holder force (F_b) is assumed to act at the outer edge of the flange and remain constant, then the approximate work required to overcome the blank-holder force is given by

$$\Delta W_b = F_b w_{max} \tag{9.5}$$

where w_{max} is the maximum lateral displacement of the flange. Sometimes, the blank-holder is modeled as a series of parallel springs located at the outer edge of the flange with the net stiffness being k [3]. Then, the work required to overcome the blank-holder force is given by

$$\Delta W_b = \frac{1}{2} k w_{\max}^2 \tag{9.6}$$

It is assumed that the in-plane shear stress $\sigma_{r\theta}$ is negligible. The work done by the membrane part of the normal in-plane stresses (σ_{rr}, $\sigma_{\theta\theta}$) is given by [3]

$$\Delta W_\sigma = -\frac{1}{2} \int_0^{2\pi} \int_{r_d}^{r_0} \left\{ \sigma_r t \left(\frac{\partial w}{\partial r} \right)^2 + \sigma_\theta t \left(\frac{1}{r} \frac{\partial w}{\partial \theta} \right)^2 \right\} r\, dr\, d\theta \tag{9.7}$$

The negative sign in the above expression assumes that the stresses (σ_{rr}, $\sigma_{\theta\theta}$) are considered as tensile. The in-plane stresses (σ_{rr}, $\sigma_{\theta\theta}$) are expressed in terms of the material, geometric and other process parameters by performing the plastic analysis of the flange. By equating the sum of ΔU and ΔW_b with ΔW_σ:

$$\Delta U + \Delta W_b = \Delta W_\sigma \tag{9.8}$$

with the help of Eqs (9.4), (9.5) (or (9.6)) and (9.7), one can obtain the wrinkling condition in terms of the process parameters: elastic-plastic material properties of the flange, the flange geometry and the blank-holder force. The condition, of course, depends on the number n of the assumed wrinkles. Yu and Johnson [3] determined the in-plane membrane stresses by the slab method and presented the wrinkling condition as a graph between the parameters $1 - r_d/r_0$ and $(\sqrt{E_0/\sigma_Y})(t/r_0)$ for different values on n. Here, σ_Y is the initial yield stress of the material. They also presented the variation of the number of wrinkles (n) with the parameter, $1 - r_d/r_0$.

Agrawal *et al.* [5] also used the energy method for the analysis of wrinkling in circular cup. They used the upper bound method, by incorporating the anisotropy of the flange material and the variation in the flange thickness, to determine the in-plane membrane stresses (σ_{rr}, $\sigma_{\theta\theta}$). Using Eq. (9.8), they minimized the blank-holder force with respect to the number of wrinkles (n) to obtain its minimum value (as a function of the material and geometric parameters of the flange) required for suppressing the wrinkling. They studied the effect of the anisotropy and the initial thickness on the minimum required blank-holder force.

The methods used by Yu and Zhang [6] and Zeng and Mahdavian [7] are similar to the energy method. Yu and Zhang [6] analyzed the flange wrinkling in circular cup by solving the differential equation for the buckling of an annular plate (but without considering the blank-holder force) by an approximate method to find the critical value of the (tensile) radial stress at the inner flange boundary. The differential equation represents the balance of forces instead of the balance of works. Zeng and Mahdavian [7] studied the flange wrinkling in circular cup by considering the balance of bending moments (instead of the balance of works) to obtain a critical condition for wrinkling. The critical condition is in terms of the material parameters (E_0, σ_Y) and the geometric parameters (r_0, r_d, t). They also studied the effect of temperature on the critical condition.

In order to provide some insight into the phenomenon of flange wrinkling, Cao and Boyce [8] and Cao and Wang [9] analyzed the wrinkling of *rectangular plates* by the energy method. The plate is subjected to compressive displacement on two edges, tensile stress on the remaining two edges and the lateral constraint in the form of blank-holder (or binder) force. The analysis is carried out in two stages. In the first stage, the work required to deform the plate when it is subjected to compressive displacement at the edges is evaluated. This is done for the flat plate as well as for the cases when the mid-plane of the plate has sinusoidal type of geometric imperfection with mode number n. Whereas Cao and Boyce [8] used the finite element (FE) software ABAQUS (with 4 noded reduced integration shell elements) to evaluate these works, Cao and Wang [9] developed analytical expressions for them.

The plate with geometric imperfection buckles and is found to require less work for its deformation. If W_0 is the work of deformation for the flat plate when it is subjected to a specified compressive displacement at the edges and W_n ($< W_0$) is the corresponding amount of work for the plate with geometric imperfection with mode number n, then the difference ($W_0 - W_n$) must be provided by the blank-holder to suppress the buckling (or the wrinkling) of the plate. Thus, the blank-holder force (F) needed to suppress the plate wrinkling can be evaluated from the following condition:

$$\int bF dw = W_0 - W_n \tag{9.9}$$

where b is the plate width and the integration is to be carried upto the maximum lateral displacement corresponding to the specified compressive displacement on the edges. The required blank-holder force, of course, becomes a function of the mode number n. Suppose the blank-holder force is specified, then condition (9) gives the critical compressive stress at which the wrinkling initiates in the flat plate. This condition is called the C-B wrinkling criterion [8]. It is represented as a graph of critical compressive stress verses the blank-holder force. Since, Cao and Wang [9] used analytical expressions for the work of deformation; they have obtained the analytical expression for the C-B wrinkling criterion. In both these works, Hill's 1948 quadratic anisotropic yield function [10] is used to model the anisotropy of the material. Further, the parametric study of the C-B wrinkling criterion is carried out with respect to the transverse tension, strain-hardening parameters and the anisotropy ratio. In both these works, it is stated that the C-B wrinkling criterion can be used to predict the wrinkling in deep drawing process. In support of this, they mentioned the observation of Hutchinson and Neale [11] that the initiation of wrinkling is most likely a local phenomenon. Further, they have reported that simulated and experimental results on the critical compressive stress in the flange wrinkling in conical cup agree with this criterion.

9.3 CONTACT ALGORITHM TECHNIQUE

In this method, the contact conditions are used to simulate the wrinkling in a deep drawing process. Whenever the contact force at a node either on the blank-die interface or on the blank-blankholder interface is found to be non-negative, the node is released from the contact and is considered free. Han and Huh [12] used the FE code CASHE to simulate wrinkling in square cup without the blank-holder force. The code uses the modified membrane FE formulation which incorporates the bending effect. They used the method of skew boundary condition to apply the contact boundary conditions. They observed a good agreement between the predicted wrinkling pattern with experimental results. They showed that the number of wrinkles depends on the blank size. Weili *et al.* [13] employed an in-house static-implicit FE code to simulate wrinkling in circular and square cups. They used the node projection method to apply the contact conditions. Further, they assumed that the gap between the die and the blank-holder (called the blank-holder gap) is greater than the blank thickness. They observed that the predicted wrinkling patterns agree well with experimental results on stamping process.

As stated earlier, this method essentially simulates wrinkling in deep drawing processes. It does not predict the critical values of any process parameter at the onset of wrinkling.

9.4 BIFURCATION METHOD: HUTCHINSON'S CRITERION

Hill [14] proposed a criterion for the existence of two solutions in terms of a functional containing the current states of stress and displacement and the fourth order elasto-plastic constitutive tensor of the material. The bifurcation occurs whenever this functional becomes zero. Hutchinson [15] specialized this criterion for sheets using the Donnell-Mushtari-Vlasov (DMV) theory of plates and shells. In this section, we discuss the Hutchinson's [15] criterion.

Let Δu_α ($\alpha, \beta = 1, 2$) and Δw be the incremental tangential and normal displacement components of the undeformed middle surface when the sheet buckles. Then, according to the DMV theory, the incremental stretching ($\Delta E_{\alpha\beta}$) and bending ($\Delta K_{\alpha\beta}$) strains in the sheet are given by [15],

$$\Delta E_{\alpha\beta} = \frac{1}{2}\left(\Delta u_{\alpha,\beta} + \Delta u_{\beta,\alpha}\right) + b_{\alpha\beta}\,\Delta w + \frac{1}{2}\,\Delta w_\alpha\,\Delta w_\beta$$

$$\Delta K_{\alpha\beta} = -\,\Delta w_{\alpha,\beta} \tag{9.10}$$

where $b_{\alpha\beta}$ is the curvature tensor of the undeformed middle surface and the comma denotes the covariant differentiation with respect to the surface coordinates. The corresponding incremental bending moments ($\Delta M^{\alpha\beta}$) and stress resultants ($\Delta N^{\alpha\beta}$) are given by [15],

$$\Delta M^{\alpha\beta} = H_{(2)}^{\alpha\beta\kappa\gamma}\,\Delta E_{\kappa\gamma} + H_{(3)}^{\alpha\beta\kappa\gamma}\,\Delta K_{\kappa\gamma},$$

$$\Delta N^{\alpha\beta} = H_{(1)}^{\alpha\beta\kappa\gamma}\,\Delta E_{\kappa\gamma} + H_{(2)}^{\alpha\beta\kappa\gamma}\,\Delta K_{\kappa\gamma} \tag{9.11}$$

where

$$H_{(i)}^{\alpha\beta\kappa\gamma} = \int_{-t/2}^{t/2} \bar{L}^{\alpha\beta\kappa\gamma}\,(\varsigma)^{i-1}\,d\varsigma \tag{9.12}$$

Here, ζ is the coordinate perpendicular to the undeformed middle surface and $\bar{L}^{\alpha\beta\kappa\gamma}$ is the plane-stress constitutive tensor relating the incremental stress $\Delta\sigma^{\alpha\beta}$ to the incremental strain ($\Delta\varepsilon_{\alpha\beta}$):

$$\Delta\sigma^{\alpha\beta} = \bar{L}^{\alpha\beta\kappa\gamma}\,\Delta\varepsilon_{\alpha\beta} \tag{9.13}$$

For the DMV theory, the equilibrium equation, in terms of the principle of virtual work, is given by [15]

$$\int_S \left\{M^{\alpha\beta}\delta K_{\alpha\beta} + N^{\alpha\beta}\delta E_{\alpha\beta}\right\} dS = \int_S T^{\alpha}\delta(\Delta u_{\alpha})dS \tag{9.14}$$

where $\boldsymbol{T}$ is the surface traction vector per unit original area and S is the domain of the undeformed middle surface of the sheet.

At the bifurcation point, two solutions exist. They are denoted by $\Delta u_{\alpha}^{a}, \Delta w^{a}$ and $\Delta u_{\alpha}^{b}, \Delta w^{b}$. Define the following difference quantities:

$$\Delta\tilde{u}_{\alpha} = \Delta u_{\alpha}^{b} - \Delta u_{\alpha}^{a}, \quad \Delta\tilde{w} = \Delta w^{a} - \Delta w^{b}, \quad \tilde{\sigma}^{\alpha\beta} = \sigma_{b}^{\alpha\beta} - \sigma_{a}^{\alpha\beta} \tag{9.15}$$

Then, for both the solutions 'a' and 'b' to exist, Eq. (9.14) reduces to:

$$\int_S \left\{\Delta\tilde{M}^{\alpha\beta}\,\Delta\tilde{K}_{\alpha\beta} + \Delta\tilde{N}^{\alpha\beta}\,\Delta\tilde{E}_{\alpha\beta} + N_0^{\alpha\beta}\,\Delta\tilde{W}_{\alpha}\,\Delta\tilde{W}_{\beta}\right\} d \;= \mathfrak{G} \tag{9.16}$$

where $N_0^{\alpha\beta}$ is the current stress resultant, $\Delta\tilde{M}^{\alpha\beta}$ and $\Delta\tilde{N}^{\alpha\beta}$ are the bending moments and the stress resultants corresponding to $\Delta\tilde{u}_{\alpha}$ and $\Delta\tilde{w}$ and

$$\Delta\tilde{E}_{\alpha\beta} = \frac{1}{2}(\Delta\tilde{u}_{\alpha,\beta} + \Delta\tilde{u}_{\beta,\alpha}) + b_{\alpha\beta}\Delta\tilde{w} + \frac{1}{2}(\Delta w_{\alpha}^{0}\Delta\tilde{w}_{\beta} + \Delta w_{\beta}^{0}\Delta\tilde{w}_{\alpha})$$

$$\Delta\tilde{K}_{\alpha\beta} = -\Delta\tilde{w}_{\alpha\beta} \tag{9.17}$$

Hutchinson [15] assumed the existence of a fictitious plane-stress comparison solid whose fourth order elasto-plastic constitutive tensor $\bar{\boldsymbol{L}}^{c}$ is assumed to be equal to $\bar{\boldsymbol{L}}$. The corresponding $\boldsymbol{H}^{c}$ tensor [to be obtained by an equation similar to Eq. (9.12)] also would be equal to $\boldsymbol{H}$. For this fictitious material, at the bifurcation point, yielding is assumed to continue both in the fundamental as well as in the bifurcated solution without any unloading. Substituting the constitutive relation (9.11) and using the equality $\boldsymbol{H}^{c} = \boldsymbol{H}$, the bifurcation condition (9.16) becomes

$$F = 0 \tag{9.18}$$

where,

$$F = \int_S \Big\{ (H_{(3)}^{\alpha\beta\kappa\gamma})^c \, \Delta\tilde{K}_{\alpha\beta}\Delta\tilde{K}_{\kappa\gamma} + 2(H_{(2)}^{\alpha\beta\kappa\gamma})^c \, \Delta\tilde{E}_{\alpha\beta}\Delta\tilde{K}_{\kappa\gamma} + (H_{(1)}^{\alpha\beta\kappa\gamma})^c \, \Delta\tilde{E}_{\alpha\beta}\Delta\hat{E}_{\kappa\gamma} + N_0^{\alpha\beta}\Delta\tilde{W}_\alpha\Delta\tilde{W}_\beta \Big\} \, dS. \quad (9.19)$$

Correia and Ferron [16] employed the Hutchinson's [15] bifurcation criterion for the prediction of wrinkling in the wall of conical cup using ABAQUS/explicit code with quadrangular shell elements. The anisotropy of the blank was modeled by the plane stress yield function proposed by Ferron *et al.* [17]. They represented the onset of wrinkling by a curve in the space of the two principal stresses, called the wrinkling limit curve.

9.5 BIFURCATION METHOD: RIKS'S CRITERION

Riks [18] obtained the bifurcation criterion for a discretized set of (non-linear) governing equations like the (non-linear) finite element equations for elastic materials. For elastic materials, the change in the total potential energy ($\Delta\Pi$) due to admissible variations of the discrete displacements (δu_i) is given by

$$\Delta\Pi\,(u_i, \delta u_i) = \frac{\partial\Pi}{\partial u_i}\delta u_i + \frac{1}{2}\frac{\partial^2\Pi}{\partial u_i \partial u_j}\delta u_i \delta u_j + \ldots \quad (9.20)$$

The first variation of the total potential energy vanishes for an equilibrium state and the second variation must be positive definite for stability of the system. The quantity $\dfrac{\partial^2\Pi}{\partial u_i\,\partial u_j}$ is called the tangent stiffness matrix and is denoted by $[K]$:

$$K_{ij} = \frac{\partial^2\Pi}{\partial u_i\,\partial u_j} \quad (9.21)$$

Thus, the stability criterion, in terms of the tangent stiffness matrix $[K]$, becomes [18]

$$\frac{\partial^2\Pi}{\partial u_i\,\partial u_j}\delta u_i\,\delta u_j = K_{ij}\,\delta u_i\,\delta u_j > 0 \quad (9.22)$$

The bifurcation criterion corresponds to the case when the tangent stiffness matrix K_{ij} ceases to be positive definite, i.e.,

$$\det\,[K] = 0 \quad (9.23)$$

The above bifurcation criterion can be extended to elasto-plastic materials by replacing the total potential energy by the integral form of the equilibrium equation (like the virtual work expression) and setting the determinant of the corresponding tangent stiffness matrix to zero.

Riks [18] also proposed that the bifurcated solution can be obtained as a linear combination of the fundamental solution upto the bifurcation state and the eigenvector of tangent stiffness matrix corresponding to the lowest eigenvalue at the bifurcation state.

Kim *et al.* [19] employed the Riks's [18] bifurcation criterion to simulate wrinkling in elliptical cup using an in-house FE code with CBR (continuum-based resultant) shell elements. They modeled the interaction of the blank-holder with the sheet not through the blank-holder force but through the contact conditions assuming a variable blank-holder gap. After the bifurcation, the bifurcated solution (i.e., the wrinkling pattern) is simulated by assuming the increment to be a linear combination of the fundamental solution (upto the bifurcation state) and the eigenvector of the tangent stiffness matrix corresponding to the lowest eigenvalue at the bifurcation state. They used Hill's 1948 anisotropic yield function [10] to model the anisotropy of the sheet. Kim *et al.* [20] carried out similar finite element analysis with CBR shell elements to study wrinkling in circular cup. Both Hill's 1948 anisotropic yield function [10] and Barlat's 1991 anisotropic yield function based on one linear transformation [21] are used by them to model the anisotropy of the sheet. They obtained the critical (average) circumferential stress, wrinkling wavelength and critical punch stroke as functions of the blank holder force. They also conducted experiments and observed that the experimental values of critical punch stroke are overestimated by the FE bifurcation analysis while the wrinkling wave length is under-predicted. They also studied the effect of material properties (like the anisotropy coefficient and the strain-hardening coefficient) and the friction coefficient on the wrinkling pattern.

9.6 BIFURCATION METHOD: HILL'S CRITERION

In this section, Hill's [14] bifurcation criterion is derived for an incremental formulation like the updated Lagrangian formulation and for a specific anisotropic elasto-plastic material. As a first step in the development of this criterion, the governing equations of the updated Lagrangian formulation problem are described. These equations are: (a) incremental strain-displacement relations, (b) incremental stress-strain relations, and (c) incremental equilibrium equation (for static problems).

9.6.1 Incremental Strain-displacement Relation

The increment size is assumed to be large. Then, instead of choosing a linear measure of the incremental deformation like the incremental linear strain tensor, one should choose an appropriate non-linear measure of incremental deformation. In this section, the incremental logarithmic strain tensor (${}_t\Delta\varepsilon_{ij}^L$) is used as a measure of the incremental deformation. It is defined by [22, 23]

$$ {}_t\Delta\varepsilon_{ij}^L = \ln\,({}_t\Delta\lambda_i)\,\delta_{ij} \quad \text{(no sum over } i\text{)} \tag{9.24}$$

where δ_{ij} is the Kronecker's delta and ${}_t\Delta\lambda_i$ are the principal values of the incremental right stretch tensor ${}_t\Delta U_{ij}$. The square of this tensor is defined by

$$ {}_t\Delta U_{ij}^2 = ({}_t\Delta F_{ik}^T)\,({}_t\Delta F_{kj}), \quad {}_t\Delta F_{ij} = \delta_{ij} + {}_t\Delta u_{i,j} \tag{9.25} $$

where ${}_t\Delta u_{i,j}$ denotes the derivative of the incremental displacement vector ${}_t\Delta u_i$ at time t with respect to the position vector ${}^t x_j$ at that time. Note that, in Eq. (9.24), the components are with respect to the principal axes of ${}_t\Delta U_{ij}$.

9.6.2 Incremental Elastic-plastic Constitutive Equation

In deep drawing, rolled sheets (without annealing) are used as blanks. Therefore, they exhibit certain anisotropy in their material behavior. For aluminium alloys, the recently proposed anisotropic yield criterion of Barlat *et al.* [24], which is based on two linear transformations, seems to capture this anisotropy more accurately. Therefore, this criterion is used in the present section to model the material anisotropy of the blank. In this section, the subscripts/superscripts denoting the time have been omitted for the sake of convenience. Then, for initial yielding, this criterion can be expressed as

$$ f(\sigma_{ij}) \equiv \sigma_{\mathrm{eq}}(\sigma_{ij}) - \sigma_y = 0 \tag{9.26} $$

where f is the yield function, σ_y is the uniaxial yield stress along the rolling direction and the equivalent stress (σ_{eq}) is defined by [24]

$$ \begin{aligned} 4\sigma_{\mathrm{eq}}^a = {} & \left|S_1^{(1)} - S_1^{(2)}\right|^a + \left|S_1^{(1)} - S_2^{(2)}\right|^a + \left|S_1^{(1)} - S_3^{(2)}\right|^a + \left|S_2^{(1)} - S_1^{(2)}\right|^a \\ & + \left|S_2^{(1)} - S_2^{(2)}\right|^a + \left|S_2^{(1)} - S_3^{(2)}\right|^a + \left|S_3^{(1)} - S_1^{(2)}\right|^a + \left|S_3^{(1)} - S_2^{(2)}\right|^a \\ & + \left|S_3^{(1)} - S_3^{(2)}\right|^a \end{aligned} \tag{9.27} $$

Here, $S_k^{(1)}$, and $S_k^{(2)}$, $k = 1, 2, 3$ are the principle values of the modified stress deviatoric tensors $\mathbf{S}^1$ and $\mathbf{S}^2$ that are obtained from the deviatoric stress tensor $\boldsymbol{\sigma}'$ by some linear transformations. In terms of the Cauchy stress tensor $\boldsymbol{\sigma}$, these transformations are given by [24]

$$ \begin{aligned} \{S^1\} &= [C^1]\,\{\sigma'\} = [C^1]\,[T]\,\{\sigma\} = [L^1]\,\{\sigma\} \\ \{S^2\} &= [C^2]\,\{\sigma'\} = [C^2]\,[T]\,\{\sigma\} = [L^2]\,\{\sigma\} \end{aligned} \tag{9.28} $$

where

$$ [C^1] = \begin{bmatrix} 0 & -\alpha_1 & -\alpha_2 & 0 & 0 & 0 \\ -\alpha_3 & 0 & -\alpha_4 & 0 & 0 & 0 \\ -\alpha_5 & -\alpha_6 & 0 & 0 & 0 & 0 \\ 0 & 0 & 0 & \alpha_9 & 0 & 0 \\ 0 & 0 & 0 & 0 & \alpha_7 & 0 \\ 0 & 0 & 0 & 0 & 0 & \alpha_8 \end{bmatrix}, $$

$$[C^2] = \begin{bmatrix} 0 & -\alpha_{10} & -\alpha_{11} & 0 & 0 & 0 \\ -\alpha_{12} & 0 & -\alpha_{13} & 0 & 0 & 0 \\ -\alpha_{14} & -\alpha_{15} & 0 & 0 & 0 & 0 \\ 0 & 0 & 0 & \alpha_{18} & 0 & 0 \\ 0 & 0 & 0 & 0 & \alpha_{16} & 0 \\ 0 & 0 & 0 & 0 & 0 & \alpha_{17} \end{bmatrix},$$

$$[T] = \frac{1}{3}\begin{bmatrix} 2 & -1 & -1 & 0 & 0 & 0 \\ -1 & 2 & -1 & 0 & 0 & 0 \\ -1 & -1 & 2 & 0 & 0 & 0 \\ 0 & 0 & 0 & 3 & 0 & 0 \\ 0 & 0 & 0 & 0 & 3 & 0 \\ 0 & 0 & 0 & 0 & 0 & 3 \end{bmatrix}, \quad \{\sigma\} = \begin{Bmatrix} \sigma_{xx} \\ \sigma_{yy} \\ \sigma_{zz} \\ \sigma_{xy} \\ \sigma_{yz} \\ \sigma_{zx} \end{Bmatrix}, \quad \{\sigma'\} = \begin{Bmatrix} \sigma'_{xx} \\ \sigma'_{yy} \\ \sigma'_{zz} \\ \sigma'_{xy} \\ \sigma'_{yz} \\ \sigma'_{zx} \end{Bmatrix},$$

$$[L^1] = [C^1]\,[T], \quad [L^2] = [C^2]\,[T]. \tag{9.29}$$

The 18 nonzero coefficients in the two transformation matrices $[C^1]$ and $[C^2]$ represent the material anisotropy. These 18 coefficients are obtained by minimizing an error function [24]. For body-centered cubic (BCC) and face-centered cubic (FCC) materials, the exponent a is 6 and 8 respectively. When all the anisotropic parameters (α_i) are equal to one and $a = 2$, the above anisotropic yield function reduces to that of the von Mises isotropic yield function.

For an isotropically hardening material, the variable yield stress depends on the hardening variable p. For the case of strain hardening, p is identified as the equivalent plastic strain $\varepsilon_{\text{eq}}^{\text{pL}}$, and is defined as:

$$p \equiv \varepsilon_{\text{eq}}^{\text{pL}} = \sum \Delta\varepsilon_{\text{eq}}^{\text{pL}} \tag{9.30}$$

where the sum is to be carried out over all the increments upto the present time and the equivalent plastic strain increment $\Delta\varepsilon_{\text{eq}}^{\text{pL}}$ is defined by

$$\Delta\varepsilon_{\text{eq}}^{\text{pL}} = \frac{\Delta\varepsilon_{ij}^{\text{pL}}\sigma_{ij}}{\sigma_{\text{eq}}} \tag{9.31}$$

Here, $\Delta\varepsilon_{ij}^{\text{pL}}$ is the plastic part of the incremental logarithmic strain tensor $\Delta\varepsilon_{ij}^{L}$. The dependence of σ_y on $\varepsilon_{\text{eq}}^{\text{pL}}$ is normally approximated by a power-law type of relationship. In the present formulation, the following Ludwik power law is used:

$$\sigma_y = H(\varepsilon_{\text{eq}}^{\text{pL}}) \equiv (\sigma_y)_o + K(\varepsilon_{\text{eq}}^{\text{pL}})^n \tag{9.32}$$

Here, $(\sigma_y)_0$ is the initial yield stress along the rolling direction and K and n are the hardening parameters.

Using the associated flow rule where the plastic potential is given by Eq. (9.26), employing the additivity of the elastic and plastic parts of the incremental logarithmic strain tensor and assuming the elastic behavior to be linear, the incremental elasto-plastic stress-strain relationship becomes [25, 26]

$$\Delta\sigma_{ij} = (C_{ijkl}^{\mathrm{EP}})\,\Delta\varepsilon_{kl}^{\mathrm{pL}} \tag{9.33}$$

where the fourth order anisotropic elasto-plastic constitutive tensor C_{ijkl}^{EP} is the inverse of the following tensor [25, 26]:

$$D_{ijkl} = -\frac{\nu}{E}\delta_{ij}\delta_{kl} + \frac{1+\nu}{E}\delta_{ik}\delta_{jl} + \frac{1}{H'}\frac{\partial\sigma_{\mathrm{eq}}}{\partial\sigma_{ij}}\frac{\partial\sigma_{\mathrm{eq}}}{\partial\sigma_{kl}} \tag{9.34}$$

Here, H' is the slope of the hardening curve Eq. (9.32) and the equivalent stress is given by Eqs (9.27)–(9.29). When the increment size is large, the variation in C_{ijkl}^{EP} over the increment is taken care of by modifying the relation (9.33) as

$${}_t\Delta\sigma_{ij} = \int_{t}^{t+\Delta t} {}^{t}C_{ijkl}^{EP}\, d({}_t\Delta\varepsilon_{kl}^{\mathrm{pL}}) \tag{9.35}$$

In the above equation, the subscripts and superscripts denoting the time have been restored.

The tensor ${}_t\Delta\sigma_{ij}$ on the left side of Eq. (9.35) must be an objective incremental stress tensor. To make it objective, it is evaluated in a material frame and the stress is updated by taking into account the material rotation [27].

9.6.3 Equilibrium Equation

For finite element application, it is convenient to obtain the Hill's bifurcation criterion from an integral form of the equilibrium equation (at time $t + \Delta t$), rather than from the incremental equilibrium equation at time t. An integral form of the equilibrium equation at time $t + \Delta t$ is given by the following virtual work expression [22]:

$$\int_{{}^{t+\Delta t}V} {}^{t+\Delta t}\sigma_{ij}\,\delta({}_t\Delta\varepsilon_{ij})\, d^{t+\Delta t}V = {}^{t+\Delta t}R \tag{9.36}$$

Here, ${}^{t+\Delta t}V$ is the domain, ${}^{t+\Delta t}\sigma_{ij}$ is the Cauchy stress tensor and ${}^{t+\Delta t}R$ is the virtual work of the external forces, all at time $t + \Delta t$. Neglecting the body forces, the expression for ${}^{t+\Delta t}R$ becomes

$${}^{t+\Delta t}R = \int_{{}^{t+\Delta t}S_t} {}^{t+\Delta t}t_i\,\delta({}_t\Delta u_i)\, d^{t+\Delta t}S \tag{9.37}$$

where ${}^{t+\Delta t}S_t$ is the boundary (with ${}^{t+\Delta t}n_j$ as the unit outward normal vector) on which the stress vector:

$${}^{t+\Delta t}t_i = {}^{t+\Delta t}\sigma_{ij}\,{}^{t+\Delta t}n_j \tag{9.38}$$

is specified and $\delta({}_t\Delta u_i)$ is the virtual incremental displacement vector at time t. Further, $\delta({}_t\Delta\varepsilon_{ij})$ represents the virtual incremental linear strain tensor corresponding to $\delta({}_t\Delta u_i)$:

$$\delta\,({}_t\Delta\varepsilon_{ij}) = \frac{1}{2}\left(\frac{\partial(\delta\,{}_t\Delta u_i)}{\partial^{t+\Delta t}x_j} + \frac{\partial(\delta\,{}_t\Delta u_j)}{\partial^{t+\Delta t}x_i}\right) \tag{9.39}$$

Since, the configuration at time $t + \Delta t$ is unknown, this expression is transformed to an integral over the known configuration at time t, i.e., over ${}^tV$ [22]:

$$\int_{{}^tV} {}^{t+\Delta t}_{\ \ t}P_{ij}\ \delta({}_t\Delta e_{ij})\ d^tV = {}^{t+\Delta t}R \tag{9.40}$$

Here, ${}^{t+\Delta t}_{\ \ t}P_{ij}$ is the 2nd Piola-Kirchhoff stress tensor:

$${}^{t+\Delta t}_{\ \ t}P_{ij} = \frac{{}^t\rho}{{}^{t+\Delta t}\rho}\ \frac{\partial^t x_i}{\partial^{t+\Delta t}x_m}\ {}^{t+\Delta t}\sigma_{mn}\ \frac{\partial^t x_j}{\partial^{t+\Delta t}x_n} \tag{9.41}$$

and $\delta({}_t\Delta e_{ij})$ is the virtual incremental Green-Lagrange strain tensor:

$$\delta\,({}_t\Delta e_{ij}) = \delta\left[\frac{1}{2}({}_t\Delta u_{i,j} + {}_t\Delta u_{j,i} + {}_t\Delta u_{k,i}\ {}_t\Delta u_{k,j})\right] \tag{9.42}$$

This virtual work expression is further simplified by decomposing ${}^{t+\Delta t}_{\ \ t}P_{ij}$ as the sum of ${}^t\sigma_{ij}$ and ${}_t\Delta P_{ij}$, decomposing $\delta({}^{t+\Delta t}_{\ \ t}e_{ij})$ into the linear and non-linear parts and neglecting the higher order terms. Then, it becomes

$$\int_{{}^tV} {}_t\Delta P_{ij}\ \delta({}_t\Delta\varepsilon_{ij})\ d^tV + \int_{{}^tV} {}^t\sigma_{ij}\ \delta({}_t\Delta\eta_{ij})\ d^tV + \int_{{}^tV} {}^t\sigma_{ij}\ \delta({}_t\Delta\varepsilon_{ij})\ d^tV = {}^{t+\Delta t}R \tag{9.43}$$

Additional simplification is achieved by approximating ${}_t\Delta P_{ij}$ as the product of the elasto-plastic constitutive tensor and the incremental linear strain tensor. Then, the virtual work expression can be written as

$$\int_{{}^tV} {}^tC^{EP}_{ijkl}\ {}_t\Delta\varepsilon_{kl}\ \delta({}_t\Delta\varepsilon_{ij})\ d^tV + \int_{{}^tV} {}^t\sigma_{ij}\ \delta({}_t\Delta\eta_{ij})\ d^tV + \int_{{}^tV} {}^t\sigma_{ij}\ \delta({}_t\Delta\varepsilon_{ij})\ d^tV = {}^{t+\Delta t}R \tag{9.44}$$

In Eqs (9.43)–(9.44), the tensors ${}_t\Delta\varepsilon_{ij}$ and ${}_t\Delta\eta_{ij}$ are given by

$${}_t\Delta\varepsilon_{ij} = \frac{1}{2}({}_t\Delta u_{i,j} + {}_t\Delta u_{j,i}),\quad {}_t\Delta\eta_{ij} = \frac{1}{2}({}_t\Delta u_{k,i}\ {}_t\Delta u_{k,j}) \tag{9.45}$$

9.6.4 Hill's Bifurcation Criterion

The Hill's bifurcation criterion is derived corresponding to the above simplified and linearized integral form of the equilibrium equation [Eq. (9.44)]. On substituting the expression for ${}_t\Delta\eta_{ij}$ [Eq. (9.45)] the equilibrium Eq. (9.44) becomes

$$\int_{{}^tV} {}^tC^{EP}_{ijkl}\ {}_t\Delta\varepsilon_{kl}\ \delta({}_t\Delta\varepsilon_{ij})\ d^tV + \int_{{}^tV} {}^t\sigma_{ijt}\ \Delta u_{k,i}\ \delta({}_t\Delta u_{k,j})\ d^tV$$
$$+ \int_{{}^tV} {}^t\sigma_{ij}\ \delta({}_t\Delta\varepsilon_{ij})\ d^tV = {}^{t+\Delta t}R \tag{9.46}$$

Let ${}_t u_i^0$ be the total displacement field upto the current level of deformation, *i.e.*, at time t. This is the fundamental solution. Suppose that bifurcation is possible at this level of deformation. Then, there are at least two solutions ${}_t\Delta u_i^a$ and ${}_t\Delta u_i^b$ of Eq. (9.46), *i.e.*, there are at least two incremental displacement fields beyond time t. The variations in the incremental displacement and strain fields can be chosen as the difference between these two solutions. Thus,

$$\delta({}_t\Delta u_i) = {}_t\Delta\tilde{u}_i \equiv {}_t\Delta u_i^a - {}_t\Delta u_i^b, \quad \delta({}_t\Delta\varepsilon_{ij}) = {}_t\Delta\tilde{\varepsilon}_{ij} \equiv {}_t\Delta\varepsilon_{ij}^a - {}_t\Delta\varepsilon_{ij}^b \tag{9.47}$$

Using this choice of $\delta({}_t\Delta u_i)$ and $\delta({}_t\Delta\varepsilon_{ij})$, the equilibrium Eq. (9.46) becomes

$$\int_{{}^tV} {}^tC_{ijkl}^{EP}\, {}_t\Delta\varepsilon_{kl}\, {}_t\Delta\tilde{\varepsilon}_{ij}\, d^tV + \int_{{}^tV} {}^t\sigma_{ij}\, {}_t\Delta u_{k,i}\, {}_t\Delta\tilde{u}_{k,j}\, d^tV + \int_{{}^tV} {}^t\sigma_{ij}\, {}_t\Delta\tilde{\varepsilon}_{ij}\, d^tV = {}^{t+\Delta t}R \tag{9.48}$$

Next, the above equilibrium Eq. (9.48) is written for each of the above mentioned two solutions ${}_t\Delta u_i^a$ and ${}_t\Delta u_i^b$. Since there is only one solution upto time t, the following things are common in the two equations:

$${}^tC_{ijkl}^{EPa} = {}^tC_{ijkl}^{EPb} = {}^tC_{ijkl}^{EP}, \quad {}^t\sigma_{ij}^a = {}^t\sigma_{ij}^b = {}^t\sigma_{ij}, \quad {}^{t+\Delta t}R^a = {}^{t+\Delta t}R^b = {}^{t+\Delta t}R \tag{9.49}$$

Taking the difference of the equilibrium Eq. (9.48) corresponding to ${}_t\Delta u_i^a$ and ${}_t\Delta u_i^b$ and using the above equalities, we get

$$\int_{{}^tV} {}^tC_{ijkl}^{EP}\, {}_t\Delta\tilde{\varepsilon}_{kl}\, {}_t\Delta\tilde{\varepsilon}_{ij}\, d^tV + \int_{{}^tV} {}^t\sigma_{ij}\, {}_t\Delta\tilde{u}_{k,i}\, {}_t\Delta\tilde{u}_{k,j}\, d^tV = 0 \tag{9.50}$$

A fictitious solid material, called comparison solid [14] is assumed to exist. For this solid, yielding is assumed to continue at the bifurcation point, without any unloading, in both the fundamental and the bifurcated solutions. Further, for this comparison solid, the fourth-order elasto-plastic constitutive tensor ${}^tC_{ijkl}^{LC}$ is assumed to be equal to ${}^tC_{ijkl}^{EP}$. Then, Eq. (9.50) leads to the following bifurcation criterion:

$$F = 0 \tag{9.51}$$

where

$$F = \int_{{}^tV} {}^tC_{ijkl}^{LC}\, {}_t\Delta\tilde{\varepsilon}_{kl}\, {}_t\Delta\tilde{\varepsilon}_{ij}\, d^tV + \int_{{}^tV} {}^t\sigma_{ij}\, {}_t\Delta\tilde{u}_{k,i}\, {}_t\Delta\tilde{u}_{k,j}\, d^tV \tag{9.52}$$

9.7 FINITE ELEMENT FORMULATION FOR THE IMPLEMENTATION OF HILL'S BIFURCATION CRITERION

In finite element method, the domain is discretized into a number of elements. In many finite element studies on deep drawing, the shell formulation has been used and thus, the domain has been the mid-surface of the sheet. In this case, a special strategy is needed to take care of the simultaneous contact of the sheet with the die on the bottom side and the punch and blank holder on the top side. On the other hand, if the sheet is considered as a 3-D domain, it is easy to handle this contact. In this section, the sheet is treated as 3-D domain, and therefore, solid finite elements are used to discretize the sheet. Further, the size of the elements in the plane of the sheet is chosen sufficiently small to avoid deterioration of the solution [28]. After the discretization, the incremental displacement field over a typical solid element is approximated as

$$\begin{Bmatrix} {}_t\Delta u_x \\ {}_t\Delta u_y \\ {}_t\Delta u_z \end{Bmatrix} = {}^t[\Phi]\ {}_t\{\Delta u\}^e \tag{9.53}$$

where the vector ${}_t\{\Delta u\}^e$ contains the incremental displacement components at the element nodes and the matrix ${}^t[\Phi]$ contains the shape functions that are known functions of the coordinates. Substitution of the approximation [Eq. (9.53)] into the virtual work expression [Eq. (9.44)] and assembly over all the elements leads to the following algebraic equation:

$${}^t[K]\ {}_t\{\Delta u\} + {}^t\{f\} = {}^{t+\Delta t}\{F\} \tag{9.54}$$

Here, ${}_t\{\Delta u\}$ represents the global incremental displacement vector and ${}^t[K]$, ${}^t\{f\}$ and ${}^{t+\Delta t}\{F\}$, respectively, denote the global coefficient matrix (at time t), global internal force vector (at time t) and global external force vector (at time $t + \Delta t$), the expressions for which are given in the thesis of Saxena [26]. Since

$${}^{t+\Delta t}[F] = {}^t[F] + {}_t\{\Delta F\}, \quad \text{and} \quad {}^t\{F\} = {}^t\{f\} \tag{9.55}$$

Eq. (9.53) can be written as

$${}^t[K]\ {}_t\{\Delta u\} = {}_t\{\Delta F\} \tag{9.56}$$

The solution of Eq. (9.56) represents only an approximate solution to the governing equations, because of the linearization and approximation involved in arriving at expression (9.44). To minimize this error, the modified Newton-Raphson iterative scheme [22, 23] is used. In this scheme, the following equation is solved iteratively:

$${}^t[K]\ {}_t\{\Delta u\}^{(i)} = {}^{t+\Delta t}\{R\}^{(i-1)} \quad \text{for } i = 1, 2, \ldots \tag{9.57}$$

where

$$\begin{aligned} {}^{t+\Delta t}[R]^{(i-1)} &= {}^{t+\Delta t}\{F\} - {}^{t+\Delta t}\{f\}^{(i-1)} \quad \text{for } i = 2, 3, \ldots \\ {}^{t+\Delta t}[R]^{(0)} &= {}_t\{\Delta F\} \end{aligned} \tag{9.58}$$

The iterations are continued till the (global) unbalanced force vector ${}^{t+\Delta t}\{R\}^{(i)}$ becomes sufficiently small.

After solving Eq. (9.57), the net global incremental displacement vector

$${}_t\{\Delta u\} = \sum_i {}_t\{\Delta u\}^{(i)} \tag{9.59}$$

is used to compute first the incremental logarithmic strain tensor using Eqs (9.24)–(9.25) and then the incremental stress tensor using Eq. (9.35). The integration in Eq. (9.35) is carried out using the Euler forward integration scheme. Finally, the Cauchy stress tensor at time $t + \Delta t$ is obtained using the updating procedure described in reference [27].

Substitution of the approximation [Eq. (9.53)] into the bifurcation criterion [Eqs (9.51)–(9.52)] and assembly over all the elements leads to the following equation:

$$_t\{\Delta\tilde{u}\}\ ^t[K^{LC}]\ _t\{\Delta\tilde{u}\} = 0 \tag{9.60}$$

where the expression for $^t[K^{LC}]$ is obtained from that of $^t[K]$ after replacing $^tC^{EP}_{ijkl}$ by $^tC^{LC}_{ijkl}$. The above equation leads to the condition

$$\det\ ^t[K^{LC}] = 0 \tag{9.61}$$

This is a discrete form of the Hill's bifurcation criterion. If unloading never occurs, $^t[K^{LC}]$ coincides with $^t[K]$ and Eq. (9.61) becomes

$$\det\ ^t[K] = 0 \tag{9.62}$$

To predict bifurcation, determinant of the coefficient matrix is checked at every Gauss point (of the plastic zone) after the convergence of each increment. Whenever Eq. (9.62) is satisfied, the state is considered as the onset of bifurcation.

Chu and Xu [29] used Hill's [14] bifurcation criterion for studying the onset of wrinkling in circular cup drawing. They analyzed only the flange region by treating it as an annular plate free at the outer edge and supported at the inner edge. Further, they used Barlat's 1991 anisotropic yield function [21], rather than Barlat's 2005 anisotropic yield function [24] [Eqs (9.26)–(9.29)] to model the anisotropy of the blank. As a result, their bifurcation criterion is similar to Hutchinson's bifurcation criterion for plates and shells [Eqs (9.18)–(9.19)] except that the fourth order (anisotropic) elasto-plastic constitutive tensor is obtained from the Barlat's 1991 anisotropic yield function [21]. In their work, they modeled the blank-holder as an elastic support of spring stiffness k. They assumed the transverse displacement $w\ (r, \theta)$ of the mid-surface of the flange as the product of a linear function or r-coordinate and some trigonometric function of θ in terms of the number of wrinkles n. They used the slab method to determine the in-plane membrane stresses (σ_{rr}, $\sigma_{\theta\theta}$). Using the bifurcation criterion, they obtained the graph of the critical drawing stress verses the ratio of the current outer radius of the flange to the inner radius of the flange.

Saxena and Dixit [30] employed Hill's [14] bifurcation criterion for predicting the maximum cup height at the onset of wrinkling in both circular as well as square cup using the finite element method. They used the governing equations (including the Barlat's 2005 anisotropic yield criterion [24] for modeling the anisotropy of the blank) as well as the finite element formulation described above. Thus, they used the discretized form of Hill's bifurcation criterion given by Eq. (9.62). An in-house 3-D FE code is developed by them based on the formulation explained earlier. They performed the parametric study of the maximum cup height (that can be drawn without the onset of wrinkling) with respect to the parameters namely blank-holder pressure, sheet thickness, die profile radius, punch profile radius, sheet anisotropy and other material properties.

9.8 AN EXAMPLE ON THE USE OF HILL'S BIFURCATION CRITERION FOR THE PREDICTION OF FLANGE WRINKLING IN DEEP DRAWING

The results from Saxena and Dixit [30] on the parametric study of flange wrinkling in circular and square cup drawing are presented in this section. The geometry of the problem for circular cup drawing is shown in Fig. 9.3 and for square cup drawing in Fig. 9.4. The X, Y and Z axes are assumed to be aligned along the rolling (RD), transverse (TD) and normal directions of the sheet respectively. The punch, die and blank-holder are assumed to be rigid. Due to the symmetry in geometry and loading, only one-quarter of the blank is considered for the analysis (Fig. 9.5). Sliding friction is assumed at the sheet-die interface. Sticking friction is assumed at the sheet-punch interface, which is a fair assumption considering the large compressive force generated at the sheet-punch interface. As the deformation progresses in the sheet, the position of the nodes and the related boundary conditions are affected. Therefore, the boundary conditions are updated in each increment. Flags are assigned to the nodes according to their position, to designate the change in the boundary conditions.

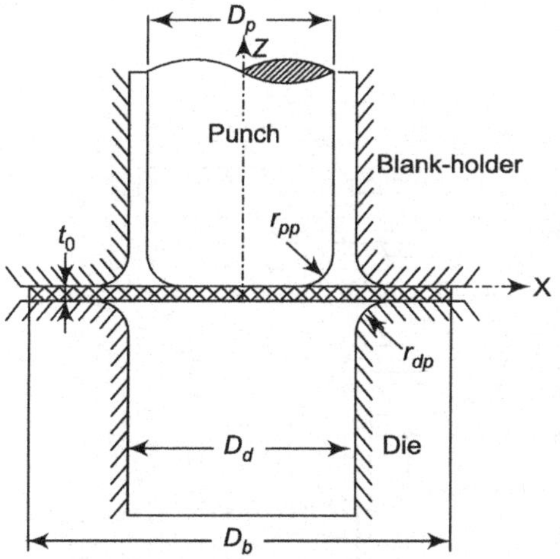

Fig. 9.3 Schematic diagram of circular cup drawing process; Symbols: D_b Blank diameter, t_0 Blank thickness, D_p Punch diameter, D_d Die opening diameter, r_{pp} Punch profile radius, r_{dp} Die profile radius [26]

The algorithm for the contact problem assumes that the tooling geometry is represented by a set of straight line segments and circular/elliptical arcs. Such a representation of tooling geometry makes it convenient to apply the contact conditions. At the beginning of each increment, the contact state from the previous increment is used. At the end of the increment, the contact state is again checked. Further, if it appears that for i^{th} contact node, the sign of contact force has changed, the node is declared to be free and if j^{th} non-contact node penetrates the punch/die/blank-holder profile, the node is projected on

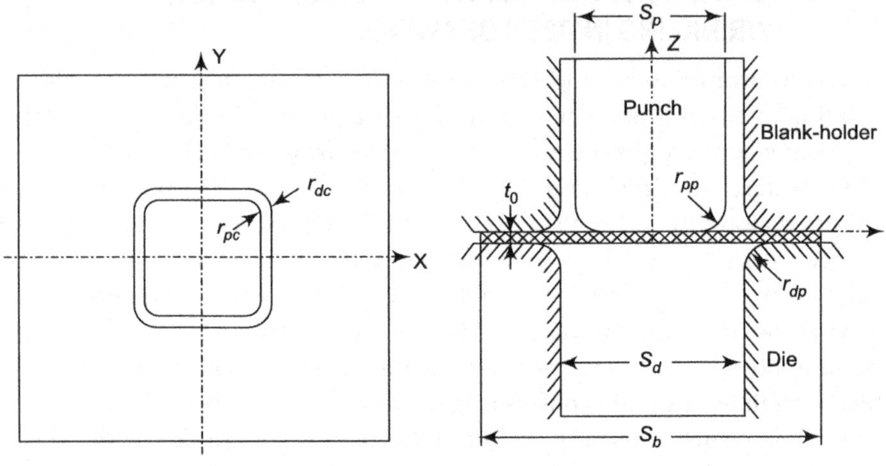

Fig. 9.4 Schematic diagram of square cup drawing process: S_b Blank size, t_0 Blank thickness, S_p Punch size, S_d Die size, r_{pp} Punch profile radius, r_{dp} Die profile radius, r_{pc} Punch corner radius, r_{dc} Die corner radius [26]

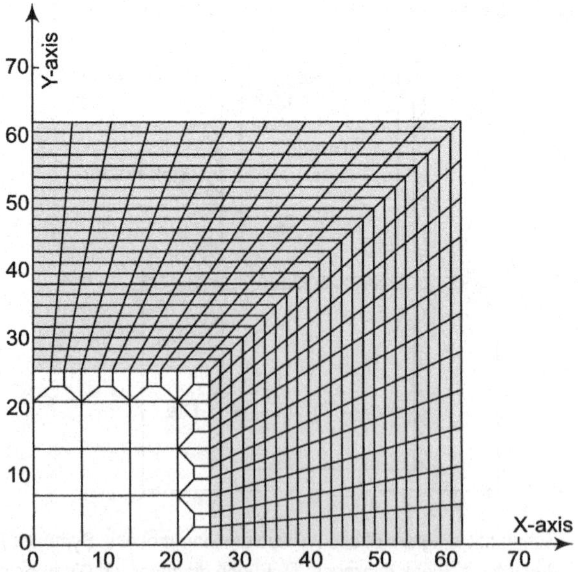

Fig. 9.5 Discretized blank for square cup drawing [26]

the profile by giving it the specified displacement. Taking into account such a change of contact state, the computational procedure of the current increment is repeated.

It is assumed that the applied blank-holder force is uniformly distributed on the top surface of the blank-holder. In order to simulate the change in the gap between the die and the bottom surface of the blank-holder (called the blank-holder gap), it is assumed that the top surface of the block-holder is connected to a series of parallel springs whose resultant stiffness if k. The value of k (assumed constant) is one of the data to be supplied for the wrinkling analysis. This value is selected so as to keep the difference between blank-holder force and the reaction force (from the sheet) on the bottom surface of the blank-holder within a specified tolerance. After each increment, the blank-holder gap is determined from the equilibrium of the blank-holder using the spring stiffness k and the resultant of the blank-holder force and the reaction force. This gap is always kept greater than or equal to the current value of the sheet thickness. If there is a change in the blank-holder gap, the computational procedure of the current increment is repeated using the new value of the blank-holder gap. The transition to the next increment is accomplished only if there is no change in the contact state and the blank-holder gap after such a recalculation.

The geometric dimensions used for the study are given in Table 9.1. The coefficient of friction used is $\mu = 0.05$. The sheet size/diameter and the coefficient of friction are kept constant in the parametric study. The materials used for the parametric study are: AA2090-T3 which is a highly anisotropic material and AA6111-T4 which is mildly anisotropic in nature [24]. The material properties for these materials are given in Tables 9.2–9.5. The number of finite elements used for the analysis is 1390. There are two elements along the thickness direction. The results are expressed in terms of the normalized cup height d/b for various values of the different process parameters where d is the cup height at the onset of wrinkling and b is the sheet-diameter for circular cup and sheet-size for square cup.

Table 9.1 Geometric dimensions [26]

Circular cup		Square cup	
Blank thickness (mm)	0.9	Blank thickness (mm)	0.9
Blank diameter (mm)	140.0	Blank size (mm)	124.0
Punch diameter (mm)	85.0	Punch size (mm)	70.0
Die opening diameter (mm)	88.0	Die size (mm)	74.0
Die profile radius (mm)	8.0	Die profile radius (mm)	5.0
Punch profile radius (mm)	7.0	Punch profile radius (mm)	8.0
		Punch corner radius (mm)	10.0
		Die corner radius (mm)	12.0

Table 9.2 Material properties [26]

AA2090-T3	
Young's modulus (E)	69 GPa
Poisson's ratio (ν)	0.33
Initial yield stress ($(\sigma_y)_0$)	280 MPa
Material exponent (a)	8
Stress-strain curve (Stress in MPa)	$\sigma_y = 646.0\ (0.025 + \varepsilon_{eq}^{pL})^{0.227}$
AA6111-T4	
Young's modulus (E)	69 GPa
Poisson's ratio (ν)	0.33
Initial yield stress ($(\sigma_y)_0$)	185.92 MPa
Material exponent (a)	8
Stress-strain curve (Stress in MPa)	$\sigma_y = 454.76 - 268.84\ \exp\ (-\ 6.45\ \varepsilon_{eq}^{pL})$

Table 9.3 Yield stress data for AA2090-T3 and AA6111-T4 [26]

	σ_0/σ_y	σ_{15}/σ_y	σ_{30}/σ_y	σ_{45}/σ_y	σ_{60}/σ_y	σ_{75}/σ_y	σ_{90}/σ_y
AA2090-T3	1.0000	0.9605	0.9102	0.8114	0.8096	0.8815	0.9102
AA6111-T4	0.9910	0.9924	0.9802	0.9602	0.9501	0.9491	0.9399

Table 9.4 Strain rate ratios for AA2090-T3 and AA6111-T4 [26]

	r_0	r_{15}	r_{30}	r_{45}	r_{60}	r_{75}	r_{90}
AA2090-T3	0.2115	0.3269	0.6923	1.5769	1.0385	0.5384	0.6923
AA6111-T4	0.8891	0.8945	0.7442	0.6131	0.6139	0.6333	0.6637

Table 9.5 Yield function coefficients for AA2090-T3 and AA6111-T4 [26]

	AA2090-T3	AA6111-T4
α_1	0.0698	1.241024
α_2	0.9364	1.078271
α_3	0.0791	1.216463
α_4	1.0030	1.223867
α_5	0.5247	1.093105
α_6	1.3631	0.889161
α_7	1.0237	0.501909
α_8	1.0690	0.557173
α_9	0.9543	1.349094
α_{10}	0.9811	0.775366
α_{11}	0.4767	0.922743
α_{12}	0.5753	0.765487
α_{13}	0.8668	0.793356
α_{14}	1.1450	0.918689
α_{15}	0.0792	1.027625
α_{16}	1.0516	1.115833
α_{17}	1.1471	1.112273
α_{18}	1.4046	0.589787

9.8.1 Effect of Blank-Holder Pressure

The blank-holder pressure is one of the important process parameters controlling the wrinkling in deep drawing. The change in the onset of wrinkling is studied for various values of the normalized blank-holder pressure. The normalized blank-holder pressure is defined as $((F_b/A)/(\sigma_y)_0)$ where, F_b is the blank-holder force and A is the current area under the blank-holder. The analysis is performed with the values of $F_b = 6, 9, 12$, and 14 kN for circular cup drawing and $F_b = 3, 6, 9$, and 12 kN for square cup drawing. For both the cases, the sheet thickness is the same (*i.e.*, 0.9 mm) as given in Table 9.1. Further, the tooling geometry dimensions are the same as given in Table 9.1. The material is AA2090-T3. Figures 9.6 and 9.7 show the normalized cup height d/b at the onset of wrinkling with respect to the normalized blank-holder pressure.

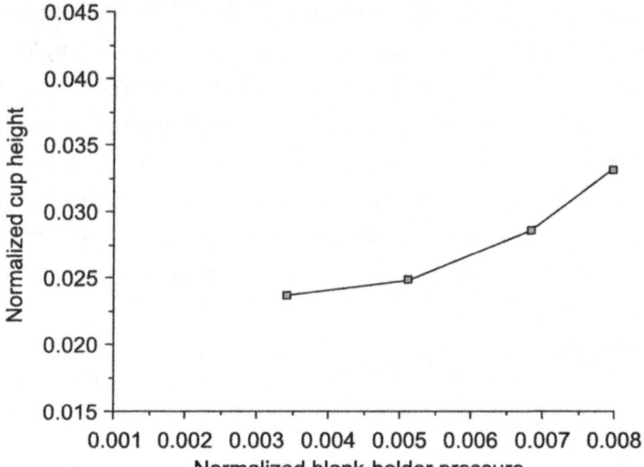

Fig. 9.6 Effect of blank-holder pressure on the onset of wrinkling for circular cup (With permission from Elsevier, copyright 2011) [30]

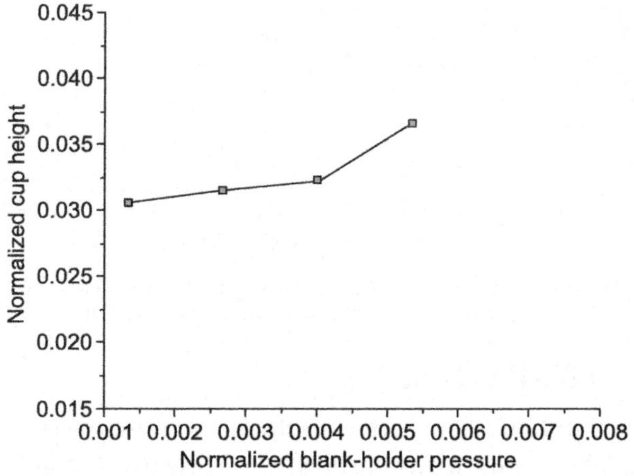

Fig. 9.7 Effect of blank-holder pressure on the onset of wrinkling for square cup (With permission from Elsevier, copyright 2011) [30]

It is observed from Figs 9.6 and 9.7 that the onset of wrinkling gets delayed with an increase in the blank-holder pressure. That is, the onset of wrinkling occurs at a higher cup height if the blank-holder pressure is increased. This is in confirmation with the fact that the higher blank-holder pressure delays the wrinkling [1]. However, the effect of blank-holder pressure is less prominent in square cup drawing.

9.8.2 Effect of Sheet Thickness

The change in the onset of wrinkling is studied for the sheet thicknesses: $t = 0.80$, 0.90, 1.0, and 1.1 mm. The analysis is carried out with the total blank-holder force of 9.0 kN for circular cup and 6.0 kN for square cup drawing. The tooling geometry remains the same. The material is AA2090-T3. Figures 9.8 and 9.9 show the variations of the normalized cup height d/b at the onset of wrinkling with respect to the normalized sheet thickness t/b for circular and square cup drawing processes respectively. The onset of wrinkling shifts towards a higher cup height with an increase in the sheet thickness. The result is in agreement with the observation of reference [31], that a thicker blank has lesser tendency to wrinkle. However, the trend of the variation of the maximum cup height (at the onset of wrinkling) with the thickness is different in square cup drawing than in circular cup drawing.

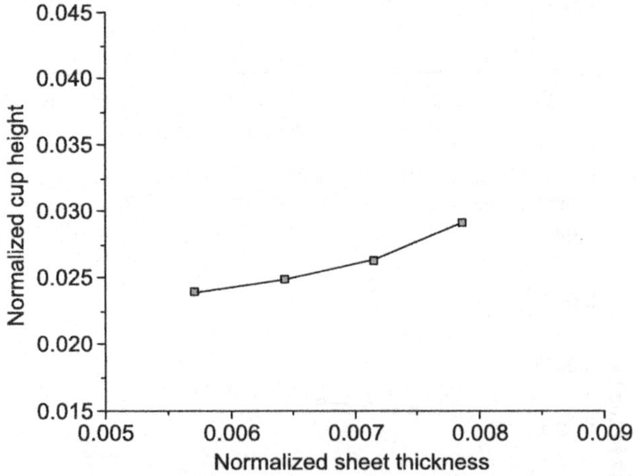

Fig. 9.8 Effect of sheet thickness on the onset of wrinkling for circular cup (With permission from Elsevier, copyright 2011) [30]

9.8.3 Effect of Die Profile Radius

The die dimension has an important influence on the onset of wrinkling in deep drawing. The change in the onset of wrinkling for circular cup drawing is studied for the die profile radii: $r_{dp} = 4.0$, 6.0, 8.0, and 10.0 mm. The total blank-holder force is 9.0 kN. For the square cup drawing, the analysis is carried out

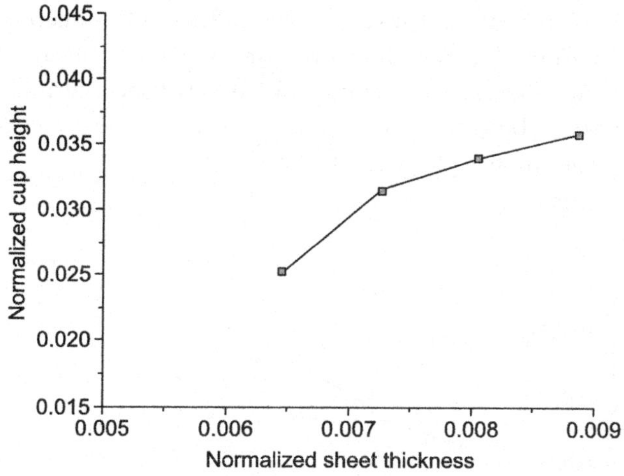

Fig. 9.9 Effect of sheet thickness on the onset of wrinkling for square cup (With permission from Elsevier, copyright 2011) [30]

for the die profile radii: r_{dp} = 2.0, 3.0, 4.0, and 5.0 mm. The total blank-holder force is 6.0 kN. For both the cases, the sheet thickness remains the same (0.9 mm). The other dimensions of the tooling geometry also remain the same. The material is AA2090-T3. Figures 9.10 and 9.11 show the normalized cup height d/b at the onset of wrinkling for various values of the normalized die profile radius r_{dp}/b for circular and square cup drawing processes respectively. It is observed that the onset of wrinkling gets delayed with an increase in the die profile radius. The reason for this is as follows.

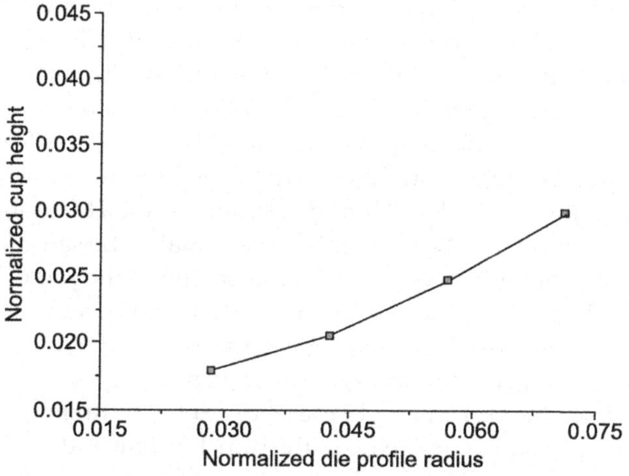

Fig. 9.10 Effect of die profile radius on the onset of wrinkling for circular cup (With permission from Elsevier, copyright 2011) [30]

The area of the flat portion of the die decreases with an increase in the die profile radius, thereby reducing the area under the blank-holder. Thus, for a constant value of the blank-holder force, the value of blank-holder pressure increases with an increase in the die profile radius. It has been shown earlier that an increase in the blank-holder pressure increases the cup height at the onset of wrinkling. Therefore, the cup height at the onset of wrinkling increases with an increase in the die profile radius. However, the effect of die profile radius is less prominent in square cup drawing.

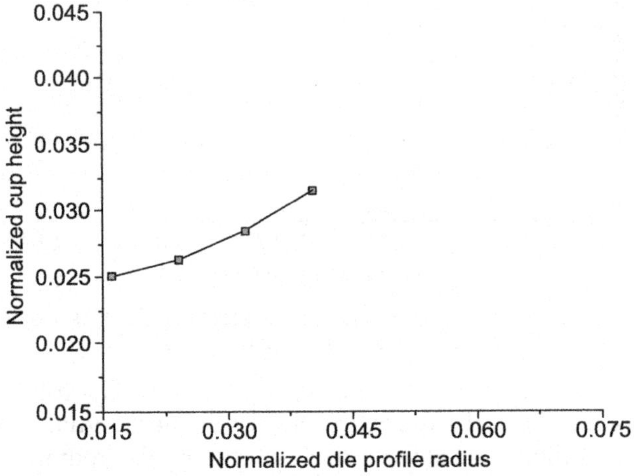

Fig. 9.11 Effect of die profile radius on the onset of wrinkling for square cup (With permission from Elsevier, copyright 2011) [30]

9.8.4 Effect of Punch Profile Radius

The punch profile also has an important influence on the onset of wrinkling in deep drawing. The change in the onset of wrinkling in deep drawing is studied for different punch profile radii in this section. The analysis for circular cup drawing process is carried out for the following values of punch profile radii: r_{pp} = 5.0, 7.0, 8.5, and 10.0 mm. The total blank-holder force is 9.0 kN. For square cup drawing, the analysis is performed for the following values of punch profile radii: r_{pp} = 4.0, 5.0, 8.0, and 9.0 mm. The total blank-holder force is 6.0 kN. For both the cases, the sheet thickness remains the same (0.9 mm). The other dimensions of the tooling geometry also remain the same. The material is AA2090-T3. Figures 9.12 and 9.13 show the variations of the normalized cup height d/b at the onset of wrinkling with respect to the normalized punch profile radius r_{pp}/b for circular and square cup drawing processes respectively. It is observed from Figs 9.12 and 9.13 that the onset of wrinkling shifts towards a higher cup height with an increase in the punch profile radius. However, the effect of punch profile radius is more significant in square cup drawing.

9.8.5 Effect of Sheet Anisotropy

The effect of sheet anisotropy on the onset of wrinkling is studied for the material AA2090-T3. The results for isotropic case are obtained by assuming that all the anisotropy parameters (α_i) in Eq. (9.29) are equal to one and

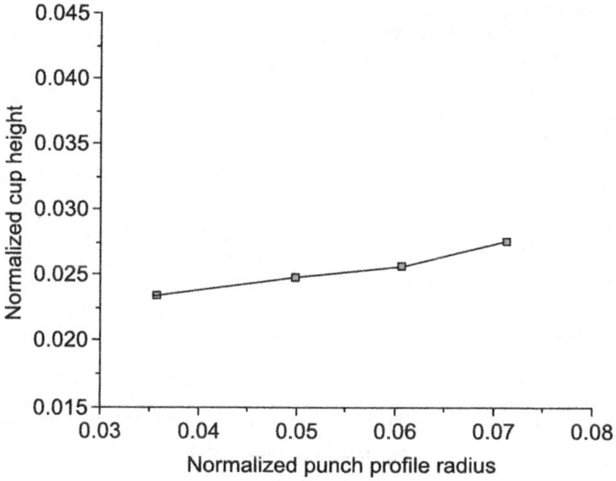

Fig. 9.12 Effect of punch profile radius on the onset of wrinkling for circular cup (With permission from Elsevier, copyright 2011) [30]

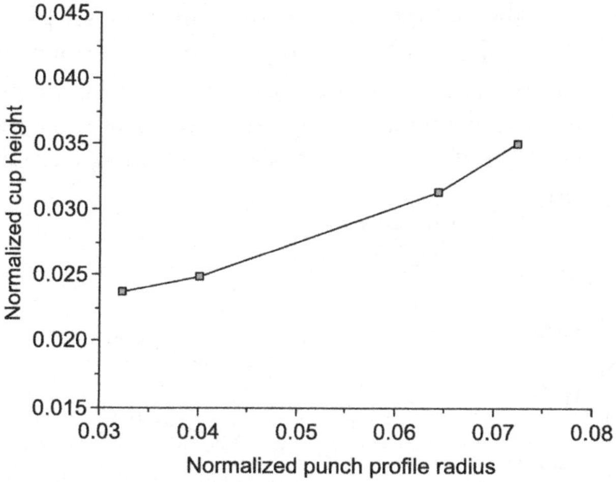

Fig. 9.13 Effect of punch profile radius on the onset of wrinkling for square cup (With permission from Elsevier, copyright 2011) [30]

the exponent (a) in Eq. (9.27) is 2. The change in the onset of wrinkling is studied for the following sheet thicknesses: $t = 0.80$, 0.90, 1.0, and 1.1 mm. The analysis is carried out with the total blank-holder force of 9.0 kN for circular cup and 6.0 kN for square cup drawing. The tooling geometry remains the same. Figures 9.14 and 9.15 show the variations of the normalized cup height d/b at the onset of wrinkling with respect to the normalized sheet thickness t/b for circular and square cup drawing respectively.

Wrinkling is facilitated by the compressive stress in the circumferential direction. This compressive stress depends on the variation of the yield stress along the circumferential direction. When the sheet is treated as isotropic, the yield stress is assumed to be the same in all directions. On the other hand, in

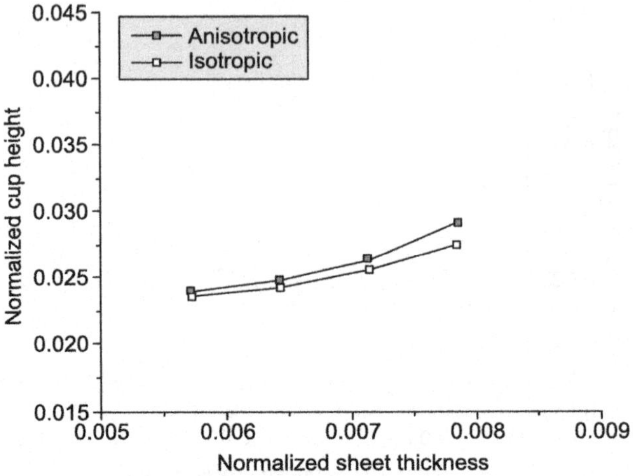

Fig. 9.14 Effect of sheet anisotropy on the onset of wrinkling for circular cup (With permission from Elsevier, copyright 2011) [30]

an anisotropic sheet, the yield stress decreases continuously through 90° from the rolling direction to achieve the minimum value in the transverse direction. Thus, the resistance of a sheet to wrinkling depends on its anisotropy. Figures 9.14 and 9.15 indicate that, this wrinkling resistance is more for the anisotropic case. Further, this effect of anisotropy is significant in square cup drawing.

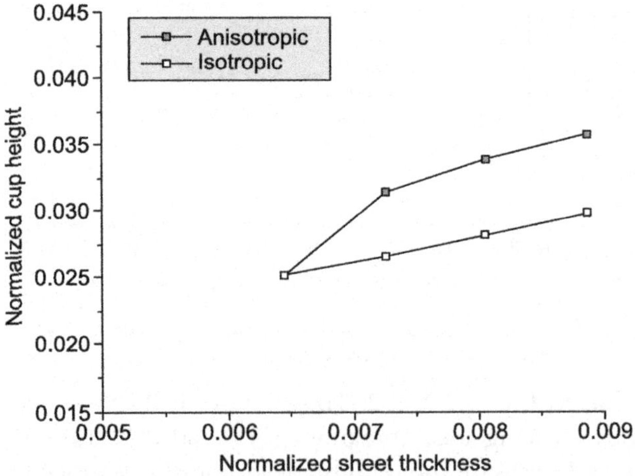

Fig. 9.15 Effect of sheet anisotropy on the onset of wrinkling for square cup (With permission from Elsevier, copyright 2011) [30]

9.8.6 Effect of Material Properties

Material properties have an important influence on the onset of wrinkling in deep drawing. The change in onset of wrinkling is studied for AA2090-T3 and AA6111-T4 materials for the sheet thicknesses: $t = 0.80, 0.90, 1.0$, and 1.1 mm. The analysis is carried out with the total blank-holder force of 9.0 kN for circular cup and 6.0 kN for square cup drawing. The tooling geometry remains the same. Figures 9.16 and 9.17 show the variations of the normalized cup height *d*/*b* at the onset of wrinkling with respect to the normalized sheet thickness *t*/*b* for circular and square cup drawing respectively.

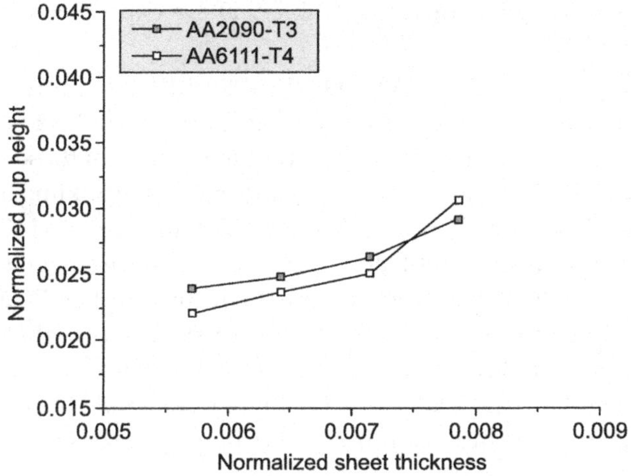

Fig. 9.16 Effect of material properties on the onset of wrinkling for circular cup (With permission from Elsevier, copyright 2011) [30]

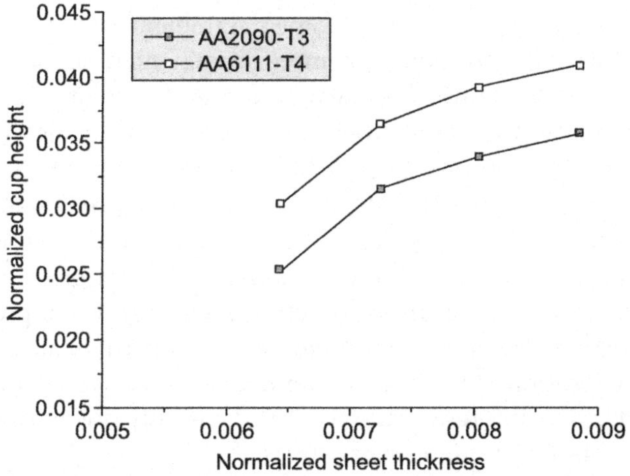

Fig. 9.17 Effect of material properties on the onset of wrinkling for square cup (With permission from Elsevier, copyright 2011) [30]

It is to be noted that, besides anisotropy, the initial yield stress $(\sigma_y)_0$ in the rolling direction, hardening coefficient *K* and the hardening exponent *n* also affect the onset of wrinkling. Therefore, to understand the reasons behind the trends of Figs. 9.16 and 9.17, parametric studies with respect to $(\sigma_y)_0$, *K* and *n* are performed. It is observed that the hardening coefficient *K* has no effect on the onset of wrinkling. But, when $(\sigma_y)_0$ is increased, the onset of wrinkling occurs at a lower cup height in both square as well as circular cup. However, when n is increased, the onset of wrinkling occurs at a higher cup height in square cup but at a lower cup height in circular cup. To study the influence of K and n on the onset of wrinkling, the stress-strain relationships of AA2090-T3 and AA6111-T4 materials (given in Table 9.2) are expressed as power laws given by Eq. (9.32).

The initial yield stress of AA2090-T3 (280 MPa) is more than that of AA6111-T4 (185.92 MPa) and the hardening exponent of AA2090-T3 (0.25) is less than that of AA6111-T4 (0.35). The combined effect of these two material properties is additive in square cup. Thus, the onset of wrinkling in square cup is observed at a lower cup height in AA2090-T3 than in AA6111-T4. However, in circular cup, the effects of $(\sigma_y)_0$ and n are opposite. In circular cup, the critical cup height for AA2090-T3 should be less than for AA6111-T4 due to larger $(\sigma_y)_0$ but should be more due to lower n. Figure 9.16 shows that the effect of n is more dominant at lower values of sheet thickness thus resulting in larger critical cup height for AA2090-T3. But at higher values of sheet thickness, the effect of $(\sigma_y)_0$ becomes more prominent leading to lower critical cup height for AA2090-T3.

9.8.7 Discussion and Conclusions

As pointed out earlier, the effects of some parameters are more significant in square cup while that of the others are more prominent in circular cup. Further, the trends of some effects are different in square and circular cups. The reasons for these differences are: (i) the difference in the flange geometry, and (ii) the difference in the way the anisotropy affects the state of the stress in the flange.

Equation (9.52) suggests that the onset of wrinkling depends on the state of stress in the flange. The significant stress components of this state of stress are the tensile stress in the radial direction and the compressive stresses in the circumferential and thickness directions. Shear stresses, present due to the friction, are smaller in magnitude. In circular cup, this state of stress is the same at every location of the circumferential stress. However, in square cup, the state of stress is different at the corners, especially due to the presence of significant amount of in-plane shear stress.

As stated earlier, the wrinkling is facilitated by the compressive stress in the circumferential direction. This compressive stress depends on the yield stress which, in anisotropic materials, varies continuously through 90° from the rolling direction. In square cup, the circumferential direction is either the rolling direction or the transverse direction. But, in circular cup, the circumferential direction varies continuously through 90° from the rolling direction. Therefore, the dependence of the compressive circumferential stress on the variable yield stress (σ_y) of the material is more complex in circular cup than in square cup.

The specific conclusions of the study of Saxena and Dixit [30] are as follows: (i) As expected, an increase in the blank-holder pressure or the sheet thickness delays the onset of wrinkling, i.e., the maximum cup height at the onset of wrinkling increases with the blank-holder pressure or the sheet thickness, (ii) The maximum cup height at the onset of wrinkling increases with an increase in the die profile radius or punch profile radius. The effect of the die profile radius is more significant in the circular cup drawing whereas that of the punch profile radius is more significant in the square cup drawing, (iii) The sheet anisotropy, in general, delays the onset of wrinkling. The effect of anisotropy is more significant in the square cup drawing in comparison to the circular cup drawing, and (iv) The material properties affect the maximum cup height at the onset of wrinkling differently in the circular and the square cup drawing processes.

SUMMARY

In this work, a study on the prediction of *flange wrinkling,* in square and circular cups is performed using finite element (FE) method. An in-house FE code, based on the updated Lagrangian formulation, is employed. An eight-noded solid element, that can handle the contact conditions between the sheet and tool effectively and can incorporate both the shear and bending effects, is used. Incremental logarithmic strain measure is used, which allows the use of a larger increment size. To make the incremental stress objective, a stress updating procedure which incorporates the incremental rotation is used. The material is assumed to be elastic-plastic strain hardening. Elastic behavior is assumed to be isotropic. Anisotropic yield criterion of Barlat *et al.* with two linear transformations (named as Yld2004-18p) is used to model the plastic anisotropy of the sheet material. The strain hardening behavior is assumed to be isotropic and modeled by a power law. The inertial and body forces are neglected. The effects of temperature and strain rate (viscoplasticity effects) on the yield strength of the material are ignored in this work. Modified Newton-Raphson method is used to solve the resultant non-linear finite element equations.

The flange wrinkling is treated as an instability of the flat configuration and thus, analyzed as a bifurcation problem. Hill's bifurcation criterion, based on his general theory of uniqueness and stability, is incorporated in the FE code to predict the onset of flange wrinkling. Further, the FE code is used to study the influence of various geometric, material and other process parameters on the maximum cup height that can be drawn without the onset of wrinkling.

REFERENCES

1. Sachs, G., *Principles and methods of sheet metal fabricating,* New York: Reinhold Publishing Company, 1966.
2. Senior, B.W., Flange wrinkling in deep-drawing operations, *Journal of the Mechanics and Physics of Solids,* Volume 4, pp.235–246, 1956.
3. Yu, T.X. and Johnson, W., The buckling of annular plates in relation to the deep-drawing process, *International Journal of Mechanical Sciences,* Volume 3, pp.175–188, 1982.

4. Timoshenko, S.P. and Gere, J.M., *Theory of elastic stability*, New York: McGraw-Hill Book Company, 1961.
5. Agrawal, A., Reddy, N.V. and Dixit, P.M., Determination of optimum process parameters for wrinkle free products in deep drawing process, *Journal of Materials Processing Technology*, Volume 191, pp. 51–54, 2007.
6. Yu, T.X. and Zhang, L.C., The elastic wrinkling of an annular plate under uniform tension on its inner edge, *International Journal of Mechanical Sciences*, Volume 28, pp. 729–737, 1986.
7. Zeng, X.M. and Mahdavian, S.M., Critical conditions of wrinkling in deep drawing at elevated temperature, *Journal of Materials Processing Technology*, Volume 84, pp. 38–46, 1998.
8. Cao, J. and Boyce, M.C., Wrinkling behavior of rectangular plates under lateral constraint, *International Journal of Solids and Structures*, Volume 34, pp. 153–176, 1997.
9. Cao, J. and Wang, X., An analytical model for plate wrinkling under tri-axial loading and its application, *International Journal of Mechanical Sciences*, Volume 42, pp. 617–633, 2000.
10. Hill, R., A theory of the yielding and plastic flow of anisotropic metals, *Proceedings of the Royal Society of London*, Volume 193, pp. 281–297, 1948.
11. Hutchinson, J.W. and Neale, K.W., Wrinkling of curved thin sheet metal, *International Symposium on Plastic Instability*, Presses Ponts et Chaussées Paris, pp. 71–78, 1985.
12. Han, S.S. and Huh, H., Modified-membrane finite-element simulation of square cup drawing processes considering influence of geometric parameters, *Journal of Material Processing Technology*, Volume 48, pp. 81–87, 1995.
13. Weili, X., Huibao, W., Yuying, Y. and Wang, Z.R., A simplified method of wrinkling simulation, *Journal of Materials Processing Technology*, Volume 121, pp. 19–22, 2002.
14. Hill, R., A general theory of uniqueness and stability in elastic-plastic solids, *Journal of the Mechanics and Physics of Solids*, Volume 6, pp. 236–249, 1958.
15. Hutchinson, J.W., *Plastic buckling, Advances in Applied Mechanics*, Volume 14, pp. 67–144, 1974.
16. Correia, J.P.D.M. and Ferron, G., Wrinkling predictions in the deep-drawing process of anisotropic metal sheets, *Journal of Material Processing Technology*, Volume 128, pp. 178–190, 2002.
17. Ferron, G., Makkouk, R. and Morreale, J., A parametric description of orthotropic plasticity in metal sheets, *International Journal of Plasticity*, Volume 10, pp. 431–449, 1994.
18. Riks, E., An incremental approach to the solution of snapping and buckling problems, *International Journal of Solids and Structures*, Volume 15, pp. 529–551, 1979.
19. Kim, J.B., Yoon, J.W., Yang, D.Y. and Barlat, F., Investigation into wrinkling behavior in the elliptical cup deep drawing process by finite element analysis using bifurcation theory. *Journal of Materials Processing Technology*, Volume 111, pp. 170–174, 2001.

20. Kim, J.B., Yoon, J.W. and Yang, D.Y., Investigation into the wrinkling behavior of thin sheets in the cylindrical cup deep drawing process using bifurcation theory, *International Journal for Numerical Methods in Engineering*, Volume 56, pp. 1673–1705, 2003.

21. Barlat, F., Lege, D.J. and Brem, J.C., A six-component yield function for anisotropic materials, *International Journal of Plasticity*, Volume 7, pp. 693–712, 1991.

22. Bathe, K.J., *Finite element procedures*, New Delhi: Prentice-Hall of India, 1996.

23. Dixit, P.M. and Dixit, U.S., *Modeling of Metal Forming and Machining Processes*, London: Springer-Verlag, 2008.

24. Barlat, F., Aretz, H., Yoon, J.W., Karabin, M.E., Brem, J.C. and Dick, R.E., Linear transformation-based anisotropic yield functions, *International Journal of Plasticity*, Volume 21, pp. 1009–1039, 2005.

25. Saxena, R.K. and Dixit, P.M., Finite element simulation of earing defect in deep drawing, *International Journal of Advanced Manufacturing Technology*, Volume 45, pp. 219–233, 2009.

26. Saxena, R.K., Studies on deep drawing defects using finite element method: earing, wrinkling and damage, *Doctoral Thesis*, Mechanical Engineering Department, IIT Kanpur, 2009.

27. Varadhan, S.N., Dynamic large deformation elasto-plastic analysis of continua, *Master's Thesis*, Mechanical Engineering Department, IIT Kanpur, 1997.

28. Lange, K., Herrmann, M., Keck, P. and Wilhelm, M., Application of an elasto-plastic finite element code to the simulation of metal forming processes, *Journal of Materials Processing Technology*, Volume 27, pp. 239–261, 1991.

29. Chu, E. and Xu, Y., An elasto-plastic analysis of flange wrinkling in deep drawing process, *International Journal of Mechanical Sciences*, Volume 43, pp. 1421–1440, 2001.

30. Saxena, R.K. and Dixit, P.M., Prediction of flange wrinkling in deep drawing process using bifurcation criterion. *Journal of Manufacturing Processes*, Volume 12, pp. 19–29, 2010.

31. Schuler, G., Metal forming handbook, Berlin: Springer-Verlag, 1998.

CHAPTER 10

Formability of Tailor Welded Blanks

R. Ganesh Narayanan*, K. Narasimhan#
*Department of Mechanical Engineering,
Indian Institute of Technology Guwahati, Guwahati 781 039, India
#Department of Metallurgical Engineering and Materials Science,
Indian Institute of Technology Bombay, Mumbai 400 076, India

10.1 INTRODUCTION

Tailor Welded Blank (TWB) is made up of blanks with similar or dissimilar thickness, materials, coatings, etc. welded in a single plane before forming. This welded blank is then formed like un-welded blanks to manufacture automotive components, with tooling and forming conditions incorporating some changes. Some of the advantages of using TWBs in the automotive sector are as follows: (1) Scrap materials from stamping industries can be reused to have new stamped products, (2) By distributing material thickness and properties, part consolidation is possible which results in cost reduction and better quality, stiffness and tolerances, (3) Provides greater flexibility for component designers, (4) Weight reduction of the product can be achieved and hence fuel consumption is lowered, (5) Improved corrosion resistance and product quality, etc. Research shows that a 1% reduction in vehicle weight can result in a reduction of fuel consumption to 0.6–1% [1]. Lot of research was conducted on TWB technology as part of Ultra Light Steel Auto Body (ULSAB) consortium, right from designing to production to post TWB forming issues [2]. Some of the applications of TWB include center pillar, bumper, front door inner, rear door inner, etc. [3, 4]. The part details of various steel TWBs are given in Table 10.1 [2]. It is clear from the table that high strength steels

(σ_{ys} = 210–550 MPa) and Ultra high strength steels (σ_{ys} > 550 MPa) are being used to make TWB components. Also, it is shown in the table that multiple welds, linear or non-linear, are possible in the manufacture of TWB components. In the case of aluminum alloys, the 5000 series for automotive inner body panels, and 6000 series for outer body panels are mainly used. Specifically, these include 6111-T4, 5754-O, 6061, 5182, 5052 and 5454. Some of the applications of aluminum TWBs include deck lids, hoods, floor and door inner panels, side frame rails, etc. [5]. With this brief background on the evolution of tailor welded blanks, advantages and applications, this chapter aims at reviewing the forming behavior of tailor welded blanks in detail. The effect of different parameters like thickness ratio, strength ratio and various weld conditions on the forming behavior of TWB like limiting strains, dome height, weld line movement, etc. is dealt with in the first few sections. Weld line movement and existing control strategies are presented. Some of the important issues on modeling TWBs and formability prediction are discussed. Finally an expert system based on neural network to predict the formability of welded blanks is described with few important results.

10.2 FORMING BEHAVIOR OF TAILOR WELDED BLANKS

It is known that the forming behavior of un-welded sheet depends on material properties, forming parameters (σ_{ys}, n, m, r, t), process parameters (blank holding force, friction coefficient, surface roughness), and microstructure of the material. In addition to the above said parameters, the forming behavior of welded blanks is affected by thickness ratio (T_{thin}/T_{thick} or T_{thick}/T_{thin}), strength ratio ($\sigma_{ys\text{-}thin}/\sigma_{ys\text{-}thick}$ or $\sigma_{ys\text{-}thick}/\sigma_{ys\text{-}thin}$) of the sheets welded, weld conditions like weld orientation, weld location, weld properties, weld microstructure, etc. While the parameters, thickness ratio, strength ratio, weld orientation, weld location depend on materials selected for welding and application of TWB, weld properties (width, yield strength, strain hardening exponent) and weld microstructure are determined by materials welded, welding techniques and welding process parameters chosen. Figure 10.1 shows the schematic description of the parameters mentioned. The effect of these parameters on the forming behavior of TWB is discussed in this section.

10.2.1 Effect of Thickness Ratio (or Difference) on the Forming Behavior of TWB

Thickness ratio is a parameter which is defined specifically for TWB, as the ratio of one blank thickness to other blank thickness, which constitute TWB. The effect of this parameter on the forming behavior is studied in few published literature. The forming behavior analyzed are forming limit strains or curve (FLC), strain distribution, dome height, deep drawability, weld line movement, etc.

Table 10.1 Steel TWB products with part specifications [2]

TWB product	Number and shape of weld line				Gauge combination (mm)	Material combination *	Property combination (σ_{ys}, MPa)
	1	2	3	4			
Body side outer	Slanting	Bi-linear	--	--	1.5/1.0/0.7	DP/DP/BH	700/700/250
Wheel inner house	Vertical	Vertical	Horizontal	--	1.4/0.6/1.1/0.6	DP/DP/DP/DP	700/500/700/500
Floor rear	Vertical	Horizontal	Horizontal	--	0.7/0.6/1.1/1.1	DP/BH/DP/DP	700/210/350/350
Rear rail	Vertical	--	--	--	1.8/1.1	DP/DP	700/500
Rocker inner	Vertical	--	--	--	1.5/0.7	DP/DP	700/700
Reinforcement B-pillar rocker rear	Horizontal	--	--	--	1.2/1.4	DP/DP	700/700
Body side outer	Slanting	Slanting	Vertical	Slanting	0.7/1.5/1.0/1.2/0.7	BH/DP/DP/DP/BH	260/700/700/700/260
Rail rear outer floor extension	Horizontal	--	--	--	0.6/1.0	BH/DP	210/500
Front door inner front	Horizontal	--	--	--	1.0/1.2	Mild/Mild	140/140
Rear door inner front	Horizontal	--	--	--	1.0/1.2	Mild/Mild	140/140

* Material combination: DP: Dual Phase steel; BH: Bake Hardenable steel; Mild: Mild Steel

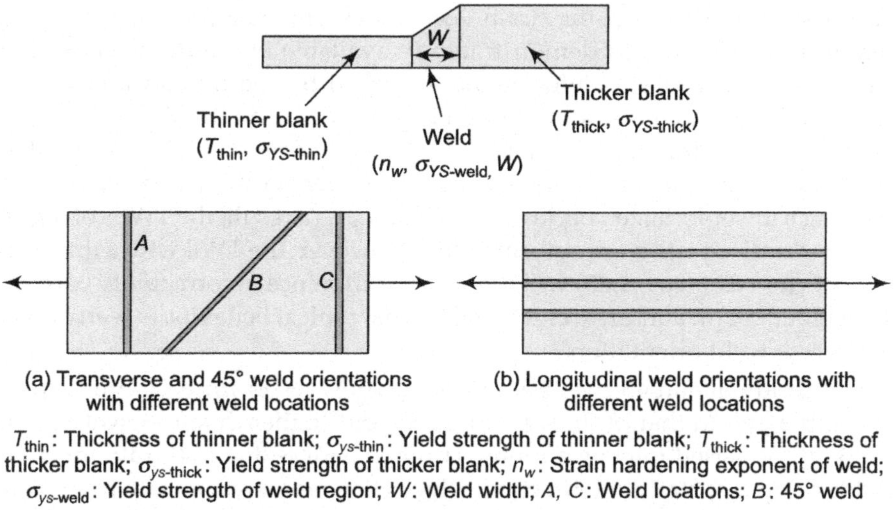

Fig. 10.1 Schematic descriptions of TWB parameters

The forming limit strains or curve (FLC) is found to decrease with increasing thickness ratio (thick/thin). Also, as the thickness ratio reduces, the forming limit of TWB tends to that of un-welded blank. Chan *et al.* [6] studied the effect of thickness ratio by performing tensile test and swift cup test. The thickness ratio selected were 2.0 (1.0/0.5), 1.67 (1.0/0.6), 1.25 (1.0/0.8) and SPCC material was chosen for the analysis. It was found from the tensile testing studies that the uniform elongation of TWB specimen is less than that of un-welded blanks. Also, with increase in thickness ratio, the uniform elongation decreases. For example, TWBs with 2.0 and 1.67 as thickness ratio possesses approximately 20% uniform elongation, while TWB with 1.25 thickness ratio posses 23% uniform elongation. Swift cup test revealed similar kind of behavior, with FLCs of TWB lower than that of un-welded blank (1 mm thick). In the case of TWB with 1.25 thickness ratio, FLC is significantly different when compared to that of other two and it moves towards FLC of un-welded sheet. Similar trend is seen with minimum major strain (%) attained during forming. The same research group has presented more data on similar formability studies in [7] and results revealed the same. The decrease in forming limit strain was related to increase in thickness ratio in [8–11] works. This was demonstrated using AA5754 material with laser welding in [8, 9] and few Al alloys and low carbon steel in [10]. Finite element simulations also predicted similar results. For instance, Vijay Bhaskar *et al.* [12] predicted same behavior for different thickness ratios using FE code, OPTRIS. Tensile test and LDH test were simulated with weld line assumption. Their study indicated that as the thickness ratio increases, the load bearing capacity and uniform elongation decreases in tensile test simulation, while progression at failure decreased in LDH test simulation. Stretch flanging simulation was performed on an industrial part for different thickness ratios (1, 1.5, 1.85) using a commercial FE code, JNIKE 3D by Nakagawa *et al.* [13] and it was illustrated that as the thickness ratio increases, the strain decreases in the thicker part. The reason behind this behavior, as demonstrated in available literature, is because of the non-uniform plastic deformation exhibited by the thinner and thicker blank in TWB which is not the case in un-welded blank. In higher thickness ratio TWBs, the thinner part deforms plastically first and much more than the thicker part. Hence, the failure occurs earlier on the thinner part, thus yielding a lower localized necking strain value. As a result, the TWBs of higher thickness ratios yield a lower formability. However, the TWB with a thickness ratio of close to 1 (say, 1.25) will reduce the difference in formability between the welded blank and un-welded blank. This typical behavior is seen only in transverse weld orientation.

Longitudinal and 45° weld orientations show lesser forming limit strains in comparison to that of un-welded blank, but higher than transverse weld TWB [9] for varied thickness ratios. This behavior is due to the alignment of weld with major straining direction, which changes the strain path from plane strain to bi-axial stretching and hence increase in limit strains is observed. Few friction stir welded Al alloys (6022-T4, 5754-O, 5182-O) were tested for formability by LDH test and OSU test [14]. Same thickness and material were welded, with weld and rolling direction parallel to the major straining direction. The results are consistent with the literature that the forming limit curve of welded samples is lower than that of un-welded samples, though the thickness ratio is 1.0. Specifically, considerable decrease in forming limit is witnessed in bi-axial stretching condition.

Some of these studies show the effect of thickness ratios on strain distribution also. Since the thinner blank deforms predominantly, thickness distribution will be non-uniform in thinner blank without much happening in the thicker blank. If the weld region is softer than the base metal, thickness change will be concentrated in the weld region, with very less concentration on thinner blank, and thicker blank the least. This behavior was demonstrated in [9], which shows that the non-uniform strain pattern and major strain peaks develop in the thinner blank. Also, the strain peaks are consistent with the earlier discussion, as thickness ratio increases, the peak strain (as observed from the strain peaks developed) decreases. For example, TWB with thickness ratio of 2 (2:1), exhibits lower peak major strain (approx. 0.2) than 1.6 (1.6:1) TWB (peak major strain = 0.55) [9]. Similar behavior is observed in [8] also. But, the behavior changes when TWB fails in the weld region with major strain as low as 0.05.

The effect of dome height at failure is similar to what is observed with limit strain. The dome height at failure of TWB decreases in comparison to that of un-welded blank. Also, with increase in thickness ratio, dome height decreases. This is due to the heterogeneity in plastic deformation exhibited by thinner and thicker blanks during forming. A few studies have illustrated this behavior [6, 13, 15, 16]. For instance, Chan *et al.* [6] performed experiments to study the thickness ratio effect on forming of SPCC material. This study shows that the dome height at failure increased with decreasing thickness ratio. The dome height at failure for different width samples also revealed the same behavior. Raymond *et al.* [15] simulated few formability tests using LS-DYNA to show the importance of modeling TWB with weld zone or without weld zone properties. Their work shows the decreasing punch travel at failure with increasing thickness ratios for LDH test simulation of 200 mm width sample with weld zone included in the model. The steel blanks also show decreased dome height at failure in Baysore *et al.* [17], where LDH tests are performed on TWBs with same material and thickness. Formability studies conducted by Miles *et al.* [14] on friction stir welded Al alloys also revealed the same, with welded blanks showing reduced dome height as that of un-welded blanks in general, but the decrease in dome height is very less with that of un-welded blank.

One issue which is quite confusing from these studies [6, 7, 15, 13] is that why and how thickness of the thicker blank is varied to study the effect of thickness ratio on the forming behavior? The basis for the question is that if the weld is perpendicular to the major straining direction, the thinner (or weaker) material will deform with absolutely negligible contribution from thicker blank and weld region (assuming harder weld zone) on the formability. Then the formability of TWB will be same as that of thinner blank, independent of thickness combination of blanks welded. So, one should change the thickness of the thinner blank, not the thicker blank, though same thickness ratios can be achieved in both the cases. In spite of varying the thicker blank thickness, the effect of thickness ratio on the forming behavior of TWB is as expected. But this depends on the forming behavior that is monitored. For example, Vijay Bhaskar *et al.* [12] has stated that the load–stroke behavior is same for different thickness ratios if the thicker blank thickness is changed in transverse weld orientation and hence they changed the thinner blank thickness to simulate the formability of TWB. This is not the case in [6, 7, 15, 13] as increase in thickness ratio reduces the limit strains, dome height at failure, etc. The possible reason for this can be the change of strain path from bi-axial stretching in lower thickness ratios to plane strain condition in higher thickness ratios, which as a result reduces the forming limit, dome height as the thickness ratio increases. This strain path change from bi-axial stretching to plane strain condition is because of the resistance provided by the thicker blank (or stronger blank) on the thinner blank (or weaker blank) during deformation. As the thickness of the thicker blank is increased, the resistance given to thinner blank is increased. So, the effect of thickness ratio is actually the effect of changing strain path as the thickness ratio increases (by changing the thicker blank thickness), and not exactly the increase in thickness difference. So, care should be taken that the thickness of thinner blank should be varied in order to study the effect of thickness ratio on the forming behavior of TWB. In the case of weld parallel to major straining direction, this issue will not arise as thinner blank, thicker blank and weld region are contributing to deformation.

The drawability parameters monitored are maximum draw ratio [18, 19], earing characteristics [20], predominantly weld zone movement and control [21–28], modeling aspects of TWB to accurately predict deep drawing of TWB [18, 20, 29–31]. In this section, effect of thickness difference of the blanks welded on the deep drawing characteristics of TWB will be briefly reviewed. In the published literature, the effect of thickness ratio is not studied much, except in few cases [21, 23, 32, 33] where the effect is studied on weld line movement only. There is no clear understanding on 'How thickness ratio affects maximum drawing ratio, formation of earing, etc.', though the understanding is simple—Increase in thickness difference deteriorates the deep drawing behavior of TWB. So, with increase in thickness ratio, limiting draw ratio should decrease, as seen with forming limit curve. The data for limiting drawing ratio, earing behavior are simply mentioned in some reports [18–20, 23] without critical analysis and explanation, in comparison to that of un-welded blanks. LDR of 1.67 to 2.0 is achieved by Ahmetoglu *et al.* [23] and 2.25 to 2.50 is seen in [18] for various thickness and material combination. In Ana Reis *et al.* [20], the effect of weld on the earing behavior of TWB is studied by taking same material and thickness. In TWB, because of the resistance given by the weld zone, ear formation is not symmetric when compared to that of un-welded blank. The effect of thickness difference on the weld line movement is understood clearly—increase in thickness ratio, increases the weld line shift. For example, Choi *et al.* [21] studied for thickness combinations of 0.8/1.2 and 0.8/1.6. SPC1 material is taken for the analysis. It is demonstrated that the weld line shift is more in the case of 0.8/1.6 thickness combination, for different weld line positions. Similar results are seen in Bhagwan *et al.* [32]. Here, different Al alloy combinations (AA 5182-O, AA 5454-O, AA 5754-O, AA 5052-O) with various thickness combinations (1.75/2.0 mm, 1.25/1.75 mm, 1.0/2.0 mm) are studied for weld line shift. It is clear from the studies that for whatever the material combination, higher the thickness difference more is the weld line shift.

This weld line movement, as demonstrated in the existing work, is due to the heterogeneity in plastic deformation achieved by the thicker (or stronger) and thinner (or weaker) blanks during forming. As a result, the thinner (or weaker) blank deforms more and hence weld zone moves towards the thicker (or stronger) blank in the cup wall region. With increase in thickness difference, the difference in deformation levels of thinner and thicker blanks increases, resulting in more weld zone shift. One important observation from [23, 33] is that the weld line not only moves but also rotates slightly (or inclination) with respect to vertical axis during drawing. Ahmetoglu *et al.* [23] presented that the weld zone turns towards the thinner blank in the flange area of about 5°, because of which the weld zone is slightly curved after deep drawing. It is seen from Pepelnjak *et al.* [33] that rotation angle of weld increases with increase in thickness difference of the blanks welded.

10.2.2 Effect of Strength Ratio (or Strength Difference) on the Forming Behavior of TWB

Strength ratio is defined as the ratio of yield strength of one material to that of another material ($\sigma_{ys\text{-thin}}/\sigma_{ys\text{-thick}}$ or $\sigma_{ys\text{-thick}}/\sigma_{ys\text{-thin}}$) welded in a single plane before forming. The effect of strength ratio on the forming behavior is similar to that of thickness ratio, with weaker material replacing thinner blank and stronger material with thicker blank. Since, the yield strength of a material cannot be varied by keeping other material properties (n, m, r, etc.) constant, welding materials of two different strengths is nothing but welding two different materials (or grades) itself. So, the intrinsic effect of strength ratio on the forming behavior of TWB cannot be studied unlike thickness ratio. The effect of strength difference on the formability of TWB is that by increasing the strength difference between the materials welded, the forming behavior is deteriorated. So, with increase in strength ratio, forming limit strain, dome height, limiting draw ratio, cup height, etc. should decrease and weld line movement should increase. This is due to differential plastic deformation levels attained by stronger and weaker material.

These issues are discussed to some extent in a few published papers [13, 18, 30, 32, 34–36]. For instance, Kusuda *et al.* [34] and Nakagawa *et al.* [13] related the effect of strength ratio with maximum forming height and limiting draw ratio. In [35], the effect of different material combinations on the limiting draw ratio and % elongation was analyzed by Geiger *et al.* The results are consistent with the basic understanding that increase in strength difference reduces the limit drawing ratio (LDR) and elongation of welded blanks. This study was performed on Al alloys AA6016 ($\sigma_{ys} = 150.8$ N/mm^2, $t = 1.1$ mm) and AA5182 ($\sigma_{ys} = 131.5$ N/mm^2, $t = 1.0$ mm) by friction stir welding. It is observed that the welded blanks show a decreased percentage elongation and LDR when compared to that of base metals. Also, TWB with AA5182 and AA6016 combination shows the least elongation and LDR, which is nothing but the effect of strength difference, as the sheet thickness are practically same. Both percentage elongation and LDR follow similar trend.

Both the thickness ratio and strength ratio are varied in some of the studies to analyze the formability of TWB [18, 30, 32, 36, 37]. The effect of different material and thickness combinations on the weld line movement is studied by Bhagwan *et al.* [32]. Since the yield strengths of Al alloys tested vary within a narrow range, the effect of thickness ratio is dominant here, and hence the effect of strength difference is not seen on weld line movement. Hayashi *et al.* [36] varied the weld location, orientation and material (EDDQ/HS; $t = 0.7$ mm) to understand the effect on weld line movement. Saunders *et al.* [30]

demonstrated the forming behavior of 1.8 mm AKDQ/2.1 mm HSLA and 0.8 mm AKDQ/1.8 mm AKDQ welded blanks by analyzing the tensile behavior, weld line shift, etc.

To summarize, increase in thickness ratio and strength ratio, deteriorates the forming behavior of TWB significantly. Forming limit strains, forming height at failure, limit drawing ratio are found to decrease with increasing thickness ratio and strength ratio. Weld line shift increases with increasing thickness or strength ratio. However, the effect depends on the weld and welding conditions viz., weld properties, location, orientation, welding techniques, parameters also, which will be discussed in next few sections.

10.2.3 Effect of Weld and Welding Conditions on the Forming Behavior of TWBs

The weld conditions refer to weld orientation, weld location, weld properties, weld microstructure, etc. The welding conditions refer to the welding techniques and parameters followed to weld the blanks. While the weld orientation and location are decided by the structural compatibility and application of TWB, weld properties and microstructure are decided by the welding techniques and materials being welded. It should be felt that once welding technique (say laser welding) and materials (say particular Steel grade or Al alloy) are fixed, weld orientation and location are the only parameters that can be varied according to the applications. In this regard, not much of work has been performed on understanding the effect of other weld conditions like weld properties and microstructure on the forming behavior of TWB. Another possible reason may be the weld region is very small, of the order of 1–2 mm in the case of laser welding (which is widely used for making TWBs) that its effect is very negligible. On the other hand, the effects of weld orientation on the formability of TWBs are studied extensively and weld location to some extent. The forming behavior studied are forming limit strains, height at failure, strain distribution, deep drawing characteristics like weld line movement, maximum draw ratio, etc. Also, the effect of welding techniques and welding parameters on the tensile and forming behavior of welded blanks has been quoted in few literatures.

The weld orientations mostly studied are 0° (weld parallel to major straining direction), 90° (weld normal to major straining direction) and 45° (weld at 45° to major straining direction). Cheng *et al.* [38] have studied the effect of transverse and longitudinal weld orientations on the tensile and forming behavior of laser welded 5754-O blanks. Same thickness (1 mm) and material is welded to obtain the effect of weld on the forming of TWB. It is found from the study that transverse TWB exhibits tensile and forming limit equivalent to that of un-welded blank, with very less forming reduction. But in the case of longitudinal weld orientation, tensile and forming limit of TWBs are drastically reduced. The reason behind this behavior is because of the difference in volume fraction of weld region in transverse and longitudinal TWB in the case of tensile testing. The forming limit strains are decided by the location of failure, which is away from the weld region in transverse TWB and failure normal to the weld in longitudinal TWB [38]. As depicted in the stress–strain curves of TWB, the UTS attained is nearly same for transverse, longitudinal and un-welded blanks. But the ductility of longitudinal TWB is decreased, when compared to that of the other two. Also, in the case of forming limit, the FLC of transverse TWB is same as that of un-welded blank, with longitudinal weld showing reduced forming limit [38]. But these results are quite different of what is said in Buste *et al.* [9]. In Buste *et al.* [9], the forming behavior of AA5754 alloy is studied for different thickness ratios (1:2, 1:1.6) using LDH test in near plane strain condition. It is seen from the analysis that longitudinal TWB has limit strain near to that of un-welded blank with transverse TWB showing reduced limit strain. The reason behind this behavior, as said in [9], is because of the change in strain path from plane strain to bi-axial stretching condition as the weld is aligned with that of major straining direction (longitudinal weld).

The percentage formability (defined by the ratio of %LDH of TWB to %LDH of thinner material) of TWBs for different welding techniques and thickness combinations are given in Tables 10.2 and 10.3. It is clear from the tables that longitudinal weld exhibits approximately 90% of the LDH of thinner base metal, with transverse TWB showing less than 70% of that of thinner base metal [9]. The behavior is same with limit strains also. This indirectly means that formability of longitudinal TWB is better than transverse TWB.

Table 10.2 NVEB welding technique [9]

Weld orientation Thickness ratio	Transverse weld	45° weld	Longitudinal weld
2:1.6 (TR=1.25)	48%*	70%*	95%
1.6:1 (TR=1.6)	52%*	55%*	95%
2:1 (TR=2.0)	60%	90%	102%#

Table 10.3 Nd:YAG welding technique [9]

Weld orientation Thickness ratio	Transverse weld	45° weld	Longitudinal weld
2:1.6 (TR = 1.25)	75%	88%	88%
1.6:1 (TR = 1.6)	70%	85%	95%
2:1 (TR = 2.0)	67%	82%	100%

TR–Thickness ratio; * indicates TWB with failure in the weld region; # lack of clamping.

Though material is same in Cheng *et al.* [38] and Buste *et al.* [9], i.e., 5754 Al alloy, the forming behavior is different. This may be due to the compounding effect of thickness difference and weld orientation in Buste *et al.* [9] analyses.

But, in Cheng *et al.* [38], the effect of thickness difference is absent as same thickness and material is welded for the study. However, studies performed by Miles *et al.* [14] on 5182-O and 5754-O alloy with friction stir welding (same material and thickness) revealed that longitudinal TWB show same forming limit in plane strain condition with slight decrease in bi-axial stretching condition. It is mentioned that practically longitudinal FSW blanks exhibit same forming limit as that of un-welded blank. This result is consistent with what is said in Buste *et al.* [9]. There is no evidence for transverse TWB in [14]. But, transverse TWB may also show same forming behavior as that of un-welded blank in this case because of the weld produced by FSW having almost same properties (σ_{ys} and n) as that of base metal. Also, it is stated by Friedman *et al.* [10] that formability of AA5182-O material (0.84/1.55 mm) drops as the weld is perpendicular to the major straining direction, with 45° welded blank showing more formability than 90° weld orientation. Kridli *et al.* [39] also showed similar forming behavior.

An important point to be noted from the formability studies performed by Buste *et al.* [9], Friedman *et al.* [10], Miles *et al.* [14], and Kridli *et al.* [39] to study the effect of weld orientation is that the lower forming limit or dome height of transverse TWB is attributed to the fact that failure occurs in the weld region or very near to the weld region. Since weld region has lower forming limit as that of base metal, lower forming limit is seen in transverse TWB. For example, in Friedman *et al.* [10] though different gauge materials are welded (0.84/1.55 mm), failure is seen in thinner material and in the weld region or near to the weld region in TWB with 90° weld. In the case of 45° weld orientation, failure occurs slightly away from the weld zone because of which forming height is increased. Also, some of the LDH data (indicated by *) given in Tables 10.2 and 10.3 are from TWB with failure in the weld region, which show lower formability when compared to that of TWB with parent metal failure. But, in the case of Cheng *et al.* [38] analysis, the welding parameters are optimized such that failure does not occur in the weld region. It is seen that failure always occurred at considerable distance from the weld region in transverse TWB, which eventually resulted in having same forming limit as that of un-welded blank [38]. In the case of longitudinal TWB, failure anyway has to occur normal to the weld region, because of which forming limit is reduced.

Ganesh Narayanan and Narasimhan [40] analyzed the effect of weld configuration (orientation, location) on the forming limit of laser welded blanks made of interstitial free steel of 0.97 mm thick. Figure 10.2 depicts the influence of weld orientation on the forming limit. It is clear from the figure that the both transverse and longitudinal welds show same forming limit as that of un-welded blanks, except in plane–strain strain path. In transverse weld orientation, the smaller influence of the weld is attributed to the facts that the weld region is not participating in deformation, and that failure occurs at 20–30 mm away from the weld zone. In longitudinal weld orientation, the smaller difference between the FLCs of un-welded blank and TWB is due to the presence of a harder and smaller weld region (less than 1 mm width). Since the weld region has a lower ductility, it participates in deformation as failure occurs normal to the weld zone. Finally the forming-limit strains of welded blanks with a weld at the geometric center of the sample are almost the same as those of the un-welded blank, irrespective of the weld orientation.

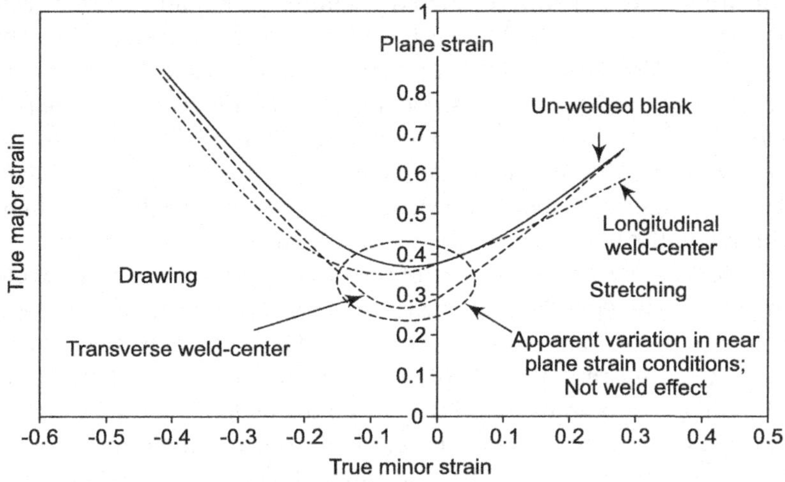

Fig. 10.2 Comparison of the FLCs of the welded blanks (transverse and longitudinal welds at the center) with that of the un-welded blank [40]; with permission from [40] copyright [2008] Sage

In summary, the effect of weld orientation on the forming limit and dome height of TWB depends on the location of failure. Because of the contradictory results of Cheng *et al.* [38] with other researchers [9, 14, 10], more critical experimentation is needed to clarify this issue like optimizing welding parameters so that weld failure can be avoided, the entire FLC can be generated for TWB to compare with that of un-welded blank, etc. One more issue which should be noted from [9] is that the limiting dome height increases with increase in thickness difference, which is just opposite to what is seen in Chan *et al.* [6, 7] and other literature. This may be due to the occurrence of failure in the weld region as the thickness ratio decreases. At higher thickness ratios, failure in the thinner base metal is witnessed (failure will move away from the weld line), and hence TWB posses the formability of thinner metal itself. But when failure occurs in the weld zone or near to the weld, as seen in lower thickness ratio TWBs, the formability will be reduced as weld region possesses lower formability than that of thinner blank.

The effect of various parameters on the forming behavior of TWB described above considers weld to be at the center of the sheet sample before deformation. But, practically weld can be located at any place on the industrial sheet part, which may affect the forming behavior of TWB viz., forming limit, weld line movement, etc. The effects of weld location on various aspects of TWB are briefly discussed here. Choi *et al.* [21] have studied the effect of initial weld line location on the weld line movement during deep drawing. It is clear from their analyses that TWB with weld at 24 mm offset show more weld line movement and weld line at center showing minimum weld line movement. This is true for any thickness combination (0.8/1.6 and 0.8/1.2) and initial shape of the blank, whether it is square or circular. Deep drawing characteristics of TWB with and without drawbeads for different weld locations (0, 12, and 24 mm from the center) are compared in [21, 22]. It is found that weld line movement is higher in the case of TWB with weld at 24 mm offset. The weld line movement is much higher when it is placed at some non-symmetric location of the product [29]. Similar results are demonstrated in [36], where EDDQ (0.7 mm) is laser welded with HS (0.7 mm) steel blanks. It is described by Hayashi *et al.* [36] that the weld line movement is because of two compounding mechanisms: (1) Primary weld line movement, where weld line moves because of the geometric change in shape and size of the blank, and (2) Secondary weld line movement, which depends on the heterogeneity of properties in the base metal and weld region, i.e., hardening degree of weld region. These mechanisms are analyzed by monitoring movement of scratched line and weld region during deep drawing. It should be noted that when the weld is placed at some offset from the center of the blank, the thinner blank contribution is more because of which the deformation exhibited by the thinner blank is higher when compared to that of thicker blank, resulting in increased weld line movement

The effect of weld line location on the forming height of AA5182 TWB (1.26:1.88 mm) is described by Kridli *et al.* [39]. The transverse weld is placed at 33.15 mm from the center and it is found that forming height of welded blank is increased when compared to that of TWB with weld at the center. As the weld location is changed, increase in thicker blank contribution is seen. As a result, formability of TWB will move towards the formability of thicker blank, which will be higher than that of TWB with weld at center as this will be closer to the formability of thinner blank. Also, the increase in thicker blank contribution may change the failure location and hence change in strain path from plane strain is expected during stretching.

Like initial weld line position increases the dome height at failure in [39], it can change the forming limit strains also. This was demonstrated by Ganesh Narayanan and Narasimhan [41] using PAMSTAMP 2G, a finite element code by using TWB made of same material and thickness combinations. Here, the transverse weld was placed at various positions at regular intervals of 5 mm from one end of the ASTM standard tensile sample. The limit strains are obtained using a novel thickness gradient based necking criterion [42]. It is found that when weld is placed at the center of the sample, the forming limit strains are higher than that of un-welded blank as shown in Fig. 10.3. The forming limit strains are decreased when the weld moves away from the center, i.e., at some offset and then increases to that of un-welded blank. The increase and decrease of limit strains for various weld positions is because of the change in strain path and movement of weld zone during deformation and is independent of weld properties (n, σ_{ys}) considered in the work. The effect of the weld location on the FLC of the transverse-welded blank can be understood from Fig. 10.4 [40]. It is clear from the figure that there is a considerable difference between the limit strains of the welded blank with a 20 and 30 mm weld offsets in the drawing region of the FLD and those of the center weld case. Approximately 10 per cent variation is seen between them. This is explained in [40] with the help of thinning gradient developed during forming.

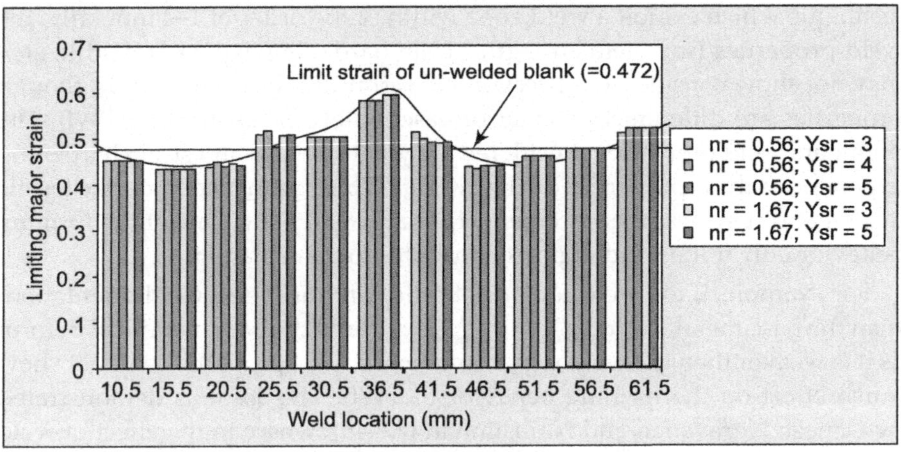

Fig. 10.3 Effect of weld locations and properties on the forming limit strains of TWB in tensile deformation [41]

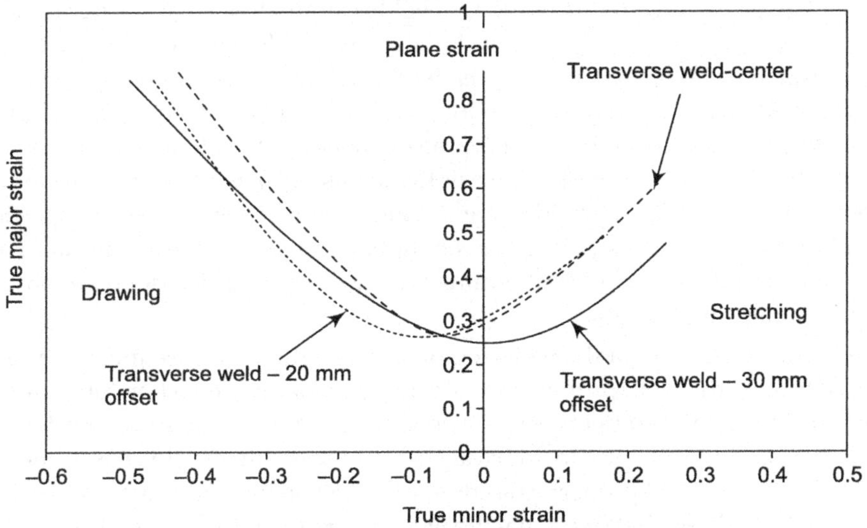

Fig. 10.4 Comparison of the FLCs of welded blanks with a transverse weld at the center and with the transverse weld at offset locations (20 mm, 30 mm) [40]; With permission from [40] copyright [2008] Sage

From the above discussion on the effect of weld orientation and location, it is clear that the initial weld line placement and orientation can affect the forming behavior of TWBs like weld line movement, dome height at failure, forming limit strains, strain distribution, etc. considerably. Under specific conditions of weld location and orientation, the forming limit strains are found to increase with weld placed at some offset from the center of the sample.

Once said that weld location and orientation affects the forming behavior, the next level of idea is to observe how the weld properties affect the forming behavior. As pointed out earlier, since laser welding is mainly used welding technique which exhibit a weld zone width of the order of 1–2 mm only, the weld properties (say yield strength, strain hardening exponent, width, etc.) may not show considerable effect on the forming behavior of TWBs, though properties are different from that of base materials constituting TWB. The dominant weld properties considered are yield strength and strain hardening exponent, as these properties can be evaluated by existing methods like tensile testing and hardness measurement. These properties will affect the forming behavior only if the weld region contributes to deformation.

For example, if the weld region is harder (in transverse weld orientation) than the base metal, i.e., $\sigma_{ys\text{-weld}} > \sigma_{ys\text{-base metal}}$, only the base metal will deform as it is weaker than the weld region and hence weld properties will not show much effect on the forming behavior of TWB. This issue is demonstrated by Ganesh Narayanan and Narasimhan [43–45], where in the effect of weld properties on forming limit strains are studied by simulations using PAM STAMP 2G. Both longitudinal and transverse weld orientations are simulated using tensile testing. In this work, the weld is simulated as the harder region, i.e., $YS_r = \sigma_{ys\text{-weld}}/\sigma_{ys\text{-base metal}} > 1$ and hence negligible effect of yield strength and strain hardening exponent is seen on the forming limit strains, irrespective of the weld locations. In the case of longitudinal weld orientation, the effect of weld yield strength and strain hardening exponent is considerable as presented in [43–45]. The limit strains increase with increasing 'n' value and it is found to decrease with increasing yield strength value. For some of the weld properties combination, the limit strains are even higher than that of base material. This increase and decrease of limit strains with weld properties depends on the mode of deformation also.

On the other hand, if the weld region is softer than the base metal, i.e., $YS_r < 1$, definitely the weld properties will have considerable effect on the forming limit of TWB. But the effect depends on other properties like strain hardening exponent and weld width. In this case, the yield strength of weld chosen should be such that the weld properties show some effect on the forming behavior, i.e., the weld should not be so soft that the weld properties effect is negligible like in TWB with harder transverse weld. Though the effect of weld properties, when weld zone is softer than the base metal, has not been studied in detail, some understanding can be obtained from Stasik and Wagoner work [46, 47].

In the case of longitudinal weld orientation (in tensile test simulation), the maximum load reached and progression till failure decreased with 'n_r' decreased from 1.3 to 1.0 for $YS_r = 3.0$. This tensile behavior can be understood from Stasik *et al.* [46] also, where the heat treatable 6111-T4 Al alloy because of softer heat affected zone, exhibit lesser ductility when compared to that of 5754-O Al alloy, in transverse loading. In 6111-T4 Al alloy, the softer weld zone possesses approximately 75% of the hardness of base material, which can be referred for yield strength of the weld zone. The 5754-O Al alloy possesses nearly same hardness in the weld zone as that of base metal. In the case of longitudinal weld orientation, the ductility of TWB will anyway decrease because of the lesser ductility (or weld 'n' value) of weld zone. The maximum load at failure (or UTS as well as yield strength) will be greater than that of the base metal, if the weld is harder than the base material like in most of the steel TWBs [40], whereas it will be less than or same as that of base metal if the weld zone is soft, as seen in [46, 47] with few automotive Al alloys. These experimental evidences reveal the significance of weld properties on the tensile behavior of TWB quantitatively. To sum up, the weld zone, softer or harder than the base material, decides the formability of TWB in the transverse loading. The weld properties are definitely an important candidate in the case of longitudinal loading of TWB, irrespective of harder or softer weld zone.

10.2.4 Compounding Effect of Factors on the Forming Behavior of TWB

Till now, the effect of various parameters, viz., thickness ratio, strength ratio, weld conditions like weld orientation, weld location, welding techniques and parameters on the forming behavior of TWBs say forming limit, strain distribution, deep drawability, failure behavior, etc. is discussed in detail. An important issue to be noted here is that these parameters may not affect the forming behavior individually. In practical experience, all these parameters can have a synergistic effect on the forming of TWB. Suppose if the effect of thickness ratio has to be studied on the forming behavior, it will be a compounding effect of thickness difference and weld conditions (weld width, weld location, weld orientation, weld properties, etc.), when compared to that of un-welded blank. In order to study the effect of weld conditions, the other parameters like thickness ratio, strength ratio have to be maintained constant, otherwise it will result in compounding effect of all the parameters. This holds good for all the parameters mentioned like strength ratio, weld conditions, etc. For example, Chan *et al.* [6, 7] studied the effect of thickness ratio on the forming limit by performing swift cup tests. It was concluded by them that the forming limit of TWB and elongation till failure decreases with increase in thickness ratio when compared to that of un-welded blank. But in reality this decrease in forming limit is actually a compounding effect of thickness ratio and weld conditions.

Similarly, studies conducted by Bhagwan *et al.* [32], Buste *et al.* [9] and few other literature can be quoted in this context. Though it is assumed that the effect of weld is insignificant because of the smaller width (of the order of 2–3 mm in laser welded TWBs), its effect depends on weld properties, weld location and weld orientation also, as explained by Ganesh Narayanan *et al.* in [43–45]. In order to understand the synergistic effect of parameters on the forming behavior, Ganesh Narayanan and Narasimhan [43] has performed DOE analysis using Taguchi and ANOVA techniques. The parameters considered in their study are: thickness ratio, yield strength ratio, strain hardening exponent ratio, weld yield strength and weld strain hardening exponent. The ratio related parameters are with respect to thinner or weaker material and thicker or stronger material. FE simulation trials were performed by following Taguchi's orthogonal array of experiments (L_{27}). Forming limit strains were monitored for each simulation experiment. The relative or percentage contribution of each parameter was obtained using ANOVA analysis. It was found from the analyses that thickness ratio show larger effect in longitudinal weld orientation, as failure takes place normal to the weld line [43]. In the case of transverse weld, yield strength ratio dominates other parameters, as it is an important material property. But, the only contradictory result is that weld properties (σ_{ys} and n) show negligible contribution, say less than 10%, in the case of longitudinal weld. The possible reason for the negligible contribution of weld region could be the dominating characteristics of other parameters if the blanks used are of different thickness and materials. The experimental [46, 47] and simulation evidences [43] are based on the findings from forming behavior of TWBs with same thickness and materials welded, wherein weld properties are supposed to show considerable effect. So, it may not be appropriate to say that weld properties do not have any effect on the forming behavior; instead it is correct to say parameters like thickness ratio, etc. dominated the weld properties, as explained in [43]. To sum up, various factors affecting the forming behavior of TWB in different proportions and prior knowledge on this will definitely help in designing the TWB process and parameters successfully.

10.3 WELD LINE MOVEMENT AND CONTROLLING METHODS

As pointed out earlier, the weld line movement is because of the heterogeneity in plastic deformation attained by the thinner or weaker material and thicker or stronger material during forming. Mostly this weld line movement is seen during deep drawing of cups and to some extent in stretching processes, if it is placed at some offset. In stretching, one can visualize the 'weld line movement' as 'bowing of weld'. Many factors show compounding effect on the weld line movement as discussed in previous sections and the deep drawing behavior deteriorates because of this. The weld line movement can be controlled by various methods like,

1. Applying differential Blank Holding Force (BHF) on thinner and thicker blanks
2. Using stepped blank holder or die and shim, to compensate for difference in thickness and strength of blanks welded
3. Using drawbeads at different locations of TWB
4. Optimum initial weld line location
5. Specially designed local adaptive controllers, specific to individual TWB parts
6. Initial shape of weld line, i.e., linear or non-linear, etc.

1. Using differential Blank Holding Force (BHF)

If the BHF acting on both base metals is the same, the thinner material can flow much more easily than the thicker material. As the thinner material is pulled towards the thicker material, it increases the stretching and shear in the thinner material. The strain becomes very high and failure occurs easily. By applying differential BHF, the higher BHF holds the thinner material and the lower BHF allows the thicker material to flow in to compensate the stretching of the thinner material and reduce the shear. Thus the weld line movement is reduced. For example, in [23, 48], this differential BH technique is demonstrated by Ahmetoglu *et al.* using deep drawing of welded blanks (0.8 and 1.8 mm thickness combination) made of AKDQ steel, with weld at center and at some offset from the center line. In order to compensate for different pressure requirements at different positions on the blank, a multi-point pressure-control technique was employed to vary the pressure around the periphery of the blank holder. To solve this issue, the die cushion of the press was replaced by a nitrogen cylinder system which consists of six nitrogen cylinders and two tanks. Three adjacent nitrogen cylinders are connected to one tank. Therefore, it is possible to set the pressure on the two halves of the blank holder at two different values. The blank-holder ring was divided into two halves using wire EDM and shims were placed under one half to create a step in the blank-holder surface. This step compensated for the thickness variation on the welded blank with the weld line located on the center line. The pressure on the sheet metal was varied locally by varying the pressure settings on the nitrogen cylinders. The blank-holder force was measured by means of six button load-cells located on top of the six nitrogen cylinders. The punch force was also measured by means of an in-line load-cell. Initially, a constant BHF of 5 tons is applied on the thinner and thicker blanks. A cup with weld line at un-symmetric location was obtained. In order to control the weld line movement, a differential BHF of 10 tons is applied on thinner side and 1 ton is applied on thicker side and this resulted in a deep drawn cup with weld line at the center, the same location where it was placed before forming.

In [49], a 2-D analytical model was proposed by Sijun He *et al.* wherein the BHF requirements in thinner and thicker blanks were predicted and validated by experiments and simulations. Here, a hydraulic pressure control system with six cylinders is equipped on a mechanical press. The hydraulic system can provide up to 2.5 tons BHF from each cylinder. A computer system is used to record the tonnage signal, the pressure signal, and LVDT signals. The experimental observation on strip drawing (0.8–1.5 mm) and box drawing reveals that weld line movement can be minimized with differential BHF of 1.2–1.4 tons in strip drawing and 2.4–7.2 tons in box drawing. A new concept for compensating the thickness fluctuation is described in [16]. According to this design concept, a prototype tooling was fabricated for TWB with straight weld line. Here, the right side of the blank holder segment was fixed to the blank holder plate. The left side BH segment is freely supported on three nitrogen cylinders. The left segment can apply different pressure according to the thickness difference and fluctuations. As a result, by compensating thickness difference, the weld line shift can be controlled by this tool design method [16]. These examples demonstrate the design of differential BH system for deep drawing of cups and controlling the weld line movement during deep drawing process. Detailed description about the segmented or differential BH system can be obtained from [49–53].

2. *Using stepped die, blank holder or punch to compensate thickness difference*

If the TWB is made of dissimilar thickness blanks, uniform blank holder pressure cannot be applied on the blank, as uniform contact between the thinner blank and the tools is absent. When dissimilar thickness blanks are used, the thickness difference should be compensated, otherwise wrinkling of thinner metal, tearing near the weld region, weld line movement will occur which will deteriorate the formability of welded blanks. This problem will not arise if same thickness blanks are used, as uniform contact is established. Compensating the thickness difference can be done in the following ways—(1) Shim can be provided, and (2) Stepped die or blank holder (or lower, upper binder) can be used. Shim is nothing but a piece of sheet inserted below or above the thinner blank as shown in the Fig. 10.5. Another method, which is the widely used in industries, is providing stepped die or blank holder (or lower, upper binder). Here, the die or blank holder is segmented into different regions, which has step near the location where weld zone contacts the die or blank holder.

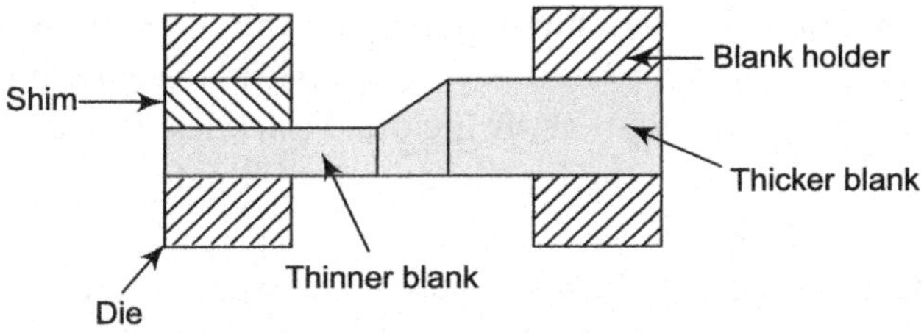

Fig. 10.5 Schematic of providing shim for thickness compensation in TWB

Following are the problems that should be taken care while designing the stepped tools during the forming of TWB.

(a) Sheet and tool tolerance should be taken care; otherwise non-uniform holding pressure may lead to compressive stresses near the weld region, resulting in failure of weld zone during forming.

(b) It should be ensured that the lower or upper tool is 'slackened' enough so that stresses can be avoided near the weld region, allowing the weld to move freely (Fig. 10.6). Depending on the welding method, the lower binder segment should be fitted with corresponding slots in which the weld can move during the forming process. In [16], mash seam welded blanks encountered a typical problem that the upper binder only made contact on the highest points on the weld line caused by the overlapped weld on both sides of the weld line during forming process. This resulted in almost no blank holder pressure on the weld region, which led to uncontrollable metal flow and considerable wrinkling in the binder flange area of the TWB. In order to solve this, slots are provided in the lower binder so that weld is pushed into the slot, during the holding process (Fig. 10.6). This leads to uniform blank holder pressure on the TWB. But these slots should be designed properly. If the slot is too deep, the weld caves-in because of the tangential compressive stress and hence missing uniform BHF in the weld region. Sometimes planishing can be performed on the welded blanks to reduce the weld thickness by simply running the welded joint through two rollers after the welding process [16].

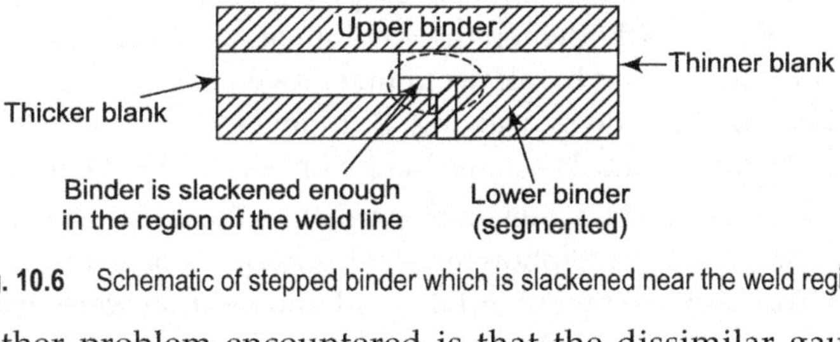

Fig. 10.6 Schematic of stepped binder which is slackened near the weld region

(c) Another problem encountered is that the dissimilar gauge sheets welded can cause sheet thickness variations greater than the allowable tolerances. Taking sheets from different batches will worsen the problem. This thickness fluctuation in dissimilar blanks will cause non-uniform blank holder pressure, which can lead to wrinkling and cracks. Experiments with intentionally included thickness fluctuations showed that fluctuations of some hundredth of millimeter cause non-uniform metal flow and bad deep drawing results [16].

(d) An important issue to be taken care is the wear and tear of slots in the tools if the weld is at some angle to the flow of metal during forming. This problem is common if the TWB has slanting weld line like the one shown in [16], where weld line subtends 20° with the punch opening line. This slanting weld line is subjected to geometric weld line movement which resulted in wear of the slot causing the mash seam weld to fail in the weld region. This type of slanting weld line may also lead to non-uniform BHF as the forming process continues and hence earlier failure of TWB is seen in [16]. The problem will be severe if the weld is non-linear through out.

3. *Using drawbeads at different locations of TWB*

The ultimate aim in controlling the weld line movement is to resist the movement of thinner (or weaker) blank in accordance with that of thicker (or stronger) blank. This results in equivalent plastic deformation levels on thinner and thicker blank, and hence weld line movement is reduced. This can be achieved by providing drawbeads or applying drawbead restraining forces at various locations of the welded blanks. For example, in Heo *et al.* [22] drawbeads of different dimensions (drawbead radius and height) are installed in the blank holder and restraining forces are applied on the thinner blank side of TWB and as a consequence weld line shift is minimized. Also, by increasing the radius and height of drawbead the weld line movement is minimized.

4. *Optimum initial weld line location*

Weld line movement can also be reduced by locating weld zone at an optimum position, such that (1) the weld line is at the same position, or (2) the weld will move to a desired position after forming. The first result can be achieved by placing weld at symmetric position of the product. For example, in Choi *et al.* [21], it is clearly shown that when the weld line is at symmetric position, i.e., at the center of the welded blank, weld line movement is minimized, in both square and circular cup deep drawing. But movement is considerable when it is located at some offset. The same behavior is seen in [36] also.

The second result can be achieved by predicting the initial weld line location in the un-deformed TWB by following some criterion. For example, in [54–56], different methods of predicting initial weld line location were proposed.

(a) *Back propagation method*: In this method, the weld line is back propagated in flat, un-deformed blank from the formed product. This gives the location of weld line in the flat blank for minimum weld line movement compared to the desired position. This was demonstrated by performing simulations of deep drawing of round cup. During simulations, the nodes in the weld zone are monitored for its movement from final product to flat, un-deformed blank, which is proposed as 'Back propagation technique' in [54]. By back propagating the nodes, one can obtain initial weld line location and design (say straight or curvilinear weld) in the un-deformed part. The same technique can be developed analytically as described in [55, 56]. Here, the initial weld location is obtained by knowing the final weld position, material properties, strain near the weld location, forming height, etc. The analytical predictions with respect to simulation results show good accuracy. This technique of predicting initial weld location depends on the thickness, material combination, and weld conditions like orientation, properties, etc. Also, the part shape—symmetric or non-symmetric, will play a vital role in determining the weld line location.

(b) *Minimum thickness strain contour method*: Here, it is stated that optimum position of weld line will correspond to the contour of minimum deformation, given by the minimum thickness strain [54]. Hence, the minimum thickness strain contour is obtained in a drawn cup. This contour is monitored from the final part to initial shape and the initial weld line position and design can be obtained. This method was demonstrated and it is proved to be a good technique for identifying initial weld location.

One can even consider major strain as a criterion like thickness strain followed in this work, as it is one of the co-ordinates to define forming limit curves. By using any of these techniques, one may end up in obtaining the initial weld as a curvilinear one as seen in back propagation technique, which is the optimum initial shape for the weld line. Otherwise, assuming straight (or linear) weld line initially will lead to increased weld line shift. The two techniques proposed in [54] are found to be better techniques to predict the initial weld line location, when compared to that of placing weld at the symmetric position.

5. Specially designed local adaptive controllers

An improved method to control weld line movement is by providing specially designed local adaptive controllers as discussed in [24–28]. This method involves restricting weld line movement by subjecting near by areas of weld region to local hydraulic pressure or clamping force. Hydraulic pressure (or clamping force) can be applied at one or two locations or along the entire weld line by designing multi local hydraulic pressure system. As deformation proceeds, the local adaptive pressure system adjusts pressure accordingly resulting in controlling weld line movement. One critical issue in this is to locate the clamping force and this depends on the part shape and size. For example, in [24], the door inner part was chosen for the study. A segmented hydraulic system was designed specifically for this purpose [24]. Also, the initial weld position has two weld lines intersecting, and region A of 2 mm thickness and region B of 1 mm thickness are generated. The clamping force locations and size are identified by FE simulations. This system was simulated for its success using LS-DYNA, FE code. The clamping force locations were simulated as fixed nodes where X and Y direction movements are restricted. It is found from the analysis that the strain pattern obtained in the simulation by adapting the clamping mechanism is well below the FLC of the material (Al 6111-T4). But in the case without clamping force, i.e., in conventional stamping process, tearing failure is seen in the intersection of weld lines. Also, the weld line movement is reduced in the case with clamping force. Relatively similar work was performed on Al 6111-T4 by Kinsey *et al.* [25, 26] and found that weld line movement can be minimized by using the local adaptive clamping mechanism. LDH tests are performed on Al 5182 TWB samples in [26] and found that failure occurs in the thinner material (1 mm), with FLC (positive major strains) of TWB lesser than that of un-welded blanks. Tearing is seen at the intersection of two weld lines in the door inner part on the thinner material side, which is consistent with what is observed in the studies, described earlier.

As said earlier, an important task in this technique is locating the clamping force on the TWB. This is well described by Kinsey *et al.* in [28] by evaluating the centroids of reaction forces in X and Y directions. The reaction forces are obtained for the fixed nodes along the weld line in X and Y directions by FE simulations. By intersecting the two area centroids on the part geometry, a proposed location for the local adaptive controllers is provided. This appropriate design of clamping mechanism lead to the increase in depth of draw without tearing failure, due to less plastic strain near the weld region. This advanced technology is also applicable to non-linear weld lines, Steel TWBs, TWBs with different materials and thickness combinations, TWBs made by other welding techniques like friction stir welding, etc.

The methods described here for controlling weld line movement can be used individually or a few techniques can be combined to obtain a good deep drawn component. Even changing the friction conditions between thinner and thicker blanks may reduce the weld line shift; for example, increasing the friction coefficient between the thinner blank and tool (say, $\mu = 0.3$) and decreasing the friction coefficient at thicker blank side (say, $\mu = 0.01$), will resist the thinner blank deformation and hence weld line shift may be minimized.

10.4 MODELLING SIMULATION OF FORMABILITY OF TWB

Modeling the formability of TWB is a critical issue because of the presence of weld region, which has relatively different mechanical properties as that of base metals. Forming behavior of TWB can be simulated by two methods—either by incorporating the properties of the weld zone (σ_{ys}, K, n, r, etc.) separately like base metals or by assuming weld region as weld line without considering properties such that weld line is just a boundary between two adjacent base metals (Fig. 10.7). The constitutive behavior of weld zone is obtained by conducting tensile tests on sub-size specimens with longitudinal weld and following rule of mixtures technique. These assumptions (weld line or zone) followed in modeling TWB play a vital role in predicting the forming behavior. Another important issue that one should consider while modeling TWB is–how to model tools like die, punch, blank holder, etc., when different thickness blanks are welded to produce TWB?. The thickness difference can be compensated by stepped die or stepped blank holder and punch as followed in some of the published work. The design and modeling of these tools, specifically in the case of un-symmetric TWB parts (as in most of the cases), needs expertise and in-depth understanding on tool design, forming behavior of TWB and modeling TWBs.

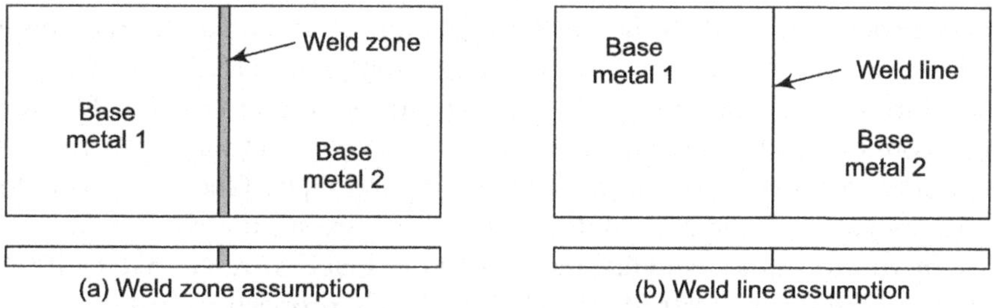

Fig. 10.7 Schematic of weld region representation during modeling of TWB formability (90° weld orientation)

Mostly TWBs are modeled without incorporating weld properties, i.e., weld line assumption is followed. The reason for this is twofold—(1) Evaluating weld properties is difficult or cumbersome, as LASER welded blanks posses weld width of the order of just 2 mm, and (2) The weld line assumption can predict the forming behavior of TWB with acceptable accuracy, as the weld region occupies very little area of the entire TWB part produced by LASER welding, say door inner, fender, bi-axial stretching (200 × 200 mm) strain path in LDH testing, etc. If weld zone assumption has to be followed, then the weld properties should be evaluated accurately like base materials. With the available methods, only few weld properties like yield strength, strain hardening exponent, strength coefficient (through tensile testing of sub-size samples with longitudinal weld and following rule of mixtures) and weld width (through micro hardness measurement, microstructure) can be evaluated. Other weld properties are assumed while modeling TWB. Though the weld line assumption can predict forming behavior to acceptable accuracy, weld region is a separate zone wherein the mechanical properties are quite different than that of base metal. So, in principle, TWB modeled with weld zone assumption should predict the forming behavior of TWB more accurately. Also, relative mechanical properties of the weld region with that of base material, say softer or harder weld zone, will affect the forming behavior prediction considerably. This issue is ruled out if the weld line assumption is followed. If the weld region is incorporated with same mechanical properties as that of base metal during modeling, one can expect that the predictions will be same as that of TWB modeled with weld line assumption or un-welded blank. But, this is practically impossible. Even if the weld properties are considered for simulation, identifying the weld shape and incorporating it in the simulations is a task by itself. This will become slightly complex if different thickness sheets are modeled, which also showed profound influence on the forming behavior prediction.

The weld line assumption is followed in many studies to predict the forming behavior of TWB like strain distribution, weld line movement control, deep drawability, etc. Few studies have aimed at including weld properties into simulations also [57–61]. In these work, the weld properties like yield strength, strength coefficient and strain hardening exponent are included in modeling weld region with some variation within or along with HAZ properties. The HAZ (4 mm) properties are given a triangular variation in the 2-D FE model which contains linear element and gradual stepped variation inside the HAZ 3-D model (containing rectangular elements) in [59] by Ghoo *et al.* The door inner panel and hemispherical dome stretching are simulated. The thickness strain distribution and weld line movement predicted correlates well with that of from experiments. In [60, 61], the HAZ and weld properties are included into the model, by knowing the yield strength of the weld zone from the relation, $\sigma_y^{weld} = (\sigma_y^{sheet} HV^{weld})/(HV^{sheet})$, where, σ_y^{weld} and σ_y^{sheet} are the yield strengths of weld and base metal, respectively. HV^{weld} and HV^{sheet} are the hardness of the weld zone and base material, respectively.

Some of the work performed on modeling issues compares the importance of weld line and zone assumption to predict the forming behavior. This issue is very important and critical from the accuracy point of view in predicting the TWB forming behavior. For example, in [62], Raymond *et al.* compared the TWB models with and without weld properties using ASTM tensile test simulation; in-plane test and LDH test simulations. Weld line movement and plastic strain are monitored for difference thickness ratios. It was found that there are a number of relatively subtle effects in which the weld is modeled and most of them are related to the constraining effect of weld region during forming. This indirectly means that weld line assumption is sufficient to model TWB. But this conclusion depends on the forming behavior that is monitored during simulation. In [62], weld line displacement is monitored, which may not be affected by the methods by which weld is modeled. Weld line moves because of the change in plastic deformation levels attained by the thinner and thicker metals during forming. Weld line or zone assumption may not affect the weld line movement prediction much. Instead, forming behavior such as load-progression curves, limit strains (or FLC) will end up in meaningful conclusions. Also, Raymond *et al.* had simulated the TWB with different thickness ratios (transverse weld). In this case, only thinner metal will deform without much happening in the thicker blank and hence it is quite obvious to expect that weld line modeling is sufficient to predict the forming behavior accurately. In Bhagwan *et al.* [31], weld line shift and equivalent plastic strains are monitored by performing LDH test simulation using LS-DYNA by including weld properties and geometry. It was found that weld geometry is more important than weld properties in predicting the weld line shift. But, this was concluded by taking only one set of weld properties (say n, σ_{ys}), without exploring much by changing them. Also, here the σ_{ys} ratio ($\sigma_{ys\text{-weld}}/\sigma_{ys\text{-base}}$) is approximately 1.06, which reflects that the weld exhibit approximately same yield strength as that of base metal. The relative property ratio plays a vital role in deciding the participation of weld region in deformation. Suppose, if the weld is softer than the base metal (say σ_{ys} ratio = 0.5), the weld properties may dominate the weld geometry effect. It is obvious, as said earlier, that for different thickness combinations, weld line assumption may be sufficient. So, one has to look into the worst case scenario of considering same material and thickness to study the importance of weld zone representation in modeling simulations.

Ganesh Narayanan *et al.* [44, 45] addressed the importance of weld region modeling by identifying the domain of weld conditions explicitly wherein weld line or zone assumption can be followed. The weld conditions considered are weld properties (n, σ_{ys}), weld orientation (transverse, longitudinal) and weld width. Tensile and LDH tests (plane strain, biaxial stretching strain paths) are simulated. Same thickness and material is modeled in the simulations. The load–stroke behavior, maximum load reached and progression at failure are monitored and comparison is done between weld line and zone assumption. It is found from the analysis that weld line modeling is not sufficient to model TWBs always and this depends on the weld conditions and mode of deformation. To summarize the deliverables from their work, representing weld region in simulations depends on whether the weld region is participating in the deformation or not. For example, if the weld region is soft and is normal to the major straining direction, failure occurs only in the weld region without much happening in the base metal. In this case weld zone representation is essential. But, if weld is harder than the base metal, then base metals absorb most of the deformation and hence weld line assumption is sufficient here. However, one cannot conclude this simply in few sentences. Suppose if the forming behavior monitored using weld line and zone assumption vary within a small range (say < 10 %) which is acceptable, one need not bother much about evaluating weld properties and including it during simulations. Instead weld line assumption can be followed. With this in mind, a criterion of 10–15% difference between the load–stroke behavior of weld line and zone assumption is assigned in [44, 45]. If the difference is less than 10–15%, weld line assumption is sufficient, otherwise weld zone assumption is essential. By following this criterion, domains are obtained for tensile and LDH tests. In transverse weld orientation, '$\sigma_{ys\text{-weld}}$' plays a vital role than 'n_{weld}' and 'width' of the weld region. In longitudinal weld orientation, the domains obtained are function of $\sigma_{ys\text{-weld}}$, n_w and weld width. It is also pointed out in [18] that weld region modeled with weld properties predict the actual shape of deep drawn cup in the weld region and base metal accurately than TWB modeled without weld properties. Similar results are seen in [63]. In the work of Zhao *et al.* [63], six different TWB models of weld parallel to bending moment, namely shell element with weld, shell element without weld, solid element with weld, and same three models for weld normal to bending moment are compared. Free bend test, stretch bend test and LDH tests are simulated and load–stroke, spring back behaviors are monitored. It is concluded from the analyses that weld region modeled with HAZ properties (by shell elements) improves the prediction of forming behavior. To summarize, the importance of weld line or zone modeling depends on the weld conditions like orientation, properties, location, deformation mode or part shape, thickness combination, etc. In general, weld line modeling is sufficient for some TWB conditions, but not always. Some of the TWB conditions demand weld zone assumption, so that forming behavior of TWB can be predicted accurately.

Tools like shim or stepped die, blank holder, punch, etc. are used in the forming of TWB. Shimming is the widely adapted method. Stepped die, blank holder, punch are also used for forming TWBs, but to produce TWB industrial parts. Not much of literature is available to describe the problems associated with modeling these tools during FE simulation. But, problems related to modeling these tools for TWB can be one among those mentioned in previous section on controlling the weld line movement. The details specifically mentioned on problems in stepped tools like providing slot near weld line, tolerance limit of sheet thickness in dissimilar TWBs, modeling tools for TWB with slanting weld line, etc. should be taken seriously while modeling tools for TWB. Meinders *et al.* [29] and Doege *et al.* [64], described how to model stepped blank holder, punch, die during the FE simulations of TWB.

10.5 EVALUATING THE CONSTITUTIVE BEHAVIOR OF WELD REGION

The constitutive behavior of weld region is important both from the formability and simulation point of view in TWB. Evaluating the constitutive behavior of weld region is not a new issue, as it is performed even before the application of TWBs. But most of application used arc welding type where weld region (weld zone + HAZ) is of the order of 15 mm, so that the weld properties can be obtained quite easily. In the case of TWB production, since LASER welding is widely used for welding, obtaining the properties of smaller weld zone (1–2 mm) accurately is really difficult and cumbersome. A few papers deal with this issue using transverse and longitudinal TWBs. In [65], the rule of mixtures technique with TWB having longitudinal weld (Fig. 10.8) is used to evaluate the stress–strain behavior of the TWB and weld region by Abdullah *et al.*

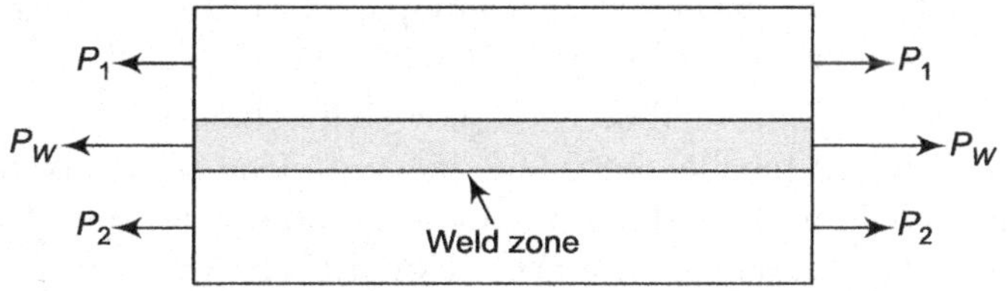

Fig. 10.8 Schematic of load sharing by base metals and weld region in longitudinal TWB

The stress–strain relationship of base metals is first obtained by tensile testing and fitting appropriate hardening law like Hollomon equation giving,

$$\sigma_1 = K_1\varepsilon_1^{n_1} \text{ and } \sigma_2 = K_2\varepsilon_2^{n_2} \qquad (10.1)$$

where 'K' and 'n' refer to strength coefficient and strain hardening exponent of base metals.

In the tensile testing of welded specimen, the total load 'P_T' on the sample is represented as,

$$P_T = P_1 + P_2 + P_W = \sigma_1 A_1 + \sigma_2 A_2 + \sigma_W A_W \qquad (10.2)$$

where 'A' refer to the cross sectional area. 'w' in Eq. (10.2) refer to weld region. Sub-scripts 1 and 2 in Eqs (10.1) and (10.2) refer to base metal 1 and base metal 2, respectively.

Substituting σ_1 and σ_2 from Eqs (10.1) in (10.2) gives,

$$\bar{\sigma}_w = \frac{P_T - (K_1\varepsilon_1^{n_1})\,A_1 - (K_2\varepsilon_2^{n_2})\,A_2}{A_W} \qquad (10.3)$$

Here, $\bar{\sigma}_w$ represents average stress in the weld region, as the hardening behavior of weld is expected to vary along its length. Longitudinal strain is assumed to be constant in the weld region during deformation, and hence

$$\varepsilon_1 = \varepsilon_2 = \varepsilon_W \qquad (10.4)$$

Substituting (10.4) in (10.3) results in the relationship,

$$\bar{\sigma}_w = \frac{P_T - (K_1\varepsilon_W^{n_1})\,A_1 - (K_2\varepsilon_W^{n_2})\,A_2}{A_W} \qquad (10.5)$$

Equation (10.5) defines the stress–strain relation in the weld region. Here, except σ_w and ε_w all other values can be obtained using sub-size tensile tests on base metal. Area of the weld region 'A_w' can be calculated by knowing the area of cross sections of base metals 1 and 2 or from weld microstructure. Measuring area of weld region is critical to the successful implementation of this method.

This method of evaluating the weld properties is followed in [65] and the stress–strain behavior (*K*, *n*) of weld region is obtained. It is clear that longitudinal TWB show higher yield strength and lower ductility than that of un-welded blank. Weld exhibits approximately twice the yield strength of base metal because of the presence of harder phase in the weld region. Since an increase in yield strength is seen in the weld region, '*n*' of weld is less than that of base metal. Also, the effect of assuming the weld area on the normalized error of *K* and *n* values is presented in [65] by Abdullah *et al.*

In Ghoo *et al.* [66], same method was followed to establish the understanding on effect of different tensile specimen size and offset weld position on the stress–strain behavior of weld region. It was found that tensile specimen size show little effect on the stress–strain behavior of TWB. Also, the offset position of weld in the longitudinal TWB showed negligible effect on the '*K*' value obtained because of the same total load absorbed by the TWBs independent of the offset values. In Auger *et al.* [67], the same methodology of rule of mixtures is used to determine the weld stress–strain behavior. Also, another method 'Plane strain compression' test is described here by which the weld properties can be evaluated. The weld region is subjected to compression test through a wedge shape tool adopted from typical plane strain compression test. By knowing data from the compression test and friction between weld and tool, weld properties can be evaluated. Cheng *et al.* [68] described the evaluation procedure for obtaining the tensile behavior of TWB and weld zone.

The global stress–strain behavior of the welded plate, behavior of individual components of the welded plate like HAZ, weld nugget can be obtained using iso-stress concept in transverse welded plate. This was demonstrated in friction stir welded plates by Liu *et al.* in [69]. Finally, the following relationship [69] is obtained which describe the global stress–strain behavior of transverse welded plate.

$$\varepsilon = \ln\left(\frac{2e^{(\sigma/K_1)^{1/n1}}\, l_{01} + 2e^{(\sigma/K_2)^{1/n2}}\, l_{02} + e^{(\sigma/K_3)^{1/n3}}\, l_{03}}{2l_{01} + 2l_{02} + l_{03}}\right) \tag{10.6}$$

The stress–strain response from this model is compared to that of experimental stress–strain behavior of Al alloy 2024-T351 [69]. It is clear from [69] that the σ–ε curve, *K*, *n* obtained from model and experiments correlate well. This demonstrates the validity of iso-stress model to evaluate the stress–strain response of transverse welded plates. Also, one can obtain the behavior of weld region and HAZ from this. The same technique can be used to evaluate the stress–strain curves of laser welded blanks, wherein one can approximate the entire transverse welded sample into three regions—base metal 1, base metal 2, and weld zone.

In [69], tensile testing of longitudinal welded sample is also performed and the results from model and experiments are compared to that of from transverse welded sample (as seen earlier). It is found from the analysis that

'*K*' and '*n*' values obtained are closely matching in both the cases. The weld properties can also be obtained by providing a reduced cross sectional area near the transverse weld region, so that failure occurs in the or near to the weld zone. This method was presented in [70, 71] by Min *et al.*, where flash welded specimens are subjected to tensile testing and v-bending operation to test its formability. One can also provide v-notch in the weld region, so that weld properties can be found directly. But, the notch should be designed properly such that intrinsic effect of weld region can be evaluated.

10.6 EXPERT SYSTEM FOR TAILOR WELDED BLANKS

The formability characteristics of TWBs is affected by weld conditions such as weld properties, weld orientation and weld location, thickness difference and strength difference between the sheets in a compounding fashion. Hence it is difficult to design the TWB conditions that can deliver a good stamped product with more or less similar formability characteristics as that of un-welded blank. In this context, few research groups have aimed at predicting the formability and deep drawability of welded blanks by using different necking theories and finite element simulations. For example, Jie *et al.* [72] studied the forming behavior of 5754-O Al alloy sheets, where in the forming limit of welded blanks with thickness ratio of 1:1.3 was experimentally evaluated and predicted using localized necking criterion based on Vertex theory. It is found from the analysis that the forming limit of the TWB is closer to thinner material FLC and the experimental and predicted FLCs correlate well with each other. Similarly, Anand *et al.* [73] investigated the limit strains of welded blanks made of IF steel of different thicknesses (0.75–1.5 mm). They have used modified Keeler criterion to predict the plane strain limit strain (FLD_0) in which average thickness and average strain hardening exponent of the blanks constituting TWB was considered. The predicted FLD_0 is comparable with that of from experiments.

Davies *et al.* [74] investigated the limit strains of aluminum alloy TWB (1:2 mm thickness), where in the FLCs predicted by Marciniak–Kuczynski (M–K) analysis are compared with the experimental results. Here the geometrical heterogeneity, i.e., the initial imperfection level, involved in the welded blank is modeled by using the strain-hardening exponent determined from miniature tensile testing together with the Hosford yield criterion, involving parameters $a = 8$ and $R = 1$ to determine a level of imperfection that exactly fits an FLD to each experimentally evaluated failure strains. The FLCs thus predicted are found to have good agreement with the experimental FLCs, except in the bi-axial stretching region. Ganesh Narayanan and Narasimhan [75] predicted the forming limit strains of laser welded blanks by using thickness gradient based necking theory incorporated into an elasto-plastic finite element code PAMSTAMP 2G®. It is found that the predictions are agreeing well in drawing region of FLD, with deviation in stretching region. Recently, Naik *et al.* [76],

Ganesh Narayanan and Naik [77] predicted the forming limit of welded blanks using the original and modified strain rate based criteria and compared with experimental results. It is observed from their work that the modified failure criterion predicts the forming limit with better accuracy. This prediction is also compared with thickness gradient based necking criterion and results are found to be encouraging. From the above discussion, it is clear that one has to follow a limit strain theory in conjunction with numerical or analytical methods to predict the forming limit strains of welded blanks for different base material and weld conditions. This is true in the case of other forming behaviors like stress–strain curve, deep drawability, earing behavior, etc. also.

It is known that the presence of thickness, strength heterogeneities and weld region deteriorates the formability of welded blanks in most of the cases. Designing TWB for a typical application will be successful only by knowing the appropriate thickness, strength combinations, weld line location and profile, number of welds, weld orientation and weld zone properties. Predicting these TWB parameters in advance will be helpful in determining the formability of TWB part in comparison to that of un-welded base materials. In order to fulfill this requirement, one has to perform lot of simulation and experimental trials separately for each of the cases which is time consuming and resource intensive. Automotive sheet forming designers will be greatly benefited if an 'expert system' is available for TWBs that can deliver its forming behavior for varied weld and blank conditions. Artificial Neural Network (ANN) is found to show better prediction of any response variable that is influenced by large number of input parameters rather than conventional way of doing experiments. Application of ANN in predicting the formability of TWB will be helpful in understanding and designing the TWB conditions that can deliver a better-stamped product.

The main objective of the present research work is to develop an 'expert system' for welded blanks that can predict their tensile, deep drawing, forming behavior under varied base material and weld conditions using different formability tests, material models, and formability criteria. It is decided to develop the expert system in conjunction with Artificial Neural Network (ANN). The data required for the expert system development is obtained through simulations only. PAM STAMP 2G a finite element code is used to generate data for varied base material and TWB conditions. The proposed expert system design for TWB forming is shown in Fig. 10.9 [78, 79]. This expert system is expected to involve three different phases. All the three phases have a design mode of operation where an initial expert system is created and put in place. The created expert system is then operated in use and update mode.

In phase 1, while the expert system is designed, a range of material properties and TWB conditions are defined within which ANN models are developed to predict the results as discussed in the earlier sections. The same phase while operated in the usage mode, the user selects base material properties and TWB conditions within the chosen range for application and prediction of formability.

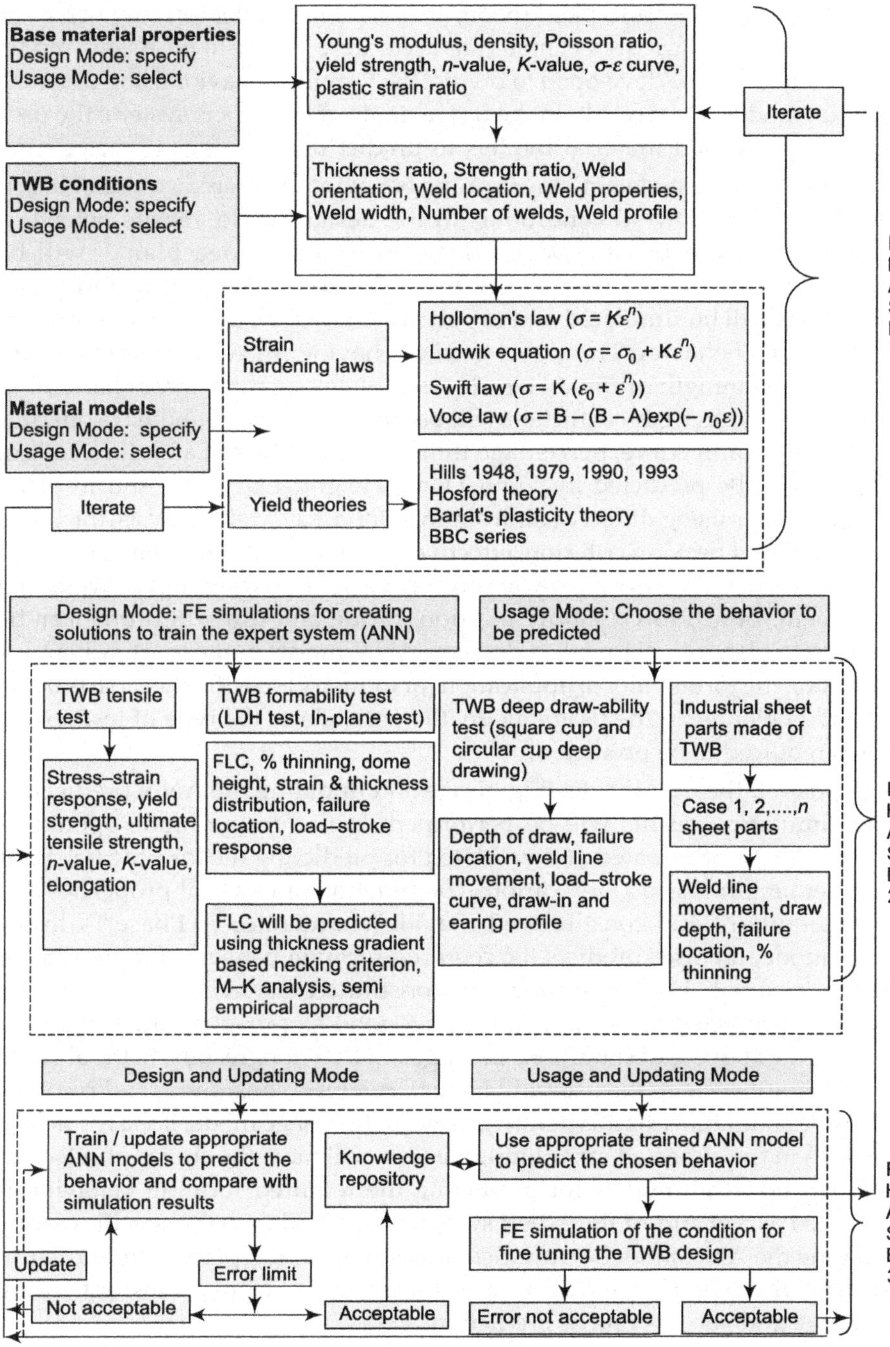

Fig. 10.9 Expert system proposed for TWB forming; with permission from [78] copyright [2009] Elsevier

In this phase, user can select different material models viz., strain hardening laws and yield theories to predict the forming behavior. There is no single strain hardening law and yield theory that can predict the forming behavior of TWBs made of varied sheet materials accurately. Hence in the design mode, ANN models will be developed to predict the forming behavior using different material models. As a result, in the usage mode of the expert system, the user can opt for desired material models to predict the forming characteristics.

Phase 2 involves selecting the forming behavior to be predicted for chosen base material and weld conditions. In the design mode, tensile behavior, formability characteristics, deep drawability of welded blanks will be simulated by standard formability tests. Different category of industrial sheet parts will be simulated and expert system will be developed to predict their forming behavior. The global tensile behavior of TWB viz., stress–strain curve, yield strength, ultimate tensile strength, elongation, strain hardening exponent and strength coefficient will be monitored. Formability properties like forming limit curve, percentage thinning, dome height at failure, failure location will be predicted by Limit Dome Height (LDH) test and in-plane stretching tests using different limit strain criteria (say M–K analysis, thickness gradient based necking criterion, effective strain rate criterion, semi empirical approach, etc.). Cup deep drawability response like draw depth, weld line movement, punch force, failure location, earing and draw-in profile can be predicted. Also it is planned to develop ANN model and expert system for predicting the formability of application (or industry) specific sheet parts made of welded blanks. In the usage mode, the user selects the type of test results that is required to be predicted.

In phase 3 the training, testing, usage and updating the ANN predictions with simulation results will be performed. In the design mode operation, various ANNs are created and validated for predicting the forming behavior (enumerated in Phase 2) for various combination of material properties and TWB conditions and constitutive behavior (enumerated in Phase 1). In the usage mode, the user predicts the required forming behavior for an initially chosen material, TWB condition and constitutive behavior. If the forming behavior predicted is not indicative of a good stamped product, the user changes the above said conditions till he gets satisfactory results. In the absence of this expert system, the user will have to run time consuming and resource intensive simulation for this iterative stage. In the usage mode, if the results are not with in the expected error limit, the user will have the choice of selecting different material models for predicting the required forming behavior as described earlier and/or the expert system is updated with the specific case by updating the ANN models to predict the case within acceptable error limits. In this way, the expert system also learns form the application cases, enhancing the range and success rate of predictions.

In this section, some representative expert system prediction like the stress–strain behavior, draw-in profile during cup deep drawing, and forming limit curve are presented. The tools required for tensile test, deep drawing test, limit dome height test simulation and modeling details can be obtained from [78–80]. The six different input parameters are varied at three different levels (decided from literature) and simulation trials were conducted as per L27 orthogonal array. The various ANN parameters like number of hidden layers, neurons, and transfer functions are optimized based on many trials to predict the outputs within the normalized error limit of 10^{-4}. Various network structures with one and two hidden layers with varying number of neurons in each layer are examined. Finally the architecture which yielded better performance is used for expert system development. In all the cases, a feed forward back propagation algorithm is selected to train the network in Matlab programming environment. Here the scaled conjugate gradient algorithm is used to minimize the error. From the available simulation data sets, 27 data sets are used to train and two intermediate data sets are utilized for testing the ANN model/expert system. The comparison between ANN predicted true stress–strain behavior and simulation results are shown in Fig. 10.10.

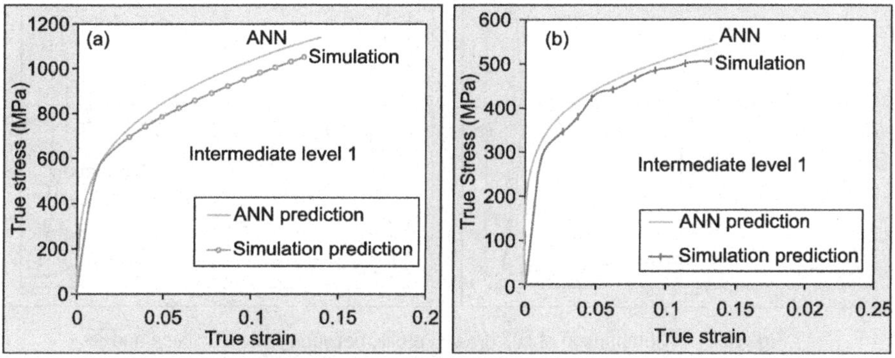

Fig. 10.10 Validating the true stress–strain behavior predicted by ANN/expert system with FE simulations; (a) Steel TWB, (b) Al alloy TWB; with permission from [78] copyright [2009] Elsevier

The strain hardening exponent (n) and strength coefficient (K) values obtained from ANN models are incorporated into Hollomon's equation ($\sigma = K\,\varepsilon^n$) for TWB made of steel and aluminum alloy base materials and true stress–strain curves are obtained [78]. It should be noted that even though Hollomon's strain hardening law is not accurate to predict the tensile behavior of aluminum alloy base material, ANN predictions are quite accurate in predicting the same.

Similarly, the comparison between ANN/expert system and simulation results of draw-in profile during square cup deep drawing is presented in Fig. 10.11. At different TWB conditions, the draw-in profile predicted by ANN model/expert system is well matched with the simulation results for both steel and Al alloy TWBs [79]. In the case of LDH test, the FLC is predicted by thickness gradient based necking criterion (TGNC). The ANN/expert system prediction is found to show excellent correlation with FLC from the criterion [Figs 10.12 (a)–(c)] for steel TWB [80, 81]. It is also found that the FLCs predicted from other failure criteria—effective strain rate, major strain rate based necking criteria, both the original and modified ones, are comparing satisfactorily with the expert system results [80, 81]. A slight deviation in the plane strain and stretching modes of deformation is seen in both the intermediate TWB conditions. The expert system can be linked with genetic algorithm to improve its efficiency. A general information about the application of expert systems in many fields including sheet metal forming can be found in the book edited by Vizureanu [82].

A detailed discussion about the weld integrity of tailor welded blanks, numerical modeling of TWB forming, its application using light weight metal alloy and advanced high strength steels can be understood from the book edited by Kinsey and Wu [83]. The potential applications of TWB in aerospace and automotive sectors is also described [83].

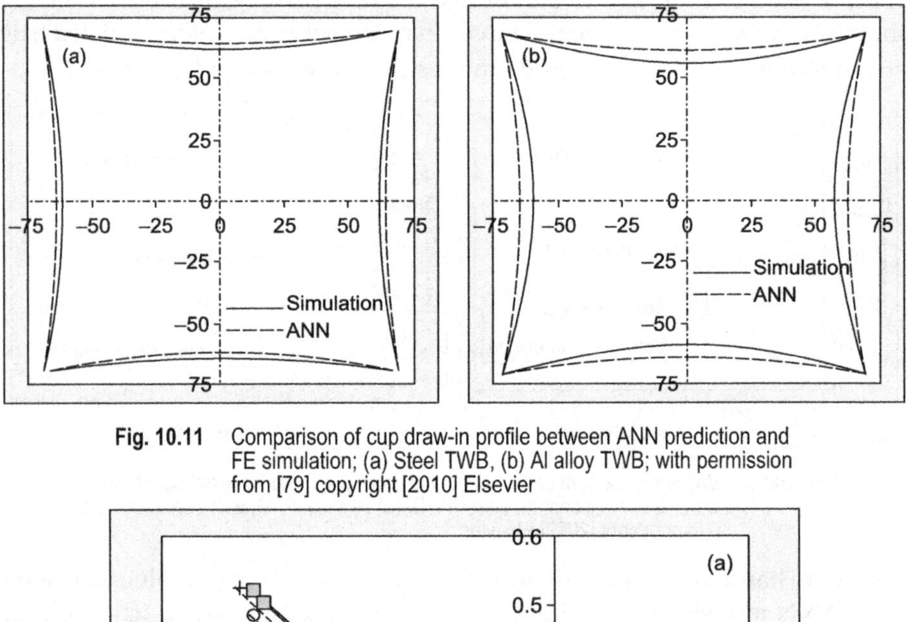

Fig. 10.11 Comparison of cup draw-in profile between ANN prediction and FE simulation; (a) Steel TWB, (b) Al alloy TWB; with permission from [79] copyright [2010] Elsevier

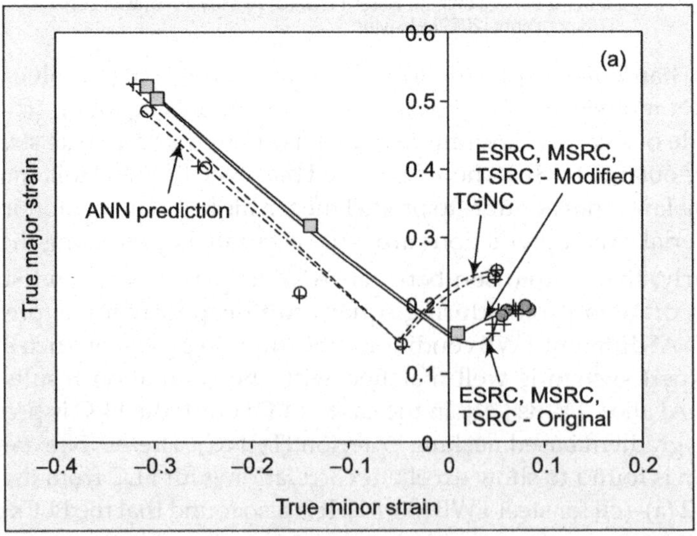

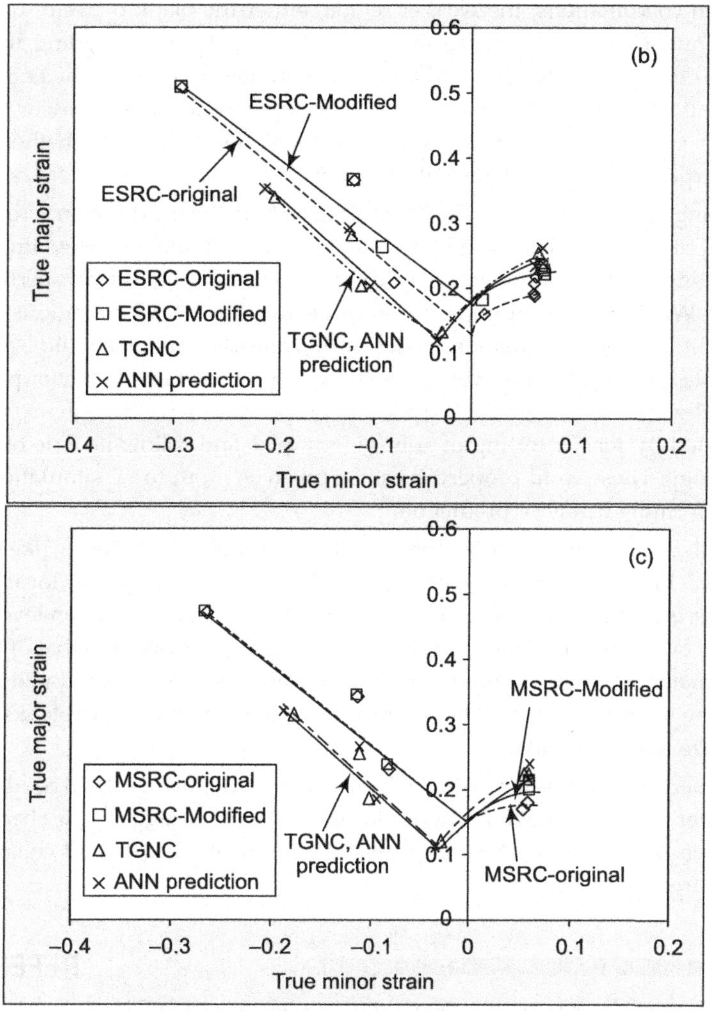

Fig. 10.12 Comparison of ANN/expert system prediction with TGNC and other failure criteria prediction—(a) intermediate level 1, ESRC, MSRC, TSRC comparison; (b) intermediate level 2, ESRC; (c) intermediate level 2, MSRC [81]

SUMMARY

The TWB parameters like thickness ratio, strength ratio, and various weld conditions affect the forming behavior differently. The weld zone may be either hard or soft, and loading may be parallel or normal to the weld region decides the formability of TWB. Assuming same material and thickness, with 'soft' weld, loading parallel to the weld may not be a problem, but normal loading will result in weld failure, because of strength differential. In the case of 'hard' weld, loading parallel to the weld can reduce the forming limit of TWB, as the weld region posses lesser ductility and no problem is seen if it is normal as it will not deform. In the case of different thickness or strength combinations, the weaker region, either the blank or weld, decides the forming limit of TWB, if the weld is normal. In the case of parallel loading, formability of TWB is decided both by the weld and base materials. In general, thickness ratio or strength ratio dominates other parameters in deciding the forming behavior, with weld properties, say σ_{ys} and n, showing significant effect depending on whether the weld zone is participating in the deformation or not.

Modeling TWB is a critical issue depending on how the weld region is represented during FE simulations. The weld region can be treated simply as a weld line, without incorporating weld properties, or as weld zone by considering weld properties during modeling. Weld line assumption is not accurate for all the TWB conditions. For some TWB conditions and deformation modes, weld zone assumption should be followed; for example, TWB with softer weld zone show accurate forming behavior predictions with weld zone modeling. The important properties of weld zone, σ_{ys}, n, K, can be evaluated by tensile testing of sub-size samples and following rule-of-mixtures methodology. These weld properties can be given as input to FE simulation models for better forming behavior prediction.

Weld line movement can be controlled by adopting techniques like applying differential BHF, draw bead restraining forces, designing special local adaptive controllers, etc. The main idea here is to control the plastic deformation level attained by thinner or weaker material, so that it will be nearly equivalent to that of thicker or stronger material, which eventually minimize weld line movement. It will be a good idea to change the friction conditions between the thinner and thicker blanks and tools to minimize weld line shift.

The expert system based on ANN will be useful to identify the TWB conditions that yields better forming behavior. This is like an index which suggests the sheet forming engineer on how to decide the best possible base material and weld configurations for defect free TWB parts.

REFERENCES

1. Pallett, R.J. and Lark, R.J. (2001). The use of tailored blanks in the manufacture of construction components, *Journal of Materials Processing Technology*, **117** (1–2); 249.
2. http://www.ulsab.org/
3. Irving, B. (1995). Welding tailored blanks is hot issue for automakers, *Welding Journal*, 49.
4. Prange, W., Schneider, C., Jaroni, U. (1994). Tailored Blanks—Contribution to an Optimized Steel Body Shell, *Proceedings of Laser Assisted Net Shape Engineering* LANE '94, Vol. 1, Meisenbach Bamberg, pp. 145.
5. Das, S. (2000). Aluminium tailor welded blanks, *Advanced Materials and Processes*, 41.
6. Chan, S.M., Chan, L.C., Lee, T.C. (2003). Tailor welded blanks of different thickness ratios effects on forming limit diagrams, *Journal of Materials Processing Technology*, **132** (1–3); 95.

7. Chan, L.C., Cheng, C.H., Chan, S.M., Lee, T.C., Chow, C.L. (2005). Formability analysis of tailor welded blanks of different thickness ratios, *Journal of Manufacturing Science and Engineering*, **127**; 743.
8. Shakeri, H.R., Buste, A., Worswick, M.J., Clarke, J.A., Feng, F., Jain, M., Finn, M. (2002). Study of damage initiation and fracture in aluminum tailor welded blanks made via different welding techniques, *Journal of Light Metals*, **2** (2); 95.
9. Buste, A., Lalbin, X., Worswick, M.J., Clarke, J.A., Finn, M., Altshuller, B., Jain, M. (1999). Prediction of strain distribution in Aluminum tailor welded blanks, *Proceedings of the 4th International conference and workshop on Numerical simulation of 3D sheet forming process*, NUMISHEET'99, Besancon, France, pp. 455.
10. Friedman, A., Kridli, G.T. (2000). Microstructural and mechanical investigation of aluminium tailor welded blanks, *Journal of Materials Engineering and Performance*, **9** (5); 541.
11. Ghoo, B.Y., Park, S.W., Keum, Y.T. (2001). New forming limit diagram of laser tailored blank, *Journal of Strain Analysis for Engineering Design*, **36**; 143.
12. Vijay Bhaskar, V., Ganesh Narayanan, R., Narasimhan, K. (2004). Effect of thickness ratio on formability of tailor welded blanks (TWB), *Proceedings of the 8th International Conference on Numerical Methods in Industrial Forming Processes*, NUMIFORM '04, Columbus, Ohio, pp. 863.
13. Nakagawa, N., Ikura, S., Natsumi, F., Iwata, Finite element simulation of stamping of laser welded blank, *SAE Paper*, 930522, pp. 734.
14. Miles, M.P., Decker, B.J., Nelson, T.W. (2004). Formability and strength of friction-stir-welded aluminium sheets, *Metallurgical and Material Transactions A*, **35A**; 3451.
15. Raymond, S.D., Wild, P.M., Bayley, C.J. (2004). On modeling of the weld line in finite element analyses of tailor-welded blank forming operations, *Journal of Materials Processing Technology*, **147**; 28.
16. Siegert, K., Knabe, E., Fundamental research and draw die concepts for deep drawing of tailored blanks, *SAE Paper*, 950921, pp. 866.
17. Baysore, J.K., Williamson, M.S., Adonyi, Y., Milian, J.L. (1995). Laser beam welding and formability of tailored blanks, *Welding Research Supplement*, 345–s.
18. Kampus, Z., Balic, J. (2003). Deep drawing of tailored blanks without a blankholder, *Journal of Materials Processing Technology*, **133**; 28.
19. Tusek, J., Kampus, Z., Suban, M. (2001). Welding of tailored blanks of different materials, *Journal of Materials Processing Technology*, **119**; 180.
20. Reis, A., Teixeira, P., Ferreira Duarte, J., Santos, A., Barata da Rocha, A., Fernandes, A.A. (2004). Tailored welded blanks—an experimental and numerical study in sheet metal forming on the effect of welding, *Computers and Structures*, **82**; 1435.
21. Choi, Y., Heo, Y., Kim, H.Y., Seo, D. (2000). Investigations of weld-line movements for the deep drawing process of tailor welded blanks, *Journal of Materials Processing Technology*, **108**; 1.
22. Heo, Y., Choi, Y., Kim, H.Y., Seo, D. (2001). Characteristics of weld line movements for the deep drawing with drawbeads of tailor-welded blanks, *Journal of Materials Processing Technology*, **111**; 164.

23. Ahmetoglu, M.A., Brouwers, D., Shulkin, L., Taupin, L., Kinzel, G.L., Altan, T. (1995). Deep drawing of round cups from tailor-welded blanks, *Journal of Material Processing Technology*, **53**; 684.
24. Kinsey, B., Liu, Z., Cao, J. (2000). A novel forming technology for tailor-welded blanks, *Journal of Materials Processing Technology*, **99**; 145.
25. Kinsey, B., Liu, Z., Cao, J., New apparatus and method for forming tailor welded blanks, *Journal of Materials and Manufacturing, SAE Paper*, 1999-01-0681, **108**; Section 5, pp. 653.
26. Kinsey, B., Viswanathan, V., Cao, J., Forming of aluminum tailor welded blanks, *SAE Paper*, 2001-01-0822, pp.1.
27. Kinsey, B., Song, N., Cao, J., Analysis of clamping mechanism for tailor welded blank forming, *SAE Paper*, 99IBECC-26.
28. Kinsey, B., Cao, J. (2001). Enhancement of sheet metal formability via local adaptive controllers, *Transactions of NAMRI/SME*, **XXIX**; 81.
29. Meinders, T., van den Berg, A., Huetink, J. (2000), Deep drawing simulations of Tailored Blanks and experimental verification, *Journal of Materials Processing Technology*, **103**; 65.
30. Saunders, F.I., Wagoner, R.H. (1996). Forming of tailor welded blanks, *Metallurgical and Material Transactions A*, **27A**; 2605.
31. Bhagwan, A.V., Kridli, G.T., Friedman, P.A. Influence of weld characteristics on numerically predicted deformation behavior of aluminum tailor welded blanks, *SAE Paper*, 02M-144, pp. 1.
32. Bhagwan, A.V., Kridli, G.T., Friedman, P.A. (2003). Formability improvement in aluminum tailor welded blanks via material combinations, *Proceedings of NAMRC XXXI*, SME paper, MF03-155, , Hamilton, Ontario, Canada.
33. Pepelnjak, T., Jurkosek, B., Kuzman, K., Jesenicnik, B., Research on the forming of tailored blanks in conventional tools, *Proceedings of Laser Assisted Net Shape Engineering 97*, LANE '97, Erlangen, Germany, pp. 767.
34. Kusuda, H., Takasago, T., Natsumi, F. (1997). Formability of tailored blanks, *Journal of Materials Processing Technology*, **71**; 136.
35. Geiger, M., Merklein, M., Artur, G. (2004). Deep drawing of friction stir welded aluminium tailored blanks, *Proceedings of International Deep Drawing Research Group 2004 Conference (IDDRG 2004)*, Sindelfingen, Germany, pp. 114.
36. Hayashi, H., Pak, S.W., Usuda, M., Metal flow of a tailored blank in square cup deep drawing, *SAE Paper*, 980447, pp. 1.
37. Koruk, A.I., Akduk, N., Cretteur, L., Corostola, J.R. Laser welded tailored blanks made of stainless steel/carbon steel, *Proceedings of Laser Assisted Net Shape Engineering 01*, LANE 2001, Erlangen, Germany, pp. 653.
38. Cheng, C.H., Chan, L.C., Chow, C.L., Lee, T.C. (2005). Experimental investigation on the weldability and forming behavior of aluminum alloy tailor-welded blanks, *Journal of Laser Applications*, **17** (2); 81.
39. Kridli, G.T., Friedman, P.A., Sherman, A.M., Formability of aluminum tailor welded blanks, *SAE Paper*, 2000-01-0772, pp. 1.

40. Ganesh Narayanan, R., Narasimhan, K. (2008). Influence of the weld conditions on the forming-limit strains of tailor-welded blanks, *Journal of Strain Analysis for Engineering Design*, **43** (4); 217.

41. Ganesh Narayanan, R. (2007). Forming behavior of tailor welded blanks, *Ph. D. thesis*, Indian Institute of Technology, Bombay, India.

42. Nandedkar V.M. (2000), Formability studies on a deep drawing quality steel, *Ph. D. thesis*, Indian Institute of Technology, Bombay, India.

43. Ganesh Narayanan, R., Narasimhan, K. (2007). Relative effect of material and geometric parameters on the forming behavior of tailor welded blanks (TWB), *International Journal of Forming Processes*, **10** (2); 145.

44. Ganesh Narayanan, R., Vijay Bhaskar, V., Narasimhan, K. (2004). Effect of weld conditions on the deformation behavior of Tailor Welded Blanks (TWB), *Proceedings of the 8th International conference on Numerical Methods in Industrial Forming Processes*, NUMIFORM '04, Columbus, Ohio, pp. 856.

45. Ganesh Narayanan, R., Narasimhan, K. (2006). Weld region representation during the simulation of TWB forming behavior, *International Journal of Forming Processes*, **9** (4); 491.

46. Stasik, M.C., Wagoner, R.H. (1998). Forming of tailor welded aluminum blanks, *International Journal of Forming Processes*, **1** (1); 9.

47. Stasik, M.C., Wagoner, R.H. (1996). Forming of tailor welded aluminum blanks, Aluminum and Magnesium for Automotive Applications, *The Minerals, Metals & Materials Society*, 69.

48. Ahmetoglu, M.A., Brouwers, D., Kinzel, G., Altan, T., Deep drawing of round cups from laser-tailor welded blanks, *Proceedings of Laser Assisted Net Shape Engineering 1994*, LANE '94, Vol. 1, Meisenbach Bamberg, p. 167.

49. He, S., Wu, X., Hu, S. J. (2003). Formability enhancement for tailor welded blanks using blanking holding force control, *Journal of Manufacturing Science and Engineering*, **125**; 461.

50. Hassan, M.A., Takakura, N., Yamaguchi, K. (2003). A novel technique of friction aided deep drawing using a blank-holder divided into four segments, *Journal of Materials Processing Technology*, **139**; 408.

51. Hassan, M.A., Suenaga, R., Takakura, N., Yamaguchi, K. (2005). A novel process on friction aided deep drawing using tapered blank holder divided into four segments, *Journal of Materials Processing Technology*, **159**; 418.

52. Gunnarsson, L., Schedin, E. (2001). Improving the properties of exterior body panels in automobiles using variable blank holder force, *Journal of Materials Processing Technology*, **114**; 168.

53. Obermeyer, E.J., Majlessi, S.A. (1998). A review of recent advances in the application of blank-holder force towards improving the forming limits of sheet metal parts, *Journal of Materials Processing Technology*, **75**; 222.

54. Thakkar, M., Date, P.P. (2005). Simulation based Control of Weld Line Movement in Tailor Welded Blanks, *Proceedings of the 6th International Conference and Workshop on Numerical Simulation of 3D Sheet Forming Process*, NUMISHEET'05, Detriot, Michigan, USA, 705.

55. Bravar, M., Kinsey, B. Analytical Determination of Initial Weld Line Position in Tailor Welded Blank Forming, *Transactions of the North American Manufacturing Research Institute of SME*, **32**; 597.

56. Kinsey, B.L., Cao, J. (2003). An analytical model for tailor welded blank forming, *Journal of Manufacturing Science and Engineering*, **125**; 344.

57. Zimniak, Z., Piela, A. (2000). Finite element analysis of a tailored blanks stamping process, *Journal of Materials Processing Technology*, **106**; 254.

58. Dry, D., Hughes, D., Owen, R. (2001). Methods of assessing influence of weld properties on formability of laser welded tailored blanks, *Ironmaking and steelmaking*, **28** (2); 89.

59. Ghoo, B.Y., Back, S.J., Keum, Y.T., Kang, S.Y. (1998). Finite element analysis of tailored sheet forming processes considering laser welding zone, *Metals and Materials*, **4** (4); 862.

60. Rojek, J., Onate, E., Piela, A., Neamtu, L. (2002). Numerical modeling and simulation of tailor welded blanks, *Proceedings of the 5th International conference and workshop on Numerical simulation of 3D sheet forming process*, NUMISHEET'02 Jeju Island, Korea, pp. 177.

61. Piela, A., Rojek, J. (2002). Experimental study and modeling tailor welded blanks, *Proceedings of the 5th International conference and workshop on Numerical simulation of 3D sheet forming process*, NUMISHEET '02, Jeju Island, Korea, pp. 225.

62. Raymond, S. D., Wild, P. M., Bayley, C. J. (2004), On modeling of the weld line in finite element analyses of tailor-welded blank forming operations, *Journal of Materials Processing Technology*, **147** (1); 28.

63. Zhao, K.M., Chun, B.K., Lee, J.K. (2001). Finite element analysis of tailor-welded blanks, *Finite Elements in Analysis and Design*, **37**; 117.

64. Doege, E., Dohrmann, H., Kosters, R. (1996). Simulation and optimization of the forming process of tailored blanks, *Proceedings of the 3rd International conference and workshop on Numerical simulation of 3D sheet forming process*, NUMISHEET'96, Dearborn, Michigan, pp. 199.

65. Abdullah, K., Wild, P. M., Jeswiet, J. J., Ghasempoor, A. (2001). Tensile testing for weld deformation properties in similar gage tailor welded blanks using the rule of mixtures, *Journal of Materials Processing Technology*, **112** (1); 91.

66. Ghoo, B.Y., Keum, Y.T., Kim, Y.S. (2001). Evaluation of the mechanical properties of welded metal in tailored steel sheet welded by CO_2 laser, *Journal of Materials Processing Technology*, **113**; 692.

67. Auger, M., Abdullah, K., Jeswiet, J.J., Wild, P., Clapham, L., Determination of weld line characteristics in tailored blanks, *SAE Paper*, 2000-01-2661.

68. Cheng, C. H., Chan, L. C., Tang, C. Y. (2005). Determination of true stress-strain curve for the weldment of aluminum laser-welded blanks, *Journal of Laser Applications*, **17**, (3); 159.

69. Liu. S., Chao, Y.J. (2005). Determination of global mechanical response of friction stir welded plates using local constitutive properties, *Modelling and Simulation in Materials Science and Engineering*, **13**; 1.

70. Min, K. B., Kang, S. S. (2000). A study on resistance welding in steel sheets for tailor welded blank Evaluation of flash weldability and formability (2nd Report), *Journal of Materials Processing Technology,* **103**; 218.

71. Min, K.B., Kim, K.S., Kang, S.S. (2000). A study on resistance welding in steel sheets using a tailor-welded blank (1st report): Evaluation of upset weldability and formability, *Journal of Materials Processing Technology,* **101** (1–3); 186.

72. Jie, M., Cheng, C.H., Chan L.C., Chow C.L., Tang C. Y. (2007). Experimental and theoretical analysis on formability of Aluminum Tailor-Welded Blanks, *Journal of Engineering Materials and Technology,* **129**; 151.

73. Anand, D., Boudreau, G., Andreychuk, P., Chen, D.L., Bhole, S.D. (2006), Forming Behaviour of Tailor (Laser) Welded Blanks of Automotive Steel Sheet, *The Canadian Journal of Metallurgy and Materials Science,* **45** (2); 187.

74. Davies, R.W., Smith, M.T., Oliver, H.E., Khaleel, M.A., Pitman S.G. (2000). Weld Metal Ductility in Aluminum Tailor Welded Blanks, *Metallurgical and Materials Transactions A,* **31** (11); 2755.

75. Ganesh Narayanan, R., Narasimhan, K. (2008). Predicting the forming limit strains of tailor welded blanks, *Journal of Strain Analysis for Engineering Design,* **43** (7); 551–563.

76. Naik, S. B., Janaki Ramulu, P., Ganesh Narayanan, R. (2010). Application of a few necking criteria in predicting the forming limit of un-welded and tailor-welded blanks, *Journal of Strain Analysis for Engineering Design,* **45**; 79.

77. Ganesh Narayanan, R., Naik, S. B. (2010). Assessing the validity of original and modified failure criteria to predict the forming limit of un-welded and tailor welded blanks with longitudinal weld, *Materials and Manufacturing Processes,* **25** (11); 1351.

78. Veera Babu, K., Ganesh Narayanan, R., Saravana Kumar, G. (2009). An expert system based on artificial neural network for predicting the tensile behavior of tailor welded blanks, *Expert Systems with Applications,* **36** (7); 10683.

79. Veera Babu, K., Ganesh Narayanan, R., Saravana Kumar, G. (2010). An expert system for predicting the deep drawing behavior of tailor welded blanks, *Expert Systems with Applications,* **37** (12); 7802.

80. Siva Krishna, K., Ganesh Narayanan, R., Saravana Kumar, G. (2010). Tailor welded blank forming limit computation by neural network based expert system, *Journal of Computer and Experimental Simulations in Engineering and Science (CESES),* Issue 8.

81. Siva Krishna, K. (2009). A expert system to predict the forming limit of tailor welded blanks, *M. Tech. thesis,* Indian Institute of Technology Guwahati, India.

82. Vizureanu, P. ed. (2010), *Expert Systems,* Intech Publishers, Croatia.

83. Kinsey, B.L., and Wu. X. ed. (2011), *Tailor welded blanks for advanced manufacturing,* Woodhead Publishing Limited, UK.

CHAPTER 11

Forming of Friction Stir Welded Blanks

Vinayak R. Malik[a], R. Ganesh Narayanan[b], Satish V. Kailas[a]
[a] Department of Mechanical Engineering,
Indian Institute of Science, Bangalore 560 012, India
[b] Department of Mechanical Engineering,
IIT Guwahati, Guwahati 781039, India

11.1 INTRODUCTION

Friction stir welding is an emerging solid state joining process. Seeing its potential, it is being implemented extensively for joining of aluminum alloys used in automotive and aerospace industries. In this context, a situation arises when the banks made by this process need to be formed. This chapter discusses on the formability aspects of friction stir welded sheets, but prior to discussing formability, we need to understand the friction star welding process in detail.

Friction stir welding (FSW) is a new process invented and patented by Welding Institute of Cambridge, UK in 1991. It is used for applications where the original metal characteristics must remain unchanged as far as possible. This process is primarily used on aluminum (originally intended for welding of aerospace alloys, especially aluminum extrusions), and most often on large pieces which cannot be easily heat treated post weld to recover temper characteristics.

FSW is an extension of conventional friction welding (FRW). In conventional friction welding, heating of interfaces is achieved through friction by rubbing two contacting surfaces, but in the FSW process, a third body is rubbed against the two surfaces to be joined in the form of a rotating non-consumable tool that is plunged into the joint. The contact pressures cause frictional heating, raising the temperature [1]. Friction stir welding (FSW) is a solid state welding process (without fusion/no melting) where a cylindrical-shouldered tool, with a profiled threaded/unthreaded probe (nib or pin) is rotated at a constant speed and fed at a constant traverse rate into the joint line between two pieces of sheet or plate material, which are butted together. The parts have to be clamped rigidly onto a backing bar in a manner that prevents the abutting joint faces from being forced apart. The length of the nib is slightly less than the weld depth required and the tool shoulder should be in intimate contact with the work surface. The nib is then moved against the work, or vice versa. The rotation action and the specific geometry of the FSW tool generate friction and mechanical working of the material. This in turn generates the heat and the mixing or stirring necessary to transport material from one side of the joint line to the other and to form a joint. This heat, along with the heat generated by the mechanical mixing process and the adiabatic heat within the material, cause the stirred materials to soften without reaching the melting point (hence cited a solid-state process), allowing the traversing of the tool along the weld line in a plasticized tubular shaft of metal. As the pin is moved in the direction of welding, the leading face of the pin, assisted by a special pin profile, forces plasticized material to the back of the pin while applying a substantial forging force to consolidate the weld metal. The welding of the material is facilitated by severe plastic deformation in the solid state, involving dynamic recrystallization of the base material. The two technical terms which need to be understood in FSW are (a) advancing side, and (b) retreating side. These require knowledge of the tool rotation and travel directions. Figure 11.1 shows the schematic of FSW. The tool rotates in the clockwise direction and moves in the direction shown (away from the reader). When the direction of the velocity vector of the tool and traverse direction are same, the side is called the advancing side of the weld, and when the direction of the velocity vector is opposite to the traverse direction, it is called the retreating side [2].

11.2 FSW PROCESS PARAMETERS

The three major process parameters identified during FSW of materials are as follows:

(i) Tool rotation and traverse speeds

The tool speeds that influence the quality of weld are (a) the speed at which the tool is rotated, and (b) the speed at which the tool is traversed (translated) along the joint. These two parameters have considerable importance and must be chosen with care to ensure a successful and efficient welding cycle. The relationship is complex between the welding speeds and the heat input during welding. In general, it can be said that increasing the rotation speed or decreasing the traverse speed will result in a hotter weld. To produce a successful weld it is necessary that the material surrounding the tool is hot enough to enable the extensive plastic flow required for welding and to reduce the forces acting on the tool. If the material is too cool then voids or other flaws occur in the stir zone and in extreme cases this may lead to damage and breakage of tool.

On the other hand, high heat input may be detrimental to the final properties of the weld. This can even induce defects due to the liquation of low-melting-point phases (just like liquation cracking in fusion welds). This leads onto the concept of a 'processing window'. This process window gives the range of process parameters that will produce a good quality weld. Within this window the resulting weld will have a sufficiently high heat input to ensure adequate material plasticity but not so high that will deteriorate the weld properties.

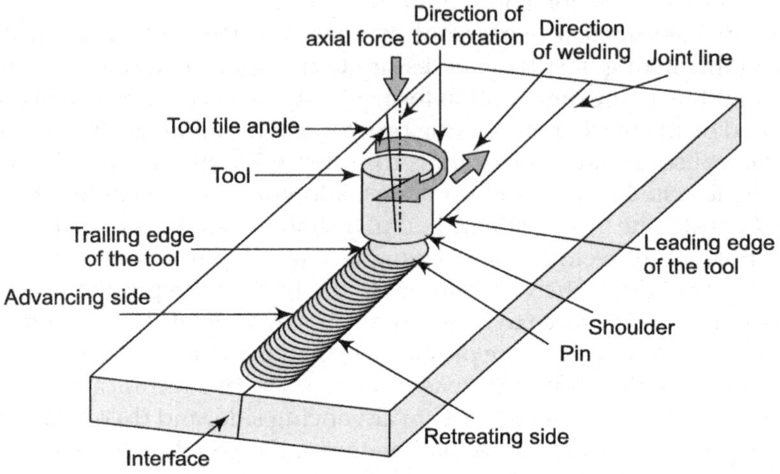

Fig. 11.1 Schematic diagram of friction stir welding (FSW) [2]; with permission from Elsevier, Copyright [2008]

To prevent tool fracture and to minimize excessive wear and tear on the tool, the welding cycle needs to be modified so that the forces acting on the tool are as low as possible. For the best combination of welding parameters it is likely that a compromise must be reached, since the conditions that favor lower forces (example: high heat input, low travel speeds) may be undesirable from the point of view of productivity and weld properties.

(ii) Tool design

The increased speed of welding is necessary for improving the production rate and keeping this process competitive compared to others. Proper design of tool plays a critical role in increasing the welding speed and improving the quality of weld. To achieve the above mentioned objective the tool material should be sufficiently strong, tough and wear resistant at the welding temperature. It should also have a good oxidation resistance and a low thermal conductivity to minimize heat loss and thermal damage to other parts of welding machine which are in vicinity. Few tools such as hot-worked tool steel (for example AISI H13) have been tried and tested for welding aluminum alloys within thickness range from 0.5 to 50 mm [3]. They proved to be perfectly all right but for more demanding applications like highly abrasive metal matrix composites, higher melting point materials such as steel or titanium, more advanced tool materials are required.

Improvements in tool design have shown to cause appreciable improvements in productivity and quality. Specifically designed tools are now available to increase the depth of penetration and plate thickness that can be successfully welded. The 'whorl' design is an example of a tool that uses a tapered pin with re-entrant features or a variable pitch thread in order to assist downward flow of material. Other designs include the Triflute and Trivex series [4]. The Triflute design has a complex system of three tapering, threaded re-entrant flutes that help in increasing material movement around the tool.

The Trivex tools use a simple non-cylindrical pin and it is observed that they reduce the forces acting on the tool during welding. A concave shoulder profile is attributed to majority of tools for the reasons such as (a) it acts as an escape volume for the material displaced by the pin, (b) it prevents material from extruding out of the sides of the shoulder, and (c) it maintains downward pressure and hence good forging of the material behind the tool. To produce additional movement of material in the upper layers of the weld a triflute tool is used which uses an alternative system with a series of concentric grooves machined into the surface.

(iii) Tool tilt and plunge depth

Plunge depth and tool tilt are critical parameters for ensuring good quality of weld (Fig. 11.2). The plunge depth is the depth of the downward extreme point of the shoulder below the surface of the plate being welded. Plunging the shoulder below the surface of the plate at the joint increases the pressure at the bottom of the tool and helps in ensuring adequate forging of the material at the rear portion of the tool. Tilting the tool by an angle of 2–4°, such that the rear portion of the tool is slightly lower than the front portion, helps in assisting the forging process required in this case of welding [4]. The plunge depth needs to be appropriately set to ensure the necessary downward pressure of

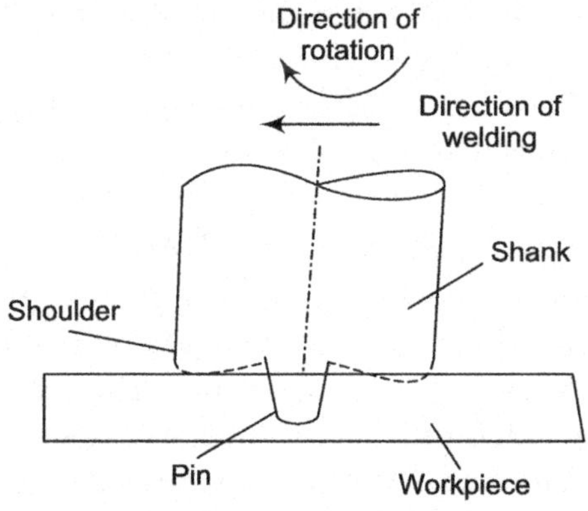

Fig. 11.2 Schematic diagram showing plunge depth and tool tilt. The movement of the tool is from right to left

the tool on the material to be welded and for full penetrations of tool probe and bottom into the material to be welded. On the other hand, if the plunge depth provided is more, the high downward force generated may lead to deflection of welding machine and may result in rubbing of the pin on the surface of the backing plate or a significant under-match of the weld thickness compared to the base material. If the plunge depth provided is less it may even result in flaws in the weld.

11.3 METALLURGY OF FRICTION STIR WELDS

Based on the microstructural characterization of grains and precipitates, the FS weld can be divided into three distinct zones viz., (a) Stirred (Nugget) zone, (b) Thermo-mechanically affected zone (TMAZ), and (c) Heat-affected zone (HAZ) as shown in Fig. 11.3.

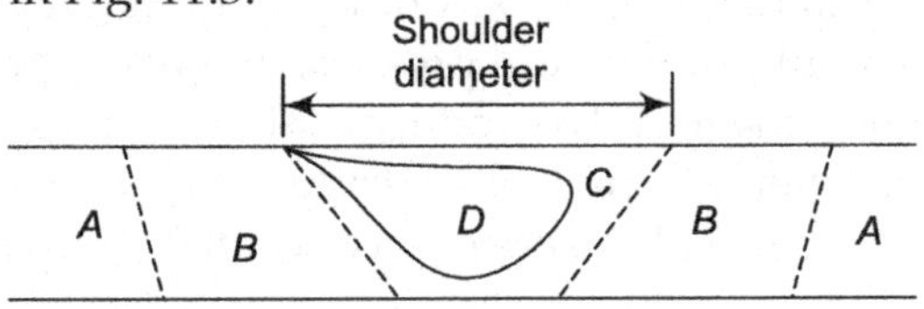

Fig. 11.3 Schematic diagram of cross-section of a typical FSW weld showing four distinct zones: (A) base metal, (B) heat-affected, (C) thermomechanically affected, and (D) stirred (nugget) zone

(i) Stirred (nugget) zone

Frictional heating and heavy plastic deformation during FSW result in generation of a recrystallized fine-grained microstructure within stirred zone. This region is referred to as nugget zone (or weld nugget) or dynamically recrystallized zone (DXZ). Under some FSW conditions, concentric rings are observed in the nugget zone. These concentric rings are referred to as the 'onion rings' in FSW [5]. There is low dislocation density in the interior of the recrystallized grains. However, from some investigations it is found that the small recrystallized grains of the nugget zone contain high density of sub-boundaries, sub-grains, and dislocations. The interface between the recrystallized nugget zone and the parent metal is relatively blurred on the retreating side of the tool, but quite sharp on the advancing side of the tool [6].

(ii) Thermo-mechanically affected zone (TMAZ)

This is a unique zone observed in FSW process. It can be termed as a transition zone, thermo-mechanically affected zone (TMAZ) between the parent material and the nugget zone. The TMAZ experiences both temperature and deformation during FSW. The TMAZ is characterized by a highly deformed structure. The grains in the parent metal are deformed in an upward flowing pattern around the nugget zone. Although the TMAZ undergoes plastic deformation, recrystallization does not occur in this zone due to insufficient deformation strain. However, dissolution of some precipitates occurs in this zone for certain alloys which show precipitates in their microstructure. The extent of dissolution depends on the thermal cycle experienced by TMAZ. The grains in the TMAZ usually contain a high density of sub-boundaries.

(iii) Heat-affected zone

Beyond the TMAZ there is a heat-affected zone (HAZ). This zone experiences a thermal cycle, but does not undergo any plastic deformation. The HAZ retains the same grain structure as the parent material. In case of heat-treatable aluminum alloys this zone experiences a temperature rise above 250°C [7]. But the thermal exposure above 250 °C exerts a significant effect on the precipitate structure which in turn affects the properties. FSW process has relatively little effect on the size of the sub-grains in the HAZ, but for certain alloys it results in coarsening of the strengthening precipitates and the precipitate-free zone (PFZ) increases by an appreciable amount. These changes in HAZ need to be considered during welding as they may bring undesirable changes in mechanical properties.

11.4 MECHANICAL PROPERTIES OF FRICTION STIR WELDS

As there is no melting involved in FSW, the quality of weld obtained is superior to other fusion welding processes which show a cast structure. FSW produces a unique microstructure within and around the stirred zone, i.e., nugget zone, TMAZ, and HAZ. The evolution of mechanical properties such as strength, ductility, fatigue, fracture toughness, and hardness during FSW are discussed in this section. Some important points are highlighted on residual stresses.

(i) Strength and ductility

Tensile properties tested on FSW specimens along (longitudinal) the weld show better strength compared to across (transverse) the weld. This is because longitudinal tensile specimens contain only fully recrystallized grains from the nugget zone, whereas transverse tensile specimens contain microstructures from all four zones, i.e., parent material, HAZ, TMAZ, and nugget zone. The strength and ductility in transverse orientation are substantially less than those in longitudinal orientation. In general, shear failures occurs in HAZ. This is because the tensile specimens in the transverse orientation cover four different microstructures. The observed ductility is an average strain over the gage length including various zones. The different zones have different resistances to deformation due to differences in grain size, precipitate size and distribution. The HAZ has the lowest strength due to significantly coarsened precipitates and the development of the precipitate free zones. Thus, during tension, strain occurs mainly in the HAZ resulting in necking and fracture. Therefore, fracture always occurs in the HAZ, resulting in a low strength and ductility along transverse orientation of the weld.

The transverse tensile strength of FSW welds and joining efficiency of FSW welds for various aluminum alloys are summarized in Mishra and Ma [8]. It is seen that the joining efficiency of FSW welds ranges from 65 to 96% for heat-treatable aluminum alloys and is 95–119% for non-heat-treatable aluminum alloy. The joining efficiency for FSW is significantly higher than that for conventional fusion welding, particularly for heat-treatable aluminum alloys [8].

Normally the strength of FSW is around 80% of the base metal which is much better than other fusion welds. The retreating side exhibits lower strength to its contrary advancing side, so failures are seen at the retreating side.

(ii) Fatigue behavior

There is a growing interest in the structural use of aluminum alloys, for applications such as automotive and railway vehicles, bridges, offshore structure topsides and high-speed ships. In all cases, welding is the primary joining method and fatigue is a major design criterion. However, as is well known, welded joints can exhibit poor fatigue properties. Thus, clear design guidelines are needed to ensure that fatigue failures are avoided in welded aluminum alloy structures. Apart from basic design of new structures, there is also increasing interest in methods for assessing the remaining fatigue lives of existing structures. A substantive work needs to be carried out in above mentioned area not only for aluminum but also for other materials which can be joined using FSW. It is important to understand the fatigue characteristics of FSW welds due to its variety of applications like aerospace structures, transport vehicles, bridge constructions, etc., where fatigue properties are critical. Substantive work is being carried out in evaluating the fatigue behavior of FSW welds, which importantly include stress—number of cycles to failure (S–N) behavior and fatigue crack propagation (FCP) behavior.

Based on the research carried out especially of S–N behavior and fatigue crack propagation (FCP) behavior of FSW welds, following points are observed. First, the fatigue strength of the FSW welds at 10^7 cycles is lower than that of the base metal, which indicates they are weaker against fatigue crack initiation.

The transverse FSW specimens have lower fatigue strength than the longitudinal FSW specimens. However, it is observed that the fatigue strength of the FSW welds is higher than that of MIG and laser welds [8]. Such studies need to be carried out for commercially important materials. A proper mechanism should be established to understand the fatigue failure in FSW in order to t improve the fatigue life of FSW components. The study should be such that it can be applied to any material joined using FSW. The reason behind better properties of FSW welds is its finer and uniform microstructure as compared to fusion (laser and MIG) welds. Second, surface quality of the FSW welds exerts a significant effect on the fatigue strength of the welds. The fatigue strength of the FSW weld decreases with increasing tool traverse speed/rotation rate (v/w) ratio due to the formation and increase of non-welded groove at the root side of the weld [8]. Skimming of layers (up to 0.5 mm) from both root and top sides removes all surface irregularities and results in fatigue strength, of both transverse and longitudinal FSW specimens, comparable to that of the base metal. Third, the fatigue resistance of friction stir welded specimens in air is inferior to that of the base metal.

In case of FCP (fatigue crack propagation), the study reveals following important observations. First, the quality of the FSW welds exerts limited effects on the d*a*/d*N*–*DK* curve, where d*a*/d*N*–*DK* gives idea about fatigue crack propagation (the data is presented in terms of crack growth rate per cycle of loading (*Da*/ *DN* or d*a*/d*N*) versus the fluctuation of the stress-intensity factor at the tip of the crack (*DKI*). *DKI* is representative of the mechanical driving force, and it incorporates the effect of changing crack length and the magnitude of the cyclic loading, the most common form of presenting fatigue crack growth data is a log–log plot of d*a*/d*N* versus *DKI*). Second, at lower loads and lower *R*-ratio of 0.1, the FCP properties of the FSW welds are superior to that of the base metal, whereas at higher loads or higher *R*-ratios of 0.7–0.8, base materials and FSW welds exhibits similar d*a*/d*N*–DK behavior. This is because of the presence of compressive residual stresses at the crack tip region in the FSW welds, which decreases the effective stress intensity at the crack front. In this case, fatigue crack propagation rates at lower loads and lower *R*-ratio are apparently reduced due to reduced effective stress intensity. However, at higher loads or higher *R*-ratios, the effect of the compressive residual stress becomes less important and similar base material and FSW d*a*/d*N*–DK curves are observed. Third, specimen geometry exhibits a considerable effect on the FCP behavior of the FSW welds. The nugget zone and HAZ of FSW welds exhibit lower fatigue crack growth rates and higher fatigue crack growth threshold, DK^{th}, at both $R = 0.1$ and 0.5, in air and in 3.5% NaCl solution, compared to the base metal. Furthermore, the FCP properties of the nugget zone are higher than those of the HAZ.

The fatigue crack growth rates in 3.5% NaCl solution for the base metal, HAZ, and nugget zone, in the intermediate and high DK regions, are about two times higher than those observed in air [8]. The crack growth behavior in the FSW joints is generally dominated by the weld residual stress, while microstructure and hardness changes in the FSW welds have minor influence. Furthermore, fatigue crack growth rates in FSW welds depends strongly on their location and orientation with respect to the weld centerline. However, if the FSW welds are mechanically stress relieved by application of 2% plastic strain, crack growth rates are almost identical to those of the base metal, irrespective of location and orientation.

(iii) Fracture toughness

No welding process is perfect and defects are bound to occur in the welds. The weld is accepted or rejected based on the intensity of defects. There are standards for acceptability of the welds pertaining to different inspection codes. The non-acceptable flaws must be repaired before the weld is put into service. Most existing codes cater toward weldments made by conventional welding techniques. FSW is generally found to produce defect-free welds. However, no established code exists so far for FSW welds. Considering the potential applications of FSW, there is a critical need for proper evaluation of the fracture behavior of the friction stir welds. The most commonly used parameters are the crack tip intensity factors (K) for linear elastic loading, and the J integral or the crack tip opening displacement (CTOD) for elastic–plastic loading.

Generally it is observed that friction stir welds exhibit better fracture toughness compared to fusion welds. The reason for higher fracture toughness associated with the FSW welds is attributed to the fracture and rounding of large primary particles by the stirring process, and the softening of the matrix. No detailed microstructure–property correlation has been established so far for the fracture toughness of FSW welds. Since there is a significant change in microstructures during FSW, it is necessary to understand the influence of microstructural characteristics on the fracture toughness of friction stir welds. For commercial precipitation-strengthened high-strength aluminum alloys, three types of particles are identified, i.e., large constituent particles (5–30 microns), dispersoids (0.2–0.5 microns), and precipitates in the nanometric size range. In the absence of constituent particles and dispersoids, the deformation behavior becomes strongly influenced by shearing of precipitates, thus, leading to strain localization. Like the conditions of strain localization in the matrix, it is found that low fracture toughness results from the presence of a narrow and soft PFZ and a large grain size.

The most critical stage in controlling fracture toughness is the control of the initiation of voids. Particles provide interfaces that are easy initiation sites for voids. Critical stress for particle cracking, is related to their size, and surface energy. The two factors viz., (i) the increase in the bonding strength between matrix and particles, and (ii) the decrease in particle size tend to increase critical stress for particle cracking, enhances the fracture toughness. Improvement in fracture toughness is often achieved by reducing iron and silicon content in aluminum alloys, thus, reducing the volume fraction and size of constituent particles. Apart from above-mentioned factors, the nature of grain boundaries influences fracture behavior of a material. A large fraction of low energy grain boundaries might toughen the material by changing the failure mode from intergranular to transgranular fracture. However, these concepts are still evolving and no quantitative relation is available to predict fracture toughness based on grain boundary character distribution.

Based on above-mentioned microstructural analyses, the fracture toughness of FSW aluminum alloys can be rationalized. FSW results in generation of a nugget zone characterized by: (a) very fine grain size (b) fine precipitates and constituent particles, (c) lower yield stress, and (d) high ratio of high-angle boundaries. Fine grain structure and small particles tend to enhance the fracture toughness of nugget zone, whereas low yield stress and high ratio of high-angle boundaries tend to reduce the fracture toughness. The overall impact of these factors is that the fracture toughness of nugget zone is higher than or comparable to that of base material, depending on the alloy chemistry and FSW parameters. The lower fracture toughness of HAZ/TMAZ region compared to the nugget zone is due to widened PFZ (precipitate free zone) and coarsened particles [8].

(iv) Hardness

Aluminum alloys can be classified into heat-treatable (precipitation-hardenable) alloys, and non-heat-treatable (solid-solution-hardened) alloys. The change in hardness in the friction stir welds is different for precipitation-hardened and solid-solution-hardened aluminum alloys. FSW creates a softened region around the weld center in a number of precipitation-hardened aluminum alloys. Such a softening is caused by coarsening and dissolution of strengthening precipitates during the thermal cycle of the FSW. If hardness is measured across the weld at intermittent distances and a graph (distance from weld centre in +ve and –ve direction on x-axis and hardness on y-axis) is generated using these hardness values, the profile thus generated is called hardness profile. Hardness profile is strongly affected by precipitate distribution rather than grain size in the weld. A typical hardness distribution across the weld of FSW 6063Al-T5 is described by Mishra and Ma [8]. Clearly, significant softening is produced throughout the weld zone, compared to the base material in T5 condition. Further, it is shown that the lowest hardness does not lie in the center part of the weld zone, but is approximately 10 mm away from the weld centerline. The hardness curve can be labeled by BM (the same hardness region as the base material), LOW (the region of lower hardness than base material), MIN (the minimum-hardness region), and SOF (the softened region).

Two kinds of precipitates are observed in the BM, LOW, and MIN regions; needle-shaped precipitates of about 40 nm in length, which are partially or completely coherent with the matrix, and rod-shaped precipitates approximately 200 nm in length, which have low coherency with the matrix. The mechanical properties of 6063Al depend mainly on the density of needle-shaped precipitates and only slightly on the density of rod-shaped precipitates [8]. The microstructure (type, size, and distribution of precipitates) in the BM region is basically the same as that in the base material, which explains the same hardness in the BM region and the base material. In the LOW region, the density of needle-shaped precipitates was substantially reduced, whereas the density of rod-shaped precipitates increases. This results in a reduction in hardness of the LOW region. For the MIN region, only low density of rod-shaped precipitates remains. Thus, not only the hardening effect of needle-shaped disappears completely, but also solid-solution-hardening effect of solutes reduces due to the existence of rod-shaped precipitates, which leads to the minimum hardness in the MIN region. In the SOF region, no precipitates are detected due to complete dissolution of the precipitates. For the solid-solution-hardened aluminum alloys, generally, FSW does not result in softening of the weld. For such alloys higher hardness is observed in the SOF region than in the base material because of the smaller grain size, higher density of sub-boundaries dislocations, strain hardening. The combined effect of the above mentioned factors leads to improved hardness of weld region in friction stir welded solid-solution-hardened aluminum alloys.

Kumar *et al.* [2] studied the role of axial load on the tensile behavior and hardness distribution of FSW blanks made of 7020-T6 material. A typical hardness distribution is shown in Fig. 11.4 for axial load of 8.1 kN and is observed that that the minimum hardness is around 120VHN, which occurs 20 mm away from the weld. The hardness minima are observed in both sides of the weld in the HAZ. It is also rationalized that in the HAZ precipitates are not dissolved in the matrix, and consequently there is no reappearance of precipitates due to post weld thermal cycle. This causes the lowest hardness and consequently fracture occurs during tensile testing in the HAZ.

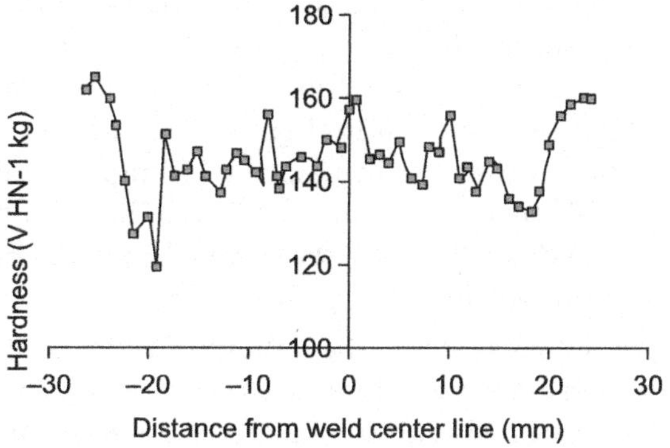

Fig. 11.4 Hardness profile across the weld region for axial load of 8.1 kN [2]; with permission from Elsevier, Copyright [2008]

(v) Residual stress

The occurrence of large residual stress influences the postweld mechanical properties, particularly the fatigue properties and its presence also shows undesirable corrosion effects if the weld is used in corrosive environment. Therefore, it is necessary to understand the residual stress distribution in the FSW welds. Generally the residual stresses are low in friction stir welds as it is a low temperature solid-state process. But, the rigid clamping used in FSW exerts a much higher restraint on the welded plates. These restraints impede the contraction of the weld nugget and heat-affected zone during cooling in both longitudinal and transverse directions, thereby resulting in generation of longitudinal and transverse stresses. Cut compliance technique, X-ray diffraction, neutron diffraction and high-energy synchrotron radiation are some of the techniques used to measure residual stresses in FSW. The residual stresses in all the FSW welds are quite low in comparison to those generated during fusion welding. Higher residual stresses are observed at the transition between the fully recrystallized and partially recrystallized regions of the weld. Generally, longitudinal (parallel to welding direction) residual stresses are tensile and transverse (normal to welding direction) residual stresses are compressive. The low residual stress in the FSW welds are attributed to the lower heat input during FSW and recrystallization accommodation of stresses. Few other observations that are made related to residual stresses in FSW welds are, (i) the longitudinal residual stresses are always higher than the transverse ones, independent of pin diameter, tool rotation rate and traverse speed, (ii) both longitudinal and transverse residual stresses exhibit an "M"-like distribution across the weld. Maximum tensile residual stresses are located at the HAZ region of the weld, (iii) residual stress distribution across the welds is similar at the top and root sides of the welds, (iv) large-diameter tool widens the M-shaped residual stress distribution and with decreasing welding speed and tool rotation rate, the magnitude of the tensile residual stresses decrease. It is observed that maximum residual stresses for various friction stir welds of aluminum alloys are below 100 MPa [9]. The residual stress magnitudes are appreciably lower than those observed in fusion welding, and also appreciably lower than yield stress of aluminum alloys. This results in substantive reduction in the distortion of FSW components and an improvement in mechanical properties.

However, for the stainless steels, the pattern of residual stress is different. This shows that the behavior of residual stress changes with material. In case of FSW of stainless steels the residual stress patterns observed are typical of most welding processes such as fusion welding, namely, high value of longitudinal tensile residual stress and very low transverse residual stress. The maximum values of longitudinal residual stress are close to the base metal yield stress, and therefore similar in magnitude to those produced by fusion welding processes in austenitic stainless steels. The longitudinal residual stress varies only slightly with depth, whereas the transverse stress varies significantly through the thickness. The sign of the transverse residual stress near the weld centerline is in general positive at the top and negative at the root. This is attributed to rapid cooling experienced by the weld root due to the intimate contact between the weld root side and the backing plate. Therefore, the distribution and magnitude of residual stress in friction stir welds are different for aluminum alloy and steel. This occurrence of difference can be related to the temperature dependence of the yield strength of material welded and the influence of final deformation by the trailing edge of the tool shoulder.

11.5 IMPORTANT PHENOMENON IN FSW

There are two important phenomenons in FSW, the knowledge of which is necessary in understanding and modeling of the process at fundamental level. They are (i) metal flow and mechanism of joining, and (ii) heat generation and heat transfer.

(i) Metal flow and mechanism of joining

The material flow during friction stir welding is quite complex and depends on the tool geometry, process parameters, and material to be welded. For optimal tool design and to obtain high structural efficient welds, the knowledge of mechanism of material flow is necessary. A number of investigations have been carried out to understand the material flow behavior during FSW. A number of approaches, such as tracer technique by marker, welding of dissimilar alloys/metals, have been used to visualize material flow pattern in FSW [10]. Understanding material flow is important for production of sound dissimilar metal welding that control the intermixing of two alloys being welded and consequent formation of new constituents which influences the weld properties. The material flow visualization study is normally utilized to analyze the mechanism of weld formation and its usefulness in improving fatigue properties and for dissimilar metal welds. Few innovative experiments had revealed that only a portion of material transported from the leading edge undergoes chaotic flow and the remaining is deposited systematically in the trailing edge of the weld [11]. Apart from experimental approaches, a number of studies have been carried out to model the material flow during FSW using different computational codes, mathematical modeling tools, simple geometrical model, and metal working model (modeled using the concept of metal forming). These attempts were made to understand the basic physics of the material flow occurring during FSW.

But the metal working model suggested seems to give results closer to experimental value. In this investigation an attempt was made to understand the mechanism of friction stir weld formation and the role of friction stir welding tool in it. This was done by understanding the material flow pattern in the weld produced in a special experiment, where the interaction of the friction stir welding tool with the base material was continuously increased. The results showed that there are two different modes of material flow regimes involved in the friction stir weld formation; namely "pin-driven flow" and "shoulder-driven flow". These material flow regimes merge together to form a defect-free weld. The etching contrast in these regimes gives rise to onion ring pattern in friction stir welds. In addition to that based on the material flow characteristics a mechanism of weld formation was also proposed [12]. The resultant microstructure and metal flow features of a friction stir weld closely resemble hot worked microstructure of typical aluminum extrusion and forging. Therefore, in this case the FSW process is modeled as a metalworking process. The process is divided into five conventional metal working zones: (a) preheat, (b) initial deformation, (c) extrusion, (d) forging, and (e) post heat/cool down [13]. In the preheat zone ahead of the pin, there is a temperature rise due to the frictional heating of the rotating tool and adiabatic heating because of the deformation of material. The thermal properties of material and the traverse speed of the tool govern the extent and heating rate of this zone. As the tool proceeds, an initial deformation zone is formed when material is heated to above a critical temperature and the magnitude of stress exceeds the critical flow stress of the material, resulting in material flow. In this zone the material is forced both upwards into the shoulder zone and downwards into the extrusion zone. A small amount of material is captured in the swirl zone beneath the pin tip where a vortex flow pattern exists. Material flows around the pin from the front to the rear, in the extrusion zone with a certain width. The width of the extrusion zone is defined by critical isotherm on each side of the tool where the magnitudes of stress and temperature are not sufficient to allow metal flow. Forging zone follows the extrusion zone where the material from the front of the tool is forced into the cavity left by the traveling pin under hydrostatic pressure conditions. Constraining the material in this cavity and application of downward forging force is done with the help of shoulder of the tool. Material from shoulder zone is dragged across the joint from the retreating side toward the advancing side. Post heat/cool zone is on the rear of forging zone where the material cools under either passive or forced cooling conditions.

The features of material flow in FSW and the joining process can be briefed into three steps as most of the material flow occurs through the retreating side and the transport of the plasticized material behind the tool forms the welded joint. The overall transport of plasticized materials during FSW is affected by three types of flows: (1) Near the tool, a slug of plasticized material rotates around the tool. The rotation of the tool and the resulting friction between the tool and the workpiece drives this movement of material, (2) Rotational motion of the threaded pin tends to push material downward close to the pin which drives an upward motion of an equivalent amount of material somewhat farther away, and (3) There is a relative motion between the tool and the workpiece. The simultaneous interaction of these three effects is responsible for the overall motion of the plasticized material and the formation of the joint.

(ii) Heat generation and heat transfer

The technique of friction stir welding (FSW) puts effective use of frictional heat for the purpose of joining metallic materials. Experiments were conducted to determine the coefficient of friction during FSW to study the interaction between the FSW tool and the base metal at various contact pressures and rotation speeds and it was found that there exists a critical temperature at which point a steep increase in the coefficient of friction is observed. Below the critical temperature the coefficient of friction is less, and above the critical temperature it reaches the maximum value. The steep increase in the coefficient of friction is found to be due to the seizure phenomenon and the contact condition during FSW between the tool and the workpiece (base metal) is found to be sticking [14]. To prevent or minimize the ill effects of welding on the mechanical properties of the weld and to increase the productivity, it is desired to minimize the heat input and increase the travel speed. At the same time it is necessary to ensure that the temperature around the tool is sufficiently high to permit adequate material flow and prevent flaws or tool fracture. When the traverse speed is increased, for a given heat input, there is less time for heat to conduct ahead of the tool and the thermal gradients are larger. At some point the speed will be so high that the material ahead of the tool will be too cold and the flow stress is too high to permit adequate material movement, resulting in flaws or tool fracture. If the 'hot zone' is too large then there is scope to increase the traverse speed and hence productivity. The variable process window is responsible for the change in total heat input and cooling rate during welding. Structural characterization of the bonded assemblies exhibits recovery-recrystallization in the stirring zone and breaking of coarser particles (phases). Dispersion of fine particles, refinement of grain size, low residual stress level, and high defect density within weld nugget contribute towards the improvement in bond strength.

Based on the heat flow and thermal profile the welding cycle can be divided into following four stages (i) Dwell: For smooth traverse of the tool, the material needs to be heated. This preheat is provided by tool rotation at appropriate speed but without linear movement in order to achieve a sufficient temperature ahead of the tool. This period also includes the plunge of the tool into the workpiece, (ii) Transient heating: As the tool starts moving along the joint there will be a transient period where the heat production and temperature around the tool will alter in a complex manner until a required steady-state condition is reached, (iii) Pseudo steady-state: It is the state where thermal field at macroscopic level remains constant around tool though there are fluctuations in heat generation, (iv) Post steady-state: It is the state where the heat reflection from the welded region leads to additional heat around the tool [14].

There are two main sources for heat generation in FSW which are (a) friction at the surface of the tool, and (b) deformation of the material around the tool. The heat generation predominantly occurs under the shoulder, due to its larger surface area, and is approximately equal to the power required to overcome the contact forces between the tool and the workpiece. The contact condition under the shoulder can be described by sliding friction, using a friction coefficient μ and interfacial pressure P, or sticking friction, based on the interfacial shear strength and torque, at an appropriate temperature and strain rate. Mathematical equations for the total heat generated by the tool shoulder during welding, i.e., Q_{total} using sliding and sticking friction approach are as given below.

$$Q_{total} = 2/3\pi P\mu\omega\,(R^3_{shoulder} - R^3_{pin}) \text{ for sliding} \qquad (11.1)$$

$$Q_{total} = 2/3\pi\tau\omega\,(R^3_{shoulder} - R^3_{pin}) \text{ for sticking} \qquad (11.2)$$

where ω is the angular velocity of the tool, $R_{shoulder}$ and R_{pin} are radius of the tool shoulder and tool pin, respectively. However, there is a difficulty while applying these equations as the values for the friction coefficient or the interfacial shear stress which are required for total heat calculations are difficult to measure in actual process. At present these parameters are used as 'fitting parameters' where the reverse calculations are performed based on measured thermal data to obtain a reasonable simulated thermal field. This reverse approach may be useful for creating process models to predict variables like residual stresses, but it is less useful for providing the insights into the process.

11.6 DIFFERENT JOINT CONFIGURATION FSW

Friction stir welding provides the advantages of being used in a variety of joint configurations. The joint configurations that can be welded are butt welds, dissimilar thickness butt welds, double sided butt, tee welds, lap penetration and lap fillet [Fig. 11.5]. The configuration which can be welded conveniently are butt and lap welds as the advantage here is that no joint preparation is required for these configurations unlike conventional welding processes. The major drawback is joining the traditional tee fillet joint which is not possible with this process and which is commonly used in many fusion welding applications. However the advantage of the FSW process can be taken by changing the product design [15].

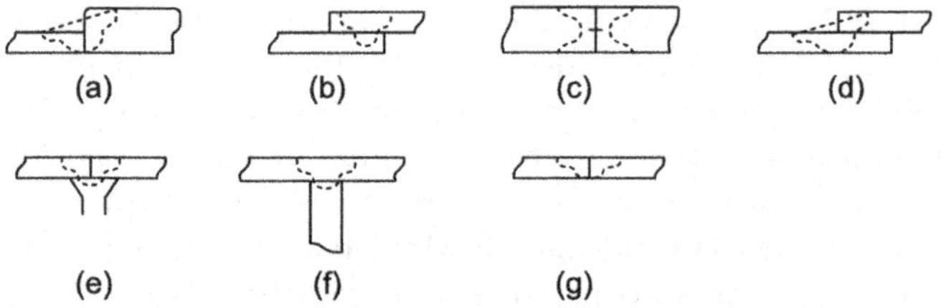

Fig. 11.5 Different joints possible using FSW (a) Dissimilar thickness butt (b) Lap penetration (c) Double sided butt (d) Lap fillet (e) Corner (f) Tee (g) Full and partial penetration butt

11.7 FSW VARIANTS

The friction stir welding process has two significant variants that can be used for other applications. The first is Friction Stir Spot Welding (FSSW), where the tool is plunged and retracted, without traversing the tool. This particular variant has parallels with the Resistance Spot Welding (RSW) process and has significant advantages over RSW, and avoids many of the difficulties of RSW of aluminum [16]. In many applications, FSSW can be readily substituted for RSW with little or no design modifications. Another variant of FSW is Friction Stir Processing (FSP), where the friction stir tool is simply traversed through the material. There is no joint in these applications. FSP can be used to significantly enhance the material properties of castings and wrought material. In castings, porosity can be eliminated. Furthermore, a significant improvement in material properties can be realized. In other applications, the material can be processed to improve the ductility of the material. This can be taken advantage of in forming or bending applications of material that typically exhibits lower ductility, such as aluminum.

11.8 MERITS AND DEMERITS OF FSW

Friction stir welding offers a variety of advantages over traditional welding processes. Problems such as porosity, solute redistribution, cracking during solidification and liquation cracking are not an issue during FSW. FSW produces a low concentration of defects and is very tolerant to variations in parameters and materials. As the joint is characterized with fine microstructure it results in number of advantages. These advantages include improved joint properties, in particular static strength and fatigue properties, low deformation of the workpieces, smaller heat affected zone. The advantages from process point of view are significant reduction in production costs and the freeing of skilled labor for other tasks. The process is also very robust, not being sensitive to environmental conditions, unlike many traditional welding processes. This is especially true for welding of lower melting point materials such as aluminum, magnesium, bronze, and copper [17]. It also yields significantly less distortion than the fusion welding processes, allowing for dramatic cost reductions in many applications. The advantage of the FSW process is, no consumables, no filler or gas shield is required, it is absolutely a safe process with no toxic fumes or spatter of molten material during welding. The other advantages include ease of automating the process as simple milling machines can be used to carry out the process which results in lower setup costs. It can be operated in all positions (horizontal, vertical, etc.), as there is no weld pool. It exhibits a good weld appearance and minimal size difference between parent material and weld, i.e., thickness under/over-matching, thus reducing the need for expensive machining after welding. In contrast, fusion welding techniques are characterized with a cast microstructure that will lead to severe degradation in the mechanical and physical properties of the joint.

On the other hand, FSW is associated with a few unique defects. Insufficient weld temperatures, due to low rotational speeds or high traverse speeds may result in inability of weld material to accommodate the extensive deformation during welding. If the material is too cool then voids or other flaws occur in the stir zone and in extreme cases this may lead to damage and breakage of tool. This also results in long, tunnel-like defects running along the weld which may occur on the surface or subsurface. Low temperatures also limit the forging action of the tool and thus reduce the continuity of the bond between the materials on each side of the weld. The light contact between the materials is termed as 'kissing-bond' in FSW [17]. This defect is quite critical as it is very difficult to detect using nondestructive methods like X-ray or ultrasonic testing. Other problem is longevity of pin, if the pin is too short or if the tool goes slightly out of the plate then the interface at the bottom of the weld will not be properly forged by the tool, resulting in a defect called insufficient-penetration of tool in material. This behaves like a notch in the material which can be a potential source of fatigue crack and reduces the fatigue resistance of weld. Other drawback is occurrence of exit at the weld hole during withdrawal of FSW tool.

11.9 APPLICATIONS OF FSW

The original application for friction stir welding (FSW) was the welding of large lengths of material in the aerospace, shipbuilding, and railway industries. Examples include large fuel tanks, for space launch vehicles, cargo decks for high-speed ferries, and roofs for railway carriages. In the last several years, the automotive industry has been aggressively studying the application of FSW in its environment. The drive to build more fuel efficient vehicles has led to the increased use of aluminum in effort to save on weight, which also improves recyclability when vehicles are scrapped [18]. The process has been commercialized in many applications, including rail car, automotive, aerospace, heavy truck, medical applications, etc. Today, the process is being transitioned into fabrication of complex assemblies, yielding significant quality and cost improvements. As the process is maturing, designers are taking advantage of the process, by designing the product specifically for the FSW process.

11.10 FORMABILITY OF FRICTION STIR WELDED BLANKS

The formability of friction stir welded blanks is important in the context of tailor welded blanks. As described elaborately in the earlier chapter, tailor welded blanks are made usually by fusion welding processes like laser welding. Recently FSW, a solid state welding process, is used for the same. In particular, it can be used to join high-strength aerospace aluminum alloys and other materials that are difficult to weld by conventional fusion welding processes. Therefore, now-a-days FSW is frequently used for making tailor welded blanks required for aerospace and automotive industries. But the weld region in the friction stir welded blanks is relatively large as compared to weld regions produced by other welding processes like laser welding, gas tungsten arc welding, etc. Also it is seen from the existing data (like hardness, strength, etc.) that, though the weld zone mechanical properties are different as compared to the base materials, they are not very different from the base materials. Therefore, the combined effect of the change in weld zone properties and dimensions on the formability of FSW blanks as a whole becomes important. It is also worth to note the changes in formability of FSW blanks with respect to TWBs made by other welding processes too. In this context, there are few research groups who have studied the formability of FSW blanks using standard formability tests and modeling methods. The theoretical results are compared and validated with experimental observations.

Some of the important results pertaining to the formability of FSW blanks are discussed in this section.

For example, Miles *et al.* [19] compared the formability of friction stir welded aluminum sheets with gas tungsten arc welded aluminum sheets under stretch forming conditions. The three aluminum alloys 6022-T4 (2.03 mm), 5754-O (1.98 mm), and 5182-O (2.03 mm) were used for formability studies. The same-thickness sheets of each alloy were welded together using both friction stir welding and gas tungsten arc welding. The FSW samples were welded at feed rates of 13 to 60 cm/min and at speeds of 1220 to 1500 rpm. The tool material is H13 steel and the tool has shoulder diameter of 10.2 mm and tool tip diameter of 1.78 mm. The gas tungsten arc welds were produced at a feed rate of 114 cm/min. The operating parameters were 50 pct AC at 200 Hz and 195 amps. The joints were compared by doing tensile test, limit dome height (LDH) test and OSU test and forming limit curves (FLC) were evaluated. The forming limit curves for FSW blanks and base materials are compared for 5182-O and 5754-O [19].

It is observed that the forming limit of base material and FSW blanks is almost same in plane strain conditions for both the 5xxx FSW blanks, while some reduction is forming limit is witnessed in the bi-axial stretching region. In the case of 6022 FSW blank, the forming limit is drastically reduced and is less in the case of bi-axial stretching region as compared to plane strain conditions. The forming limits of the FSW blanks decrease rapidly as the strain paths move toward biaxial tension. This is consistent with the formability results where the lowest dome heights for FSW 6022-T4 were obtained with a fully clamped specimen that produced biaxial or near biaxial strain states during failure. For the 5182-O and the 5754-O alloys, there were no significant differences in any of the average tensile properties of the FSW and GTAW samples. For 6022-T4, the yield strength and ultimate tensile strength drops significantly in the welded specimens, although the FSW specimens retain more strength than the GTAW specimens. In addition, the FSW samples had a greater average elongation than the GTAW samples, by about 50%. In this alloy, the fracture location was always in the HAZ for the FSW samples, while it was always in the weld for the GTAW samples.

For LDH and OSU test the two 5xxx series alloys perform about the same regardless of the welding process used, while the 6022-T4 has a lower LDH for gas tungsten arc welding than for friction stir welding. The FSW specimens failed in the HAZ and not in the weld for the LDH test, but several of the GTAW samples failed in the weld. For the 6022-T4, the differences between the friction stir welding and gas tungsten arc welding results are not as great when the OSU test was performed. This is because the OSU test causes stretching along the weld and not across it, resulting in a plane-strain fracture perpendicular to the weld.

Rodrigues *et al.* [20] assessed the formability of TWB joined by two different conditions of welding named 'hot weld' (HW) and 'cold weld' (CW) by deep drawing axi-symmetric cups. The 'hot weld (HW)' is generated by a tool having conical shoulder with welding speed, rotation speed, and tool tilt angle of 180 mm/min, 1800 rpm, and 2.5° respectively. The 'cold weld (CW)' is generated by a tool having scrolled shoulder with welding speed, rotation speed and tool tilt angle of 320 mm/min, 1120 rpm, and 0°, respectively. The deep drawing tests were performed using circular TWB specimens, with 200 mm diameter. No rupture was observed in the welds for both types of TWBs, which confirms the good plastic deformation behavior of both the even-matched (HW) and under-matched (CW) welds. Though no rupture was registered, strong wrinkling occurred for the HW blanks. The punch force–displacement curves for the TWBs were monitored and it is observed that the punch force evolution for the CW blank is very similar to that of the base material. In the case of even-matched HW blanks, the punch force behavior is very close to that of the base material, until approximately 20 mm punch displacement. After this, strong wrinkling started at the cup flange region, which was accompanied by a strong decrease in the punch force relative to the base material. The main cause for wrinkling is the deformation taking place in the zone between the die and the punch, where the workpiece is unsupported and tangential compressive stresses arise promoting buckling and folds. The absence of wrinkling in the case of the under-matched CW blanks, which were drawn under the same blank-holder force used for the HW blanks, can be explained by the plastic deformation of the softer weld material that avoids buckling.

Miles *et al.* [21] evaluated the forming behavior of the dissimilar welds of the different combination base materials like 5182-H16, 5754-O, 6022-T4 and 6111-T4 using LDH test and OSU test and finally generated forming limit diagrams experimentally. The dome testing was effective for discriminating between an acceptable weld and a bad weld. This test causes biaxial strain in the weld area, resulting in stretching across the weld and subsequent splitting in the weld nugget, if the welding parameters are not good. The OSU plane-strain testing showed that, when stretching is done only along the weld (without any transverse stretching), the ductility of the welded dissimilar-alloy pairs are same as that of the base materials. The FLD of 5182/5754 appears to have a relatively flat FLC, exhibiting similar formability both in plane-strain and in biaxial stretching conditions. This is caused by the strength difference in the two alloys, where more strain occurs in the weaker alloy, as is seen in the shifting of the weld line during testing. This results in relatively more minor strain at a given major limit strain. The FLDs for 5182/6022 and 5754/6022 have curves that are maximum at strain-path corresponding to plane strain and that slope toward minimum values as strain ratios approach biaxial tension. This behavior is caused by the localization of strain in the HAZ of the 6022 alloy, especially as biaxial-strain conditions cause stretching across the weld, and is similar to that of monolithic, friction-stir-welded 6022.

Kim *et al.* [22] measured FLD experimentally for base materials utilizing the hemispherical dome stretching test. However, FLC of weld zones were calculated utilizing Hill's bifurcation theory and the M–K theory, based on isotropic hardening rigid-plasticity with the isotropic yield function. The forming limit criterion based on Hill's bifurcation theory was applied to the drawing side of FLD (or negative minor strain) and the M–K model was followed on the stretching side of FLD (or positive minor strain). For the M–K theory, the defect size was obtained such that the M–K theory would provide the same limit strain as Hill's theory in the plane-strain strain path (with the minor strain zero). In their work, the same materials with the same thickness (similar gauges, SG) and different thickness (dissimilar gauges, DG) were welded together. The thinner part of all (DG) materials was in the retreating side of the FSW. The four different automotive sheets namely 6111-T4, 5083-H18, 5083-O and DP590 steel sheets, each having one or two different thicknesses were used for the analyses [22].

For formability studies, three tests with varied complexity were performed: the simple tension test with three loading directions namely—type I: along the weld, type II: across the weld and type III: at an angle of 45° to the weld for simple tension stretching, the hemisphere dome stretching test (or the FLD test) for biaxial stretching and the cylindrical cup deep drawing test over a cylindrical punch for draw forming. Three different simulation methods were selectively utilized for performance comparison as—model-A: the perfect welding condition (the weld zone property is ignored), model-B: the measured average mechanical properties of the weld zone with uniform average weld zone thickness, and model-C: the average weld zone property with varying weld zone thickness. In simple tensile test, load-engineering strain profiles were obtained up to failure and engineering strains at the maximum load and failure were measured [22].

For numerical simulations, the reduced four node shell elements S4R with five integration points through thickness were used for specimens with parallel (Type I) and vertical (Type II) weld lines. For the sample with 45° weld line (Type III), triangular elements S3R with five integration points through thickness were used within the gauge length and S4R was utilized for the rest. In order to investigate the formability of FSW sheets in biaxial stretching deformation, hemispherical dome stretching (HDS) tests were performed with six welded samples: 6111-T4 (SG, DG), 5083-H18 (SG, DG), AA5083-O (SG) and DP-steel (SG). The hemispherical dome stretching (HDS) test was carried out on a 50 ton double action hydraulic type press. The punch speed was 1.5

mm/s and blank holding force was applied just enough to completely clamp the blank, which was about 200 kN. The lubricant WD-40 was applied on the punch only. The punch load profiles during forming were monitored and the limit dome heights (LDH) at the failure were measured. The failure onset locations and patterns were also observed and compared with simulation results. The FLDs of different FSW sheets (both SG, DG) predicted using Hollomon and Voce's strain hardening laws are compared with each other for analysis. The failure pattern and location is also compared for all the FSW blanks in plane-strain and bi-axial stretching strain paths. The load–stroke curves during HDS tests were also obtained from experiments and simulations for all the FSW blanks and are compared [22].

The cylindrical cup drawing (CCD) test was carried out on the 50-ton double action hydraulic type press. The punch speed was set at 1.5 mm/s and tool dimensions are: punch diameter = 50 mm; punch corner profile radius = 6 mm; die opening diameter = 53.68 mm; die corner profile radius = 8 mm. The failure for 5083-O (SG) sample during experimentation occurred normal to the weld region at the bottom of the cup wall. Similarly, the simulated failure was initiated at the weld zone of the cup wall bottom and propagated to the base along the bottom rim rapidly. In the case of DP590 (SG) sample, the experimental and simulated failure lines were generated perpendicular to the weld line at the weld zone of the cup wall bottom [22].

Miles *et al.* [23] compared the forming behavior of friction stir welded and laser welded dual phase 590 steel sheets of both same and different thickness using transverse tensile test, LDH test and OSU test. The LDH tests were performed to find out some acceptable welding condition and using these conditions FSW were carried out which were compared with the laser welded joints. The dissimilar gauge tension tests (1.6/1.0 mm) showed that transverse tensile ductility in the friction stir welded specimens could reach the same level of performance as the laser welded specimens. When stretching along the weld was carried out using the plane strain OSU formability test, the friction stir welded blanks had better formability than the laser welded blanks by 20%. It is observed from the experiments that the FSW sample failed at an average punch stroke of 36.1 mm compared to 30.1 mm for the laser welded sample [23]. The improved ductility of the FSW blank is due the lower peak hardness in the weld region as compared to laser welded blank.

Lee *et al.* [24] studied the forming behavior of tailor welded blanks both experimentally and numerically. When the formability performance of the weld samples are compared to that of the base sheets, the improved ductility of the weld promoted the performance for AA5083-H18, AA5083-O, and AZ31 TWB sheets. However, the weld strength was inferior for all welded sheets, except AA5083-O and hence the TWB samples were vulnerable to the strain localization. Therefore, aligning the weld zone properly in an optimized angle to avoid strain localization and to take advantage of the improved (or almost sustained) weld zone ductility would be important in the process design.

Gan *et al.* [25] analyzed the formability of 5083-H18 and 6111-T4 FSW sheets by tensile tests and one-dimensional analysis of tensile test in both rolling direction (RD) and transverse direction (TD). The temperature measurements and stress–strain behavior were monitored during the FSW process. The strain hardening behavior of 6111 base material and the weld zone are same in both the transverse and longitudinal rolling directions. The hardening behaviors were fit into equations and the resulting stress–strain behavior is also compared. The equation based stress–strain behavior agrees well with the experimental curves [25]. A one-dimensional analysis of tensile behavior is also performed to study the formability and the results are compared with the experiments. Since 1-D analysis is done, the rate effects and 3-D effects are not included and hence the tensile behavior is valid till necking only. It is seen that the simulated stress–strain behavior and experimental results correlates well with each other for both the 6111 and 5083 FSW blanks [25].

The deep drawing analysis of FSW blanks made of similar and dissimilar Al sheets was performed by Leitão *et al.* [26]. The TWBs made of AA 5182-H111 and AA 6016-T4 were formed to round cups. The tool with a threaded probe of 3 mm in diameter and 0.9 mm long and the scrolled shoulder of 14 mm diameter is used for welding. The tool was moved at 320 mm/min travelling speed and 1120 rpm rotational speed. Both the similar (A5182–A5182 and A6016–A6016) and dissimilar (A5182–A6016) TWBs were made. It is concluded that the formability of the TWBs, made of similar material, is influenced by the difference in mechanical properties between the weld and the base materials, and also, by the initial size of the blanks. By using a 200 mm diameter TWB, non-defective cups were formed, while defective cups were formed if 180 mm diameter TWB was used. In the case of dissimilar TWBs, the presence of small defects at the weld root of the dissimilar welds induced failure of some of the blanks during the formability tests. This is seen in TWBs with diameter of 200 and 180 mm tested with 8 kN blank-holder force. However, it was possible to draw the 180 mm diameter blanks under higher blank-holder forces (16, 20, and 32 kN) which proves that the rupture resulted exclusively from the presence of the defects. The dissimilar TWBs have good formability behavior as compared to similar TWBs and base material [26].

An important issue during the formability prediction of friction-stir welded blanks is the accurate implementation of weld zone details like the mechanical and geometrical properties of nugget zone and HAZ. It is relatively easy to evaluate the global mechanical properties of weld zone (including nugget zone and HAZ as a single entity) as compared to the evaluation of NZ and HAZ properties separately. In this context, Zadpoor *et al.* [27] analyzed the effects of the implementation of the weld details on the accuracy of the failure prediction, strain distribution, and springback behavior of FSW TWBs for the limiting dome height (LDH) test and the S-rail problem. The effects of implementing the weld detail on the simulation time are also considered. The M–K model is used to predict the forming limit of FSW blanks. It is observed from the simulation results that neglecting the mechanical properties of the weld zone can result in 38% over-prediction of the dome height in the LDH test and 27% overestimation of the terminal punch travel in the S-rail simulation. In the case of LDH test, there is not much difference in the punch travel obtained from the models without HAZ and with NZ and HAZ separately. Similar pattern is observed in the case of 'stamping' stage of spring back simulation. However, the simulation time of the 'springback' stage increased more than 120% with implementation of the weld details in the FEM models. It is concluded by them that the weld details implementation is not significant if springback is not simulated. The implementation of weld zone details is still important while predicting the strain distribution and onset of failure [27].

Similar analysis was performed by Perumalla *et al.* [28–30] wherein the predictions from single zone model (without separate NZ and HAZ) and double zone model (NZ and HAZ are modeled as separate entities) are compared to identify the domain of weld conditions in which single zone assumption is valid. This is done by comparing the load–punch stroke behavior of FSW blanks with single zone and double zone models. A criterion based on the difference between the load–punch stroke behavior of single zone and double zone model is defined beyond which the double zone model is valid. It is found that multiple domains wherein double zone model is valid are possible due to parametric changes in weld conditions and such changes affect the local and global deformation influencing the model representation. When the weld conditions are equivalent to reference model weld conditions, it represents single zone model. If the conditions are very different from the reference model, it breaks down the single zone model and double zone is found to be valid. Figures 11.6 and 11.7 show the domain of weld conditions for single zone and double zone model validity for FSW blanks made of similar thickness sheets (1.5:1.5 mm). The failure locations also affect the weld zone representation based on the critical error percentage difference between reference model (single zone) and double zone model parameters [28–30].

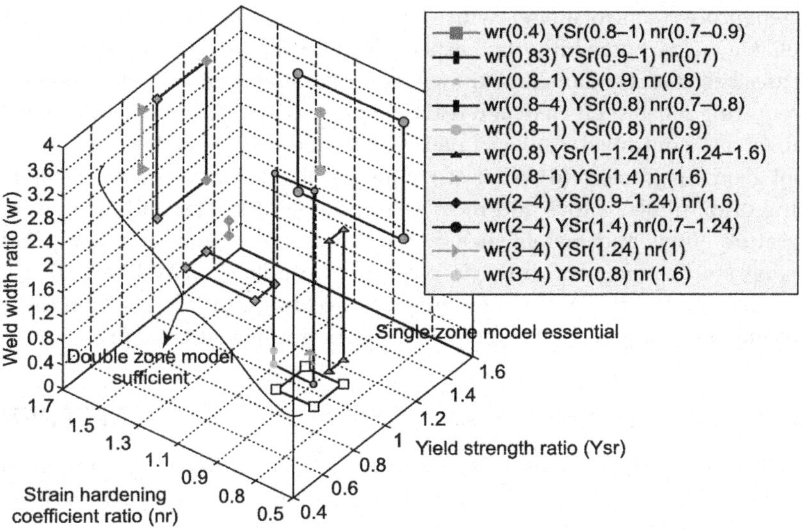

Fig. 11.6 Domains of weld conditions for representing the double zone modeling validity with maximum load at failure as criterion (longitudinal weld orientation) [29]; with permission from [29] copyright [2011] Sage

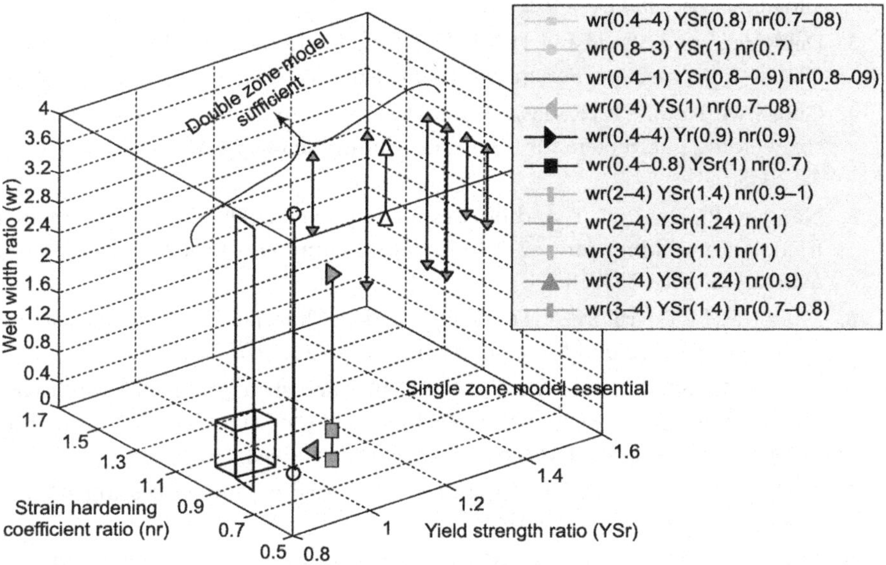

Fig. 11.7 Domains of weld conditions for representing the double zone model validity with displacement at failure as criterion (longitudinal weld orientation) [29]; with permission from [29] copyright [2011] Sage

SUMMARY

Despite considerable interests in the FSW technology in past, the basic understanding of the process is lacking. The fundamental knowledge of the FSW process and the knowledge of evolution of the structure and properties needs to be combined to build intelligent process control models with a goal to achieve, defect free, structurally sound and reliable welds. Some important aspects include: (a) Tailoring weld structure and properties based on fundamental knowledge still remains an important milestone in FSW requiring reliable, efficient and real time process modeling so that process can be optimized and controlled. (b) Exact nature of material flow needs to be understood. (c) Tool geometry design. (d) Wear of welding tool. (e) Micro structural stability. (f) Welding of dissimilar alloys and metals. (g) Joining of harder and/or high melting temperature alloys and metals such as steel, nickel, titanium, etc. using FSW. (h) Corrosion studies after welding. (i) Work on variants of FSW like FSP (Friction Stir Processing) and FSSW (Friction Stir Spot Welding). (j) Process Monitoring. (k) Identifying newer applications where FSW can be applied successfully.

REFERENCES

1. Thomas, W.M, Nicholas, E.D., Needham, J.C., Murch, M.G., Temple-Smith, P., and Dawes, C.J. (1991). Friction stir butt welding, *International Patent Application* No. PCT/GB92/02203.
2. Kumar, K. and Kailas, S.V. (2008). On the role of axial load and the effect of interface position on the tensile strength of a friction stir welded aluminium alloy, *Materials and Design*, **29**; 791.
3. DebRoy, T. and Bhadeshia, H.K.D.H. (2010). Friction stir welding of dissimilar alloys—a perspective, *Science and Technology of Welding and Joining*, **15** (4); 260.
4. Ghosh, M., Kumar, K., Kailas, S.V., and Ray, A.K. (2010). Optimization of friction stir welding parameters for dissimilar aluminum alloys, *Materials and Design*, **31**; 3033.
5. Nandan, R., DebRoy, T. and Bhadeshia, H.K.D.H. (2008). Recent advances in friction-stir welding-process, weldment structure and properties, *Progress in Materials Science*, **53**; 980.
6. Murr, L.E., Li, Y., Trillo, E.A., Flores, R.D., and McClure, J.C. (1998). Microstructures in Friction-Stir Welded Metal, *J. Mater. Process. Manuf. Sci.*, **7**; 146.
7. Somasekharan, A.C. and Murr, L.E. (2005). Microstructural Details of Friction Stir Weld Interfaces, Friction Stir Welding and Processing III, in K.V. Jata, (Ed.), *The Minerals, Metals and Materials Society*, Warrendale, PA, p. 261.
8. Mishra, RS. and Ma, ZY. (2005). Friction stir welding and processing, *Material Science and Engineering*, **R 50**; 1.
9. Murr, L.E. (2010). A review of FSW research on dissimilar metal and alloy systems, *Journal of Materials Engineering and Performance*, Published online: 2 February, DOI:10.1007/s11665-010-9598-0.
10. Flores, R.D., Murr, L.E., Shindo, D.J., and Trillo, E.A. (2001). Friction-Stir Welding of Metals and Alloys: Fundamental Studies of Solid-State and Intercalated Flow, *J. Mater. Process. Technol.*, **117** (3); Sect. 1A.
11. Kumar, K. and Kailas, S.V. (2010). Positional dependence of material flow in friction stir welding: analysis of joint line remnant and its relevance to dissimilar metal welding, *Science and Technology of Welding and Joining*, **15** (4); 305.

12. Kumar, K. and Kailas, S.V. (2008). The role of friction stir welding tool on material flow and weld formation, *Materials Science and Engineering A*, **485**; 367.
13. Reynolds, (2000). Visualisation of material flow in autogenous friction stir welds, *Science and Technology of Welding and Joining*, **5** (2); 120.
14. Kumar, K., Kalyan, C., Kailas, S.V., and Srivatsan, T. S. (2009). An investigation of friction during friction stir welding of metallic materials, *Materials and Manufacturing Processes*, **24**; 438.
15. Lohwasser, D. and Chen, Z. (2010), *Friction stir welding—From basics to applications*, Chapter 5, Woodhead Publishing, p. 118.
16. Arbegast, W.J. (2006). Friction stir welding: After a decade of development, *Welding Journal*, March, 28.
17. DebRoy, T. and Bhadeshia, H.K.D.H. (2010). Friction stir welding of dissimilar alloys—a perspective, *Science and Technology of Welding and Joining*, **15**(4); 260.
18. Murr, L.E., Li, Y., Trillo, E.A., and McClure, J.C. (2000). Fundamental Issues and Industrial Applications of Friction-Stir Welding, *Mater. Tech. Adv. Performer. Mater*, **15** (1); 37.
19. Miles, M. P., Decker, B. J. and Nelson, T. W. (2004). Formability and strength of friction-stir-welded aluminum sheets, *Metallurgical and Materials Transactions A*, **35a**; 3461.
20. Rodrigues, D.M., Loureiro, A., Leitao, C., Leal, R.M., Chaparro, B.M., and Vilaça, P. (2009). Influence of friction stir welding parameters on the microstructural and mechanical properties of AA 6016-T4 thin welds, *Materials and Design*, **30**; 1913.
21. Miles, M.P., Melton, D.W. and Nelson, T.W. (2005). Formability of friction-stir-welded dissimilar aluminum alloy sheets, *Metallurgical and Materials Transactions A*, **36a**; 3335.
22. Kim, D., Lee, W., Kim, J., Chung, K.H., Kim, C., and Okamoto, K. (2010). Wagoner, R.H., Chung, K., Macro-performance evaluation of friction stir welded automotive tailor-welded blank sheets: Part II—Formability, *International Journal of Solids and Structures*, **47**; 1063.
23. Miles, M. P., Pew, J., Nelson, T. W., and Li, M (2006). Comparison of formability of friction stir welded and laser welded dual phase 590 steel sheets, *Science and Technology of Welding and Joining*, **11** (4); 384 .
24. Lee, W., Chung, K.H., Kim, D., Kim, J., Kim, C., Okamoto, K., Wagoner, R.H., and Chung, K. (2009). Experimental and numerical study on formability of friction stir welded TWB sheets based on hemispherical dome stretch tests, *International Journal of Plasticity*, **25**; 1626.
25. Gan, W., Okamoto, K., Hirano, S., Chung, K., Kim, C., and Wagoner, R. H. (2008). Properties of friction-stir welded aluminum alloys 6111 and 5083, *Journal of Engineering Materials and Technology*, **130**; 031007–1.
26. Leitão, C., Emílio, B., Chaparro, B.M., and Rodrigues, D.M. (2009). Formability of similar and dissimilar friction stir welded AA 5182-H111 and AA 6016-T4 tailored blanks, *Materials and Design*, **30**; 3235.

27. Zadpoor, A.A., Sinke, J. and Benedictus, R. (2009). Finite element modeling and failure prediction of friction stir welded blanks, *Materials and Design*, **30**; 1423.

28. Perumalla Janaki Ramulu, Nilesh. B., and Ganesh Narayanan, R. (2010). Weld zone representation during friction stir welded blanks formability prediction with equal thicknesses, *Proceedings of the Third International and Twenty Fourth All India Manufacturing Technology*, Design and Research conference, Andhra University, Visakhapatnam, India, December 13–15, 989.

29. Perumalla Janaki Ramulu and Ganesh Narayanan, R. (2011), Weld zone representation during the formability prediction of friction stir welded blanks with similar thickness sheets, *Journal of Strain Analysis for Engineering Design*, **46**; . 456.

30. Perumalla Janaki Ramulu and Ganesh Narayanan, R. (2011). Comparing the weld zone representation methods during the formability prediction of friction stir welded blanks, *The 14th International ESAFORM Conference on Material Forming*, Queen's University, Belfast, April 27–29, *AIP Conference Proceedings*, **1353** 213.

CHAPTER 12

Probabilistic and Mesoscale Approaches to Model Interfacial Phenomena During Hot Flat Rolling

Sumitesh Das
sumitesh.das@tatasteel.com
Research and Development Division
Tata Steel Limited, Jamshedpur – 831007, India

12.1 INTRODUCTION

Intricate interactions between process variables in metal forming make it a challenge to the modeling community. Metal forming processes such as hot rolling of steels usually begin by heating the workpiece to 1150–1250°C in a reheating furnace to obtain a homogenous temperature throughout the workpiece. However, during the deformation process, a spatial temperature distribution is created within the workpiece. This is because heat is lost through the exposed surfaces to the surrounding atmosphere by conduction, convection, and radiation. Also, a small amount of heat is added internally in the form of deformational heating. The temperature distribution results in the development of a heterogeneous microstructure that produces a variation of mechanical properties. There is also a surrounding environment that may manifest through the tool–workpiece interface or the air/water–workpiece interface. Heat transfer in the former case is governed by conduction to the tool, convection via the entrapped air or lubricant or water and, to a lesser extent, radiation. Heat transfer in the latter case is dominated by convection and radiation. The workpiece surface is therefore the region that is most strongly dominated by interacting temperature dependent phenomena. In both the above interfaces, a heat transfer coefficient (h) is used to characterize the heat transfer phenomena.

The derivation of h within the roll bite is done in two steps. The first step is to embed a thermocouple at a depth of 1.5–2.5 mm below the strip surface and another at the specimen centerline [14]. Figure 12.1 shows the thermocouple locations in a typical laboratory step up. In Fig. 12.1, the thermocouples locations are marked by the filled circles and named A and B. These are shown outside the roll gap for sake of clarity. Heat transfer takes place when the locations marked by A and B move into the roll bite. The second step is to use either an analytical or numerical model to solve the equation

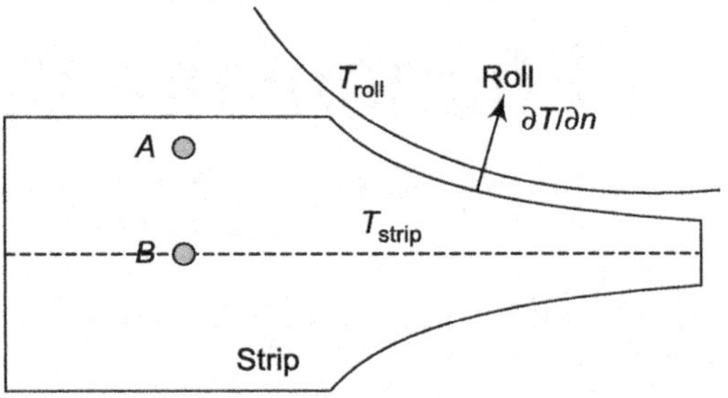

Fig. 12.1 Schematic illustration of an instrumented strip used to determine h. A and B are the surface and center thermocouples embedded in the strip. The dotted horizontal line is the strip centerline. T_{strip} and T_{roll} are the strip and roll temperatures that vary spatially within the strip and roll

$$-k_{strip}\frac{\partial T}{\partial n} = h(T^t_{strip} - T^t_{roll}) \tag{12.1}$$

where k_{strip} is the thermal conductivity of the strip, $\frac{\partial T}{\partial n}$ is the direction of heat flow, h is the heat transfer coefficient, T^t_{strip} and T^t_{roll} are the strips and roll temperatures at a particular time instant. The value of h is then iterated such that the predicted temperature distribution matches closely with the experimentally observed profile.

Although the use of a single parameter h is widespread, it is a highly idealized representation of a complex set of interactive phenomena. Early researchers [14] studied the development of the thermal profile assuming a constant value of h ($h_{constant}$). Their models used two-dimensional (2D) finite difference (FD) methods to determine the thermal gradient. Harding [14] used a value of 2.06 kW/m^{2}K at 700°C, and a value of 5.1 kW/m^{2}K at 1100°C in his simulations. However, Devadas *et al.* [10] showed from their laboratory rolling experiments on steels that h is dependent on process variables, in particular, rolling speed, initial workpiece temperature and workpiece material composition. Kumar *et al.* [21] used first order polynomial expressions relating *h* to the speed of rolling. The coefficients used in these expressions were obtained by curve fitting to the experimental data of Devadas *et al.* [10].

Kumar's formulation is independent of the temperature and material composition of the workpiece. Pietrzyk and Lenard [34] and Chen *et al.* [5] reported the sensitivity of *h* to normal pressure in addition to the workpiece temperature for low carbon steels, stainless steels and microalloyed steels. Based on their findings, Pietrzyk and Lenard [34] used a first order pressure and temperature based formulation to calculate *h*. The sensitivity of *h* to normal pressure is such that the peak value of *h* is coincident with the local pressure peak within the roll bite.

Table 12.1 shows reported values of *h* gleaned from the literature; the numerical values vary widely. The main reason for this wide variation appears to be the lack of a proper control and characterization of the workpiece and tool surfaces, in particular, the presence of asperities and oxide scale. Wilson and Sheu [46] presented experimental data on the behavior of asperities under compressive loads. Their experiments were conducted on aluminum alloy workpieces. The benefit of these measurements is that they are uncorrupted by scales (unlike in steels where there is a substantial scale formation). They measured the width of the flattened asperity crests and showed that the fractional area, defined as the ratio of the apparent contact area to the real contact area, asymptotically approaches unity at larger strains. The presence of asperities, therefore, reduces the actual area of contact between the workpiece surface and the tool surface. The reduction in available area results in a high thermal resistance and a temperature jump across the two contacting surfaces. Sutcliffe [43], using a slip-line field approach, showed that the crushing rate of commercially pure copper asperities depends on the mean normal pressure on the surfaces and the fractional area. Williamson and Majumdar [45] assumed a fractal distribution of the asperities and investigated the influence of surface deformations on contact conductance for dissimilar metals in contact. From their compression experiments on a variety of metals, in particular aluminum–stainless steel and copper–stainless steel, they concluded that the rise in the value of *h* depends

Table 12.1 Reported values of the heat transfer coefficient (h) for hot rolling of steels

Research group	Work piece	h (kW/m^{2}K)
Stevens *et al.* [41]	Carbon steels	18–38
Murata *et al.* [33]	Carbon steels	20–81
Sellars [40]	304 Stainless steels	200
Devadas and Samarasekera [9]	Low carbon steels	17–51
Pietrzyk and Lenard [34]	Low carbon steels	5–50
Xin *et al.* [47]	Low carbon steels	185
Li and Sellars [25]	Plain carbon steels	33–37
	Low alloy steels	33–46
	316 stainless steels	33–62
	Silicon steels	28–47

on the behavior of asperity penetration and deformation and is a non-linear function of the normal pressure.

However, the reduction in the contact area is complicated by the presence of oxide scales; especially during the hot rolling of steels. This is because the thermal conductivity of the oxide scale is some 10–15 times smaller that that of the steel substrate. Therefore, even a thin layer of scale on the surface of a hot workpiece has a strong influence on the rate at which heat flows through the surface[1]. During the deformation process, both the thermal and geometrical properties of the scale determine the thermal gradient within the workpiece and tool. Under certain combinations of temperature and strain rate, through-thickness brittle fracture is observed, resulting in direct contact between the workpiece and the tool. This is because the metal extrudes through the crack, thereby producing a sharp increase in the value of h. Under other circumstances, the oxide layer behaves as a ductile material and covers the workpiece surface. Qualitatively, the thicker the scale, the lower the overall value of h and the higher is the workpiece surface temperature.

Since the oxide scale presents a significant potential barrier to heat transfer during the hot rolling of steels, its effective characterization is important in deriving meaningful values of h. However, very few mathematical models combine an analysis of heat transfer and deformation of the tool–workpiece interface taking into account the evolving effect of oxide scale behavior. In most models, the oxide is positioned at pre-determined locations on the strip surface. These oxide scales may be fixed [11] (Fletcher, 1998) or sliding [18]. Krzyzanowski and Beynon [18] used a value of 37 kW/m^2K for strip positions with no scales, a value that is decreased to 20 kW/m^2K for strip locations covered by scale. However, depending on local temperatures and adhesive properties, scales can either remain fixed and undergo plastic deformation or suffer brittle cracking during deformation. When brittle cracking occurs, the fresh metal extrudes through the cracks and comes in direct contact with the tool. Li [23] showed from his compression and hot rolling tests that the percentage of fresh metal coming in contact with the tool rises asymptotically with reduction, irrespective of the scale thickness. Based on this experimental finding, this oxide layer is modeled as a two-layer zone of contact created by the fracture of the oxide scale, with an effective value of h derived as a function of the fractional area of contact using an analogy of resistances in series. Results [25, 27] show that h is a function of processing parameters (temperature, pressure, operating speed), material parameters (workpiece, tool, oxide) and scale and surface dynamics. However, the spatial gradients in temperature within the oxide layer and the scale dynamics (scale cracking) are not considered explicitly in that work. Specifically, fracture of

[1] When iron oxidizes in air at high temperatures, it grows a scale consisting of layers of wüstite (FeO), hematite (Fe_3O_4) and magnetite (Fe_2O_3). The relative thicknesses are in the ratio of roughly 95:4:1 at 1000 °C.

the oxide scale, resultant crack spacing and crack width were not taken into consideration.

The development of more sophisticated models is thwarted mainly by the lack of data of the oxide scale behavior and the necessary conditions for its failure. More so, reliable data of the scale behavior is elusive, particularly in the regime experienced in hot rolling. Krzyzanowski and Beynon [18, 19] showed from their hot tensile experiments that a brittle-ductile regime could be defined to characterize the scale behavior. For certain grades of mild steel, brittle fracture ensues below 850°C. Above this temperature, the scale is ductile and the weakest link is the metal–scale interface. As the percentage of alloying elements such as Si and Mn increases and Cu decreases, the regime for the onset of brittle fracture shifts up to 910°C. Based on these observations, Krzyzanowski and Beynon [18] modeled the scale effects explicitly using 2D finite element (FE) model of the scale layer. In their simulations, h is kept constant at 30 kW/m^2K. Due to the predominantly brittle nature of the scale cracking, they treated the scale failure as a linear elastic fracture and they modeled it as a series of pre-bonded rafts that separate when the strain exceeds a given critical debonding strain. Crack nucleation (the phenomenon that determines the crack widths) is pre-determined by bonded rafts that separate when the critical strain for brittle fracture is exceeded. To model the wide spatial distribution (10~100 μm) of crack widths that have been seen in laboratory rolling experiments Krzyzanowski and Beynon [18, 19, 26] had to use about 10^4 finite elements, a requirement that entails high computational costs.

The quantification of h clearly depends on the surface characteristics of oxide and asperities, in particular the spatial and temporal development of oxide cracks. However, the study of the interacting basic principles governing this development is not yet sufficiently advanced for realistic and accurate fully-deterministic modeling. Furthermore, the rolled oxide surface also lacks regularity. Accordingly, the paper presented here accepts these unknowns, and models these effects through random variables. It involves two approaches. Firstly, it describes a numerical experiment that visualizes the tool–workpiece interface phenomena as a random process driven by process variables. The source of randomness is partly due to an ignorance of the values of the physical parameters operating under the roll bite and also a consequence of the underlying small-scale physical interactions. This ignorance is mainly due to the difficulty in conducting realistic experiments to investigate the fundamental phenomena governing the spatio-temporal evolution of h. Coupling this with the observations of strong statistical variations in the scale cracking phenomena provides an ideal candidate for probabilistic modeling.

The chapter implements these variations using a Probability Distribution Diagram (PDD). It uses the concept of a phantom oxide layer to mimic the presence of an oxide layer between the strip and roll surfaces. The algorithm is implemented using subroutines that can interact with any commercial finite element codes, e.g., ABAQUS^{TM}. At any instant, the possible configurations of the strip surface are evaluated with reference to the PDD, resulting in an instantaneous value of the heat transfer coefficient (h_{PDD}). Sources of randomness such as instantaneous temperatures of the strip and roll surfaces, instantaneous temperature and thickness of the oxide layer and distribution of asperities are at the modeller's discretion and require an experienced guess.

Greater insight into the behavior of the oxide scale results from modeling the oxide layer within the roll bite at a microstructurally-based length scale. This is the basis of the second approach presented here. The internal length scale should both capture the microscopic behavior responsible for the macroscopic behavior and provide suitable input to model the global behavior. This mesoscopic view is achieved here as an entity comprised of uniformly distributed Cellular Automata (Neumann, 1966) (CA) cells attached to each FE gauss integration point. This array of CA cells represents a local material region in the oxide scale. It initiates and propagates a damage variable within the oxide scale. The benefit of such a combined Cellular Automata Based Finite Element (CAFE) approach is two-fold. Firstly, microstructural variables can be easily built in. Secondly, the computation is greatly simplified because the number of finite elements is reduced 100 fold compared with the number used by other researchers [20].

As mentioned earlier, comparison of predicted temperature values for material points close to the roll surface by actual experiments is elusive. The paper presents two numerical models that attempt to fill up this gap in experimental data by simulating realistically experimental conditions and provides a methodology to characterize the surface behavior within the roll bite during rolling. Results from both the probabilistic and mesoscale models clearly show the effect of the surface behavior on the development of h and its influence on the development of the temperature gradient within the underlying strip.

12.2 PROBABILISTIC MODEL FORMULATION

Heat transfer and friction are length-based concepts related intrinsically with an area (3D) macro FE model or line (2D) macro FE model. Typical finite element models identify this length with the elemental dimension; for example, this length could be that between the corner nodes of a typical element. The approach adopted here is based on the probability of finding the strip surface in a specific configuration or state at any instant of time in the roll gap.

Figure 12.2(a) shows the different probable configurations (*A*, *B*, *C*, *D*, *E*, *F* and *G*) of a strip surface relative to the roll surface in the early stages of rolling. The exaggerated undulations show the roughness of the roll and strip surfaces. The different configurations of the workpiece surface node are identified as follows.

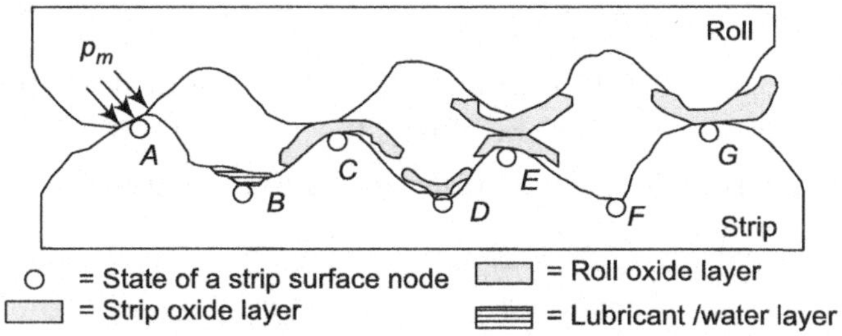

Fig. 12.2 (a) Initial strip–oxide–roll configuration at the start of rolling. The figure is not to scale as the strip and roll roughness is of the order of 0.1~0.3 microns and the oxide scale thickness is in the range of 10~300 microns. The different states of the contacting surfaces are classified as

State **A**: Strip surface is in contact with the roll surface and is under pressure.

State **B**: Strip surface is in contact with lubricant/water.

State **C**: Strip surface is in contact with strip oxide and is under pressure.

State **D**: Strip surface is in contact with the strip oxide layer.

State **E**: Strip surface is in contact with the strip oxide and a pressurized roll oxide.

State **F**: Strip surface is in contact with air.

State **G**: Strip surface is in contact with a pressurized roll oxide layer.

- State **A**: In this state, the surface is in contact with the roll and is under pressure. This configuration represents the ideal state in the absence of oxide scales and with relatively smooth surfaces at the interface. There are no gaps and the heat transfer coefficient is that for solid bodies under pressure. The friction for metal-to-metal contact is assumed to model this configuration. Under low-pressure conditions, the model uses the Amonton–Coulomb law with an appropriate friction coefficient. However, if the friction stress exceeds a particular value, i.e., the shear stress, then it is switched to Tresca's law with its shear factor m. A lower limiting friction coefficient of 0.2 is assumed for metal-to-metal contact with smooth interfaces. The upper limiting value of the shear factor is determined from geometrical constraints of initial and final strip thicknesses, roll diameter and rolling reduction [7]. The contact coefficient (h_c) determines the contact between surfaces under pressure and is proposed by [23]

$$h_c = h_i \frac{k_h}{R_{\text{av}}} \left[1 - \exp\left(-c_p \frac{p_m}{HV} \right) \right]^n \tag{12.2}$$

where h_i (= 0.405), c_p (= 0.3) and n (= 1.5) are constants [23] for hot rolling, p_m is the contact pressure and HV (Vickers Diamond Pyramid) is the surface hardness of the softer of the two contacting solids. R_{av} is the combined standard deviation of the surface heights of the two surfaces expressed [23] as

$$R_{av} = \sqrt{R_{strip}^2 + R_{roll}^2} \tag{12.3a}$$

where R_{strip} (= 0~0.15 μm) and R_{roll} (= 0.3~0.8 μm) are the standard deviations of the surface heights of the strip and roll surfaces.

k_h is the harmonic mean thermal conductivity and is given as

$$k_h = \frac{2k_{strip}k_{roll}}{k_{strip} + k_{roll}} \tag{12.3b}$$

where k_{strip} (= 22 + 0.01091 {T_{strip} − 500} W/mK for mild carbon steels) [50, 52] and k_{roll} (= 28 W/mK for tool steels) are the thermal conductivities of the strip and roll, respectively. The hardness for the plastically deforming strip (the softer material of the two) is calculated as

$$HV_{strip} = 3\sigma_y\left(\varepsilon, \dot{\varepsilon}, T\right) \tag{12.4}$$

where σ_y is the strip yield stress at a particular strain, strain rate and temperature within the roll bite.

- State **B**: In this state, the strip surface is in contact with lubricant/water. This represents the state occurring at the valleys or exposed peaks of the strip surface. The heat transfer is mainly dictated by convection and radiation [52] and is expressed as

$$h_{conv-red} = b_{conv}\,(T_{strip} - T_{atmosphere}) + c_{rad}\,(T_{strip}^4 - T_{atmosphere}^4) \tag{12.5}$$

where b_{conv} (≈ 0.0125 kW/m^{2}K for mild carbon steels) and c_{rad} are convection and radiation constants, respectively. c_{rad} is a product of the Stefan–Boltzmann constant (= 5.67051 × 10^{-8} W/m^{2}K^4) and the emissivity (= 0.84) [53].

- State **C**: In this state, the strip surface is in contact with the strip oxide layer and is under pressure. The limiting base value for metal–oxide contact friction is set lower than that for metal–metal contact friction. A reasonable assumption is that as metal–metal contact friction for smooth interfaces is 0.2, a reasonable lower limit metal–oxide friction would be 0.1. The overall heat transfer coefficient is defined by the interface between the oxide scale and strip and oxide scale and roll. The roll does not come in contact with the strip. In this state, the overall heat transfer coefficient is

$$\frac{1}{h} = \frac{1}{h_{strip-oxide}} + \frac{1}{h_{oxide}} + \frac{1}{h_{oxide-roll}} \tag{12.6}$$

where $h_{strip–oxide}$ and $h_{oxide–roll}$ are the pressurized contact heat transfer coefficients of the strip–oxide and oxide–roll interfaces whose values come from the use of Eq. (12.2). For the oxide–roll interface, the oxide is taken to be the softer material and its hardness is expressed as [23]

$$HV_{\text{oxide}} = 7075 - 5.387\ T_{\text{oxide}} \tag{12.7}$$

where T_{oxide} is the oxide temperature in Kelvin and HV_{oxide} is the hardness of the oxide layer in MPa.

The heat transfer coefficient of the oxide, h_{oxide} is

$$h_{\text{oxide}} = C_o \frac{k_{\text{oxide}}}{\delta^o_{\text{oxide}}} \tag{12.8}$$

where C_o is a constant (≈ 1), k_{oxide} is the thermal conductivity of the oxide layer and δ^o_{oxide} is the initial thickness of a uniform oxide layer [51].

- State **D**: In this case, the strip surface is in contact with the strip oxide layer and is not under pressure. This is a special case of **C** where the strip is in contact with the oxide but is not in contact with the roll. Equation (12.8) models this state.
- State **E**: In this state, the strip surface is in contact with the strip oxide layer and a pressurized roll oxide layer. This is a special case of state **C**. This state is simplified if the roll oxide and strip oxide are combined together to form an effective oxide layer and the interface between the oxide layers is neglected. Under such circumstances, Eq. (12.6) models the heat transfer.
- State **F**: In this state, the strip surface is in contact with air and the heat transfer coefficient is that for air. For the confined space where humidity, temperature and pressure dominate, the heat may be higher.
- State **G**: In this state, the strip is in contact with a pressurized roll oxide layer. This is similar to states **C** and **E** and its treatment is similar to those. In the simplified formulation, the roll oxide is similar to strip oxide.

Figure 12.2(b) illustrates a typical state of the strip surface at an instant within the roll bite. It is possible that a new state develops under certain combinations of temperature, reduction and scale thickness. This is designated as É in Fig. 12.2(b). É is a special case of states **A** and **E** due to the brittleness of the oxide layer below a critical temperature (T_{crit}). Above this temperature, the oxide layer is ductile and deforms plastically. If the oxide layer is thick enough, a special case of **É** forms that does not allow the strip surface to come in contact with the roll surface even under high reductions.

The Probability Distribution Diagram (PDD) shown in Fig. 12.3 summarizes these phenomena under the assumption that the variations are linear. This is qualitatively correct since it incorporates the expected dimensions and changes with reduction that experimental observations demand. Further information on the shapes of these distributions that emerge from detailed experiments on roll bite interactions can easily be incorporated into a modified diagram.

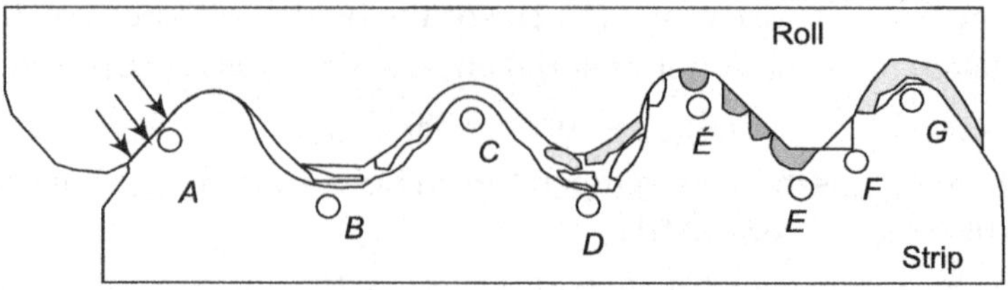

Fig. 12.2 (b) Intermediate strip–oxide-roll configuration at an instant within the roll bite. A new state É develops due to the brittleness of the oxide layer below a critical ductile-brittle transition temperature

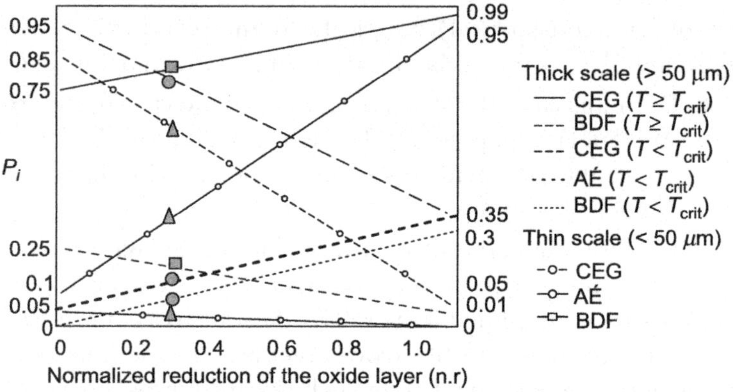

Fig. 12.3 Probability Distribution Diagram (PDD) of a particular state (BDF, CEG and AÉ) as a function of reduction of the oxide layer. A linear variation with reduction is assumed for each of the states. Data from metallographic examination of rolled strip surfaces are used to get the most appropriate representation of each of the above states

The diagram lumps states **B, D** and **F** together to represent an effective probabilistic state defined as a fraction of the side of the element in contact with air, lubricant or water. For thin scales, most of the strip is covered with scale at the point of entry to the roll bite. Since the only possible route of contact with the air or lubricant or water is after cracking takes place at entry to the roll-bite or at the edges due to pre-spread, the PDD ascribes a very low probability of 0.05 at the start of rolling. As rolling proceeds, the underlying steel substrate deforms plastically and fills up the pores in the oxide layer. Almost all the air pores are eliminated and the corresponding probability decreases to 0.0. On the other hand, for a thick scale, brittle fracture takes place due to chilling and the steel cannot come into direct contact with the roll bite. The PDD then increases the initial probability of 0.0–0.3. This is the case when the operating temperature is below the critical temperature for the ductile to brittle transition of the oxide scale. This temperature is in the range of 850–910°C for mild steel strips [20]. Above the critical temperature, that is, above 910°C, a different scenario exists. Since most of the oxide is now ductile and can be easily deformed, its surface area increases under shear loading. Accordingly, the PDD decreases the initial value of 0.25–0.01.

Without any data on the individual state properties, it is sensible to combine states **C, E** and **G** to form an effective probabilistic state defined as the fraction of the side of the strip element both in contact with oxide and under pressure. This is the form shown in Fig. 12.3. For thin scales, rapid chilling of the oxide layer takes place and brittle cracking ensues. The initial probability, assumed to be 0.85, therefore falls rapidly toward the final value of 0.05 assumed at the exit from the roll bite. For thicker scales, although the outer layers are chilled and also fracture, the inner layers may deform plastically and remain covering the steel substrate. This has prompted the assumption of an initial value of 0.95 and final value of 0.35. The drop is less steep than for thin scales. Above the critical temperature as mentioned earlier, the oxide scale deforms plastically, and covers the whole steel surface such that states **A** and **É** never occur. In these circumstances the PDD of Fig. 12.3 has an initial value of 0.75, increasing to 0.99 at exit. For thicker scale, the same argument holds good. Under state **CEG**, friction works between the roll and the oxide layer and is transmitted via the oxide layer to the strip surface. The value is calculated as a reduced fraction of the assumed lower and upper limits of 0.15 and 0.2 for metal-to-oxide contact. This value is multiplied by the corresponding probability factor to calculate the instantaneous friction factor.

Finally, the PDD of Fig. 12.3 combines states **A** and **É** to form an effective state in which the roll is in direct contact with the strip. For thin scales below the critical temperature, the probability increases rapidly. So the initial value of 0.1 increases rapidly to 0.95. For thick scales, although the roll experiences the presence of the steel substrate, the substrate does not extrude sufficiently to touch the roll surface. So the probability rises more gently with the initial value of 0.05 increasing only to 0.35. For thick and thin scales above the critical temperature range, the model here assumes that states **A** and **É** never occur. Table 12.2 shows the initial and final values for each of the above states.

Table 12.2 Initial and final values of the probable states used in the present work. The initial and final values are for a non-porous scale. For porous scale, an additional 15% is added to states **AÉ** and subtracted from states **CEG**. In the present work, scale thicknesses below 50 μm are "thin" while those above 50 μm are thick. In addition, scales above 200 μm are porous

	$T_{oxide} < T_{crit}$						$T_{oxide} > T_{crit}$	
State	**BDF**		**CEG**		**AÉ**		**BDF**	**CEG**
Thickness of oxide scale	Thin	Thick	Thin	Thick	Thin	Thick	Independent of thickness	
Initial values	0.05	0.0	0.85	0.95	0.1	0.05	0.25	0.75
Final values	0.0	0.3	0.05	0.35	0.95	0.35	0.01	0.99

Experiments [4, 20] show that under certain conditions of temperature and compositions of the substrate, the growing oxide scale contains pores of varying degrees. These pores promote brittle fracture faster compared to a solid oxide layer. This means that a porous oxide layer allows steel to come in contact with the roll earlier, i.e., state **AÉ** is achieved ahead in the roll gap. For the thicker oxide scales of 220 μm, the PDD model implements this by using higher values of state **AÉ** and lower values for state **CEG** at any given reduction of the oxide scale. This means that for the 220 μm porous scale, the initial and final values of state **AÉ** are 0.0575 and 0.4025, respectively, while for a 100 μm relatively non-porous scale, the initial and final values of state **AÉ** are 0.05 and 0.35, respectively, as shown in the table.

At any overall state, the sum of all contributing probabilities sum to unity, i.e., $\sum P_i = 1.0$. For example: if a normalized reduction of 0.3 is considered, then the following cases might arise.

- Thin scale with T_{oxide} less than T_{crit}

$$P_{\mathbf{BDF}} = 0.035,\ P_{\mathbf{CEG}} = 0.61 \text{ and } P_{\mathbf{AÉ}} = 0.355$$

This case is shown by the filled triangle in Fig. 12.3.

- Thick scale with T_{oxide} greater than T_{crit}

$$P_{\mathbf{BDF}} = 0.09,\ P_{\mathbf{CEG}} = 0.77 \text{ and } P_{\mathbf{AÉ}} = 0.14$$

This case is shown by the filled circles in Fig. 12.3.

- Thick/Thin scale with T_{oxide} greater than T_{crit}

$$P_{\mathbf{BDF}} = 0.178,\ P_{\mathbf{CEG}} = 0.822$$

This case is shown by the filled squares in Fig. 12.3.

The instantaneous heat transfer coefficient (h_{PDD}) results from the product of the heat transfer coefficient for a particular configuration with the corresponding probability factor. This is expressed as

$$h_{PDD} = P_{\mathbf{BDF}}\, h_{conv-rad} + P_{\mathbf{CEG}}\, h + P_{\mathbf{AÉ}}\, h_c \tag{12.9}$$

12.3 MESOSCALE MODEL FORMULATION

Modeling of material characteristics involves a hierarchy of length scales. At one extreme, atomistic simulations such as molecular dynamics (MD) exploit the interactions between atomic species in order to interpret the effects at a higher length scale. At the other extreme, macroscopic simulations such as finite elements (FE) work on the principle of local material homogeneity. Specifically, material homogeneity is assumed at a length scale smaller than the element size. Typically, atomic simulations involve scales of nanometers whilst finite element simulations sample micrometer scales. Since the FE mesh is not a strict division of the macroscopic domain and is an abstract demarcation of the domain into elements, single crystal (equivalent to single grains) and polycrystalline FE models are available that sample a scale lower than the micrometer scale by using a fine FE mesh, a development that rapidly increases the number of elements. The main drawback of using too fine a FE mesh in either of the above models is that in a *N*-dimensional space with *X* finite elements along each direction, the scaling order of FE is X^N. In addition, single crystal models do not consider the effect of neighboring grains and so are unrealistic. On the other hand, FE polycrystal models do not specify the nature of interfaces between grains. In such strategies, the FE mesh is narrowed down to the size of the feature in question. This feature may be the grain, grain boundary, oxide geometry, second phase particles, or a myriad of measures of internal metallurgical substructure. This makes the stand-alone FE method very computationally expensive as the size of the feature scales down.

These considerations pose two important questions. Firstly, what is a suitable scale at which numerical models are sufficient in detail and are also appropriate in terms of computational time? The answer lies in identifying an abstract scale in between the micro and macro scales, the "mesoscale". Secondly, can additional numerical techniques be used in conjunction with a conventional FE representation to model material heterogeneity and related properties? This work attempts to answer this using Cellular Automata, an approach that offers flexibility in length scales and which is used to model the mesoscale. The framework of the mesoscale model is a coupling of Cellular Automata based Finite Elements (CAFE) with each entity suitably scaled to represent the essential mechanics at its individual length scale.

A meso-geometry is a representative mathematical abstraction of a physical system containing sufficient details of metallurgical structures so as to obtain the material properties at the macroscopic scale. Metallurgical features used to characterize deformation include grain sizes (and associated grain boundaries), dislocations (mobile and immobile), second phase particles (as in bi-phase/multi-alloyed systems) and defects (represented by microcracks or microvoids). During deformation, these features evolve at different length and time scales; the fastest associated with the smallest length scales. Specifically, the average grain size is of the order of 10~100 μm (typical of hot working of steels and aluminum alloys), while dislocations, microcracks and microvoids are of the order of 0.1~1000 nanometers. An underlying functional relationship exists across these length scales that determine the behavior of a particular metal under deformation.

In the CAFE model with an average grain size of 10~100 μm, the first level meso-geometry is of the order of 40~400 μm. The internal structure of this first level meso-geometry evolves with the deformation process and the generation and dissolution of metallurgical substructures (microcracks, microvoids, cells, and subgrains) are addressed by a second level meso-geometry, the order of which is 0.1~1 μm. Any feature below this length scale (dislocations for example) is held as an averaged value at that particular spatial location. This means that the spatial and temporal distribution of metallurgical structures at any instant of the deformation process needs to be mapped as discrete variable values held at specific sites.

CAFE is built in six distinct steps. These are:

1. Defining the metallurgical structure using CA.
2. Connecting the continuum with microstructure and adapting the FE input deck.
3. Linking CA and FE.
4. Approximating the strain space within CA.
5. Physical averaging to FE.
6. Updating CA using the transformation matrix.

Each of the above steps is elaborated in the following sections.

12.3.1 Defining the Metallurgical Structure Using CA

CAs, first introduced by von Neumann [44] as idealizations of biological systems, are mathematical idealizations in which space and time are discrete. Heselbarth and Göbel [15] (1991) used CA to model static recrystallization following deformation. Their model used a probability based nucleation criterion to populate the CA domain. The growth of these nuclei is based on the surrounding (neighborhood) temperature field that is obtained either from a heat-flow computation or from temperature measurements. Rappaz and Gandin [36] used a similar approach to simulate the grain growth during solidification of as-cast structures. For a better approximation of the enthalpy field associated with the liquid solid transition, a finite element program 3-MOS was used; thereby introducing the term CA-FE [12]. Makarov *et al.* [31] modeled damage propagation using a CA based approach. The local stress values required to propagate the damage was obtained from a standalone FE program. Matic and Geltmacher [32] also used a similar approach; the stress values being supplied by a standalone FD program with damage being propagated using the technique of node-splitting. However, in the above models, there is no link to the underlying microstructure and its response to the loading pattern.

As mentioned earlier, CA consists of a uniformly (or non-uniformly) spaced array, usually infinite or periodic in nature, with single or multiple discrete variables within each cell. Each CA cell maps a region in the actual material and the values of its variables completely specify the underlying material state. On the other hand, the CA cell possesses no intrinsic length and time scale. This spatio-temporal independence means that the length and time scale of a single CA cell corresponds to the particular entity under consideration. The present work exploits this feature to capture two levels of microstructural entity.

Figure 12.4(a) shows a 2D spatial domain of dimensions l_{CA} by b_{CA} subdivided into a number of cells, N_x and N_y, along the X and Y-*axes*. At this stage, these axes are a more general system of ordering the CA cells in mutual connectedness of which the Cartesian axes would be a subset. The CA cell size is given by the algebraic product of (l_{CA}/N_x) and (b_{CA}/N_y) and is linked to the size of the metallurgical feature in question. As mentioned earlier, metallurgical features such as grains are often of the order of microns, so the physical CA cell size corresponds to the micron scale in the examples studied later. This means that the region marked by $(l_{CA}/N_x, b_{CA}/N_y)$ in Fig. 12.4(b) is of the order 5~40 μm. The present work treats this scale as the first level CA.

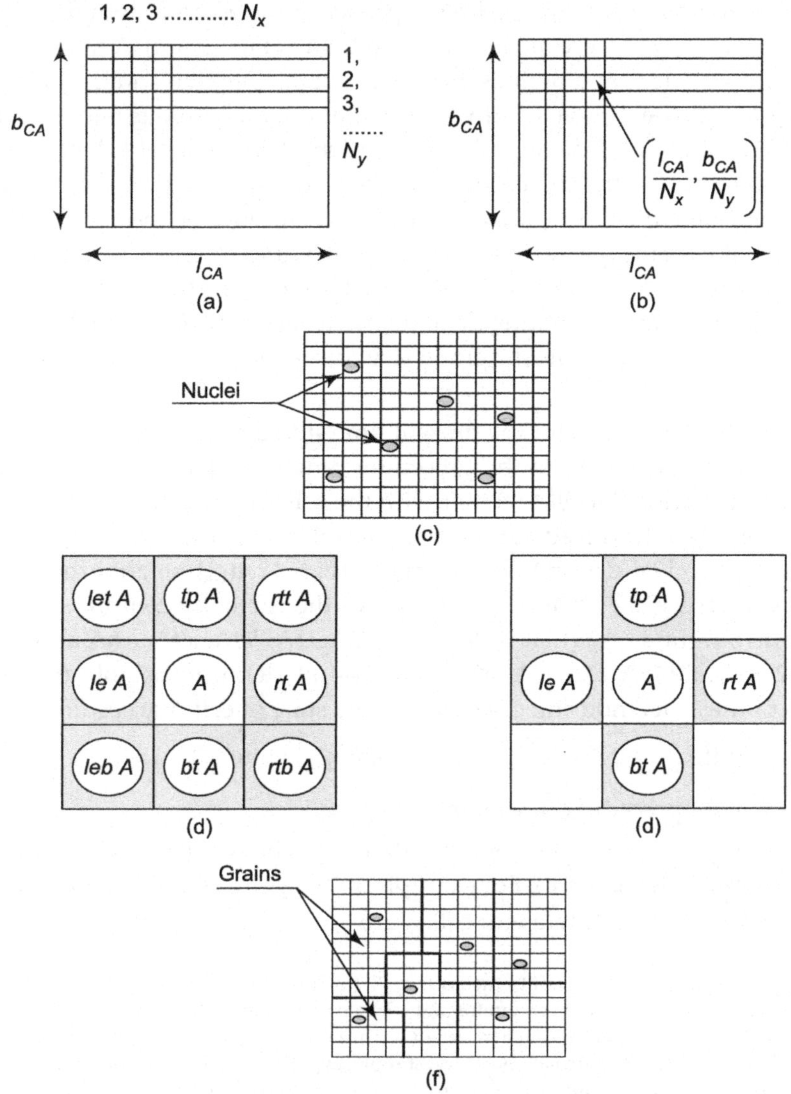

Fig. 12.4 Microstructure representation on a 2D domain using cellular automata. (a) Definition of a 2D domain. (b) Smallest unit definition (N_x and N_y are the number of divisions along the X and Y axes) (c) Nucleation (d) Moore's neighborhood (e) Neumann neighborhood (f) Grain boundaries using transition rules

To start the simulation of a microstructure, grain nuclei are randomly scattered over the domain of Fig. 12.4(a). A second-level array of CA cells of 0.1~5 μm size represent these nuclei. The small circles in Fig. 12.4(c) represent these nuclei. All nuclei can be assumed to be present at the start of the simulation as is the case with site-saturated nucleation or they can continue to appear during the simulation as a function of the deformation variables. The present work employs the first approach and uses a random function[2] to generate and locate these nuclei on the domain of Fig. 12.4(a). A numerical tag identifies a nucleus and defines its unique spatial position and its orientation with respect to the chosen X and Y coordinate system. A unique integer number generated randomly[3] defined the orientation. All first-level CA cells except those occupied by the nuclei have a "zero" orientation that represents the matrix within which the grain can grow. The presence or absence of nuclei is an internal variable for the first-level CA cell. Other internal variables are the spatial location of the CA cell and the orientation of the microstructure it will represent. This form of linking a microstructural entity to a spatial geometric entity allows embedding different structure–geometry relationships in a multi-level CA.

The state of a CA cell for the present time step depends on the states of the neighboring CA cells from the preceding time step. The neighborhood of a cell includes the cell itself together with the cells sharing an edge or a corner of the cell. As introduced earlier, the variables are propagated using either Moore's [Fig. 12.4(d)] or Von Neumann [Fig. 12.4(e)] neighborhoods. The cell "A" in Figs. 12.4(d) and 12(e) indicates the currently computed cell. The cells surrounding "A" [that is, *tp*A, *bt*A, *le*A, *rt*A, *let*A, *rtt*A, *leb*A and *rtb*A of Fig. 12.4(d) and *tp*A, *bt*A, *le*A, *rt*A of Fig. 12.4(e)] define its neighborhood. The states of these neighboring cells modify the state of cell A according to

$$\text{State}_A^{t+1} = f(\text{State}_{leA}^t, \text{State}_{rtA}^t, \text{State}_{tpA}^t, \text{State}_{btA}^t) \tag{12.10}$$

where t is the preceding time step.

State_A represents the state of cell A that might be zero (no nuclei or dead) or unity (with nuclei or alive). For example, in the game of life [6], the transition rule used to change the state of cell A are:

---

[2] The random generation is a call to the Fortran 77 function *r and* (i) where "i" is the seed. This function is a linear congruential generator and is based on the formulae *rand* (i + 1) = (*rand* (i) *b* + *a*) *mod* m. "*b*" and "*a*" are positive integers called the multiplier and increment respectively. "m" is the modulus (usually it's the machine's largest representable integer ~ 2^{32}). For a given seed, the function returns a series of numbers uniformly distributed between 0 and $2^{32} - 1$. A different seed gives a different sequence of random numbers. The random function can be scaled as required in defining grain orientations or dislocation densities.
The same function can be used to obtain a "normal" gaussian distribution with a pre-specified mean and standard deviation. (further details are at www.nr.com).

[3] Usually, a wide variation, say 1–2000, is used to specify the orientation of the CA cells. This way, there is a lower probability of getting grains of similar orientation.

- Cell A remains alive in the next time step if either two or more of its neighbors are alive.
- Cell A dies of overcrowding in the next time step if more than three of its neighbors are alive.
- Cell A dies of loneliness in the next time step if fewer than one of its neighbors is alive.
- Cell A becomes alive in the next time step if exactly two of its neighbors are alive.

Simple transition rules similar to those above result in patterns of immense complexities. To start the simulation of the microstructure, the velocity (v_{gb}) of the moving front of each nucleus is expressed in terms of the spatial dimension (dX or dY) moved per unit time (t). This approach relates the microstructural evolution to physical times. Davies [8] assumed a mean grain interface migration rate given by

$$v_{gb} = \frac{dX}{dt} = \frac{dY}{dt} = C_o t^{-n} \tag{12.11}$$

where C_o is a constant (= 2) and n is equal to 0.6.

At each step, the nuclei spread over cells that have a zero orientation. The simulation assumes an isotropic grain growth and the simulation stops when the moving fronts of two nuclei meet. This interface forms the grain boundary having a mis-orientation defined by the difference of orientations across the boundaries. This is the form of Fig. 12.4(f), which illustrates the structure–membership relationship of the first level CA.

12.3.2 Adapting the Finite Element (FE) Code

The FE method approximates any domain using elements. The type and shape of elements used depend on the domain type (2D or 3D) and shape (linear or non-linear) and are defined by the number and location of the nodes. Figure 12.5 shows some commonly used 2D and 3D elements.

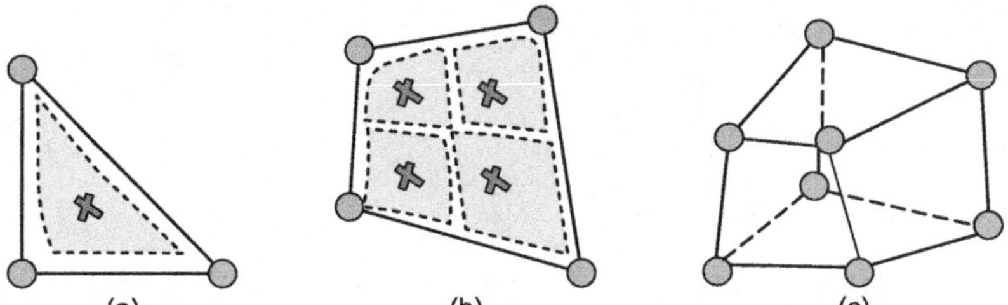

Fig. 12.5 Commonly used (a) 2D triangular (b) 2D quadrilateral (c) 3D solid elements. Circles indicate the location of the nodes and the crosses indicate the integration points. The grey regions are the probable areas of influence by each integration point

Nodes also carry the information regarding the state of the domain as nodal variables—these may either be pre-defined boundary or internal fields generated during the simulation. Field gradients across nodes are approximated within the master element at integration points located at a distance of 0.577 from the nodes [49] (Barlow, 1976). For a real element in a Lagrangian framework where the material moves with the deformed mesh, the shape functions define the location of the integration point. An abstract area of influence (indicated by shaded regions in Figs. 12.5(a) and 12.5(b)) surrounds each integration points (indicated by crosses in Figs. 12.5(a) and 12.5(b)). In a deforming mesh, this area corresponds to the actual state of the deforming material. The algorithm connects an array of CA cells with the FE integration point areas by exploiting the information regarding the nodal locations and the element connectivities.

Figure 12.6(a) shows a typical input file for a FE simulation. The nodes are numbered N1, N2, ..., Nn where n is the number of nodes. Each of these nodes has a spatial definition in terms of its coordinates along the X, Y and Z directions defined in a global Cartesian coordinate system. Linking the nodes in sequence gives the element connectivities. This is shown for one four-noded element (E1) defined by the nodes N1, N2, N3 and N4 in Fig. 12.6(b). The following steps define the regions (spheres of influence) of each integration point:

1. Determine the mid points M12, M23, M34, M14 of each edge for element E1.
2. Determine the midpoint (DM) obtained by the intersection of the diagonals for element *E*1.

Figure 12.6(c) shows the areas (I2, I3, I3 and I4) influenced by each of the four integration points. The assumption of an averaged material point information implies that the same conditions apply, i.e., a variable has the same value at any point within each area. This is where the present modeling approach seeks

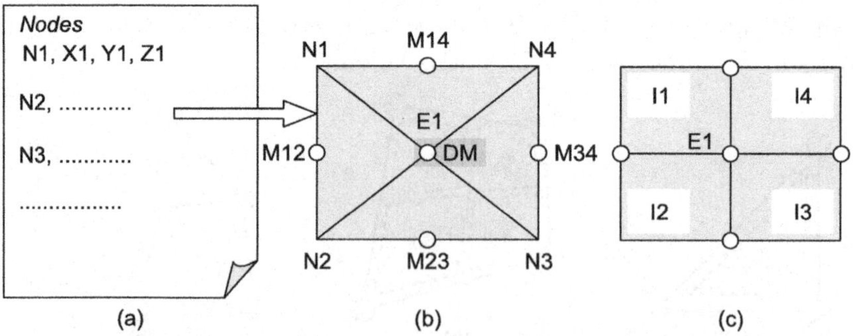

Fig. 12.6 The finite element input deck modification. (a) Typical input file used in a FE simulation. N1, N2, N3, N4 are the nodes and X1, Y1, Z1 are the spatial locations of node N1. E1, E2, etc., are the elements. (b) Determination of midpoints and diagonals for a particular element E1. (c) Determination of regions of influence by integration points (I1, I2, I3 and I4) for a particular element E1

to exploit the capability of CA at a suitably placed length scale. This length scale is that of a mesoscale such that there are at least 5–20 grains at a single integration point of the finite element.

12.3.3 The Triangulation Technique to Form the Link Between CA and FE

A link between a CA cell and a finite element is then done by asking a purely geometrical question as to which CA cell is encompassed by which finite element. Figure 12.7(a) shows this schematically. Then for each CA cell, the algorithm checks if the

- *x coordinate of the CA cell lies between* X_{min} *and* X_{max} *of the finite element.*
- *y coordinate of the CA cell lies between* Y_{min} *and* Y_{max} *of the finite element.*

If the CA cell satisfies both of the above criteria, then the algorithm classifies the cell as IN else OUT. Figure 12.7(b) shows this algorithm schematically.

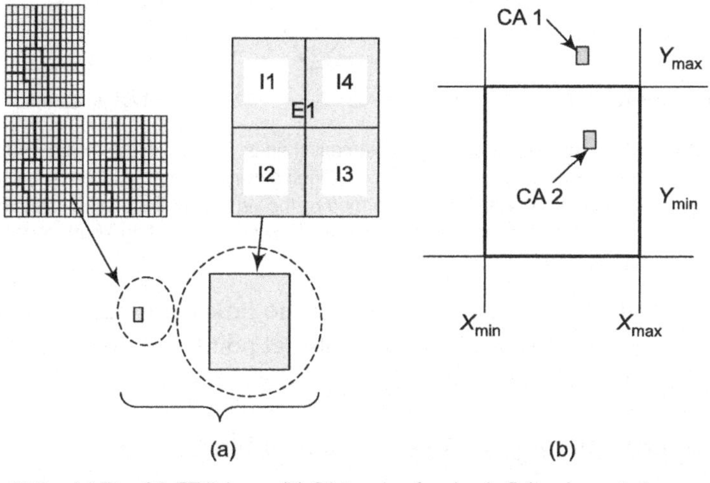

Fig. 12.7 (a) The CA-FE linkage. (b) CA tagging for simple finite element shapes such as rectangles. In the figure, CA-1 is OUT and CA-2 is IN the element

However, if the FE shape is that as shown in Fig. 12.8(a), the results of above algorithm are wrong; a further iteration is required. To achieve this, the work proposes a triangulation technique. This technique involves the division of a finite element into two triangles. The areas formed by the CA cell with each of the three corners are compared to the area of the original triangle. In doing so, three cases arise

1. The sum of the areas (A1 + A2 + A3) is greater than the area of the original triangle (A). Figure 12.8(b) shows this configuration. The algorithm classifies this CA cell as OUT of the particular FE.
2. The sum of the areas (A1 + A2) is equal to the area of the original triangle (A). The area (A3) is equal to zero. Figure 12.8(c) shows this configuration. The algorithm classifies this CA cell as IN the particular FE.

3. The sum of the areas (A1+A2+A3) is less than the area of the original triangle (A). Figure 12.8(d) shows this configuration. The algorithm classifies this CA cell as IN the particular FE.

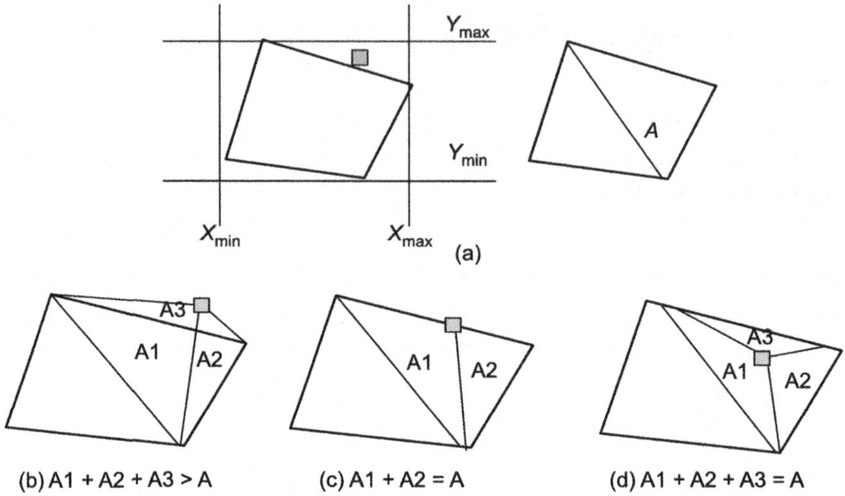

Fig. 12.8 (a) CA tagging for other finite element shapes. Division of the element into two triangles, area of each is A. (b) CA tagging using the triangulation technique. A1, A2 and A3 are the areas formed by the vertices of the original triangle and the CA cell. This is the case where the CA cell is OUT of the element. (c) and (d) CA cell is IN the element

All CA cells follow the above procedure to link with the FE. This form of tying the CA and FE ensures that the material point remains associated with the given integration point throughout the deformation.

12.3.4 Approximating the Strain Space within CA

The response to an imposed macroscopic strain of an ensemble of mesoscopic features including grains, grain boundaries, second phase carbides and inclusions determines the local strain and stress fields. Mesoscale studies for a duplex stainless steel show a strain inhomogeneity across different phases [16, 35]. Specifically, the strain in the ferrite is about 1% higher than the overall strain, while the strain in the austenite is 1% lower. In addition, grain boundaries [42, 48] and second phase inclusions [30] act as dislocation sinks by impeding their movement and result in higher strains. However, a precise quantification in terms of macroscopic strain, strain rate, and temperature is not known. The present work acknowledges this lack of quantitative information by adopting a feature-driven biased distribution of the macroscopic strain.

Figure 12.9(a) is a schematic representation of a typical microstructure. It deals with grains, grain boundaries and second phase carbides or inclusions.

The small circles in the figure show the location of carbide particles. A suitably scaled CA captures this approximated microstructure and links it to the integration point of a finite element [Fig. 12.9(b)] using the procedure outlined in the previous section.

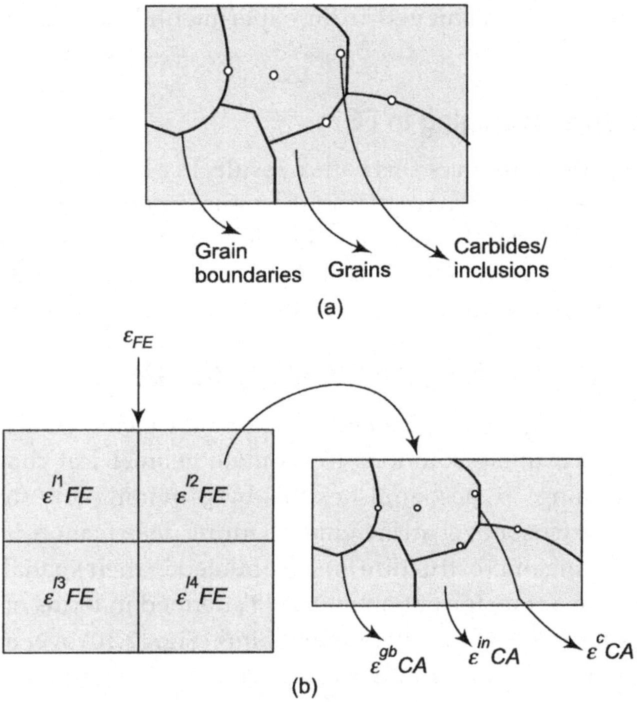

Fig. 12.9 (a) Schematic representation of entities in a typical microstructure showing grains, grain boundaries, and carbides/inclusions. (b) Mapping elemental strains to feature-based strains at the CA level. ε_{FE} is the macroscopic strain. ε^{J1}_{FE}, ε^{J2}_{FE}, ε^{J3}_{FE} and ε^{J4}_{FE} are the integration point strains, ε^{gb}_{CA}, ε^{in}_{CA} and ε^{c}_{CA} are the grain boundary, grain interior and carbide/inclusion strains at the CA level, respectively

The strain at a CA cell is a function of the macro-strain at each integration point expressed as:

$$\varepsilon^{i}_{CA} = f^{i}\varepsilon^{j}_{FE} \tag{12.12}$$

where ε^{i}_{CA} is the strain at each CA cell and i is the property of the CA cell expressed in terms of microstructural features such as grain interiors, grain boundaries, carbide particles, etc., ε^{j}_{FE} is the macro-strain at a particular integration point, with j being the integration point number (j = 1, 2, 3 or 4 for a 4 noded linear finite element).

f^{i} is the scaling factor whose value depends on the microstructural feature. In the present model, f^{i} takes values of 1, 1.2 and 3 for grain interiors, grain boundaries and carbide particles. Thus, the strain distribution is homogenous for CA cells forming part of the grain interior whilst CA cells that form part of the boundary are at a higher strain.

Here, the transfer of information from the macro- to the micro-scale adopts a feature-based approach. This is good to a first approximation and a better representation can be achieved from experimental data as they become available.

12.3.5 Physical Averaging to FE

A strain space distribution of this nature results in a stress space distribution within the CA. The stress space distribution is an internal variable of the first level CA. The FE integration point stress is an average of the stresses at all "active" CA cells. The issue of "active" cells arises in oxide scale modeling where cells "fail" or become "inactive" upon exceeding a critical strain.

12.3.6 Updating CA Using the Transformation Matrix

As mentioned earlier, Lagrangian FE simulations are incremental formulations that derive approximate solutions to estimate geometrical changes with a change in straining. FE accomplishes this by providing an estimate of the deformation rate tensor over an increment. During deformation, it is necessary that the underlying microstructure is also updated in their spatial definitions. The CA cells use a transformation matrix [T] defined in terms of the present and expected locations of the integration points (Fig. 12.10) over an increment to update the microstructure associated with the integration point and is expressed as

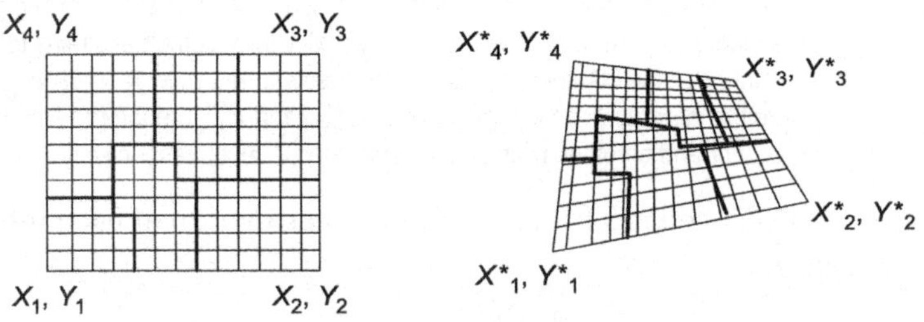

$$\begin{bmatrix} X_i & Y_i & 1 \\ X_j & Y_j & 1 \\ X_k & Y_k & 1 \end{bmatrix} = \begin{bmatrix} T_{11} & T_{12} & T_{13} \\ T_{21} & T_{22} & T_{23} \\ T_{31} & T_{32} & T_{33} \end{bmatrix} \begin{bmatrix} X_i^* & Y_i^* & 1 \\ X_j^* & Y_j^* & 1 \\ X_k^* & Y_k^* & 1 \end{bmatrix}$$

Fig. 12.10 The Lagrangian formulation for the CAFE model. T_{ij} is the transforming matrix. The subscripts $\{i, j, k\}$ are taken in an order such that (X_4, Y_4), (X_1, Y_1), (X_2, Y_2) are used for transforming integration point I_1; similarly (X_1, Y_1), (X_2, Y_2), (X_3, Y_3) are used for transforming integration point I_2; (X_2, Y_2), (X_3, Y_3), (X_4, Y_4) for integration point I_3; (X_3, Y_3), (X_4, Y_4), (X_1, Y_1) for integration point I_4

$$[O]^{t+1} = [T][O]^t \tag{12.13}$$

where $[O]$ is the matrix containing the material coordinates at time t.

As mentioned earlier, the CAFE model associates a material region with each integration point. The work uses Eq. (12.13) to update the instantaneous location of all CA cells that comprise the material region around an integration point. However, the assumption of a homogenous deformation in the presence of grain boundaries is a simplification. The present work assumes that the magnitude of movement of the grain boundaries in relation to the overall deformation is negligible, and so does not register on the mesoscale.

12.4 APPLICATION TO OXIDE DAMAGE MODELLING

Figure 12.11(a) shows a typical meso-structure with grains of different orientations. Point A of Fig. 12.11(a) illustrates a crack front that propagates as a brittle fracture only through grains (intragranular fracture) or boundaries (intergranular fracture) based on their respective orientations. An idealized representation of this meso-structure identifies the CA cells that form part of the grain boundary and those that form part of the grain interior. Figure 12.11(b) shows a schematic representation of the discretised oxide layer domain with 10 finite elements. The CA cell array is shown for one of the finite elements. The present work studies random crack initiation due to the inherent irregularity of the process itself. The filled circles in Fig. 12.11(c) show this. The crack nucleus is half the distance between two successive CA cell centroids. A spectrum of crack nuclei sizes can also be considered with the smallest crack nuclei being the distance between two successive CA cell centroids. Although this method can be used to model the nucleation of cracks within the oxide layer, this is not considered here and the initiation of cracks is restricted to the top layer only. Since experiments show that the propagation of the crack is predominantly downwards towards the strip, the Moore's neighborhood of Fig. 12.12(a) needs suitable modification. This modified CA neighborhood is shown in Fig. 12.12(b) in which a three-cell neighborhood is used to propagate the crack front. Any of the cells represented as *leA*, *tpA* and *rtA* can be a part of the grain boundary. For the cell A currently being computed, the transition rule defining the changeover from a healthy cell to a defective one is whenever

- any of the cells (State_{leA}, State_{tpA}, State_{rtA}) have failed in the preceeding time step [schematically illustrated by ***F*** in Fig. 12.12(c)] and the strain in cell A exceeds the critical strain required for fracture at the present time step.

Crack front propagation is restricted to the CA cells and is based on their initial state, that is, whether they are a part of a grain boundary (for grain boundary dominated failure) or in other cases not reported here, whether they contain second phase particles or voids that coalescence to propagate a ductile type fracture. The directionality of the crack front is material dependent and evolves naturally within the modeling strategy according to the details of the transition rules used. For this work, the directionality follows implicitly by restricting the crack front to the cells one level below the present failed CA cell. Figure 12.11(d) shows this schematically.

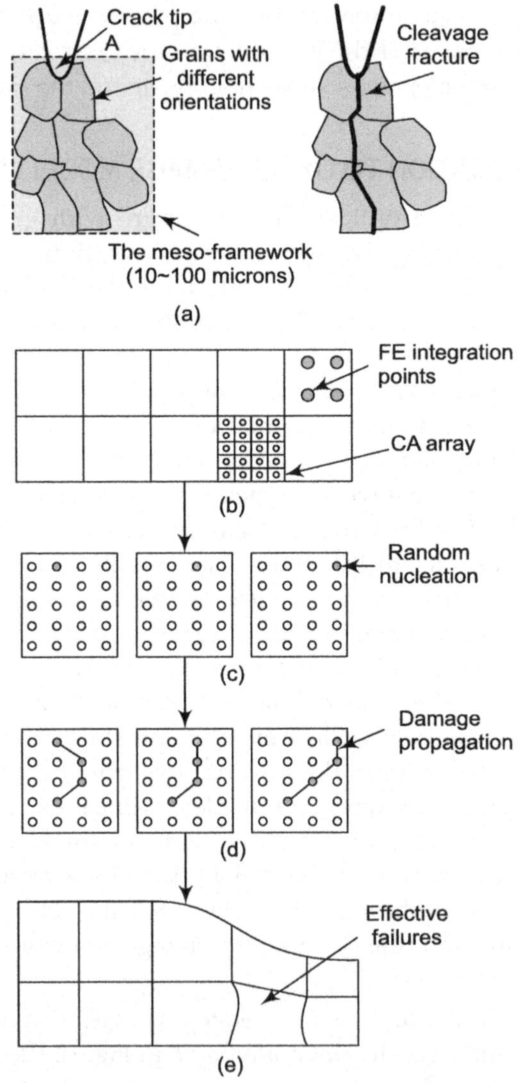

Fig. 12.11 (a) Application of the CAFE formulation to model oxide cracking phenomena. A schematic showing the physical phenomena of oxide brittle fracture. (b) CAFE formulation and its incorporation into the finite element mesh. (c) Random nucleation sites in the CA grid. (d) Propagation of the damage front as the deformation proceeds. (e) Effective failure of the overlying finite element

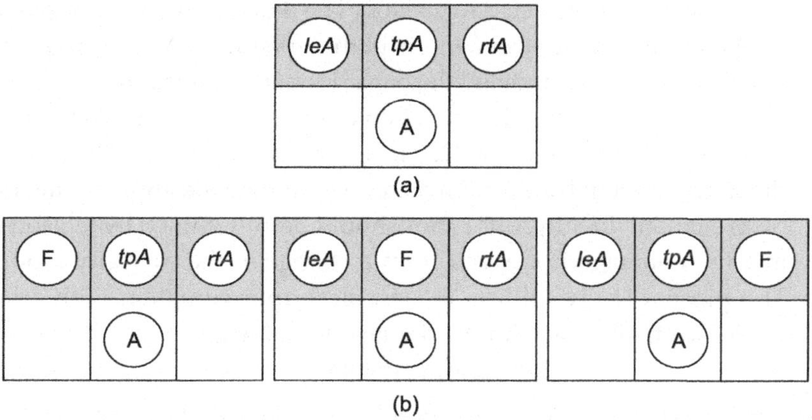

Fig. 12.12 (a) Modified Moore's neighborhood. The CA cell being computed is represented by A. leA, tpA and rtA are the CA cells to the left, top and right of the CA cell A. (b) Situations in which the CA cell A fails. The cell marked F has failed in the preceding time step

The above methodology can capture any crack fronts that propagate simultaneously through the material. When this happens, the corresponding encompassing finite element fails faster. Figure 12.11(e) shows this schematically. The propagation of the crack front is particularly sensitive to two factors.

- The number of CA cells aligned favorably with respect to the global fracture direction.
- Regions of high strain.

The first factor follows from an assumed random Gaussian distribution of potential nuclei over all the CA cells before the start of the deformation. The second factor evolves during the deformation process itself. Two indices help make the necessary characterization.

- An initial failure index (IFI) defined as the ratio of the area occupied by potential crack nuclei sites (includes second phase carbides, inclusions and voids) to the total area occupied by the first level CA cells within a finite element. A high ratio indicates a greater susceptibility to fracture while a lower value indicates a greater stiffness.
- A final failure index (FFI) defined as the ratio of area occupied by the "failed" or "inactive" first level CA cells to the total area occupied by the first level CA cells within a finite element. The FFI is the internal variable associated with the CA cell and links to the overlying finite element at each time increment. The FFI controls the strength of the finite element at each time increment and is therefore responsible for the macroscopic material behavior. Rice [38] defined a scalar quantity called "damage" for characterizing failure in steels, copper and aluminum

alloy specimens subjected to tension. The damage evolves linearly with increment in plastic strain and is path dependent. An effective stress–strain relationship modeled the behavior of the damaged material such that the effective stress is a function of the instantaneous change in the damage variable[4]. Lemaitre [22] achieved this by reducing the rigidity of the finite element by a certain factor to simulate the approaching death of an element. Ideally, this factor should be determined by tension tests in which damage is derived from a change in the elasticity modulus. The present study utilizes the simplest approximation with a linear decrease of stiffness with the damage variable given by

$$E^{t+1}(T) = E^{t}(T)\,(1 - FFI^{t})^{5} \tag{12.14}$$

where the superscript '$t+1$' is the present time step and FFI^t is the final failure index at the previous time step.

Li and Sellars [29] show that brittle fracture dominates at low reductions of 10–20% and thick scales of 160~370 μm. This is similar to thin scales of 22~40 μm that undergo brittle failure at higher reductions. This means that a linear elastic fracture model (LEFM) can be adopted for these classes of rolling simulations and the *FFI* can be linked to a decrease in the elasticity modulus. The critical strain technique [39] gives the critical strain for brittle fracture. For oxide scales on a substrate that is under tension, the critical strain is

$$\varepsilon_{cr} = \left(\frac{2\gamma(T)}{F_s^2 \pi E(T) c}\right)^{0.5} \tag{12.15}$$

where F_s takes values of 1.12, 1 and $2/\pi$ for a surface notch of depth c, for buried notch of width $2c$ and for a semicircular surface notch of radius c, respectively. c varies between 20~100% of the oxide thickness.

---

[4] Beremin [1] simplifies the cleavage fracture in structural components into a two-step process and deals only with slip-initiated cleavage. In ferritic steels, microcracks originate or are present in the carbide particles at the grain boundaries. These propagate into macrocracks through the grain when a critical strain is achieved. In such cases, the *IFI* (and *FFI*) is defined as the number of failed CA carbide particles to the total number of carbide particles [3]. In this work, propagation of the crack into the grain takes place only when two or more cracked carbide particles are found with a radius. As a first approximation, a radius of 40 μm (~1.5 times the average grain diameter of the ferritic steels) was assumed.

The above definition works for structural components where failure takes place once the largest flaw has failed. Robertson and Manning [39] pointed out that for scales, the presence of an underlying substrate allows further straining (and cracking) even after the largest flaw has failed. They proposed a failure criterion in terms of crack propagation rather than the initiation of flaws (used in structural failures). Krzyzanowski and Beynon [20] adopted a similar approach. The present *IFI* (and *FFI*) definition takes into account both the crack initiation and propagation in the sense that every crack nucleus has a potential to form a crack if the correct circumstances prevail. And these crack nuclei propagate into the bulk CA cells—the decrease in strength being defined in terms of the area of the "inactive" CA cells.

[5] $E(T) = 240[1 - 0.00047(T - 25)]$ GPa for an undamaged oxide. T is the oxide temperature [39].

$\gamma(T)$ is the temperature-dependent surface fracture energy and is the most difficult parameter of the theory to set because quoted values can vary over one order of magnitude for a given oxide. The plane strain fracture toughness K_{IC} relates $\gamma(T)$ to $E(T)$ through

$$\gamma(T) = \left(\frac{K_{IC}^2}{2E(T)}\right) \tag{12.16}$$

The stress intensity factor K_{IC}, is a function of process variables such as strain, strain rate and temperature. For the range of temperatures that are of interest in hot rolling, it is convenient to use the relationship of Krzyzanowski and Beynon [18] obtained using a polynomial fit[6] to the experimental data of Hancock and Nicholls [13].

Equation (12.15) depends on the material properties of surface fracture energy and rigidity and the geometrical distributions of F_s and c[7]. Also, Eq. (12.15) is valid only for oxide scales on a substrate that is under tension; the values of $\gamma(T)$, the shape factor F_s and the flaw size c under mixed loading conditions, as experienced within the roll gap, are virtually unknown. It is common to assume the critical strain under mixed loading conditions to be a multiple of the critical strain under tensile loading. For example: Krzyzanowski and Beynon [18] assume that the critical strain under shear loading is twice that under tensile loading. The CAFE model defines c and ε_{cr}, as

- c is a gross measure of cracking through the scale thickness and corresponds to the maximum distance travelled by a surface crack along the scale thickness. In the upper limit, c is equal to the instantaneous oxide thickness, δ_{oxide}.
- The operating critical strain (ε_{cr}) is a random multiple[8] of the critical strain under tension.

Li and Sellars [26] observed that once the oxide scale fails, fresh steel rapidly comes in contact with the roll. The present model cannot simulate separation of the finite elements. Therefore, a simple replacement technique represents the important interfacial changes that happen as fresh steel extrudes to the surface in the roll-bite. When the original oxide CA cells fail, i.e., the FFI changes, CA cells with the material properties of the underlying substrate replace these. In this way, the finite element changes its material behavior. Figure 12.13 illustrates the change in the elastic modulus as a function of the normalized internal variable *FV2* (damage) at different oxide temperatures.

---

[6] $K_{IC} = a_o + a_1T + a_2T^2 + a_3T^3 + a_4T^4 + a_5T^5$ where $a_o = 1.425$, $a_1 = -8.897 \times 10^{-3}$, $a_2 = -8.21 \times 10^{-5}$, $a_3 = 3.176 \times 10^{-7}$, $a_4 = -5.455 \times 10^{-10}$, $a_5 = 3.437 \times 10^{-13}$. The equation is valid for the temperature (T) range of 20–820°C [13] and is for wüstite (FeO).

[7] As a rough estimate, $\varepsilon_{cr} = 0.0014$ for $\delta_{oxide} = 50$ μm, $\varepsilon_{cr} = 0.001$ for $\delta_{oxide} = 100$ μm, $\varepsilon_{cr} = 0.00067$ for $\delta_{oxide} = 225$ μm. In this calculations, $c = 0.2\ \delta_{oxide}$, $F_s = 1.12$, $\gamma(T)/E(T) = 0.04 \times 10^{-9}$/m. These strains are encountered just at entry into the roll gap.

[8] The operating critical strain varies between 1.5 to 2 times the critical strain in tension.

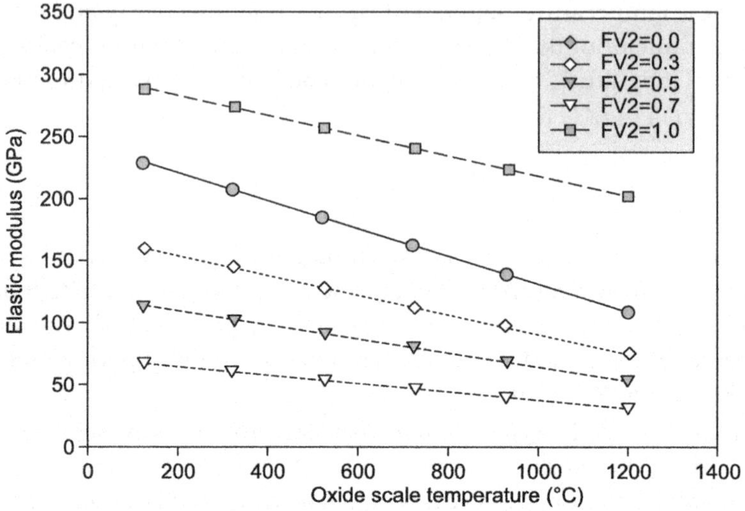

Fig. 12.13 Variation in elastic modulus as a function of field variables. Two field variables, oxide scale temperature and damage (FV2), are used to modify the elastic modulus

12.5 MESOSCALE MODELLING

Figure 12.14 shows the 2D FE mesh for the oxide layer, strip and roll. An array of four noded isoparametric fully thermo-mechanically coupled elements form the FE model of strip, roll and the oxide layer. The work studies the behavior of two oxide thicknesses of 225 μm and 110 μm; these are typical of industrial and laboratory hot rolling of steels. For the 225 μm layer, each oxide finite element is of size 225 μm by 250 μm. For the 110 μm layer, each oxide finite element is of size 110 μm by 125 μm. In addition, the model assumes an initial strip temperature of 1000°C, an initial roll temperature of 30°C and an initial oxide temperature of 1000°C.

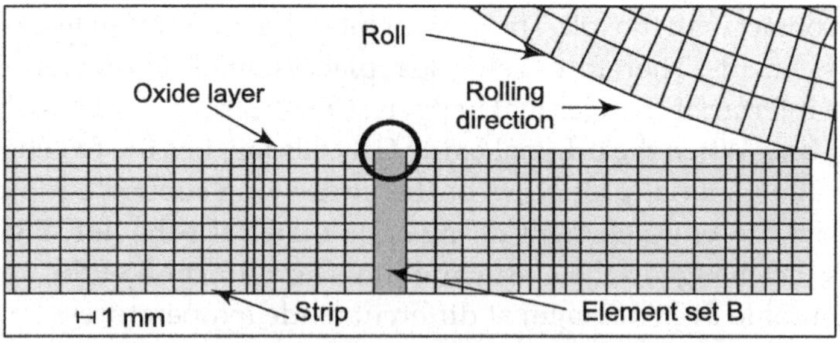

Fig. 12.14 2D finite element configuration for the CAFE formulation

The CA mesoscale geometry involves laying first level CA cells of 11 μm by 11 μm within each finite element of the oxide layer. Figure 12.15 shows the metallurgical entities modeled at the CA scale for an oxide finite element. The unfilled circles represent second level CA cells containing the crack nuclei (microvoids, micro inclusions, etc.).

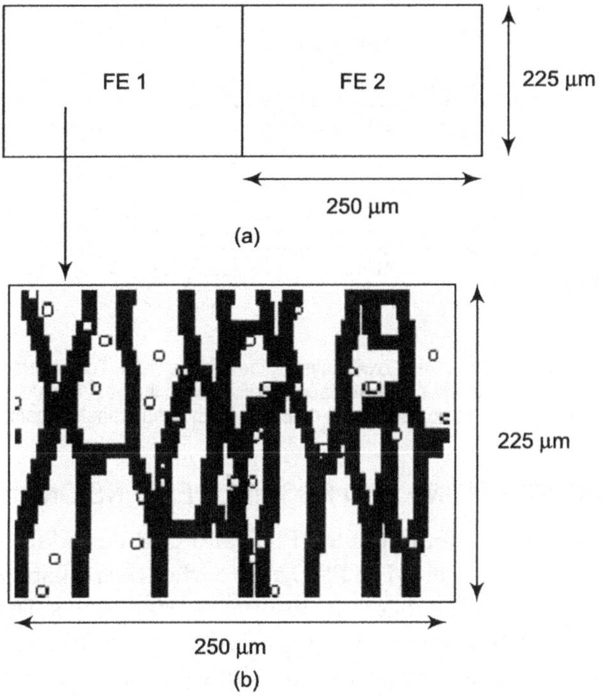

Fig. 12.15 (a) Oxide layer finite element sizes. (b) Metallurgical entities associated with an oxide finite element

The CAFE model monitors the evolution of crack nuclei into cracks and analyses its effect on heat transfer at the oxide–roll (Interface A in Fig. 12.16) and oxide–strip (Interface C in Fig. 12.16) interfaces. The final failure index, FFI, modifies the interface properties as the cracks develop. To achieve this, the model assumes lower and upper bounds on the heat transfer coefficients. Results from the PDD model show that for an initial oxide thickness of 225 μm, an effective heat transfer coefficient of 15–20 kW/m^{2}K results whilst for a lower oxide thickness of 100 μm, a higher heat transfer coefficient of 30 kW/m^{2}K is more appropriate. The rise in the heat transfer coefficient in the 100 μm case comes from the break-up of the scale layer leading thereby to enhanced strip–roll contact. Based on these results, a linear function relates *FFI* to h_{CAFE}.[9]

$$h_{CAFE} = h_{lower} + FFI\,(h_{upper} - h_{lower}) \tag{12.17}$$

where h_{lower} and h_{upper} are the lower and upper bounds on the heat transfer coefficient and are equal to 10 and 30 kW/m^{2}K, respectively. The model assumes the similar heat transfer coefficient to operate at both the A and C interfaces.

[9] The linear relationship is used here as a first approximation due to the lack of experimental data.

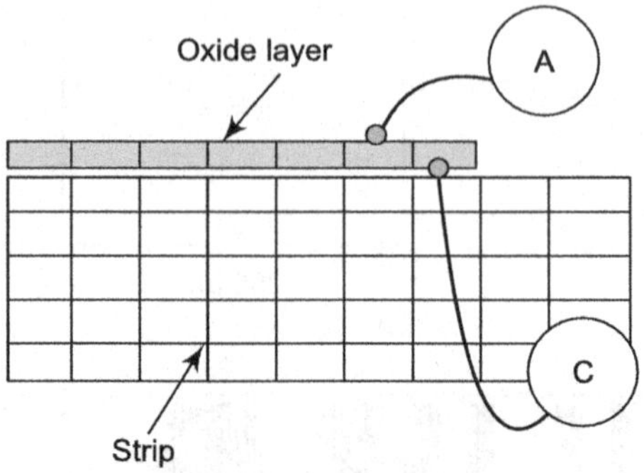

Fig. 12.16 The inclusion of a "real" oxide layer in the CAFE model. Two interfaces need to be defined. The roll–oxide interface is defined with the heat transfer coefficient (h_A). The oxide–strip interface is defined with the heat transfer coefficient (h_C)

12.6 CONCLUSIONS AND POSSIBLE EXTENSIONS

The work showed the potential of the PDD and CAFE models to incorporate interface variables effectively. The PDD approaches the roll-strip interface as a "black" box driven by experimentally observed surface states. These states vary with processing conditions (e.g., rolling speed, temperature, reduction), state of the oxide (viz. compact or porous) and surface conditions (viz. roughness). CAFE, on the other hand, is a "white" box model that seeks to incorporate the metallurgical details of the oxide and evaluate the effects of processing conditions on its behavior. CAFE represented the oxide grain structures and crack nuclei using arrays of CA cells and modeled the evolution of the cracks at the CA level using the strain and temperature data obtained from an overlying finite element. The effects of these cracks were analyzed in terms of the final failure index (FFI) that was related to heat transfer coefficient. The PDD and CAFE models can be incorporated into the standard finite element codes, e.g., ABAQUS using user-subroutines.

The initial and final values for a state in the PDD is a modeling input and requires an educated guess. In metal forming operations such as rolling, the exact limits of each state are difficult to quantify. An alternative was to use the CAFE model to elicit these limits and feed each state in the PDD. However, an explicit modeling the oxide layer requires an accurate representation of the oxide layer itself in terms of the oxide species, their distribution and its compactness, a set of reliable thermo-mechanical data and a prior knowledge of the behavior of the strip–oxide and oxide–roll interfaces. To start with, CAFE modeled the oxide layer as a single metallurgical entity with the thermal and mechanical properties of wüstite (FeO). Within the CAFE framework, the interface effects were shown in terms of the FFI with the bounds for each interface variable derived from the PDD model. This forms the groundwork to incorporate further details in the CAFE model particularly in terms of the interface behavior to complete the PDD-CAFE data transfer route.

12.6.1 Possible Extensions

The present understanding of the roll–strip interface in hot rolling is just beginning to unfold. Fundamental enquiries regarding the causal effects of friction and heat transfer is necessary. In particular, the physical and chemical aspects of adhesion and spallation.

The PDD model offers a first approximation of the relationships between the causal effects of the oxide scale. The PDD incorporated the reduction of the oxide thickness, the geometric parameters of the oxide thickness and the processing conditions of temperature. The PDD may be extended to model effects of chemical composition in form of a chemical state driven PDD or a PDD based on the substrate chemical composition and the oxide species formed on that substrate.

The CAFE model may be extended to model the actual adhesion and spallation activity itself. For this, a layer of CA cells may be introduced between the oxide CA cells and the underlying finite element. These interfacial CA cells are fed with the chemical composition of the substrate and the oxide species. This in turn, transfers to an adhesion index where 0 means "zero" adhesion and 1 is "perfect" adhesion. A similar formulation may be adopted for the oxide on the roll.

A related issue is that of lubrication in the rolls and its effects in the roll gap. The PDD can accommodate this need easily by the state BDF. Depending on the type of lubricant used and its position on the roll bite, the initial and final states may be adjusted. The CAFE model is ideally suited to model the effect of lubrication. Depending on the type of lubricant and its accessibility into the roll gap, some CA cells can be modified into lubricant CA cells with the thermal properties of that of a lubricant.

REFERENCES

1. Beremin, F.M. (1983). A local approach for cleavage fracture of a nuclear pressure vessel steel, *Metallurgical Transactions,* **14A**; 2277–2287.
2. Beynon, J.H. and Sellars, C.M. (1992). Modelling microstructure and its effects during multipass hot rolling. *ISIJ International,* **32**; 359–367.
3. Beynon, J.H., Das S., Howard, I.C., and Shterenlikht, A. (2002), Extending the local approach to fracture: Methods for direct incorporation of microstructural effects into finite element models of fracture. *New and Emerging Computational Methods: Applications to Fracture, Damage and Reliability, Pressure Vessels and Piping Conference (ASME-PVP 2002),* Vol. 438, ed. F. W. Brust, Vancouver, Canada, pp. 229–237.
4. Burke, D.P. and Higginson, R.L. (2000). Characterisation of Multi-Component Scales By Electron Back Scattered Diffraction (EBSD). *Scripta Metallurgica et Materialia,* **42**, 277–281.

5. Chen, W.C., Samarasekera., I.V., Kumar., A. and Hawbolt., E.B., 1993. Mathematical modelling of heat flow and deformation during rough rolling. *Ironmaking and Steelmaking,* **20** (2); 113–125.
6. Conway, J. (1976). *On Numbers and Games,* New York: Academic Press, Inc.
7. Das, S., Palmiere, E.J. and Howard, I.C. (2001). The effect of friction on deformation during rolling as revealed by the embedded pin technique. *Materials Science and Technology,* **17** (7), 864–873.
8. Davies, C.H.J. (1996). The Cellular Automata simulation of static recrystallisation, in Beynon, J.H., et al. (eds.), *Proceedings of the 2nd International conference on modelling of metal rolling processes,* Institute of Materials, London, 178–184.
9. Devadas, C. and Samarasekera, I.V. (1986). Heat transfer during hot rolling of steel strip. *Ironmaking and Steelmaking,* **13**; 311–321.
10. Devadas, C., Samarasekera, I.V. and Hawbolt, E.B. (1991). The thermal and metallurgical state of steel strip during hot rolling. *Metallurgical Transactions-A,* **22A**; 309–320.
11. Fletcher, J.D. (1998). Process Simulation of the hot rolling of thin gauge strip by finite element modelling, *Ph.D. Dissertation,* The University of Sheffield.
12. Gandin., Ch.A. and Rappaz., M. (1994). A coupled finite element—cellular automaton model for the prediction of dendritic grain structures in solidification process. *Acta. Metall.,* **42** (7); 2233–2246.
13. Hancock, P. and Nicholls, J.R. (1988). Application of fracture mechanics to failure of surface oxide scales. *Materials Science and Technology,* **4**; 398–406.
14. Harding, R.A. (1976). Temperature and Structural changes during hot rolling. *Ph.D. Dissertation,* The University of Sheffield.
15. Hesselbarth, H.W. and Göbel., I.R. (1991). Simulation of recrystallisation by Cellular Automata, *Acta Metall.,* **39** (1); 2135–2143.
16. Hoc, T., Rey, C. and Viaris de Lasengo, P. (2000). Mesostructure of the localisation in prestrained mild steel, *Scripta Metallurgica et Materialia,* **42**; 749–754.
17. Hollander, F., 1970. Mathematical models in metallurgical process development. *Iron and Steel Institute,* **13**; 46–78.
18. Krzyzanowski., M. and Beynon, J. H. (1999a). The tensile failure of mild steel oxide under hot rolling conditions. *Steel Research,* **70** (1); 22–27.
19. Krzyzanowski, M. and Beynon, J.H. (1999b). Finite element model of steel oxide during tensile testing during hot rolling conditions. *Materials Science and Technology,* **15**; 1191–1198.
20. Krzyzanowski., M. and Beynon, J.H. (2000). Modelling the boundary conditions for thermo-mechanical processing—oxide scale behaviour and composition effects. *Modelling and Simulation in Materials Science and Engineering,* **8**; 927–945.
21. Kumar, A., Samarasekera, I.V. and Hawbolt, E.B. (1992). Roll bite deformation during the hot rolling of steel strip. *Journal of Materials Processing and Technology,* **30**; 91–114.
22. Lemaitre, J. (1985). A continuum damage mechanics model for ductile fracture. *Journal of Engineering Materials and Technology,* **109**; 83–89.

23. Li, Y.H. (1996). Modelling of boundary conditions and their effects in hot forging and rolling, *Ph.D. Dissertation*, The University of Sheffield, UK.

24. Li, Y.H. and Sellars., C.M. (1996a). Modelling surface temperatures during hot rolling of steel, in Beynon, J.H., et al. (Eds.), *Proceedings of an International Conference on Modelling of Metal Rolling Processes-3*. The Institute of Metals, pp. 178–186.

25. Li, Y.H. and Sellars, C.M. (1996b), Evaluation of interfacial heat transfer and its effects on hot forming processes, *Ironmaking and Steelmaking*, **23** (1); 58–61.

26. Li, Y.H. and Sellars, C.M. (1998). Final Report for DTI-NPL Project MMP6/M6, Department of Engineering Materials, The University of Sheffield.

27. Li, Y.H. and Sellars., C.M. (1999a). Modelling surface temperatures during hot rolling of steel, in J. H. Beynon *et al.* (Eds.), *Proceedings of an International Conference on Modelling of Metal Rolling Processes-3*. The Institute of Metals, pp. 178–186.

28. Li, Y.H. and Sellars, C.M. (1999b). Effect of chemical composition on interfacial heat transfer coefficient during hot rolling of steels. *Proceedings of the* 6th *ICTP, Advanced Technology of Plasticity*, 111, pp. 1973–1978.

29. Li, Y.H. and Sellars, C.M. (2000). Experimental investigations of cracking and deformation behaviour of oxide scales during hot flat rolling of steel, Report No. 0023, IMMPETUS, The University of Sheffield.

30. Luo., C. and Ståhlberg., U. (2001). Deformation of inclusions during hot rolling of steels. *Journal of Materials Processing and Technology*, **114**; 87–97.

31. Makarov., P.V. et al. (1999). Modelling of development of localised plastic deformation and pre fracture stage in meso volumes of heterogeneous media, *International Journal of Fracture*, **100**; 121–131.

32. Matic., P. and Geltmacher., A.B. (2001). A Cellular Automaton based technique for modelling mesoscale damage evolution. *Computational Materials Science*, **20**; 120–141.

33. Murata, K., Morise, H., Mitsutsuka, M., Naito, H., Komatsu, T., and Shida, S. (1984). Heat transfer between metals in contact and its application to protection of rolls. *Transactions of the Iron and Steel Institute of Japan*, **24** (9); B309.

34. Pietrzyk, M. and Lenard, J.G. (1989). A study of thermal–mechanical modelling of hot flat rolling. *Journal of Materials Shaping Technology*, **7** (2); 117–126.

35. Pinna, C., Beynon., J.H. and Sellars., C.M. (2000). Experimental investigation and micromechanical modeling of the hot deformation of duplex stainless steels, in Martin, P. et al. (Eds.), *Proceedings on Mathematical Modelling in Metals Processing and Manufacturing*, Conference of Metallurgists, pp. 1–16.

36. Rappaz., M., Gandin., Ch.A. (1993). Probabilistic modelling of microstructure formation in solidification processes. *Acta Metallurgica et Materialia*, **41** (n2); 345–360.

37. Rappaz, M., Gandin, Ch.A., Desbiolles, J.L., and Thévoz, Ph. (1996). Prediction of grain structures in various solidification processes, *Metallurgical and Materials Transactions*, **27**; 695–705.

38. Rice, R.W. (1977) in MacCrone R.K. (Ed.), pp. 200–208. *Treatise in Materials Science and Technology*, Academic Press, Inc.

39. Robertson, J., and Manning, M.I., (1990). Limits to adherence of oxide scales. *Materials Science and Technology*, **6**; 81–91.

40. Sellars, C.M. (1985). Computer Modelling of Hot Working Processes. *Materials Science and Technology*, **1**; 325–332.

41. Stevens, P.G., Ivans, K.P., Harper, P. (1971). Increasing work roll life by improved roll cooling practice. *Journal of the Iron and Steel Institute*, **209**; 1–11.

42. Sun., S., Adams, B.L. (2000). Observations of lattice curvature near the interface of a deformed aluminium bicrystal. *Philosophical Magazine A*, **80**; 9–25.

43. Sutcliffe, M.P.F. (1988). Surface asperity deformation in metal forming processes. *International Journal of Mechanical Sciences*, **30** (11); 847–868.

44. Von Neumann, J. (1966). *Theory of Self Reproducing Automata*, in Banks, A.W. (Ed.), University of Illinois, Urbana.

45. Williamson, M. and Majumdar, A. (1992). Effect of surface deformations on contact conductance, *Transactions of the ASME*, **114**; 802–809.

46. Wilson, W.R.D. and Sheu, S., (1988). Real area of contact and boundary friction in metal forming. *International Journal of Mechanical Sciences*, **30** (7); 475–489.

47. Xin, P., Aizawa and T. Kihara, J. (1991). Roll pass evaluation for hot shape rolling processes. *Journal of Materials Processing Technology*, **27**; 163–178.

48. Zhang, Z.F., Wang, Z.G., Hu, Y.M. (2000). Effect of grain size on grain boundary strengthening of Copper bi-crystals under cyclic loading. *Materials Science and Technology*, **16**; 157–162.

49. Barlow, J., (1976). Optimal stress locations in finite element models, *International Journal of Numerical Methods for Engineering*, **10**, 243-251.

50. Dadras, P. and Wells, W.R. 1984. Heat transfer aspects of Non-isothermal axi-symmetric upset forging. *Journal of Engineering for Industry, Transactions of ASME*, **106**, 187–195.

51. Torres, M and Colas, R. 2000. A model for heat conduction through the oxide layer of steel during hot rolling. *Journal of Materials Processing Technology*, **105**, 258–263.

52. *SLIMMER Manual*, 1992, Dept. of Engineering Materials, The University of Sheffield, UK.

53. Caglayan, H. and Beynon, J.H. 1992, SLIMMER: Sheffield Leicester Integrated Model for Micro structural Evolution in Rolling, *Proceedings of an International Conference of Modeling of Metal Rolling Process*, Institute of Materials, London, pp. 274–282.

CHAPTER 13

Epilogue

U.S. Dixit
Department of Mechanical Engineering,
Indian Institute of Technology Guwahati, Guwahati-781039, India

13.1 INTRODUCTION

The area of metal forming is quite wide. In this book, basic fundamentals of metal forming processes and their modeling are described. The book also describes the application of newer techniques of modeling the metal forming processes as well as the technology of newer forming processes. The purpose of this chapter is to briefly describe some metal forming processes that have not been described elsewhere in the book and provide directions for future research and development. As the topics are described briefly, the reader is directed to a number of references.

13.2 MICROFORMING

Starting from the last decade, demand for miniature components has increased considerably due to their application in micro-electro-mechanical systems (MEMS) and micro-system technologies (MST). One of the most popular methods for fabricating small components is LIGA, which is a German acronym for Lithographie, Galvanoformung, Abformung, meaning lithography, electroplating, and molding, respectively. Other popular methods are bulk micromachining, surface micromachining and meso-scale machining. In bulk micromachining, selective etching inside the substrate is carried out. On the other hand, in surface micromachining, deposition and selective etching is carried out layer by layer on top of a substrate. This process is suitable only of building 2.5D (two-and-a-half-dimensional) structures. The meso-scale machining includes focused ion-beam machining, micro-milling, micro-turning, laser machining and micro-electro discharge machining. These manufacturing processes are expensive.

Small components can also be fabricated by plastic deformation and the process is called microforming. Geiger *et al.* [1] defined the microforming as the production of parts or structures with at least two dimensions in the sub-millimeter range. More precisely we can define the microforming as the process of manufacturing a part or feature by plastic deformation, whose at least one orthogonal view can be enclosed in a square of 1 mm size. Some examples of micro-formed components are IC-carrier, fasteners, micro-screws, lead frames, sockets and various kinds of connecting elements [2]. Compared to other processes, microforming has the following advantages:

- It is an economical process for mass production.
- The production rate is high.
- High accuracy and good surface finish can be obtained.
- The mechanical properties of the micro-formed parts are better.

However, it faces the following challenges:

- Fabrication of tooling for microforming requires special materials and precision manufacturing.
- The tool gets worn and sometimes may break during the process.

We can think of micro version of all conventional metal forming processes. Due to size effect, the existing knowledge cannot be applied as it is. Also the precision requirement is more. Some of the micro-forming processes are micro-extrusion, micro-forging and micro-rolling.

As the size of a feature to be manufactured becomes small, the underlying physics of the manufacturing also changes. The laws of the metal forming applicable for manufacturing a normal sized product or feature are no longer applicable for forming smaller components or features. The size has a major influence on flow stress and friction [3]. Vollertsen [4] grouped the size effect into three categories: (i) density size effect, (ii) shape size effect and (iii) microstructure size effect.

Density size effect occurs when the number of integral value of a material feature per unit volume remains more or less constant. Therefore, with reduction of size, the absolute value of material feature decreases. This affects the material properties. The material features can be small pores, dislocation lines, interface areas or grains. The reduction in the absolute number of features in a component affects material properties. For example, if the number of grains reduce, the individual orientation of grains becomes important. Due to orientation effect, the variation of properties among micro-components will be more than among bigger sized component. The strength of the brittle material is expected to increase with decreasing size due to reduction of absolute number of cracks in the micro-component. The density size effect usually occurs in the samples of size range 1–10 mm.

Shape size effect occurs due to change in the shape of component. This effect has been explained by surface layer model theory [2]. As the shape of the component changes, the ratio of surface area to volume changes. The grains on the surface are loosely held compared to grains in the interior, because of lesser number of grains surrounding the surface grains. Hence, as the ratio of surface area to volume increases, the flow strength of the material decreases. Chen and Tsai [5] conducted experiments on two cylindrical specimens with dimensions: (i) 5 mm diameter, 7.5 mm height and (ii) 2 mm diameter and 3 mm height. They observed that the hardness decreased with decreased size. Also, as the ratio of surface to volume increases, the adhesion of a small sized component to a gripper is more than a normal sized component. This effect is prominent in the sample size range of 100 μm to 1 mm.

Microstructure size effect occurs when some micro-structural features remain invariant as the size changes. One example is strain gradient plasticity. Each material has a characteristic length. As the object size changes, the ratio of object size to characteristic length also changes. When the object size becomes of the order of characteristic length of the material, strain gradient effect becomes significant. This effect is prominent in the sample size range of 10–100 μm.

Size also has a profound influence in the frictional behavior of the object. This effect was studied in detail by Engel *et al.* [6]. For extrusion process, Engel and Eckstein [2] found that the value of friction factor increases as the size of the billet decreases. In an extrusion process, friction factor increased by 20 times for reduced size when using extrusion oil as lubricant [1]. This behavior is explained by the open and closed lubricant pockets model. The billet surface is not smooth and contains pockets. The pockets that are not connected to edges of the specimen can retain the lubricant during the process. Such pockets are called closed pockets. The pockets that are connected to edges are called open pockets and are unable to retain the lubricant when pressure is applied. In a smaller component the proportion of closed to open pockets is less. Therefore, the lubricant is not retained effectively. Due to this, pressure on the peaks of the billet surface increases and they get flattened. This increases the friction force. The open and closed lubricant pocket model is confirmed by the fact that in dry lubrication the friction does not vary significantly with size.

13.3 SEVERE PLASTIC DEFORMATION (SPD) PROCESSES

It has been observed that ultrafine grained (grain size 100 to 600 nm) and nanostructured (grain size <100 nm) metals and alloys exhibit increased strength, hardness and ductility. These materials also provide an opportunity for superplastic deformation. Severe plastic deformation (SPD) is a process to convert conventional macro-grained metal or alloy into submicrometer or nanoscale material by using extreme plastic straining. The process is environmentally clean. There are a number of different types of SPD processes as discussed in the paper of Azushima *et al.* [7] and the book edited by Altan [8]. Four most commonly used SPD processes are (i) High Pressure Torsion (HPT) (ii) Equal Channel Angular Extrusion (ECAE) (iii) Cyclic Closed-Die Forging (CCDF), and (iv) Accumulative Roll Bonding (ARB). Typical strains in this process are of order of 10, due to which a 2–3 times increase in strength is observed.

Figure 13.1 shows a schematic of HPT process. During this process, the material in the form of disc experiences shear deformation between a fixed and rotating anvil. A large amount of pressure is applied during the twisting of the disc. After the process, the size of the disc remains unaltered. It is possible to adjust cumulative strain, applied pressure and deformation speed continuously and obtain ultrafine grained materials with a fairly uniform structure. The process is limited to manufacturing small sized disc.

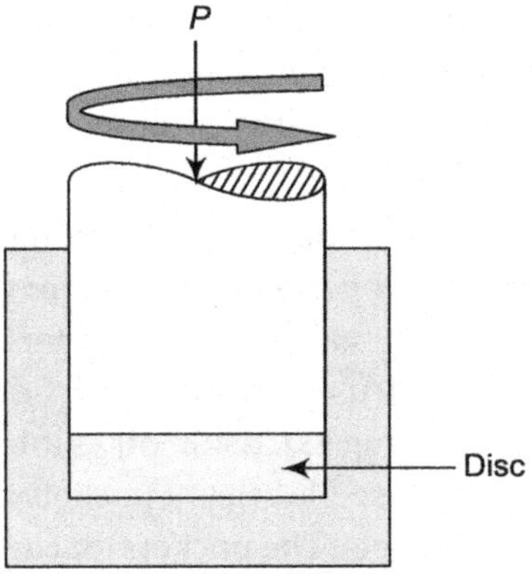

Fig. 13.1 A schematic of high pressure torsion process

Figure 13.2 shows a schematic of ECAE process. Here, a billet is passed though two intersecting closed channels of the same cross-section meeting at an angle. The billet is deformed by shear, but its cross-section remains same. The billet is extruded through these channels in several passes. Although this process is applied to solid metals, it may also be used for the consolidation of metallic powders.

Figure 13.3 shows a schematic of CCDF process. Here, a billet of height H and width W is compressed till its height becomes W and width H. The billet is now rotated by 90°, such that its height is again H and width W. It is compressed again. The approximate equivalent strain after n compressions is given by

$$\varepsilon = \frac{2n}{\sqrt{3}} \ln\left(\frac{H}{W}\right) \tag{13.1}$$

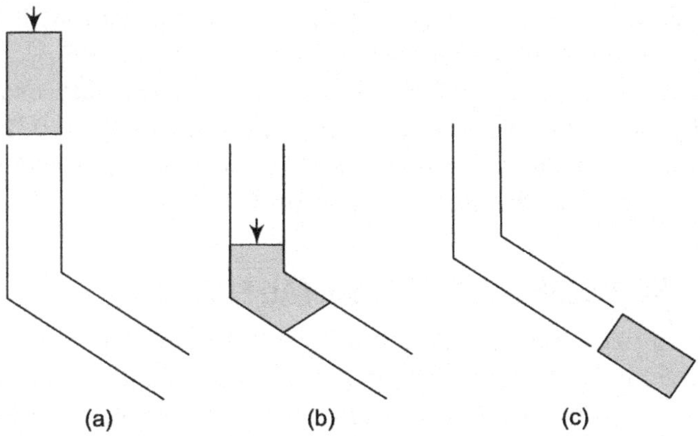

Fig. 13.2 A schematic diagram of equal channel angular extrusion process: (a) billet about to enter the channel (b) billet in the channel (c) billet after passing through the channel

In this process, the strain distribution is not uniform, which is a disadvantage.

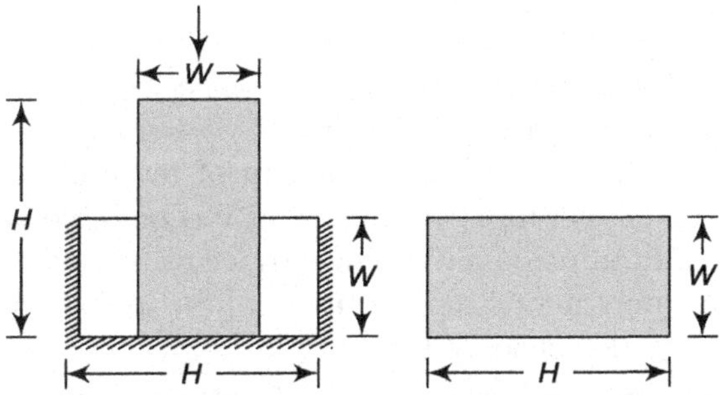

Fig. 13.3 A schematic of cyclic closed die forging process

Figure 13.4 shows a schematic of ARB process, in which stacking of sheets and conventional roll bonding are repeated. Two strips are stacked together and passed through two counter rotating cylinders. The sheets get pressed and deformed. It is important to surface treat two sheets to increase the bond strength. The deformed length of the rolled material is sectioned into two halves. The sectioned sheets are again surface treated and stacked. The stacked sheets are again passed through the roll and sequence of operations is repeated. Usually, in one pass, a 50% reduction is carried out. In that case approximate strain in n passes will be $0.8n$.

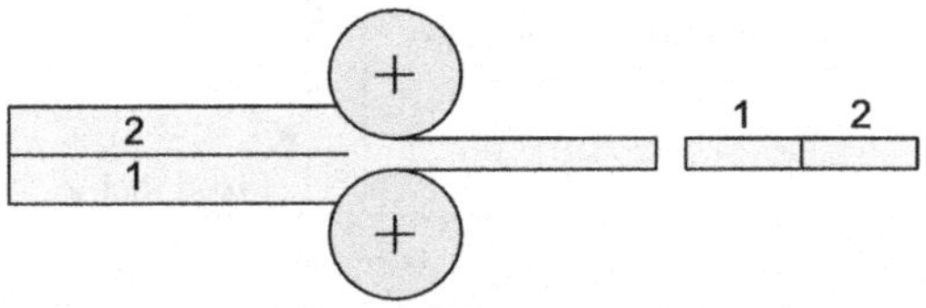

Fig. 13.4 A schematic of accumulative roll bonding process

A combination of several processes is also being practiced. For example, Mani *et al.* [9] have developed a process by modifying ECAE process. In this process, billet coming out of the channels is twisted. The authors named this process as Torsional-Equal Channel Angular Pressing (T-ECAP). This process is capable of imposing larger strain than that of the ECAE process in one single pass. Also, the strain distribution is more uniform.

13.4 INCREMENTAL SHEET METAL FORMING

As the name implies, the incremental sheet metal forming involves deforming the sheet metal in increments. The process traces back to its origin in metal spinning and incremental deformation by a black smith's hammer. In one sense, even rolling can be considered as an incremental forming process. However, the present notion of incremental sheet forming (ISF) or Asymmetric Incremental Sheet Forming (AISF) is somewhat different. Emmens *et al.* [10] define the incremental sheet forming as a family of sheet metal forming processes where the deformation is highly localized, without drawing in of material from a surrounding area and using a fully clamped blank where the final shape is determined by the *xyz* movement of some tool part without the need for a die.

The incremental sheet metal forming process has been reviewed by Jeswiet *et al.* [11]. Almost in all forms of the process, a hemispherical-head tool is used to deform the material. The movement of the tool is controlled by a CNC controller. By proper tool path planning, it is possible to produce a very complex shape without using a die. The surface roughness of the product is dependent on the incremental displacement of the tool. At high draw angle, orange peel effect is also observed.

In the early version of incremental sheet forming process, a support acting as a partial or full die was also used. As the sheet is pressed at two points, it is called Two Point Incremental Forming (TPIF). A schematic diagram of TIPF process is shown in Fig. 13.5. A rotating tool is traversed on a rigidly clamped

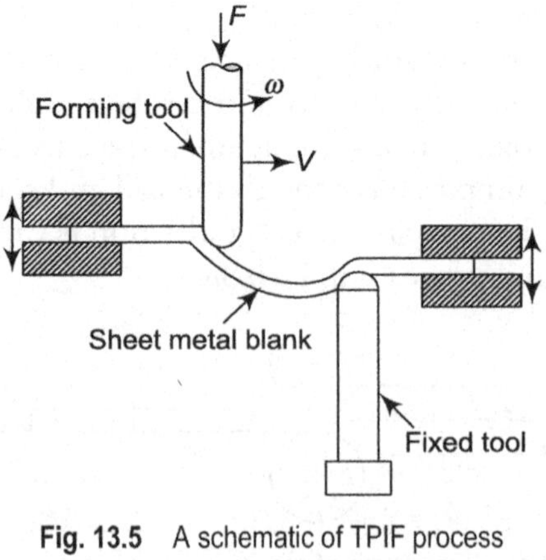

Fig. 13.5 A schematic of TPIF process

blank on a well-planned path. In multi-pass process, the path can be traversed several times by changing the depth of the tool. This process can be carried out on a CNC milling machine.

TPIF is not a truly die-less process. Figure 13.6 shows a truly die-less process called Single Point Incremental Forming (SPIF). Here, except at the rigid supports, the sheet is not supported anywhere. There are several advantages of this process. The complex shapes can be produced without the need of any die. The process can be performed on a conventional CNC machine and the size of a part is limited by the size of machine alone. The power and maximum force are not the limitations as the final form can be produced in a number of passes involving small vertical movement of the tool. The process is economical for small batch size production or rapid prototyping. The process does have some limitations. The part accuracy is of the order of mm and time required is much longer in comparison to processes like deep drawing.

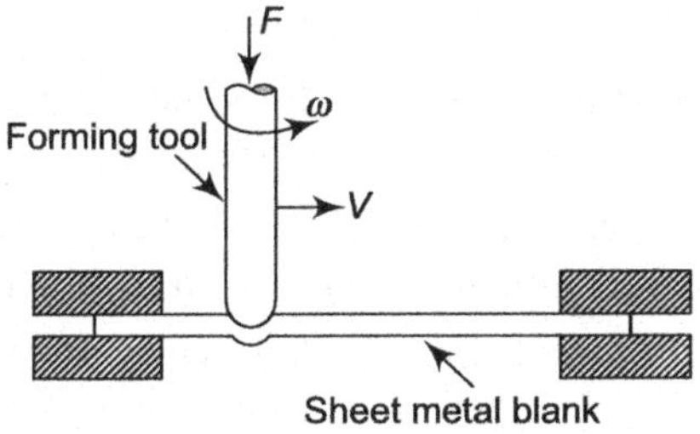

Fig. 13.6 A schematic of TPIF process

Heating of the material reduces the flow stress thus reducing the forming forces. A high speed rotation of the tools generates a lot of frictional heat. The heating can also be carried out by a laser beam.

The incremental sheet forming process is still not a popular production process. However, several researchers have demonstrated its potential of becoming an important industrial process. The main limitation of the process seems to be its slow speed in comparison to conventional processes. The geometrical accuracy and forming time is highly dependent on the tool path and the strategies for generating the proper tool path are being developed [12, 13].

13.5 APPLICATION OF LASER IN METAL FORMING

Laser has been extensively used for machining and welding. It also finds application in sheet metal forming. Laser can be used to assist a conventional sheet metal forming process, to complement a forming process or to replace a conventional forming process. The following subsections briefly describe these applications.

13.5.1 Heating by Laser during Forming

Heating of the metal reduces its flow stress and planar anisotropy, thus improving its formability. A sapphire tool can be used which is transparent to laser light. This helps to increase the temperature of the local region of the workpiece. Gillner *et al.* [14] has reported that in deep drawing of a high grade steel of thickness 150 μm with a 4 mm diameter punch, the limiting drawing ratio can be increased from 2 to 2.3 with laser heating. However, a substantial part of energy is lost in the tool parts through absorption, reflection and heat conduction.

Samm *et al.* [15] has listed the following advantages of laser assisted micro-forming:

- The laser energy input and thus the resulting temperature in the work-piece can easily be controlled via the current of a diode laser.
- Local heating of selected areas of the work-piece is possible limiting the heating zone.
- Needed temperature gradient can be achieved by controlling the power of the laser.
- The absorption of laser radiation allows short process times that cannot be accomplished with heat transfer from pre-heated tools.
- Material properties can be manipulated by controlling the down cooling via laser power.

13.5.2 Hybrid Processing

The laser processing can be combined with conventional processing. The initial shape can be created by mechanical tools and it can be corrected by laser. Compared to laser forming, conventional forming processes are fast and large deformations can be obtained in one step. Hybrid forming potentially combines the speed of conventional forming with the accuracy of laser forming [16]. Conventional forming can also be combined with laser bending for producing tight joints. Laser cutting can be performed before and/or after forming. Laser heat can also be used to stress-relieve the formed parts.

13.5.3 Laser Forming

Laser forming (LF) is a process that forms the sheet metal by means of stresses induced due to external heat instead of external force [17] It is used to deform the sheets made up of high thermal expansion coefficient materials such as stainless steels, light alloys of aluminum, magnesium and titanium. By proper path planning of the laser beam, it is possible to produce a complex form. Figure 13.7 shows schematic of a laser bending process. Here a laser beam is scanned across the width of a cantilever beam. As a result, the sheet gets bent with bend line as the line of traverse of the laser beam. In a single pass, only a small bend angle is produced. Multi-pass operation can produce larger bend angle.

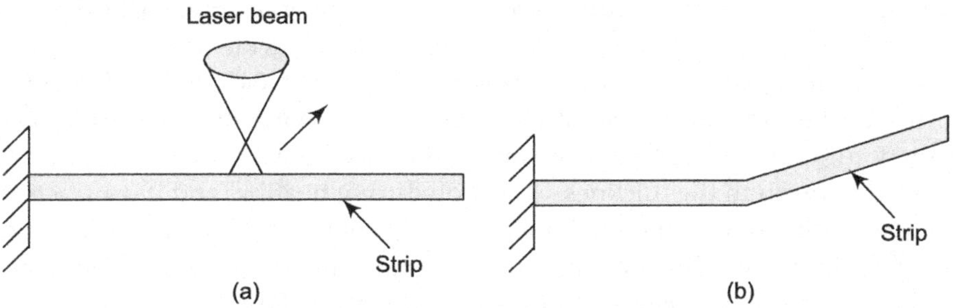

Fig. 13.7 A schematic diagram of laser forming

Laser forming is a complex transient process that involves thermodynamics, heat transfer, elastoplastic mechanics and metallurgy to control the deformation of the metal workpiece. Three mechanisms of laser forming have been proposed [18] temperature gradient mechanism, buckling mechanism, and upsetting mechanism. The temperature gradient mechanism (TGM) is the most widely reported laser bending mechanism. In TGM, the diameter of heated area is nearly equal to sheet thickness. TGM occurs in two stages *viz.* heating of the workpiece using laser beam and natural convection cooling. During heating phase, a laser beam quickly scans over the work piece which produces a steep temperature gradient in the workpiece across its thickness. As a result, a differential thermal expansion occurs through the thickness direction. The material expands in the heated zone and the whole shape of the material bends away from the beam. This is called as 'counter-bending'. During cooling the material contracts again in the upper layers and further plastically deforms. It leads to the local shortening of the upper layers and bending occurs in the opposite direction, thus obtaining a positive bend angle. Buckling mechanism (BM) is another mechanism used in laser forming to manufacture three dimensional complex components. It is generally used to deform thin metal sheets. In BM, the deformation is achieved by employing low feed rate and larger beam diameters (approximately ten times of the sheet thickness) to avoid the steep temperature gradient across the sheet thickness unlike in TGM. Due to heating, thermal compressive stresses are developed in the sheet which results in a large amount of thermo-elastic strain that in turn causes local thermo-elastic–plastic buckling of the material. The direction of the bending angle is not defined by the process itself as it is for TGM. The part can bend in either the positive or negative directions depending on a number of factors such as laser parameters, pre-bending orientation of the sheet, pre-existing residual stresses, direction in which any other elastic stresses are applied, internal stresses and external or gravitation forces. The upsetting mechanism (UM) is a shortening (or thickening) mechanism. In UM, the process parameters are similar to BM, but the dimension of the heated area is much smaller compared to sheet thickness. The thermal gradient across the work piece thickness is low due to comparatively slow scanning of laser beam over the work surface. During UM, the growth of the buckle will not take place as the geometry of the work piece is stiffer, instead, a shortening of material in the heated area and simultaneous thickening along the laser irradiated path occur. In the upsetting mechanism, the sheet metal is almost constantly expanded through the thickness direction during heating, and it contracts in the width direction of the sheet metal during cooling. Thus, the thickness of the plate increases. This heating process is repeated across the whole width of the material to change the thickness of the plate. This mechanism can be used to form a plane sheet into a specially formed part with a proper heating strategy. Aligning, adjustment and rapid prototyping are also possible. In practice a combination of these mechanisms is operative.

13.6 MODELLING OF THE EVOLUTION OF THE MICROSTRUCTURE

In metal forming, the quality of the finished product is important. The quality of the product is dependent on the microstructure. Hence, one is interested in determining the evolution of microstructure during metal forming. Microstructural models can be based on the physics or they can be empirical models based on experimental observations. The models can also be a combination of physically-based models and empirical models. Microstructural models cover, the dynamic and static phenomena. Dynamic phenomena arise at the beginning of deformation and may continue through the deformation. They include hardening, dynamic recovery and dynamic recrystallization. Dynamic phenomena leave the material in an unstable state and promote static and metadynamic recovery and recrystallization, which are then followed by grain growth.

Microstructural evolution can also be studied by Cellular Automata (CA) technique. The CA approach can be used to describe the growth kinetics and microstructure evolution in the time and space scale. The introduction of computers has made the CA modeling technique, a very simple and easy tool for describing various physical processes. For further details one can refer [19].

13.7 DIRECTIONS FOR FUTURE RESEARCH

In a review paper, Dixit *et al.* [20] has envisaged certain areas that will dominate the research in the modeling of metal forming and machining processes. These are briefly summarized here.

1. **Incorporation of metallurgical aspects in the theory of plasticity:** Classical theory of plasticity is applicable to ductile continuum materials and does not have provision to incorporate finer features of material behavior. In future more advanced theories like crystal plasticity will find increasing applications in the modeling of metal forming.
2. **Modeling of the evolution of microstructure during processing:** High speed computers offer the possibility of simulating the microstructural evolution. It will help in controlling the quality of the product.
3. **Modeling of micro and nano manufacturing processes:** The plasticity theory developed for micro-fabrication deviates from reality for micro and nano-fabrication. There is a scope to develop theories for modeling these processes. Some of the processing techniques require the understanding of material behavior at the atomistic scale that requires different approach of modeling. One such approach is Molecular Dynamics (MD) simulation [21].
4. **Multi-scale modeling:** The use of different models at several smaller length scales with attempts to transfer information between them has come to be known as 'multi-scale material modeling' [22]. In near future, this technique is expected to gain momentum in the modeling of manufacturing processes.
5. **Comparison of processing techniques in terms of product quality:** There is a need to carry out evaluation of various processes based on their impact on product quality. In future, this is expected to be an important area of applied research.
6. **Development of experimental techniques for obtaining the material parameters:** It is difficult to obtain some material parameters of the model through experiments. Extensive research efforts are required on hardware as well as software side. With the advent of powerful computational techniques, it will be possible to accurately simulate many experiments.
7. **Development of hybrid computational methods:** In future hybrid computational methods comprising physics-based and AI-based models are expected to gain momentum.

Apart from these areas concerned with the modeling of metal forming processes, the research is required in the following areas:

- Processing of materials having low formability
- Forming of very intricate and small components
- Development of processes providing more productivity and quality
- Development of sustainable forming processes
- Development of fully automatic and adaptive forming processes.

13.8 CONCLUSIONS

The description in this book started with the introduction and basic fundamentals of the metal forming processes. Subsequently, the recent developments in well-established and industrial metal forming processes were described. The last chapter provides a brief description of the processes and issues that need to be yet developed for making them industrially competitive. Some directions for future research have also been provided.

REFERENCES

1. Geiger, M., Kleiner M., Eckstein, R., Tiesler, N., and Engel. U. (2001). Microforming, *CIRP Annals—Manufacturing Technology*, **50** (2); 445–462.
2. Engel, U., and Eckstein, R. (2002). Microforming—from basic research to its realization. *Journal of Materials Processing Technology*. **125–126**; 35–44.
3. Geiger, M., Geibdorfer S. and Engel, U. (2007). Mesoscopic model: advanced simulation of microforming process, *Production Engineering Research and Development*, **1**; 79–84.
4. Vollertsen, F. (2008). Categories of size effects. *Production Engineering Research and Development*, **2**; 377–383.
5. Chen, F.K. and Tsai, J.W. (2006). A study of size effects in micro-forming with micro-hardness tests. *Journal of Materials Processing Technology*. **177**; 146–149.
6. Engel, U., Messner, A. and Tiesler, N. (1998). Cold forging of microparts—effect of miniaturization on friction, in J.L Chenot et al. (eds.), *Proceedings of the 1st ESAFORM Conf. on Materials Forming*, Sophia Antipolis, France, p. 77–80.
7. Azushima, R., Kopp, A., Korhonen, D.Y., Yang, F., Micari, G.D., Lahoti, P., Groche, J., Yanagimoto, N., Tsuji, A., Rosochowski, A., and Yanagida, (2008). Severe plastic deformation (SPD) processes for metals, *CIRP Annals-Manufacturing Technology*, **57** (2); 716–735.
8. Altan, B. (2006), *Severe Plastic Deformation: Towards Bulk Production of Nanostructured Materials*, Nova Publishers, New York.
9. Mani, B., Jahedi, M. and Hossein Paydar, M. (2011). A modification on ECAP process by incorporating torsional deformation, *Materials Science and Engineering: A*, **528** (12); 4159–4165.
10. Emmens, W.C., Sebastiani, G. and van den Boogaard, A.H. (2010). The technology of Incremental Sheet Forming—A brief review of the history, *Journal of Materials Processing Technology*, **210** (8); 981–997.
11. Jeswiet, J., Micari, F., Hirt, G., Bramley, A., Duflou, J., and Allwood, J. (2005). Asymmetric Single Point Incremental Forming of Sheet Metal, *CIRP Annals—Manufacturing Technology*, **54** (2); 88–114.
12. Malhotra, R., Bhattacharya, A., Kumar, A., Reddy, N.V., and Cao, J. (2011). A new methodology for multi-pass single point incremental forming with mixed toolpaths, *CIRP Annals—Manufacturing Technology*, In Press, DOI: 10.1016/j.cirp.2011.03.145.

13. Malhotra, R., Reddy, N. V. and Cao J. (2010). Automatic 3D Spiral Toolpath Generation for Single Point Incremental Forming, *J. Manuf. Sci. Eng.*, **132**; 061003 (10 pages).

14. Gillner, A., Holtkamp, J., Hartmann., C., Olowinsky., A., Gedicke., J., Klages., K., Bosse, L., and Bayer A. 2005. Laser applications in microtechnology. *Journal of Materials Processing Technology* **167**; 494–498.

15. Samm, K., Terzi, M., Ostendorf, A., and Wulfsberg, J. (2009). Laser-assisted micro-forming process with miniaturised structures in sapphire dies. *Applied Surface Science* **255**; 9830–9834.

16. Magee, J., and De Vin, L.J. (2002). Process planning for laser-assisted forming. *Journal of Materials Processing Technology*, **120**; 322–326.

17. Geiger, M., F. Vollertsen, and G. Deinzer. 1993. Flexible straightening of car body shells by laser forming. *International Congress and Exposition*, Detroit. MI, USA: Publ by SAE, 930: 279.

18. Geiger, M. and F. Vollertsen. 1993. The mechanisms of laser forming. *CIRP Annals*, 42 no. 1: 301–304.

19. Janssens, K.G.F., Raabe, D., Kozeschnik, E., Miodownik, M.A., and Nestler, B., (2007), *Computational Materials Engineering: An Introduction to Microstructure Evolution*, Elsevier, London.

20. Dixit, U.S., Joshi, S.N. and Davim, J.P. (2011). Incorporation of material behavior in modeling of metal forming and machining processes: A review, *Materials & Design*, **32** (7); 3655–3670.

21. Komanduri, R. and Raff, L.M. (2010), Molecular Dynamics (MD) simulations of machining at the atomistic scale, in V.K. Jain (ed.), *Introduction to Micromachining*, Narosa Publishing House, New Delhi.

22. Campbell, G.H., Foiles, S.M., Huang, H., Hughes, D.A., King, Wayne E., Lassila, D.H., Nikkel, D.J., de la Rubia, T.D., Shu, J.Y., and Smyshlyaev, V.P. (1998). Multi-scale modeling of polycrystal plasticity: a workshop report, *Materials Science and Engineering A*, **251**; 1–22.

List of Contributors

1. **Dr. Arbind Kumar Singh**

   Department of Civil Engineering

   Indian Institute of Technology Guwahati, India 781 039

   Dr. Arbind Kumar Singh is an Associate Professor at Department of Civil Engineering, Indian Institute of Technology Guwahati. His main area of research is plasticity.

2. **Dr. R. Ganesh Narayanan**

   Department of Mechanical Engineering

   Indian Institute of Technology Guwahati, India 781 039

   Dr. R. Ganesh Narayanan is an Assistant Professor at Department of Mechanical Engineering, Indian Institute of Technology Guwahati. His area of research includes Sheet metal forming, Cold forging, Joining and forming, Tailor welded blanks, Formability and forming limit, and Computational applications in metal forming.

3. **Dr. Uday S. Dixit**

   Department of Mechanical Engineering

   Indian Institute of Technology Guwahati, India 781 039

   A Ph.D. from IIT Kanpur, Dr. U.S. Dixit is currently Professor in the Department of Mechanical Engineering at Indian Institute of Technology Guwahati. He has about 5 years of experience in industry and 15 years of experience in academia and has contributed several papers to journals of international repute.

4. **Dr. Satish V. Kailas**

   Department of Mechanical Engineering

   Indian Institute of Science, Bangalore, India 560 012

   Dr. Satish V. Kailas is a Professor in the Department of Mechanical Engineering, Indian Institute of Science, Bangalore, India. He got his Masters and Ph.D. from the Indian Institute of Science in 1989 and 1994, respectively. His fields of interest include tribology, metal forming and eco-friendly lubricants.

5. **Dr. R. K. Uyyuru**

   FMC Kongsberg Subsea

   Kongsberg, Norway

   Rama Krishna Uyyuru is working as Senior Material Engineer with FMC Kongsberg Subsea AS (part of FMC Technologies Inc.), Kongsberg, Norway. He got his Ph.D. from NTNU, Trondheim, Norway, while he did his M.E. (Metallurgy) from IISc, Bangalore, India. His fields of interest include material selection, tribology in manufacturing, manufacturing (forming and welding) simulation and engineering analysis.

6. **Mr. M. Shimjith**

   GE Global Research Center, Bangalore, India.

   M. Shimjith is a Research Associate at the GE Global Research Center, Bangalore, where he has been working for past 2 years. He received his Bachelor's degree in 2005 and worked for over 5 years at Indian Institute of Science as Research Assistant. His principal line of research is Tribology, Erosion and Hot Corrosion studies of Gas Turbine materials.

7. **Mr. Vinayak R. Mallik**

   Department of Mechanical Engineering

   Indian Institute of Science, Bangalore, India 560 012

   Vinayak R. Malik is currently a Research Scholar under Prof. Satish Vasu Kailas in the Department of Mechanical Engineering, Indian Institute of Science, Bangalore. He completed his Bachelors and Masters degrees from Gogte Institute of Technology, Belgaum. His areas of interest are Material Science, Metal Forming, and Joining of Engineering Materials.

8. **Dr. Santosh Kumar**

   Department of Mechanical Engineering

   IIT-BHU Varanasi 221 005, India

   A Ph.D. from IIT Kanpur, Dr. Santosh Kumar is a professor in the Department of Mechanical Engineering, IIT-BHU. His main area of research is extrusion. He has published several papers and executed a number of sponsored projects.

9. **Dr. Jay S. Gunasekera**

   Department of Mechanical Engineering

   Ohio University, USA

   Dr. Jay S. Gunasekera is Professor of Mechanical Engineering at Russ College of Engineering and Technology, Ohio University. He was the chair of the Department of Mechanical Engineering from 1991 to 2006.

Dr. Gunasekera has advised over 90 graduate students, many of who are now leaders in their respective fields.

10. **Mr. Arjaan Buijk, Managing Director, Simufact-Americas LLC, USA**

    11685 Spicer Drive, Plymouth, MI 48170-4347, USA

    After receiving his MS in Aerospace Engineering from the Technical University of Delft, he joined MSC.Software in 1988. In 2008, he joined Simufact, and is responsible for the Americas region. Over the course of 20+ years in the industry, he has gained specific expertise in the field of metal forming simulation, using the software products Simufact. forming and Simufact.welding.

11. **Dr. Gary Huang, Manager, Simufact-Americas LLC,**

    11685 Spicer Drive, Plymouth, MI 48170-4347, USA

    After receiving his PhD in Civil Engineering Department from the University College of Swansea, UK, he joined Rockfield Software Ltd, Swansea in 1989. In 1998, he started working for MARC Corp and later MSC.Software Corp. For 9 years, he was a developer and project manager of MSC.SuperForm. In 2008, he joined Simufact-Americas, as manager of the solver development team.

12. **Dr. Gabriel (Barton) McBain, Director**

    Simufact India Pvt. Ltd., Bangalore, India

    Dr. Gabriel McBain studied Metallurgical Engineering and Materials Science at the RWTH Aachen University with a focus on metal forming sciences and numerical modelling. Worked as senior consultant in the field of numerical modelling of metal manufacturing processes. Currently he is the director of Simufact India Pvt. Ltd., Bangalore.

13. **Dr. Takashi Kuboki**

    Department of Mechanical Engineering & Intelligent Systems,

    University of Electro-Communications, Chofu_shi, Tokyo, Japan

    Dr. Kuboki is an Associate Professor in Department of Mechanical Engineering & Intelligent Systems, University of Electro-Communications, Chofu_shi, Tokyo. He works in modeling of metal forming process using Finite Element Method. He has about 11 patents, more than 20 journal papers and more than 10 conference papers.

14. **Dr. P. P. Date**

    Department of Mechanical Engineering, IIT Bombay, Powai, Mumbai 400076, India

    Dr. P.P. Date is a Professor at Department of Mechanical Engineering, IIT Bombay. He works on metal forming, Mechanical metallurgy, Sheet metal formability, AI applications to Metal forming. He has published many papers in journals and conferences of international repute.

15. **Dr. K. Narasimhan**

Department of Metallurgical Engineering and Materials Science

Indian Institute of Technology Bombay, Mumbai 400 076, India

Dr. K. Narasimhan is a Professor at Department of Metallurgical Engineering and Materials Science, IIT Bombay. He works on Sheet metal forming, Formability, Superplasticity, Modelling and Simulation. He has completed several sponsored projects and consultancy work. He has authored several papers in international journal and conferences.

16. **Dr. Prakash M. Dixit**

Department of Mechanical Engineering

Indian Institute of Technology, Kanpur – 208 016, India

Dr. P.M. Dixit is a Professor at Department of Mechanical Engineering, Indian Institute of Technology, Kanpur. His areas of interest include Metal Forming, Large Deformation Elasto-Plastic Impact/Contact Problems, Ductile Fracture, and Finite Element Method. He has guided several students at Ph. D., M.Tech. level. He has published several journal papers and books of international repute.

17. **Dr. Ravindra K. Saxena**

Department of Mechanical Engineering

Sant Longowal Institute of Engineering and Technology , Longowal – 148 106, India

Dr. Ravindra K. Saxena is an Associate Professor in Department of Mechanical Engineering at Sant Longowal Institute of Engineering and Technology, Longowal. His research area is sheet metal forming.

18. **Dr. Sumitesh Das**

Research and Development, Tata Steel, Jamshedpur, India

Dr. Sumitesh Das holds a post-graduate degree in Manufacturing Science from IIT Kanpur and a PhD in Engineering Materials from The University of Sheffield, UK. His research interests include: multi-level modelling using a variety of modelling tools, incorporating experimental data into modelling, and transfer of models into practical usage at operations. He was also recognized as Young Engineer by INAE. He presently heads the global research programs at Global R&D of Tata Steel.

Index

S

T

Made in United States
Orlando, FL
04 November 2026

90200625R00321